ESO ASTROPHYSICS SYMPOSIA
European Southern Observatory

Series Editor: Philippe Crane

Jeremy R. Walsh   Ivan J. Danziger (Eds.)

Science with the VLT

Proceedings of the ESO Workshop
Held at Garching, Germany,
28 June – 1 July 1994

 Springer

Volume Editors

Jeremy R. Walsh
Ivan J. Danziger
European Southern Observatory
Karl-Schwarzschild-Strasse 2
D-85748 Garching, Germany

Series Editor

Philippe Crane
European Southern Observatory
Karl-Schwarzschild-Strasse 2
D-85748 Garching, Germany

ISBN 978-3-662-22397-0   ISBN 978-3-540-49215-3 (eBook)
DOI 10.1007/978-3-540-49215-3

CIP data applied for

© Springer-Verlag Berlin Heidelberg 1995
Originally published by Springer-Verlag Berlin Heidelberg New York in 1995.
Softcover reprint of the hardcover 1st edition 1995
Typesetting: Camera ready by author/editor
SPIN: 10481240   55/3142-543210 - Printed on acid-free paper

Preface

At its meeting in May 1993, the ESO Scientific and Technical Committee (STC) set up a small working group to examine the scientific priorities for the VLT. In particular, this working group was charged with recommending new VLT instruments based on scientific precedence. At that time there were only four approved VLT instruments and with twelve foci available (excluding Coudé), there was scope for a wide range of future instrumentation. The Working Group on Scientific Priorities for the VLT met during the latter half of 1993 and circulated a questionnaire. The initial work of this group is reported by Vigroux (ESO Messanger, No. 74, p. 28, 1993). It became clear that a Workshop would be the best way both to poll astronomical community interest in possible new instruments and to provide a forum for discussing science projects for the VLT.

The Working Group, whose members were L. Vigroux (Chair), K.S. de Boer, B. Fort, R.-P. Kudritzki and B. Marano, together with S. D'Odorico, J. Wampler and J.R. Walsh from ESO, formed the organising committee for the workshop. I.J. Danziger headed the local organizing committee. The workshop, held during a very warm week in late June 1994 at Garching, led to these proceedings. The success of the conference was assured by the organisational assistance of Christina Stoffer from ESO. Also from ESO, Britt Sjoeberg and Hans-Jürgen Kraus provided invaluable help and Edouard Pomaroli assisted with the tape recording of the Panel Discussions. We are very grateful to the Max-Planck Institute for Plasma Physics for allowing use of their auditorium.

In the week before the conference, Antoine Duquennoy, who was due to present a contributed paper, was tragically killed in a traffic accident. J.-M. Marriotti very kindly wrote his contribution for the proceedings. We dedicate this volume to his memory knowing that, had he lived, he would have been a very active contributor to science with the VLT and VLTI.

The large number of reviews, as well as oral contributions and poster papers, has precluded publication of all papers in a single volume. Therefore the poster papers and contributions on telescopes and instruments are published in a supplement volume by ESO, whose contents are listed following the contents list for this volume.

This was the first workshop dedicated specifically to science opportunities with the VLT and is therefore very broad in scope. We look forward to many more VLT conferences and perhaps to another with the same title, not "Science (to be performed) with the VLT" but "Science (done) with the VLT".

Garching, January 1995
J. R. Walsh
I. J. Danziger

Contents

Contents

Poster Paper Supplement
(Separately published by ESO)

List of Participants

CESARSKY, Catherine   CEA Saclay, Service d'Astrophysique
cesarsky@sapvxg.saclay.cea.fr

CHAKAVEH, Sepideh   Universitäts-Sternwarte, Gottingen
sepi@usw008.dnet.gwdg.de

CHALABAEV, Almas   Observatoire de Grenoble
chalabae@gag.observ-gr.fr

CHARBONNEL, Corinne   Observatoire Midi-Pyrenees, Toulouse
corinne@obs-mip.fr

CHINCARINI, Guido   Osservatorio Astronomico di Brera
chincarini@astmim.astro.it

COHEN, Martin   University of California, Berkeley,
Radio Astronomy Laboratory
cohen@bkyast.berkeley.edu

COLLESS, Matthew   Mt. Stromlo & Siding Spring
Observatories
colless@mso.anu.edu.au

COUDÉ DU FORESTO, Vincent   Observatoire de Paris, DESPA, Meudon
foresto@megasx.obspm.fr

CRANE, Phillipe   ESO, Garching
crane@eso.org

CRISTIANI, Stefano   Osservatorio Astronomico di Padova
cristiani@astrpd.astro.it

CROCKER, James   ESO, Garching
jcrocker@eso.org

CRUZALÈBES, Pierre   Observatoire de la Cote d'Azur, Grasse
cruzalebes@ocar01.obs-azur.fr

CUBY, Jean-Gabriel   ESO, Garching
jcuby@eso.org

D'ODORICO, Sandro   ESO, Garching
sdodoric@eso.org

DANZIGER, John   ESO, Garching
jdanzige@eso.org

DE BERGH, Catherine   Observatoire de Paris, DESPA, Meudon
debergh@mesiob.obspm.fr

DE BOER, Klaas   Sternwarte der Universität, Bonn
deboer@astro.uni-bonn.de

DE GRAAUW, Thijs   SRON, Laboratory for Space Research,
Groningen
thijsdg@sron.rug.nl

DE ROBERTIS, Michael   York University, Dept. of Physics &
Astronomy
mmdr@sol.yorku.ca

DEJONGHE, Herwig   Universiteit Gent, Sterrenkundig
Observatorium
Herwig.Dejonghe@rug.ac.be

DENNEFELD, Michel — Institut d'Astrophysique, Paris
dennefel@iap.fr

DI SEREGO ALIGHIERI, Sperello — Osservatorio Astrofisico di Arcetri
sperello@arcetri.astro.it

DJORGOVSKI, George — Caltech, Pasadena
george@deimos.caltech.edu

DRAVINS, Dainis — Lund Observatory
dainis@astro.lu.se

DUSCHL, Wolfgang — Universität Heidelberg, Institüt für
Theoretische Astrophysik
wjd@platon.ita.uni-heidelberg.de

ENARD, Daniel — ESO, Garching
denard@eso.org

FELENBOK, Paul — Observatoire de Paris, DAEC, Meudon
felenbok@obspm.fr

FERLET, Roger — Institut d'Astrophysique, Paris
ferlet@iap.fr

FORT, Bernard — Observatoire de Paris, DEMIRN
fort@mesioa.obspm.fr

FOY, Renaud — Observatoire de Lyon
foy@obs.univ.lyon1.fr

FRANÇOIS, Patrick — Observatoire de Paris, Paris
patrick@iap.fr

FRANSSON, Claes — Stockholm Observatory
claes@astro.su.se

FRICKE, Klaus — Universitäts-Sternwarte, Gottingen
kfricke@uni-sw.gwdg.de

FULLERTON, Alex — Universitäts-Sternwarte, Munchen
fullerton@usm.uni-muenchen.de

FUSI-PECCI, Flavio — Osservatorio Astronomico di Bologna
flavio@astbo3.astro.it

GENZEL, Reinhard — MPI für Extraterrestrische Physik,
Garching
genzel@mpe-garching.mpe.de

GIACCONI, Riccardo — ESO, Garching
rgiaccon@eso.org

GILMOZZI, Roberto — ESO, Garching
rgilmozz@eso.org

GROSBØL, Preben — ESO, Garching
pgrosbol@eso.org

GUSTAFSSON, Bengt — University of Uppsala, Astronomical
Observatory
Bengt.Gustafsson@astro.uu.se

HAMILTON, Donald — MPI für Astronomie, Heidelberg
hamilton@mpia-hd.mpg.de

HAMMER, Francois   Observatoire de Paris, DAEC, Meudon
hammer@gin.obspm.fr

HEBER, Ulrich   Astronomisches Institüt, Universität
Erlangen-Nurnberg, Bamberg
heber@sternwarte.uni-erlangen.de

HENSBERGE, Herman   Royal Observatory, Brussels
herman@astro.oma.be

HEWETT, Paul   Institute of Astronomy, University of
Cambridge
phewett@mail.ast.cam.ac.uk

HUBIN, Norbert   ESO, Garching
nhubin@eso.org

ILLINGWORTH, Garth   UCO/Lick Observatory
gdi@lick.ucsc.edu

IYE, Masanori   National Astronomical Observatory of
Japan, Tokyo
iye@optik.mtk.nao.ac.jp

JACOBY, George   KPNO, Tucson
jacoby@noao.edu

JØRGENSEN, Henning   Copenhagen University, Astronomical
Observatory
henning@astro.ku.dk

JORISSEN, Alain   Universite Libre de Bruxelles
ajorisse@astro.ulb.ac.be

KÄUFL, Hans Ulrich   ESO, Garching
hukaufl@eso.org

KISSLER, Markus   ESO, Garching
mkissler@eso.org

KOBAYASHI, Naoto   National Astronomical Observatory of
Japan, Tokyo
naoto@merope.mtk.nao.ac.jp

KOESTER, Detlev   Universität Kiel, Institüt für
Theoretische Physik
supas027@astrophysik.uni-kiel.d400.de

KOTILAINEN, Jari   Tuorla Observatory, University of Turku
jarkot@polaris.cc.utu.fi

KRAUTTER, Joachim   Landessternwarte, Heidelberg
jkrautte@hp2.lsw.uni-heidelberg.de

KUDRITZKI, Rolf-Peter   Universitäts-Sternwarte, Munchen
kudritzki@usm.uni-muenchen.de

LA FRANCA, Fabio   Universita di Padova, Dip. di Astronomia
lafranca@astrpd.astro.it

LABHART, Lukas   University of Basel, Astronomical
Institute
labhardt@urz.unibas.ch

LAFON, Jean-Pierre	Observatoire de Paris, DASGAL, Meudon lafon@mesiob.obspm.fr
LAGAGE, Pierre-Olivier	CEA Saclay, Service d'Astrophysique lagage@sapvxg.saclay.cea.fr
LE FÈVRE, Olivier	Observatoire de Paris, Meudon leFevre@gin.obspm.fr
LÉGER, Alain	Institut d'Astrophysique Spatiale, Université de Paris XI leger@iaslab.ias.fr
LEIBUNDGUT, Bruno	ESO, Garching bleibund@eso.org
LÉNA, Pierre	Université & Observatoire de Paris lena@megasx.obspm.fr
LINDNER, Ulrich	Universitäts-Sternwarte, Gottingen ulindner@eden.uni-sw.gwdg.de
LUCY, Leon	ESO, Garching llucy@eso.org
MAGAIN, Pierre	Institut d'Astrophysique, Liege magain@astro.ulg.ac.be
MARANO, Bruno	Osservatorio Astronomico di Bologna marano@astbo3.bo.astro.it
MARIOTTI, Jean-Marie	Observatoire de Paris, DESPA, Meudon mariotti@mesioa.obspm.fr
MATHEZ, Guy	Observatoire Midi-Pyrenees, Toulouse mathez@obs-mip.fr
MATTEUCCI, Francesca	Osservatorio Astronomico di Trieste matteucci@atlantis.oat.ts.astro.it
MAZZALI, Paolo	Osservatorio Astronomico di Trieste mazzali@hpw002.oat.ts.astro.it
MELLIER, Yannick	Observatoire Midi-Pyrenees, Toulouse mellier@obs-mip.fr
MENDEZ, Roberto	Universitäts-Sternwarte, Munchen mendez@usm.uni-muenchen.de
MEYLAN, Georges	ESO, Garching gmeylan@eso.org
MILEY, George	Sterrewacht Leiden miley@strw.LeidenUniv.nl
MINNITI, Dante	ESO, Garching dminniti@eso.org
MONNET, Guy	Observatoire de Lyon monnet@orion.univ-lyon1.fr
MOORWOOD, Alan	ESO, Garching amoor@eso.org
MUELLER, Thomas	Universität Wurzburg, Astronomie mueller@astro.uni-wuerzburg.de

NORDSTRÖM, Birgitta	Copenhagen University, Astronomical Observatory birgitta@astro.ku.dk
OCH, Susanne	ESO, Garching soch@eso.org
ORTOLANI, Sergio	Osservatorio Astronomico di Padova ortolani@astrpd.astro.it
PACINI, Franco	Osservatorio Astrofisico di Arcetri pacini@arcetri.astro.it
PASIAN, Fabio	Osservatorio Astronomico di Trieste pasian@oat.ts.astro.it
PATAT, Ferdinando	ESO, Garching fpatat@eso.org
PEDERSEN, Holger	Copenhagen University, Astronomical Observatory holger@astro.ku.dk
PELLEGRINI, Silvia	ESO, Garching spellegr@eso.org
PETITJEAN, Patrick	Institut d'Astrophysique, Paris petitjean@iap.fr
PICAT, Jean-Pierre	Observatoire Midi-Pyrenees, Toulouse picat@obs-mip.fr
QUINN, Peter	Mt. Stromlo & Siding Spring Observatories pjq@mso.anu.edu.au
QUIRRENBACH, Andreas	MPI für Extraterrestrische Physik, Garching quirrenbach@mpe-garching.mpg.de
RAUCH, Michael	OCIW, Pasadena mr@ociw.edu
REINHEIMER, Thorsten	MPI für Radioastronomie, Bonn tr@specklec.mpifr-bonn.mpg.de
REIPURTH, Bo	ESO, La Silla reipurth@eso.org
RENZINI, Alvio	Universita di Bologna, Dip. di Astronomia alvio@alma02.cineca.it
RICH, Michael	Columbia University, Dept. of Astronomy, New York rmr@cuphyd.phys.columbia.edu
RIDGWAY, Stephen	NOAO, Kitt Peak ridgway@noao.edu
RODONÒ, Marcello	Universita di Catania, Instituto di Astronomia mrodono@astrct.ct.astro.it

RODRIGUEZ-ESPINOSA, José   Instituto de Astrofisica de Canarias
jre@ll.iac.es

RODRIGUEZ-ULLOA, Jesus   ESO, Garching
jrodrigu@eso.org

ROTH, Martin   Astrophysical Institüt Potsdam
mmroth@aip.de

ROUAN, Daniel   Observatoire de Paris, DESPA, Meudon
rouan@obspm.fr

RUDER, Hanns   Universität Tubingen, Theoretische &
Astrophysik

RUPPRECHT, Gero   ESO, Garching
grupprec@eso.org

SCARAMELLA, Roberto   Osservatorio Astronomico di Roma,
Monteporzio
kosmobob@astrmp.astro.it

SCHINDLER, Sabine   MPI für Extraterrestrische Physik,
Garching
sas@mpa-garching.mpg.de

SCHNEIDER, Jean   Observatoire de Paris, Meudon
schneider@mesiob.obspm.fr

SCHOBER, Hans Josef   Karl-Franzens-Universität Graz, Institüt
für Astronomie
schober@bkfug.kfunigraz.ac.at

SECCO, Luigi   Osservatorio Astronomico di Padova
secco@astrpd.astro.it

SETTI, Giancarlo   Istituto di Radioastronomia, Bologna
setti@astbo1.bo.astro.it

SHAVER, Peter   ESO, Garching
pshaver@eso.org

SINACHOPOULOS, Dimitrios   Observatoire Royal, Brussels
dimitris@astro.oma.be

SPYROMILIO, Jason   ESO, Garching
jspyromieso.org

STECKLUM, Bringfried   Max-Planck-Gesellschaft AG, Jena
pbs@physik.uni-jena.de

STIAVELLI, Massimo   Scuola Normale Superiore, Pisa
mstiavel@astro.sns.it

STRASSMEIER, Klaus   Universität Wien, Institüt für
Astronomie
strassmeier@astro.ast.univie.ac.at

SURDEJ, Jean   Space Telescope Science Institute
surdej@stsci.edu

SWINGS, Jean-Pierre   Institut d'Astrophysique, Liege
jpswings@vm1.ulg.ac.be

TACCONI-GARMAN, Lowell   MPI für Extraterrestrische Physik,

	Garching
	lowell@mpe-garching.mpg.de
TAMMANN, Gustav	University of Basel, Astronomical Institute
	tammann@urz.unibas.ch
TARENGHI, Massimo	ESO, Garching
	mtarengh@eso.org
THEODORE, Bertrand	ESO, Garching
	btheodor@eso.org
THIMM, Guido	ESO, Garching
	gthimm@eso.org
TINNEY, Chris	Anglo-Australian Observatory, Epping
	cgt@aaoepp.aao.gov.au
ULRICH, Marie-Helene	ESO, Garching
	mulrich@eso.org
VAN DER HULST, Thijs	Kapteyn Astronomical Institute, Groningen
	vdhulst@astro.rug.nl
VAN WINCKEL, Hans	Katholieke Universiteit Leuven, Instituut voor Sterrenkunde
	hans@ster.kuleuven.ac.be
VAUTERIN, Paul	Universiteit Gent, Sterrenkundig Observatorium
	vauterin@izar.rug.ac.be
VERBUNT, Frank	Institute of Astronomy, Utrecht
	verbunt@fys.ruu.nl
VETTOLANI, Giampaolo	Istituto di Radioastronomia, Bologna
	vettolani@astbo1.cnr.bo.it
VIGROUX, Laurent	CEA Saclay, Service d'Astrophysique
	vigroux@sapvxg.saclay.cea.fr
VILLAR-MARTIN, Montserrat	ESO, Garching
	mvillar@eso.org
VON DER LÜHE, Oskar	ESO, Garching
	ovdluhe@eso.org
WALSH, Jeremy	ESO, Garching
	jwalsh@eso.org
WAMPLER, Joe	ESO, Garching
	jwampler@eso.org
WARD, Martin	University of Oxford
	mjw@astro.ox.ac.uk
WATERS, Rens	SRON, Laboratory for Space Research, Groningen
	rens@sron.rug.nl
WEIGELT, Gerd	MPI für Radioastronomie, Bonn
	weigelt@mpifr-bonn.mpg.de

WEST, Richard	ESO, Garching rwest@eso.org
WIEDEMANN, Gunter	ESO, Garching gwiedema@eso.org
WILLIGER, Gerard	MPI für Astronomie, Heidelberg williger@mpia-hd.mpg.de
WOLTJER, Lo	Observatoire de Haute-Provence
ZAMORANI, Giovanni	Osservatorio Astronomico di Bologna 37929::zamorani
ZIEBELL, Manfred	ESO, Garching mziebell@eso.org
ZIJLSTRA, Albert	ESO, Garching azijlstra@eso.org
ZINNECKER, Hans	Universität Heidelberg, Institut für Theoretische Astrophysik hans@astro.uni-wuerzburg.de

Part 1

The Solar System and the Extra-Solar System Planets

The Solar System and the VLT

C. de Bergh

Observatoire de Paris, 92195-Meudon Cedex, France

Abstract. After considering the different methods that can be used to study Solar System objects, this paper reviews some recent work on the Solar System that has been accomplished by ground-based measurements. The possibilities offered by the VLT and instrumentation already planned are then considered, as well as other needs not fulfilled by the planned VLT instrumentation and which will not be covered by space observations in the near future.

1 Introduction

The Solar System includes, in addition to the Sun and the Earth, cold objects of very different nature. These objects have, as seen from the Earth, various brightnesses and spatial extents which means that different instrumental capabilities are required to study them. Besides the eight known extra-terrestrial planets that include five extended bright objects (Mercury, Venus, Mars, Jupiter and Saturn), and three faint planets with apparent sizes less than 3 arcsec (Uranus, Neptune and Pluto), there are sixty-one known planetary satellites that are (except for the Moon) faint objects with apparent diameters less than 1.7 arcsec, one bright extended ring sytem around Saturn and three faint ring systems (around Jupiter, Uranus and Neptune). The Solar System also includes a very large number of small objects: more than 6500 asteroids whose orbital properties are sufficiently well known than they can be classified, more than 1000 identified comets and at least seventeen trans-Neptunian objects. Many more asteroids have been detected, and many new ones are discovered every year. Most of the asteroids are faint and their apparent size is less than 0.7 arcsec in diameter. A few comets are very bright and can be very extended when they get close to the Sun, but many are faint and rather small. The class of trans-Neptunian objects has been identified recently. These objects, whose exact nature is not yet known, orbit the Sun beyond Neptune. The visual magnitudes of the objects discovered so far are between 22 and 24.

While some of these Solar System objects (essentially the brightest planets) have been extensively studied from space, important progress has been accomplished recently by ground-based investigations, thanks to progress in technology and telescope size. Ground-based studies are absolutely essential to prepare space missions that can carry only limited instrumentation and operate for limited periods of time. After reviewing the different methods of investigation and some recent ground-based studies, we will see what could be the possible main objectives for a VLT program on the Solar System, given our current state of knowledge and the possibilites offered by space missions and Earth-orbit satellites. We will also briefly examine whether or not current planned instruments for the VLT could allow us to carry out such a program.

2 The Different Methods of Investigation

Various techniques can be used to study Solar System objects. The classical methods used in astronomy are also used here: imaging; photometry; polarimetry; spectroscopy; spectrophotometry; spectro-imaging; radiometry. This can be done by remote-sensing either from the Earth, from a balloon, from an airplane or from Earth-orbit satellites. It can also be done from a spacecraft in orbit around a planet or during a fly-by, which is specific to Solar System studies. Other approaches include observations of occultations, either solar, spacecraft or stellar occultations. These are used to study planetary atmospheric composition, rings, size and shape of objects. Solar System objects can also be studied by radar observations made either from the Earth or from space. These radar studies have so far concerned a few satellites, the surfaces of Mars, Venus and Titan, the rings of Saturn, and a few asteroids and comets. In addition, from a spacecraft, there are unique possibilites to do in-situ measurements of the atmospheres and solid surfaces.

Earth-orbit satellites have been used, or are being used, to study the Solar System. Among them are the astronomical satellites IRAS, IUE, HST. Furthermore, there have been a few satellites, essentially UV satellites, specific for planetary observations. The Kuiper Airborne Observatory has also been used for observations of comets, Mars, Venus and the giant planets. In the near future, the ISO satellite will allow important progress to be made on solar system objects. Later on, the planned sub-millimetric FIRST satellite should open new windows particularly interesting for planets and comets, and the planned NASA-SIRTF satellite should provide a nice complement to ISO observations. The Hubble Space Telescope will, in the future, be equipped with new instruments that will be very useful for planetary studies, and, in particular, a near-IR camera.

Concerning the space observations, Venus and Mars have been extensively explored from space, with orbiter and in-situ observations. Jupiter, Saturn, Uranus and Neptune have been explored only during fly-by missions. No probe has ever been sent to Pluto. Concerning the smaller bodies, only two asteroids (Gaspra and Ida) and three comets (Halley, Giacobini-Zinner and Grigg-Skjellerup) have been visited by a spacecraft. Future space missions include the arrival of the NASA-GALILEO spacecraft to the Jupiter system in December 1995 and the NASA-ESA CASSINI/HUYGENS mission to the Saturn system, with a launch in 1997. For Mars, the next step is an intensive Mars exploration with, ultimately, meteorological networks, rovers and sample return, that will start with a NASA-MARS SURVEYOR mission to replace MARS OBSERVER. At the beginning of the next century, there will be the ESA-ROSETTA mission to a comet and several asteroids. Plans also include several small NASA Discovery missions to planets, comets and asteroids, and a Pluto-Fast-Flyby mission.

In spite of the possibilites offered by space exploration, much can be done from the Earth. Earth-based observations can benefit from larger telescopes and more complex instrumentation than space observations. In addition, they are absolutely essential to study phenomena that vary over long periods of time.

3   Ground-Based Exploration - Recent Progress

Planetary objects are relatively cold bodies. Their radiation consists essentially of reflected sunlight in the visible and near-infrared ($\lambda < 3\ \mu$m), and thermal radiation at longer wavelengths emitted by the planetary surface or the atmosphere when heated by the Sun. This thermal radiation peaks, for planets, at wavelengths between 10 and 100 μm, depending on their distance from the Sun, their albedo and their rotation rate.

Recent progress in Earth-based studies of planets, satellites, asteroids and comets has come, for a large part, from infrared and millimetric studies, mainly because of important improvement in the sensitivity of the detectors and instruments at these wavelengths, and access to larger telescopes. We will not consider here studies made in the millimetre, since they are outside the scope of this paper. Progress has also come from three other factors: the possibility to use large telescopes on very good sites (high-altitude); to work at higher spectral resolution than from a spacecraft; and to study phenomena that vary over long periods of time.

We have selected here some examples of recent studies that have been performed by ground-based observations in the visible and in the infrared.

a) **examples of photometric and imaging studies:**
   - detailed studies of the rings of Saturn and Uranus by stellar occultations in the visible. A resolution as high as 3 km can be obtained for the Uranus rings by measurements made from the Earth. The occultations allow study of the long term dynamical evolution of the rings (changes in ring widths, precession, etc.);
   - detection of a tiny atmosphere around Pluto (the pressure at the surface of Pluto is probably no more than a few microbars) by photometry in the visible observed during a stellar occultation in 1988 (see the review by Stern, 1993);
   - studies of the Pluto-Charon system during mutual eclipses and occultations by photometry in the years 1985-1990. Such opportunities occur only every 124 years. These studies, which are difficult, have provided information not only on the masses and radii of the two objects but also on albedo variations at the surface of Pluto (see Stern, 1993);
   - numerous studies of the rotational properties, diameters, direction of the rotation axes and shapes of asteroids, thanks to accurate photometric measurements. Some studies have also started, in particular at the ESO 3.6-m telescope, that make use of adaptive optics systems, allowing spatial resolution on objects less than 0.7 arcsec diameter in apparent size (Saint-Pé et al., 1993);

– observation from ESO of an outburst of Halley's comet when the comet reached 14.3 AU (West at al., 1991);

– the detection of many new asteroids, faint comets, and also the detection of trans-neptunian objects (see West and Hainaut, these proceedings);

– studies of Io's volcanic activity by photometry in the visible and IR, and by imaging, particularly around 5 μm where the hot spots, which can have temperatures in excess of 600K, radiate strongly;

– observations of the spectacular collision of comet P/Shoemaker-Levy 9 with Jupiter in July 1994 that provided excellent results from imaging in the near and thermal infrared (measurements made only from the Earth), and visible. Ground-based observations were made from numerous sites all over the world (see The Messenger, September 1994).

b) examples of high-resolution (R > 1000) spectroscopic studies:

– study of cometary gaseous emission at different distances from the nucleus. We take, as an example, the measurements of Arpigny et al. (1994) that concern C_2 emission observed at the center of a comet and at 800 km from the center (see Figure 1);

– deep probing of the atmosphere of Venus by near-IR spectroscopy. The lower part of the atmosphere of Venus is hidden by the clouds and difficult to study. There have been a few in-situ studies with mass-spectrometers and gas chromatographs, but they provided contradictory results on the abundances of some of the compounds, in particular water vapor. Thermal radiation from deep atmospheric levels can be detected at wavelengths shorter than 3 μm in a few spectral windows by remote sensing of the night side of the planet. The spectra of Venus in these infrared windows are extremely complex, and both high resolution (resolutions up to 28,600 have been used) and the access to broad spectral ranges covering entire atmospheric windows (as available with Fourier Transform Spectrometers) have been key factors in retrieving accurate information on the abundance and vertical distribution of compounds detected in the deep atmosphere of the planet like CO, H_2O, HDO, SO_2, OCS and HF (see, e.g., Bézard et al., 1990 and Bézard et al., 1993);

– the study of the aurorae of Jupiter, Saturn and Uranus by IR spectroscopy. Emission due to H_3^+ that is formed at very high altitudes was discovered in 2 and 4μm spectra of Jupiter, Saturn and Uranus at resolutions higher than a few thousands. FTS and grating spectrometers have been used for these studies. In addition, very high resolution FTS spectra of Jupiter (R=115 000) have allowed the natural width of the 4μm H_3^+ lines to be measured (Drossart et al., 1993);

– discovery of new compounds such as AsH_3 in the deep tropospheres of Jupiter and Saturn by IR spectroscopy in the 5 μm range at resolutions of the order of 21 000. This was made using Fourier Transform spectrometers which provided the high resolution and broad spectral coverage required to identify all the other absorbers and to define the proper atmospheric model

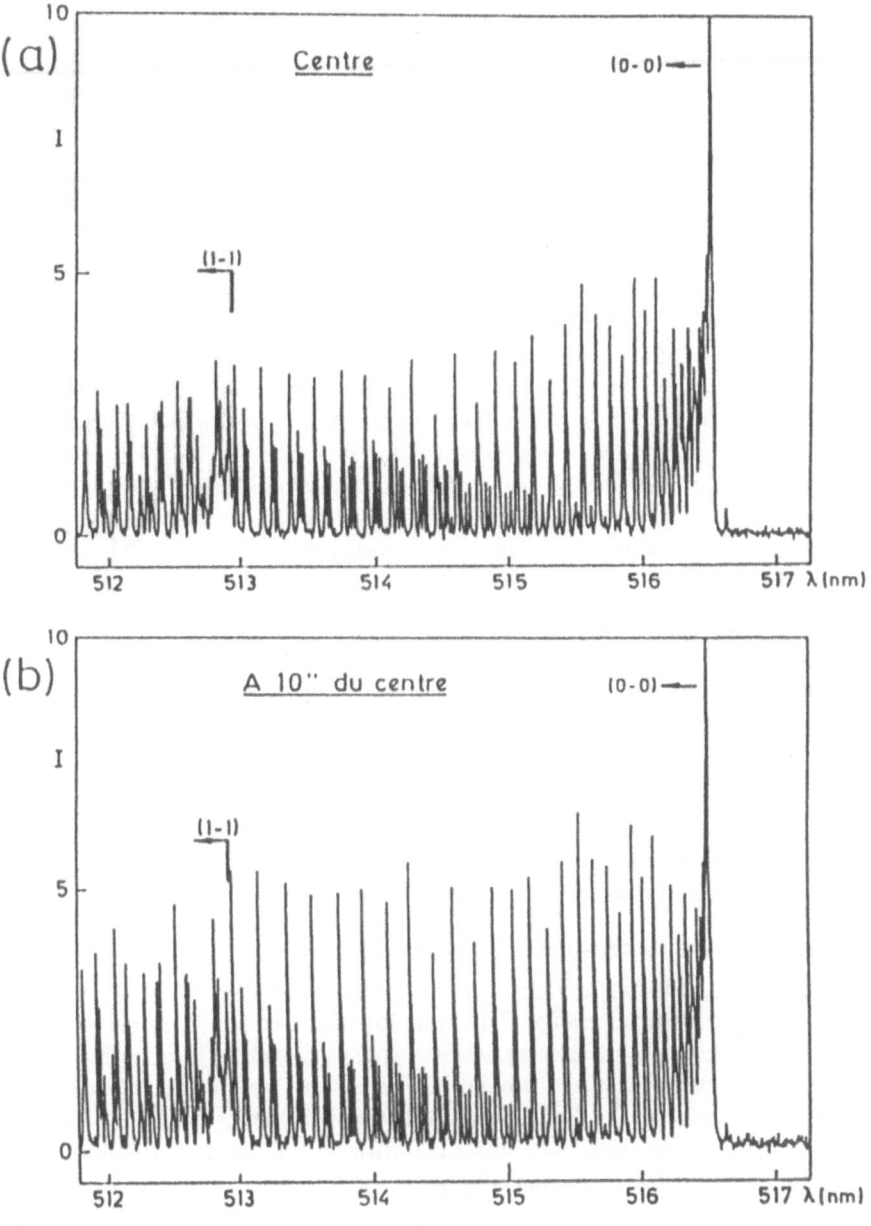

Fig. 1. The (0-0) and (1-1) C$_2$ Swan bands in comet Aarseth-Brewington observed at high spectral resolution (R= 60 000) with a 3 arcsec aperture. a) The spectrum of the comet at its center. b) The spectrum of the comet 10 arcsec from the center. The rotational intensity distribution differs in the two spectra, indicating differences in temperature. These spectra were obtained at the Observatoire de Haute-Provence 1.5 m telescope (Arpigny, 1994).

needed to retrieve the abundance of the newly observed species (see, e.g, Bézard et al., 1989);

– studies of important isotopic ratios (^{12}C/^{13}C, D/H, ^{14}N/^{15}N) from 10μm spectroscopy of the giant planets and Titan. Most of them have been made with the cryogenic echelle array spectrometer IRSHELL at the IRTF telescope at resolutions of about 10 000. These studies also provide information on hydrocarbon abundances and stratospheric temperatures (e.g. Orton et al., 1992).

c) examples of low-resolution (R < 1000) spectroscopic studies:

– determination of the nature of the ices at the surfaces of Triton by 1.4 to 2.5 μm spectroscopy. Triton was intensively observed by Voyager 2 in 1989, but no direct information on its surface composition was obtained as Voyager 2 did not carry instruments well adapted for that. CH$_4$ ice had been detected from Earth in 1979 and N$_2$ ice in 1984. Recent ground-based measurements using a cooled-grating spectrometer (CGS4) on the UKIRT telescope at a spectral resolution of about 300 have revealed the presence of CO and CO$_2$ ices at the surface of Triton (see Cruikshank et al., 1993);

– visible and infrared spectroscopy of asteroids to study their composition. Recent studies concern outer-belt asteroids (Cybeles, Hildas, Trojans) and members of dynamical families (see, e.g., Binzel and Xu, 1993). Spectral resolutions of the order of a few hundreds are generally sufficient for these studies;

– observations of the 3.4μm emission, which has now been detected in a number of comets in spectra with resolutions up to 500. Part of this emission is certainly due to methanol, another part to other unidentified carbon compounds (Davies et al., 1993; Bockelée-Morvan et al., 1994).

d) examples of spectro-imaging studies:

– study of O$_2$ emission produced in the upper atmosphere of Venus, around the 90-100 km altitude range. This emission observed at 1.27 μm shows very important spatial and time variations. Its study can help in retracing the complex circulation at the probed altitudes. Spectro-imaging studies have been carried out at the AAT with a grating spectrometer used in a scanning mode (Allen et al., 1992) and at CFHT with an FTS coupled to an infrared camera (Maillard and Simons, 1992);

– study of the effects of the comet S-L 9 collision on Jupiter. Strong 3.5 μm H$_3^+$ emission was observed at impact sites in the Southern hemisphere and also at similar latitudes and longitudes but in the Northern hemisphere. The observations were made using IRSPEC at ESO at resolutions $\sim$ 2 000 and by scanning over the whole disk of Jupiter (see The Messenger, September 1994).

4 Current Needs

In the case of planets, there is a clear need to better study the atmospheres of Uranus and Neptune. Relatively limited information on the atmospheres of these planets was obtained from the Voyager 2 exploration. Better ground-based spectroscopy and spectro-imaging would help in addressing directly or indirectly current problems concerning these planets such as the incredibly dynamical state of Neptune's meteorology and the lack of significant interior heat from Uranus. Such studies will require larger telescopes and better spatial resolution than presently available. Indeed these planets are rather faint and have small apparent sizes (see Table I).

More generally, for a better understanding of the complex photochemistry occuring in the atmospheres of the planets and Titan, we need to search for other compounds that may be present in relatively small amounts but still play

Table 1

Object	Diameter (km)	Maximum apparent size (arc-sec)	Maximum visual magnitude
Jupiter	142,984	47	-2.7
Saturn	120,536	19	0.67
Mars	6,794	18	-2
Uranus	51,118	3.9	5.5
Neptune	49,528	2.3	7.8
Ganymede	5,262	1.7	4.6
Callisto	4,800	1.6	5.6
Io	3,630	1.2	5
Europa	3,138	1	5.3
Titan	5,150	0.8	8.3
Ceres	913	0.7	7.4
Vesta	501	0.5	5.9
Pallas	523	0.4	8.1
Triton	2,700	0.18	13.5
Pluto	2,300	0.1	15.1

an important role. This requires very high spectral resolution in the infrared, as well as access to large spectral ranges. For Mars, ground-based studies are needed, in complement to the space missions, for a study of seasonal and long-term effects (changes in the atmospheric pressure, in the extent and composition of the polar caps, onset of storms, formation of clouds. etc.).

There is now considerable interest in bodies with tenous atmospheres that are controlled by the surface temperature or the volcanic activity, such as Pluto, Triton, 2060 Chiron and Io. Until recently, we knew very little about these bodies. Current work includes the search for gaseous CH_4 and CO in the atmospheres of Pluto and Triton, and the search for absorptions by gaseous SO_2 in the spectra of Io. These searches require studies in the infrared and they must be made at high spectral resolution ($> 20\,0000$) as the lines are intrinsically narrow. Spatial resolution would also be extremely important. In the case of Io, for instance, there is indeed clear evidence from millimetric and UV studies for a non-uniform atmosphere over the disk of the satellite. To obtain some spatial resolution on the objects, very sensitive spectrometers coupled with adaptive-optics systems would be required on a very large telescope. In addition, fiveμm imaging at high spatial resolution would be essential to better monitor Io's volcanic activity.

Other programs of interest include the study of faint emission on bright objects such as the distribution of the faint H_3^+ emission outside the auroral regions on Jupiter. This emission, discovered in 1992 in FTS spectra (Marten et al., 1994), is hard to study because it is quite faint. A complete spatial mapping of this emission and temporal monitoring would provide important information on the ionosphere of Jupiter. High spatial resolution can now be achieved (see Figure 2 and Ballester et al. 1994). A higher spectral resolution than obtained so far in spectro-imaging mode would be required to better study this emission.

Stellar occultations are very important to study the rings, their thickness and dynamical evolution, as well as to study the tenous atmospheres of Pluto, Triton or Chiron or the upper atmospheres of the giant planets. However, occultations by sufficiently bright stars are rare. With larger telescopes, one would have access to more stars, which would increase significantly the number of opportunities.

For asteroids, there is a need to study fainter objects for classification (small asteroids in the outer edge of the main belt, or further away; magnitudes higher than 16). Information on their shape can be retrieved from their photometric light-curves. Higher sensitivity and higher spatial resolutions in the infrared and in the visible are required to study the mineralogy and surface properties of asteroids or small satellites. High spatial resolution imaging studies are now being made in the visible for some asteroids with the HST. To do much better in the visible would require the VLTI. In the near-infrared, we will have to rely on ground-based observatories equipped with adaptive optics systems until a near-infrared camera is installed on the HST. What is essential for these near-infrared studies is utilize circular variable filters in order to be able to isolate signatures of solids.

For comets, it would, for example, be very important to identify the compounds responsible for the extra emission near 3.3 μm. This would require a

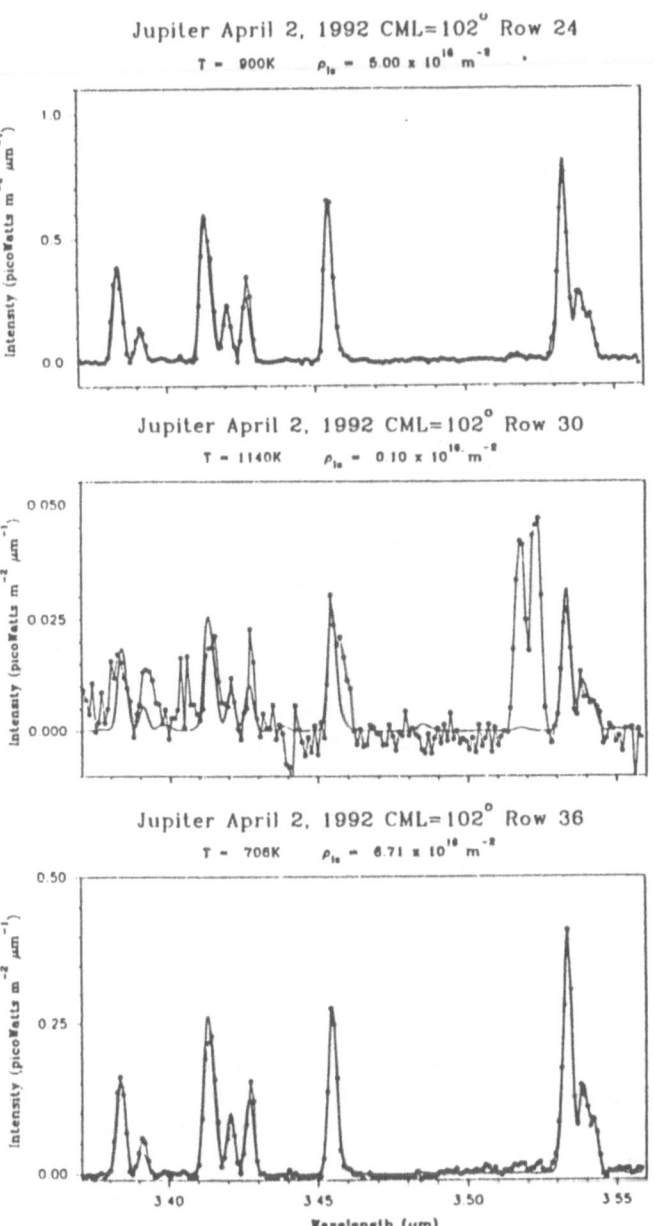

Fig. 2. Spectra of Jupiter recorded with the CGS4 array on the UKIRT telescope (resolution: 1000 to 1300). The three spectra correspond to the 102° System III longitude on the planet and to three different latitudes: 75° North, 0° and 67° South (from top to bottom). The emission detected (apart from two unidentified features at 3.517 and 3.522 μm) is from lines of H_3^+. Spectra at higher resolution and signal-to-noise would be needed to retrieve information on the temperatures and column densities for the emission detected at the low latitudes (Ballester et al., 1994).

spectral resolution higher than 500 (as achieved so far) in order to better separate gaseous from solid emission. Spectro-imaging of inner comae at moderate spectral resolutions, as has been started at CFHT to study gaseous emission with the Tiger field-integral spectro-imager (Festou, private communication), needs to be pursued and extended to fainter comets. More sensitive two-dimensional spectro-imagers in the visible and access to larger telescopes will be needed. Higher spectra resolution would also be required in order to study the fine structure of the molecular bands (see Figure 1). Resolutions of the order of 5 000 to 10 000 in a spectro-imaging mode may become possible in the future. To work at still higher resolutions in a complete two-dimensional imaging mode, as would be necessary for a refined study of the C_2 production and equilibrium as well as for many other species (see the reviews of Festou et al., 1993a and 1993b), may be difficult to achieve. Larger telescopes will offer the possibility to study more comets. Indeed, only the very bright ones have been studied in detail so far. Larger telescopes will also allow improved studies of the activity of comets when they are very far from the Sun (see West and Hainaut, these proceedings). Imaging and spectroscopic studies of fainter members of the asteroid-comet population will also help us in better understanding the relationship between asteroids and comets.

A very important new aspect of Solar System research is to complete the inventory of the Solar System. In particular, it is essential to search for more Kuiper belt objects and to study their nature. This requires more sensitive photometry and the capability to do routinely spectroscopy of objects with magnitudes higher than about 22 (West and Hainaut, these proceedings).

5 The Instrumentation Required

What type of instrumentation is needed to carry out all these programs? The needs concerning the very distant objects of the Solar System can be found in West and Hainaut (these proceedings). The most obvious needs and specificities for brighter objects are summarized here. There is a strong need for infrared and visible spectro-imagers. The range of spectral resolutions needed for spectro-imaging is very large: from a few thousands for the study of gases to a few hundreds for the study of solids. Most spectro-imaging studies concern fields of view less than 50 arcsec. in diameter. Exceptions are cometary studies and studies of the Io torus. Spectroscopy combined with adaptive optics capabilites is essential. Spectroscopy combined with an interferometer such as the VLTI would be extremely valuable. It is very important to have access to the 10 and 20 μm spectral ranges, both for imaging and for spectroscopy. It is also important to do spectroscopy over broad spectral ranges. Very high-resolution spectroscopy (up to about 10^5) is needed at all wavelengths for the study of bright comets and planetary atmospheres.

In the list of instruments under study for the VLT, the CRIRES (VLT high-resolution IR echelle spectrometer) and MIIS (Mid Infrared Imager/Spectrometer) instruments are the most important for planetary studies. CRIRES is also important for cometary studies. The possibility offered by the VLT to do simultaneous observations in different wavelength ranges by using different instrumentation on the different telescopes is very interesting. Possible studies of Solar System objects with the VLTI are relatively limited. They include diameters and shapes of asteroids, localization of hot spots on Io and studies of some cometary nuclei (if the coma is faint enough, and if the comet is not too far away from the Earth since the nuclei are generally less than 10 km in diameter).

What more would be needed at the VLT? It is important to develop adaptive optics as much as possible and install it on the VLT. The coupling of adaptive optics devices with spectrometers is essential. The minimum spectral range to be covered in the infrared is the 1 to 5 μm region. A spectrometer for the 10-20 μm region would also be very important. Indeed, as we have seen, this is a spectral range very favorable to the study of planetary atmospheres. Simultaneous access to a larger spectral range than currently offered by the 10μm IRSHELL instrument (about 3 cm^{-1}) would be important. Fabry-Perot imaging systems would also be very useful for more refined studies of planetary atmospheric emission or of the ions and neutrals in the Io torus. A spectro-imaging system equivalent to the FTS coupled with an infrared camera as currently exists at CFHT for the 1 to 2.5 μm spectral range, or something equivalent, would be extremely valuable. It should be cooled and coupled to a camera extending to at least 5 μm.

6 Conclusions

In summary, possible Solar System studies to be carried out with the VLT include the study of far-away or faint objects (such as faint comets, asteroids and Kuiper-Belt objects) by imaging and by spectroscopy in the visible and in the near-infrared. It includes also spectro-imaging of extended objects that are intrinsically faint (Uranus, Neptune, Io, faint comets) or have faint spatially variable features (such as weak H_3^+ emission outside the auroral zones of Jupiter, auroral emission on Saturn, weak cometary or Io torus emission). Spectro-imaging with higher spatial resolution than has been possible so far is required. There is a clear need for adaptive optics coupled with spectrometers. It is also essential in many of these studies to have a broad spectral coverage. Access to fields of view up to about 50 arcsec is necessary. Concerning the wavelength ranges of interest, it is clear that the infrared range is the most important and that the access to the 8-12 and 20 μm windows is essential. A few programs could make use of the VLTI. The best adapted instruments among the ones under study for the VLT are CRIRES and MIIS.

14

References

Allen, D.A., Crisp, D., Meadows, V. (1992): *Nature* **359**, 516

Arpigny, C. (1994): in "Molecules and Grains in Space", AIP Conference Proceedings, ed. I. Nenner, p.205

Ballester, G.E., Miller, S., Tennyson, J., Trafton, L.M., Geballe, T.R. (1994): *Icarus* **107**, 189

Bézard, B., Drossart, P., Lellouch, E., Tarrago, G., Maillard, J.P. (1989): *Astrophys. J.* **346**, 509

Bézard, B., de Bergh, C., Crisp, D., Maillard, J.P. (1990): *Nature* **345**, 508

Bézard, B., de Bergh, C., Fegley, B., Maillard, J.P., Crisp, D., Owen, T., Pollack, J.B., Grinspoon, D. (1993): *Geophys. Res. Letters* **20**, 1587

Binzel, R.P. and Xu, S. (1993): *Science* **260** 186

Bockelée-Morvan, D., Brooke, T.Y., Crovisier, J. (1994): submitted to Icarus.

Cruikshank, D.P., Roush, T.L., Owen, T.C., Geballe, T.R., de Bergh, C., Schmitt, B., Brown, R.H., Bartholomew, M.J. (1993): *Science* **261**, 742

Davies, J.K., Mumma, M.J., Reuter, D.C., Hoban, S., Weaver, H.A., Puxley, P.J., Lumsden, S.L. (1993): *Mon. Not. Roy. Astron. Soc.* **251**, 1022

Drossart, P., Maillard, J.P., Caldwell, J., Rosenqvist, J. (1993): *Astrophs. J.* **402**, L25

Festou, M.C., Rickman, H., West, R.M.: *Astron. Astrophys. Reviews*, **4**, 363

Festou, M.C., Rickman, H., West, R.M. (1993b): *Astron. Astrophys. Reviews*, **5**, 37

Maillard, J.P., Simons, D. (1992): in Proceedings of an ESA Workshop on "Solar Physics and Astrophysics at Interferometric Resolutions", Paris (France), February 1992, p. 205

Marten, A., de Bergh, C., Owen, T., Gautier, D., Maillard, J.P., Drossart, P., Lutz, B.L., Orton, G.S. (1994): *Planetary and Space Science*, **42**, 391

Orton, G.S., Lacy, J.H., Achtermann, M., Parmar, P., Blass, W.E. (1992): *Icarus* **100**, 541

Stern, S.A. (1992): *Ann. Rev. Astron. Astrophys.* **30**, 185

Saint-Pé, O., Combes, M., Rigault, F. (1993): *Icarus* **105**, 271

West, R.M., Hainaut, O., Smette, A. (1991): *Astron. Astrophys.* **246**, L77

Very Distant Objects in the Solar System

Richard M. West[1], Olivier R. Hainaut[12]

[1] European Southern Observatory, Karl-Schwarzschild-Strasse 2, D-85748 Garching, Germany
[2] Institute for Astronomy, University of Hawaii, 2680 Woodlawn Drive, Honolulu, HI 96822, U.S.A.

Abstract. We discuss photometric and spectroscopic observations of very faint and distant minor objects in the Solar System that will become possible with the ESO Very Large Telescope (VLT). The proposed studies are based on extrapolations from recent, related pilot programmes at the performance limits of present observational facilities, notably the 3.58-metre ESO New Technology Telescope with associated instrumentation (EMMI and SUSI). The VLT observations will most certainly have a major impact on future studies of physical processes in hitherto inaccessible reaches of the outer solar system and, thus, on the continued investigation of its origin and early evolution.

1   Current Observational Status

The outer reaches of the solar system, here defined as the region beyond heliocentric distance $r = 10$ AU, are populated by several different classes of *known* objects: 1) the *major planets* (Uranus, Neptune); 2) their apparently quite diverse *moons* (e.g., Miranda, Triton); 3) an increasing number of comparatively large *"minor planets"* (at this moment naturally divided into the more nearby "Centaurs" like (2060) Chiron, and 14 transneptunian objects discovered during 1992–94); 4) a small number of *comets* (all long-period objects detected around their immediately preceeding perihelion passage); and 5) *charged and neutral particles of the solar wind* (detected by Pioneer and Voyager S/C). In addition, there is most likely some *dust* near the main plane (possibly already imaged by IRAS), as well as many, as yet *undetected comets*.

The exact status of Pluto and Charon within this scheme is unclear. However, there is now a tendency to consider that these two objects, possibly also the neptunian moon Triton, as well as most of the "Centaurs" and the new transneptunian objects, all originate in the "Kuiper Belt", i.e., the outermost part of the flattened solar nebula, first hypothesized by Kuiper (1951) at $r \sim 50 - 100$ AU.

Just a few years ago, the only possible observations of objects in the outer solar system were those of the major planets, their moons and Pluto/Charon, and once in a while a particularly bright, long-period comet on its way out. They included for instance regular monitoring of the atmospheres of Uranus and Neptune in various optical and IR wavebands, medium-resolution spectroscopy of Pluto, Charon and the brighter moons, and broad-band CCD imaging of the faint comets. Now, however, due to progressively better astronomical instrumentation, in particular in terms of sensitivity and angular resolution, it has become possible to perform more detailed observations of these objects and, not least, to detect and subsequently observe in more detail many smaller objects in the outer solar system.

Regular observations are now done of even very distant comets (Meech 1993, Hainaut *et al.* 1994a, Hainaut 1994). These studies have led to important new insights into the behaviour of these objects, for instance the unexpected, very violent outburst of P/Halley at $r = 14.3$ AU (West *et al.* 1991; Sekanina *et al.* 1992) and the continued activity of comet Cernis ((1983 XII) at 24 AU (Meech *et al.* 1994). At the same time, more detailed observations of the five known "Centaurs", (2060) Chiron, (5145) Pholus, (5335) Damocles, 1993 HA2 and 1994 TA, have become possible (cf. Meech and Buie, 1994).

Table 1. The known transneptunian objects (October 1994)

Object	a (AU)	e	i deg	V mag	Orbit MPC	Arc
1992 QB1	43.887	0.070	2.189	23.8	22971	2 opp.
1993 FW	43.872	0.047	7.741	23.4	23870	2 opp.
1993 RO	39.334	0.198	3.720	23.2	23982	2 opp.
1993 RP	39.329	0.114	2.570	24.6	23493	2 days
1993 SB	39.421	0.321	1.929	23.3	[1]	2 opp.
1993 SC	34.391	0.185	5.164	22.5	23863	2 opp.
1994 ES2	45.269	0.012	1.036	24.8	23653	86 days
1994 EV3	43.130	0.043	1.626	23.7	23653	86 days
1994 GV9	42.184	0 [2]	0.056	23.4	23653	23 days
1994 JQ1	43.306	0 [2]	3.837	23.5	23654	23 days
1994 JR1	35.264	0 [2]	3.814	22.6	23654	26 days
1994 JS	36.540	0 [2]	15.422	22.8	23653	27 days
1994 JV	35.250	0 [2]	18.081	23.6	23653	25 days
1994 TB	31.716	0 [2]	10.228	22.2	[3]	2 days

[1] IAUC 6085; another possible, but unstable orbit: $a = 42.9$, $e = 0.42$
[2] Eccentricity assumed
[3] Announced in MPEC 1994-T02 on Oct. 5, 1994

One of the most important recent developments, though, is the discovery of no less than 14 transneptunian objects in just over two years, beginning with 1992 QB1 in August 1992, cf. Table 1 that represents the situation in early October 1994. They represent an entirely new population of objects in the solar system. It is, however, unfortunate that due to lack of astrometric observations only five of them have so far been observed at two oppositions and have therefore been "secured"; the others have only been observed over very short orbital arcs and are most probably "lost".

2 Observational Possibilities with the VLT

The *faintest* solar system object so far observed is P/Halley at 18.8 AU and V = 26.5 mag (Hainaut *et al.* 1994b); the *most distant* is 1994 ES2 at ~ 45 AU (Table 1), and *the most distant and active* object is comet Cernis at 24.6 AU and V = 25.5 (Meech *et al.* 1994). The *highest angular resolution* achieved during imaging of a distant solar system object is ~ 0.04 arcsec (Pluto and Charon; Albrecht *et al.* 1994). The VLT will be able perform observations significantly beyond these limits, both because of its large photon collecting area and its unsurpassed interferometric resolution (≈0.001 arcsec).

For VLT observations of individual, faint objects, we show in Fig. 1 the predicted exposure times necessary to reach an object of a given size and heliocentric distance (i.e., of a given magnitude) with the FORS instrument at an 8.2-metre VLT Unit Telescope. The upper part (Fig. 1a) corresponds to imaging (S/N = 5) and the lower (Fig. 1b) to spectroscopy (S/N = 10, R = 600 at $\lambda = 550$ nm). It is thus possible, within a reasonable, total exposure time (say, 2 hours), to securely detect an object of 500-km diameter at $r = 165$ AU, or to obtain a somewhat weakly exposed spectrum of an object of 125-km diameter at $r = 40$ AU.

In the survey mode, it should be possible to cover with the same telescope and FORS about four 6×6 arcmin fields twice per night at S/N = 5 to V = 27.5 mag. Assuming a transneptunian object density of $\sim 50 - 500$ per square degree at $V \leq 27.5$ mag (Marsden, private comm.), the predicted discovery rate is $2 - 20$ new objects per night, cf. the extensive discussion of this subject in the paper by Hainaut *et al.* (1994a).

3 Specific Projects

We now mention some of the possible VLT programmes specifically concerned with the most distant objects in the solar system.

For the outer planets and their larger moons, the high angular resolution of the VLT will allow very detailed *optical and IR images* to be obtained (nominal resolution ~ 100 km at the distance of Neptune). *Near-IR spectroscopy*, in particular in the "organic" wavebands, also appears particularly promising for these objects. More projects are mentioned in the review paper by C. de Bergh in this volume.

For the transneptunian and Kuiper belt objects, the following main questions may now be addressed by means of dedicated surveys and observations of individual objects:

Density/distance/orbits: It is of interest to learn the distribution of these objects with heliocentric distance and their orbits, as well as their dynamical history. Do their numbers increase outwards ? Is there a maximum at a certain distance ? Are there "gaps" due to orbital resonances ? How stable are the orbits, i.e., how long have the objects been at this distance ? This would also cast more light on the distribution of matter in the proto-solar disk.

18

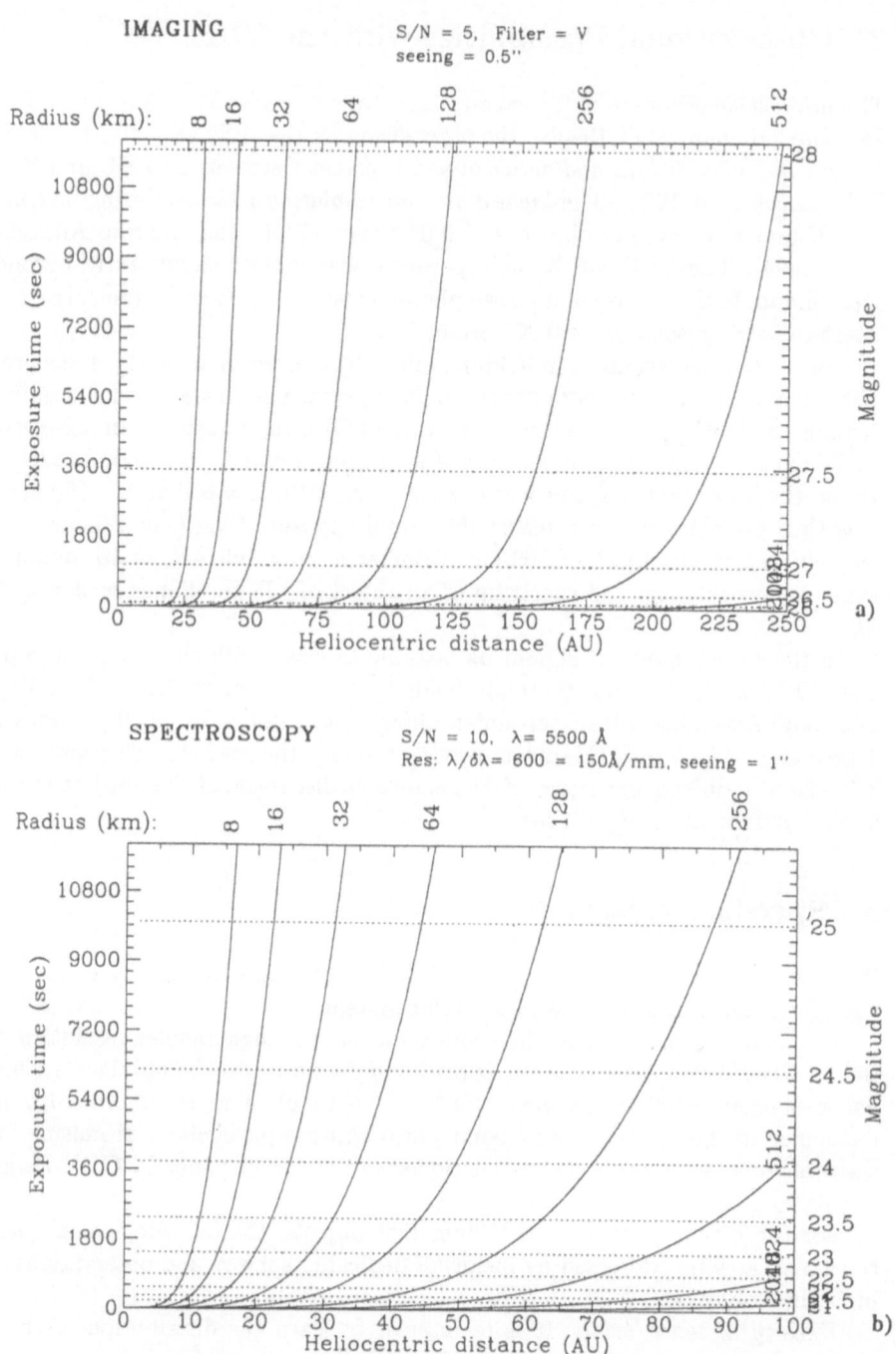

Fig. 1. Observing limits with an 8.2-metre VLT Unit Telescope and FORS. Exposure time vrs. heliocentric distance r and object radius (km), needed to obtain **a.** a V-image at S/N = 5, and **b.** an R=600, S/N = 10 spectrum at 5500 Å. In both cases, the assumed albedo = 0.04, i.e., equal to that of the nucleus of P/Halley.

Number/size/shape/rotation: Whenever possible, it will be important to determine the (radiometric) size and the rotational properties of individual objects. Are the larger objects more spherical than the smaller ones ? Do the smaller rotate faster than the larger ? This may contain information about their origin, in particular whether some of them are fragments of larger objects that collided at an earlier epoch.

Surface composition: Multi-band photometry and/or low-resolution spectroscopy will make it possible to identify the main surface components. Are they all similar, or are there particular object "types" (as in the main belt) ? Are they similar to other minor bodies, in particular to the short-period comets (that are supposed to originate in the Kuiper Belt) ?

Atmosphere/coma/activity: While (2060) Chiron and some of the distant comets have a coma, most of the known distant, minor objects apparently have not. Is this because of compositional differences or is it due to their recent orbital histories ? Will major outbursts be observed ? If so, at which distances – this may hint at the causes and help to identify the responsible constituents.

4   Conclusions and Desiderata

We conclude that the VLT will be able to perform unique and extremely interesting observations of distant solar system objects, in particular the fainter ones which may be detected in large numbers by this new instrumental facility. In this way the VLT will provide a major contribution to the exploration of the outer solar system and thus to the study of its origin and early evolution.

However, there are two important provisos connected to this. First of all, it would be most useful to have at least one instrument at one of the 8.2-metre unit telescope foci with a *much larger sky field than presently foreseen*, possibly by means of a focal reducer combined with a very large detector in the form of a state-of-the-art CCD mosaic. This would greatly enhance the VLT's ability to survey reasonably large sky areas in order to discover new objects of this type. Furthermore, in order to avoid excessive light losses because of image smearing, it is of paramount importance that the VLT is able to perform *accurate off-set guiding* during the imaging and spectral observations of moving objects.

References

Albrecht, R., Barbieri, C., Adorf, H.-M., Corrain, G., Gemmo, A., Greenfield, P., Hainaut, O., Hook, R.N., Tholen, D., Blades, J.C., Sparks, W.B. (1994), ApJ. Let., submitted

Hainaut, O.R. (1994), "Selected Observations of Solar System Minor Bodies", Ph.D. Thesis, Université de Liège and ESO.

Hainaut, O.R., West, R.M., Smette, A., Marsden, B.G. (1994a), A&A 289, 311

Hainaut, O.R., West, R.M., Marsden, B.G., Smette, A., Meech, K.,J. (1994b), A&A, in press, ESO preprint 1030

Kuiper, G. (1951), in "Astrophysics, a Topical Symposium", Hynek, McGraw-Hill Eds., New York, p. 357

Meech, K. (1993), in Proceedings of Lenggries "Workshop on the activity of distant comets", Huebner *et al.* Eds., Southwest Research Inst., San Antonio, p. 12–20

Meech, K.J., Buie, M.W. (1994), submitted to Icarus

Meech, K., Hainaut, O.R., Marsden, B.G., Smette, A., West, R.M. (1994), in preparation

Sekanina, Z., Larson, S., Hainaut, O., Smette, A., West, R.M. (1992), A&A 263, 376

West, R.M., Hainaut, O., Smette, A. (1991), A&A 246, L77

How to Search for Extra-Solar Planets with the VLT/VISA?

A. Léger[1], J-M. Mariotti[2], D. Rouan[2], J. Schneider[2]

[1] Institut d'Astrophysique Spatiale, CNRS, bat 121, Université Paris-Sud, F-91405 Orsay
[2] Observatoire de Paris, CNRS, place J. Janssen, F-92195 Meudon

Abstract. We review the motivation for searching for Extra-Solar planets, and point out that we *really do not know* if the Solar System is a common place or a rare stellar system. A weak version of the Anthropic Principle prevents us from applying the Copernicus Principle to deduce that there are many other planetary systems similar to ours. We consider the different possibilities to search for exo-planets: (i) direct methods, where one has to beat the huge contrast between the stellar flux and that of the planet (10^{10} in the visible for the Sun/Earth system); (ii) indirect ones, where the stellar motion or photometry are studied, looking for perturbations due to possible planets. The indirect methods can be of different types: differential astrometry, radial velocimetry, (pulsar) timing and precise photometry. We show that the VLT and VLTI have remarkable astrometric and velocimetric potential for searching for Jovian planets around a large sample of stars of many types, including solar-like ones. We conclude that *Europe should not miss the start of a key field for 21st century Astrophysics* and should begin to build the appropriate instruments.

1 Introduction: Why search for Extra-Solar Planets?

The motivation to search for extra-solar planets (exo-planets) are two-fold. First, we would like to know how the formation of planetary systems depends on various factors. The second problem is quite different in nature: how frequent is Life in the Universe? The first question is purely astrophysical while the second, in addition to its (pre-)biological aspects, offers an occasion for astrophysics to have contact with the deepest questions on the connection of Mind with the physical Universe.

1.1 Astrophysical Reasons

For several years, an increasing number of circumstellar disks has been detected around other stars. It is generally believed that they may evolve toward planetary systems. In the unique case of the Beta Pic dust disk (Smith and Terrile, 1984), this detection is direct. In all other cases the detection is indirect. The disks are detected either by infrared or by radio excesses in the spectrum of the central star, either in Herbig Ae/Be stars or around T Tauri stars. For T Tauri stars, 50% of them have been found to have evidence for a disk (Beckwith and Sargent 1993).

Although we do not understand in detail the formation of the Solar System, the simulation of the growth of planets from planetesimals (Wetherill, 1994) shows that the formation of planets is the natural and final outcome of dust disks.

We would like to know the answer to several questions:

- - how frequent are other planetary systems?
- - are they similar to ours?
- - how do they form and evolve?
- - how do these features depend on the type of the central star (mass, age, binarity etc.)?
- - are there regions of the Galaxy where their formation is more frequent?

All these questions require statistical studies, which in turn demand the detection of as many extra-solar planets as possible.

1.2   The Presence of Life in the Universe

By "life", two different notions are meant. First of all, by primitive life we mean any, low entropy content, self replicating, chemical activity. Signs of such an activity can be found in extra-solar planets by spectroscopic studies (Sect. 2.2). The second aspect, far more important but more delicate, can be formulated as "is there a physical activity associated with ther complex organisms in the Universe? " In the present paper we present only the first steps towards this goal, but this latter aspect *is in the background of everyone's questioning.*

• A Conflict between two Principles

The outcome of the search for life outside the Earth is quite open because we are facing a conflict between two general principles which are usually guidelines:

– the Copernicus principle states that "the Earth is a common place in the Universe". Its generalisation implies the existence of many other such places, including other habitable planets;

– a reasoning similar to that leading to the Anthropic principle warns us that "the mere fact that Earth has to be habitable for us to ask the question may bias the problem". If there were a single such planet in the Galaxy, it would be the Earth because it is "the" place where life has developed and evolved so that intelligent beings can live and ask themselves such questions.

The Copernicus principle applies to questions that have no implications upon the possibility for mind to occur. An example is the position of the Sun in the Galaxy, possibly excluding its very center, but it does not apply to parameters that are necessary for life to occur.

The straightforward conclusion is that *observations are required.*

• The Drake equation

Drake (1965) has estimated the number of contemporary technological civilisations in the Galaxy as the product of:

– the number of stars formed per year (1);
– the probability of having a habitable planet around a star (2);
– the probability that life appears when conditions are favourable (3);

– the probability that evolution leads to a technological civilisation (4);

– the mean lifetime of such civilisations (5).

In this equation, several factors are completely unknown. If one tries to make (middle) educated guesses for each of them, the large number of stars in the Galaxy ($N \sim 10^{11}$) would lead to a large number of such civilisations. But then, one faces the Enrico Fermi paradox, "where are they?", implying "why are they not here already?". Interstellar travel, at velocities $c/30$ to $c/10$, seem not impossible for Human civilisation in a few centuries. Hence, extra-solar civilisations, born (much) earlier than ours, should have travelled to us and possibly colonised us.

One or several factors in the Drake equation can be much lower than first guessed and produce bottlenecks. Then, it is of the highest interest to measure those factors that are accessible to our present and near-future technology. This is described further as the "step by step" approach.

• The SETI approach

The different programs of Search for Extraterrestrial Intelligence (SETI) such as HRMS, formerly supported by NASA, META, SERENDIP and OSURO attempt a short cut, searching for signals emitted, intentionally or not, by other technological civilisations that would reach us. Signals have been emitted from the Earth and are presently detectable within a sphere of radius about 50 lightyears, unless *all* of them have switched to non-radiating means of communication (e.g. optical fibers). These searches would be of the greatest interest if successful, but the difficulty is that we have no way to estimate the probability of success.

• The "Step by Step" approach

An alternative to SETI is an approach where one tries to measure the factors in the Drake equation which are accessible to our telescopes in the near and middle future, namely (1), (2) and (3). It is our opinion that both approaches should be pursued in parallel, when one considers their implications for Science and, more generally, for Mankind.

As stated in the TOPS (Burke 1992) and COMPLEX (Burns 1994) reports, we live in a special epoch: *for the first time*, we are able to *give objective answers* to questions that Humanity has been asking for centuries (e.g. Epicurus, 3rd cent BC), such as "are we the only beings with mind in the Cosmos?"

In this review, we concentrate on a question related to factor (2) of the Drake equation: "are there numerous planets around nearby stars?" and more specifically "is the Solar System a special stellar system or a commonplace one?" It leads to the search for: (i) Jovian exo-planets; and (ii) telluric ones.

The central difficulty to directly detect an exo-planet is the huge contrast between the fluxes of the star and the planet. This contrast is 10^{10} in the visible for the Sun/Earth system (Angel et al 1986). The different detection methods can be divided in two categories: (i) the direct ones that try to overcome this huge handicap; and (ii) the indirect ones, where the star is observed and perturbations in its photometry or motion, due to possible planet(s), are searched for.

2   Direct Detection Methods

• Imaging in the visible

Various direct imaging methods in the visible have been proposed (Angel ,1994; Malbet et al. 1994; Labeyrie, 1994). These programs would be very spectacular if successful. However, they would give access only to a very limited sample of star systems which would be a drastic handicap if no detection were made.

• IR Detection and spectroscopy from space

The main hope to directly detect and spectroscopically study a reasonable sample of planetary systems is from space in the mid-IR. This wavelength domain has two major advantages: (i) the contrast between the star and the planet is reduced by almost 3 orders of magnitude from the situation in the visible (Angel et al 1986); and (ii) the $5 - 20\mu m$ range contains characteristic bands of key atmospheric constituents. The emission of the Earth's atmosphere however prevents such observations from the ground.

Several proposals have been made in recent years. The SISTER project (Bély et al, 1992) is based on a Fizeau interferometer and aims to image exo-planets, including telluric ones. A project has been proposed to ESA for its Horizon 2000+ program, named DARWIN, which is still more ambitious. It aims at detecting telluric exo-planets and, in case of success, to take the IR spectrum ($5 - 20\mu m$, $R = 20$), searching for H_2O, CO_2 and O_3 spectral signatures. The presence of this last molecule being an indicator of photosynthetic activity that produces O_2 which is needed to make large amounts of O_3, and therefore points to a carbon chemistry based life. If successful, such a mission would permit an estimate of factor (3) in the Drake equation.

These projects are long-term ones and this justifies the working out of (simpler) indirect methods from the ground and space.

3   Indirect Methods

These methods do not attempt to separate the planet(s) light from the stellar radiation and search for features that can characterize the presence of planet(s) around the star.

3.1   The Star Photometry

If the orbital plane of a planet is edge on, or close to it, at some point during the planet's revolution, it will transit in front of the star disk and make a partial eclipse of the latter (Fig. 1).

The probability for occurrence of transits is small, but not vanishing for planets close to the star, i.e. telluric ones. Considering the Solar System observed from a random direction: in 0.5% of cases an observer would see Earth transits; and in an additional 2.2%, transits of either Mercury, Venus or Mars would be observed. The change in the stellar flux is $(R_{pl}/R_*)^2$. In the case of the Earth/Sun system, the change in flux is 1.10^{-4} and the transit lasts about 10 hours.

Space-based photometric missions in the visible are presently under consideration for other purposes. For example the ESA mission STAR, is planned to make photometry with a *rms* precision of about 10^{-5} for integration times of one hour. This could make a significant number of detections, if a large number (5,000 - 10,000) of stars is continuously monitored during 2-3 years.

From the ground, photometry cannot be made so accurately, but differential measurements with an *rms* of 10^{-3} should be possible. This would allow the search for transits of larger planets around smaller stars (e.g. Jupiters/M dwarfs). The stellar systems have to be carefully chosen (e.g. eclipsing binaries or stars with known spinning axes) to enhance the probability of having the correct orientation for their planetary planes, because only a smaller number of objects can be monitored (Schneider & Chevreton, 1990).

The problem of possible astrophysical artefacts that could mimic the transit of a planet, e.g. stellar spots in a slowly rotating star, has to be considered as well as the stellar photometric noise that could bury the planetary signal. Fortunately, planet transits have specific features (achromaticity, recurrence, etc.) that should help their characterization.

Note that, with direct observations in the IR from space, the search for transits is one of the very few methods that could detect *telluric planets* in the near to middle future. It is not suitable for searching for Jovian planets (low geometrical probability, infrequent transits) and is complementary to the following methods.

Fig. 1. Transit of a planet in front of a star

Fig. 2. A star and a planet orbiting around their center of mass I

3.2 Star Dynamics Studied by Differential Astrometry

The planet(s) and star orbit around their center of mass I (Fig 2). The latter has an uniform motion if other bodies are far away. The star describes an ellipse around I. The semi-major axes a^*, a, masses M_*, m, angular separation θ and star distance D are related by:

$$a_* M_* = am \;, \qquad \theta \equiv \frac{a_*}{D} = 500 \; \frac{a/5.2AU}{D/10pc} \times \frac{m/m_{Jup}}{M_*/M_\odot} \; \mu arcsec \qquad (6)$$

θ is the quantity that can be measured by differential astrometry between the target star and a (distant) reference star whose angular motion on the sky can be considered as uniform. Relation (6) indicates that the method is sensitive to massive and distant planets, i.e. Jovian ones. In the paper by Quirrenbach (these proceedings), the actual motion of the Sun is shown; t is complex but dominated by Jupiter and Saturn (Burke, 1992, Fig. 3-2).

The reflex motions of the Sun, if seen from 10 pc, would have amplitudes:

$Jupiter/\odot \longrightarrow \pm 500 \mu arcsec$

$Saturn/\odot \longrightarrow \pm 270 \mu arcsec$

$Earth/\odot \longrightarrow \pm 0.3 \mu arcsec$

The interferometric mode of the VLT, even restricted to VISA, *is remarkably suited to differential astrometric measurements*, provided a modification of the interferometric system is scheduled that allows for an increase of the effective field of view up to 15-30″, and makes possible simultaneous fringe measurements for the target star (required magnitude $K < 12$) and a reference star ($K \le 17$) (von der Lühe et al. these proceedings). Around a randomly chosen point in the sky, there is a 90 % probability to find a $K \le 17$ star at less than 17″. The actual measurement to get the angular $\Delta\theta$ vector between the two stars is that of a differential delay path that has to be added to the reference star. Real time fringe tracking is performed on the (bright) target star and the fringes of the reference star are searched for, with some integration time, which allows the latter to be fainter.

A 3-telescope VISA configuration is needed to measure the two components of $\Delta\theta$. Von der Lühe et al (these proceedings) have shown that given the Paranal atmospheric turbulence conditions, 30 min integration time and a 100 m interferometer baseline, angular measurements should have rms errors:

$$\sigma = 26 \mu as \quad \text{for 17as star separation,} \qquad\qquad \sigma =$$
$8\mu as \quad$ for 5as $\quad -$

Such measurements would open up a large number of accessible stellar systems. For instance, if we require, per point, a 5σ measurement of the ellipse semi-major axis, Jupiter-mass planets would be detectable around solar type stars up to

$10pc \times 500\mu arcsec/(26\mu arcsec \times 5) \simeq 40pc.$

There are about 2000 G stars within this distance (with magnitude $K \le 8$, which well fulfils the brightness requirement) and a total of 6000 F+G+K stars.

This would give the possibility to carefully select a sample of $\sim$200 stars to be studied. In fact, if systematic errors can be eliminated, the random error on individual measurements can be smaller because a trajectory is determined by many points, leading to larger accessible distances (see Quirrenbach, these proceedings, for details). The ESA mission GAIA ($20\mu arcsec$ absolute astrometry) should also be very valuable in the further future if it flies long enough.

The differential astrometric method of detecting Jovian planets is basically *free from astrophysical artefacts*. This can be realised when noting that the Jupiter/Sun center of mass, I, is at $1.1R\odot$ from the Sun center. Clearly, spots and chromospheric eruptions are unable to move its photometric barycenter sufficiently to prevent a good determination of the Sun to I distance.

In addition, the measurement of a_*, the period of the orbit, gives a from Kepler's 3rd law and the knowledge of M_*, *yields directly the planet mass m, independently of the orbital plane angle i* with the sky plane.

3.3 Star Dynamics Studied by Velocimetry

● Typical numbers

When the inclination angle i is different from zero, a radial velocity (RV or v) is induced by the presence of a planet. This reflex motion can be measured by the Doppler effect on stellar spectral lines. The extremum RV is:

$$v = \pm 13 \ \frac{(m/m_{Jupiter})\sin i}{(a/5.2AU)^{1/2}(M_*/M_{\odot})^{1/2}} \ ms^{-1} \qquad (7)$$

This velocity is independent of the system distance but the stellar flux can limit the accuracy of the measurement (photon noise). Values for Jupiter and the Earth around the Sun, for $\sin i = 1$, are:

$$Jupiter \longrightarrow v = \pm 13 ms^{-1}, \quad P = 12yrs$$
$$Earth \longrightarrow v = \pm 0.1 ms^{-1}, \quad P = 1yr$$

The global RV of the Sun is dominated by Jupiter (Burke, 1992, Fig. 3-5).

● The Technical Challenge

The difficulty of measuring such velocities can be appreciated from a single figure. To measure a RV with an rms precision of $2ms^{-1}$ with a $R = 100,000$ spectrometer and a detector oversampling of 2, requires the position of the stellar lines to be determined at the level of 1.310^{-3} detector pixels. The measurements need long term stability as they have to be compared at 10 or 20yr intervals.

● Methods

The techniques presently considered are:

– Pulsar timing: the precision of the measurement, after (delicate) Earth motion corrections, is fantastic. When expressed in term of RV, it is typically $0.1mms^{-1}$! However, it applies only to very specific stellar systems, a priori, not favourable for life to develop.

– Fabry Pérot interferometer: used by Mc Millan et al. (1993)

– Gas cell reference: a gas is used that has many transitions in the visible in order to give an absolute wavelength reference. The stellar flux travels through the cell, i.e. with the same optical path. HF cells have been used by Walker et al (1992) and I_2 cells by Marcy et al (1992), Hatze et al (1993) and Kürster et al (1994). The presently announced accuracy is $5 - 15ms^{-1}$.

– CORAVEL: a correlation is made between the star spectrum and a reference one that can be shifted in frequency, until a maximum is obtained, indicating the actual Doppler shift. An absolute standard lamp is required (Mayor et al, 1992). With the new stable ELODIE spectrometer, a precision of $10 - 14ms^{-1}$ is presently obtained that could be improved on in the future.

– LASER monitored Fabry Pérot: Connes (1994) is developing an instrument that could be adapted to most spectrometers. The absolute standard is a Laser and the system should reach the photon noise limit.

• Problems

The keys points to obtain the required long term high stability are still a matter of investigation. It appears that:

– systematic error sources have to be tracked down. They come from thermal changes in the spectrometer, bad focussing of the telescope, atmospheric pressure changes, etc.; – astrophysical processes that can mimic planets such as magnetic cycles, chromospheric activity, special stellar oscillation modes, etc. have to be considered carefully (see next Section).

4 First Results

• **Planets around pulsars**

After a false announcement by another group, Wolszczan & Frail (1992) have claimed the discovery of two Earth-size planets ($3.4M \oplus / \sin i$ and $2.8M \oplus / \sin i$) at $0.36Au$ and $0.47Au$ distance from the $6.2ms$ pulsar $PSRB1257 + 12$. This claim was confirmed 2 years later and a third Moon-size planet ($0.015M \oplus / sini$) was found closer to the pulsar ($0.19AU$) by Wolszczan (1994). This discovery is now well established and raises many intriguing questions. The harsh local conditions point to inhabitable planets, but this point need to be investigated further.

• **Are astrophysical artefacts irreducable problems?**

We have seen that the astrometric method is basically free from artefacts, so the question applies to the RV method. A false detection has been claimed on γ Cephei, where stellar processes have mimicked the presence of a planet (Walker et al, 1989 and 1992). However this was measured on a giant star, and it seems that the situation is much safer with Main Sequence stars. The monitoring of Sun light reflected by the Moon, which gives a good averaging of the whole disk emission, has been performed by Mc Millan et al (1993). They claim to have found no spurious variation at the level $\pm 4ms^{-1}$ over a 5 yrs period that includes the Solar Cycle 22 maximum. It demonstrates that the Sun activity would not prevent a $\pm 13ms^{-1}$ detection. In addition, the actual effect of the solar activity can be still much lower because the accuracy was technically limited.

- **Intriguing first results**

Table 1. Accessible Jovian planets $(T_{planet}/ \sim 130K)$ when photon limited, 15min integration, 8m telescope, 50% sky visibility, $\sin i = 0.87$, only single stars.

Stellar type	P (yr)	v (ms⁻¹)	Number	Remark
F	25	1.4	120	too long orbital periods, too short-life stars
G	10	2.2	130	OK
K	4	3.9	70	OK
M_{0-4}	1.3	7.8	45	telluric planets habitable?

Walker et al. (1993, 1994) and Marcy & Butler (1992) have studied 21 and 65 main sequence F, G and K stars during 12 and 3 years respectively. Their data show *no evidence* of planets with $M \sin i$ equal or larger than 1-3 Jupiter-mass ... This encourages us to push further the observations by decreasing the mass limit detection (down to Saturn-sized planets?) and increasing the number of studied stars. This is important, more especially as studies by Wetherill (1994) have concluded that, in the Solar System, Jupiter was necessary for life to develop on Earth because of its action on the cometary flux impact on our planet. If Jupiter was not there, he conclude that the impacting flux would have been 3 orders of magnitude larger than the actual one, throughout the whole Earth history. This would probably have been quite unfavourable for life to develop.

5 Why VISA and VLT?

5.1 Differential Astrometry

VISA, in a configuration with 3×1.8 m telescopes, that does not need to be changed, is an *ideal tool to search for Jovian exo-planets*, provided some minor modifications are made and the construction of adapted instrumentation is done, has been indicated above. Quirrenbach (these proceedings) suggests a key program using VISA during 300 observing nights spread over 10 yrs. It would build an unbiased sample of 200 stars and could give real answers to the frequency of occurrence of Jovian planets around different types of star.

However, it is (probably) more tricky than the RV approach because interferometer tuning has been, up to now, a delicate job. This justifies the parallel development of the RV approach which has its own intrinsic advantages, e.g. the independence of the RV amplitude on the star distance.

5.2 Radial Velocity

From the discussion in sections 3 and 4, it follows that we need measurements of $2ms^{-1}$ RV to detect planets somewhat smaller than Jupiter, e.g., with a mean value $\sin i = 0.87$, $m_{Jup}/6$ or $m_{Saturn}/1.8$. In addition, we need rather fast observations to build a large star sample (~ 200). Then, photon noise becomes a problem and a 8m telescope is really useful. Table 1 gives the number of accessible stars. It shows that we would have access to a total of 200 favourable stellar systems .

• <u>How many runs?</u>

Considering the 12 yr period of Jupiter/Sun, 2 runs per year per object seem suitable during 1 or 2 decades. To study a sample of 200 stars with individual integration times of 10 mn, would require: $200 stars \times 2 runs \times 10 min \times (8 hours/night)^{-1} \times 3(overheads) = 25\ nights/year$ on one 8 m telescope.

• <u>Instrument requirements.</u>

The required high resolution and large spectral range of the spectrometer point to an échelle spectrometer with *very good stability* and on-line absolute calibration.

Among planned instruments, UVES seems the most suitable one as it has an I_2 cell that can be put in front of the entrance slit. The highest care should be taken to attain a high stability (no thermal shocks, no vibrations, reproducible focusing of the telescope, etc.).

The participation of Marcy, who has a long experience with I_2 cells, would be a valuable advantage. Recent campaigns at the ESO $1.4m$ CAT/CES by Kürster et al. (1994), where long term velocity precision of $4 - 7 ms^{-1}$ has been obtained, is encouraging.

The goal being $1 - 2ms^{-1}$ long term precision, if the checks with UVES were to conclude that such a stability is out of reach, one should start studies for a dedicated instrument.

Connes "Absolute Accelerometer" (1994) that should be photon noise limited is also an interesting approach. Its feasibility should be checked in the coming 2-3 years at the OHP ELODIE Spectrometer.

6   Comparison of Differential Astrometry and Radial Velocimetry Techniques.

Table 2.	Differential Astrometry	Radial Velocity
sensitive to	$\dfrac{m}{M_*} \times \dfrac{a}{D}$	$\dfrac{m \sin i}{[a\, M_*]^{1/2}}$
determ. quant.	m, i, a	m sini, a
advantage	- no foreseen artefact - yields directly m	- fast ? - D dependence via star flux only
disadvantage	- long and delicate tuning of interferometer	- possible astrophysical artefacts - yields m sini

7   Conclusions

– Differential Astrometry and Radial Velocimetry appear to be complementary techniques and they can be scheduled for the VISA and the VLT respectively.

– They can detect only Jovian planets, but studies by Wetherill (1994) point to the need of such planets for local telluric planets to be habitable. If this appears to be correct, these measurements would directly address the estimate of factor (2) in the Drake equation (sect 1.2)

– We conclude that the search for exo-planets and their caracterization will be a key subject for the 21st century Science. *Europe should not miss it.* In the USA, this program used to be called Towards Other Planetary System (TOPS). At ESO, we have very interesting capabilities with the VLT and VISA provided that we build the proper instruments.

Practically, we strongly plead for :

1 - The building of VISA, in a 3-telescope version, even with a limited number of telescope locations, provided it includes two (almost) orthogonal long baselines.
2 - The modification of the VISA effective field of view to allow the astrometric measurement of two stars as distant as 17″ proposed by von der Lühe et al.(these proceedings) and the construction of a specific instrument that can make the precise angular measurements (differential delay line).
3 The construction of UVES with special attention to its absolute calibration (I_2 cell), long term stability and light entrance conditions (such as telescope focusing). If these instruments are built, we should have the capabilities to play a major role in the long term studies of Bio-Astronomy in the next century, with the corresponding *great impact on the Public.*

References

Angel, J.R.P., Cheng A.Y.S. & Woolf, N.J. (1986), Nature 232, 341

Angel, J.R.P. (1994), Nature 368, 203

Beckwith, S.V.W. & Sargent, A.I. (1993) in Protostars & Planets III, eds. Levy E. & Lunine J., Univ. of Arizona Press p. 521

Bély, P.Y., Burrows, C.J., Roddier, F. Weigelt, G. & Bernasconi, M.C. (1992) in ESA Colloquium 13-16 October ESA SP-354

Burns, J.A. (1994), COMPLEX, Nat. Acc. of Sc. available from ftp.nas.edu/pub/reports

Burke, B.F. (1992), TOPS NASA report of the Solar Syst. Expl. Div., p XIV.

Connes, P. (1994), Astrophys. Sp. Sc. 212, 357 & 110, 211 (1985)

Drake, S. (1965), in Current Aspects of Exobiology, ed Mamikunian, G. & Briggs, M.H. Pergamon Press.

Epicurus (-300), letter to Herodotitus in Lettres et Maximes, Presses Universitaires de France, Paris, 1987 p.105

Hatzes, A.P. & Cochran, W.D. (1993), Ap J 413, 339

Kürster, M., Hatzes, A.P., Cochran, W.D., Pulliam, C.E., Dennerl, K. & Döbereiner, (1994), The Messanger, ESO, June, p.51.

Malbet, F., Shao, M., Yu, M.S. (1994), SPIE 2201, 1135

Marcy, G.W. & Butler, R.P. (1992), PASP 104,270

Mayor, M., Duquennoy, A., Halbwachs, J.L. & Mermilliod, J.C. (1992), IAU Colloq. 135, ASP conf;. series 32, 73

McMillan, R.S., Moore, T.L., Perry, M.L. & Smith, P.M. (1993), Ap J 403, 801

Schneider, J. & Chevreton, M. (1990), A&A 232, 251

Schneider, J. (1994), in First Circumstellar Habitable Zones Conference, Conference, ed. L. Doyle (Travis Press), in press

Smith, B. & Terrile, J. (1984), Science, 226, 421

Walker, G. A., et al. (1989) Ap J Let 343, L21

Walker, G. A., et al. (1992), Ap J Let 396, L91

Walker, G.A.,(1993), comm. at the 2nd Int.Conf. on Planetary Systems, Dec 13-15, Hawaii, USA

Walker, G. A., et al. (1994), (preprint)

Wetherill, G. (1994b),Astroph. & Space Sc., 212,23

Wolszczan, A. & Frail, D.A. (1992) Nature, 355, 145

Wolszczan, A. (1994), Science 264, 538.

Astrometric Detection and Investigation of Planetary Systems with the VLT Interferometer

A. Quirrenbach

Max-Planck-Institut für Extraterrestrische Physik, Postfach 1603, D-85740 Garching, Germany

Abstract. The VLT Interferometer is capable of doing astrometry with ~ 10 microarcsecond precision over fields of a few arcseconds by measuring the differential delay between the target star and a (possibly faint) reference object in the K band. With this precision, it is possible to detect planets through the reflex motion of the central star at considerable distance; Jupiter could be found out to $\sim 1\,\mathrm{kpc}$. The masses of any planets detected can be determined, independently of the inclination of the orbit. In a "key program", which would use about 300 observing nights over 10 years on VISA (1.8 m telescopes only), 200 stars could be observed, with 20 to 30 measurements for each star. As a result of this program, an inventory of giant planets around stars of various spectral types could be obtained, including the multiplicity and mass function of the planets. Smaller planets (down to ~ 10 Earth masses) could be found around a few nearby stars. Inclusion of pre-main-sequence objects in the target list would give additional information about the formation of planetary systems.

1 Introduction

Among the most interesting open problems in astronomy is the question whether planetary systems are common around main sequence stars. Since direct imaging of planets close to much brighter stars is extremely difficult (e.g. Angel 1994), most attempts to detect them have been made with indirect methods. The most promising approach seems to be the search for the motion of the star around the planetary system's center of mass, either by looking for changes of the radial velocity (e.g. Cochran and Hatzes 1994), or by astrometric methods in the optical (Dekany et al. 1994) and radio (Lestrade et al. 1994) wavelength ranges. The sensitivity of narrow-angle astrometry in the optical and near IR can be substantially increased, if a long-baseline interferometer is used instead of a single telescope (Shao and Colavita 1992). Pilot experiments with the MkIII interferometer on Mt. Wilson have shown the validity of the underlying theory and the feasibility of the proposed methods (Quirrenbach et al. 1994, Colavita 1994). Astrometric observations with the VLT Interferometer will thus be a powerful method to detect planetary systems, and to characterize them once they have been found.

2 Astrometry with the VLT Interferometer

To appreciate the potential of interferometric astrometry, it is important to note that ground-based narrow-angle astrometry is limited by seeing effects. Furthermore, if the separation θ between target and reference star is sufficiently small, the astrometric error σ due to atmospheric turbulence scales as $\sigma \propto B^{-2/3} \theta$, where B is the telescope diameter, or the baseline length of the interferometer (Lindegren 1980). Based on measurements of turbulence and wind speed profiles on Paranal, the astrometric error expected for a half-hour integration with the VLTI is $\sim 10\,\mu$as for $\theta = 10''$ (von der Lühe, Quirrenbach, and Koehler 1994). Since most instrumental errors cancel to first order for differential astrometric measurements, the requirements on the instrumental stability necessary to reach the limit imposed by the atmosphere are not more stringent than those necessary for interferometric imaging. The baseline vector, for instance, has to be known with $\sim 50\,\mu$m precision, which can be achieved without an extensive metrology system by periodic observations of a set of stars with good sky coverage (Hummel et al. 1994). Nevertheless, the implementation of narrow-angle astrometry requires a thorough understanding of the instrumental effects that relate to delay variations (vibrations, drifts etc.); the efforts spent in this area will contribute significantly to the overall success of the VLTI.

The target stars in a program aimed at searching for planets will normally be fairly bright and can therefore be used to cophase the interferometer; the limiting magnitude should be of order $K = 12$ for 1.8 m telescopes (ESO VLT Panel 1989). The astrometric reference stars can be much fainter, since they are not used to cophase the instrument. With the numbers from VLT (1989), and taking into account the signal-to-noise required to measure the position with $10\,\mu$as precision, a limiting magnitude of $K = 17$ is obtained for the reference stars. The VLTI delay lines have been designed to give an unvignetted field of view of $8''$; for the VISA array it can be made as large as $\sim 20''$ by means of simple beam expander optics at the entrance of the delay lines. This field of view is sufficiently large to give a high probability for finding a suitable astrometric reference for any randomly chosen target star. These and other technical issues related to the measurement of the differential delay between the target and reference stars are discussed by von der Lühe, Quirrenbach, and Koehler (1994).

3 Observations of Extrasolar Planetary Systems

The outstanding sensitivity of the VLTI for the detection of planets is apparent from Figure 1, which shows the reflex motion of the Sun due to the presence of the planets as observed over a range of distances. Jupiter could be found out to 1 kpc, Uranus at a distance of 200 pc, and a planet with 10 Earth masses can be detected around nearby stars. Unlike searches for radial velocity variations, the sensitivity of astrometric methods does not depend on the inclination of the planetary orbit; thus increasing the detection probability and giving a tremendous advantage for statistical studies. Furthermore, the mass of each planet detected astrometrically can be determined with high precision, provided only that the basic parameters (mass and distance) of the central star are known. It will thus be possible to take an inventory of giant planets around stars of various spectral types, including multiplicity, distribution of the orbits, statistics on orbital resonances, and mass function of the planets. This database will contain a wealth of information about the formation and evolution of planetary systems, and provide a completely new framework for the interpretation of the history of our own Solar system. The inclusion of peculiar stars and pre-main-sequence objects in the target list will provide an additional means of studying the process of star formation and the genesis of planets. It would be interesting, for example, to see whether the detection rate of giant planets is affected by the presence of IR excesses or other indicators of circumstellar disks.

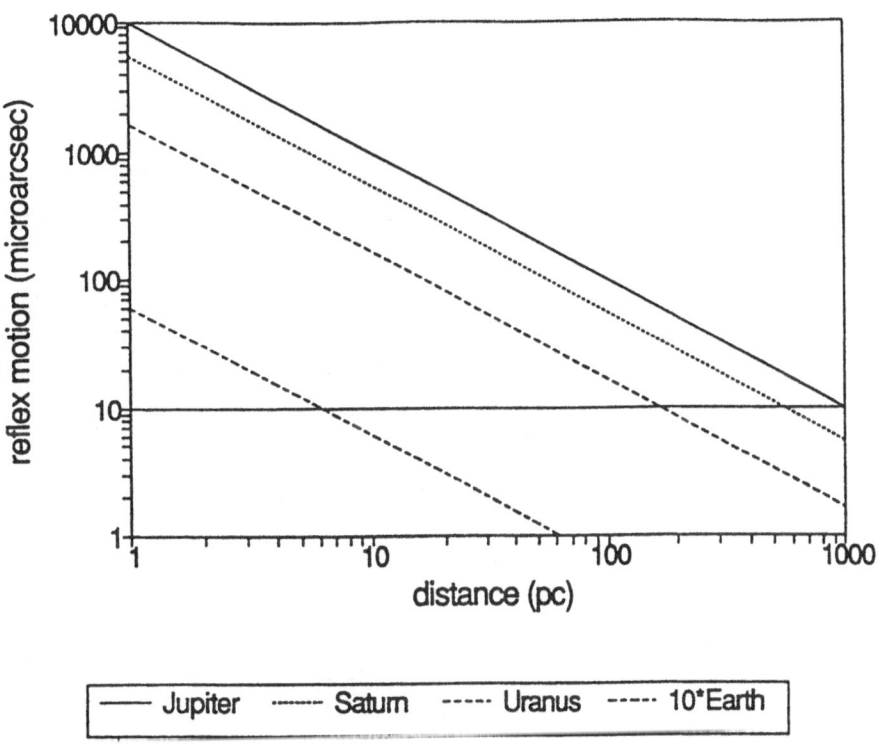

Fig. 1. Reflex motion of the Sun, due to the presence of giant and Earth-like planets, as a function of distance. The horizontal line indicates the detection limit anticipated for the VLTI (10 μas).

4 Observing Strategy

The search for extrasolar planets is inherently a long-term project, which requires a homogeneous data set and consistent data reduction; it will therefore most easily be implemented within a "key program" or similar framework. This program would use the VISA (1.8 m) telescopes only. The optimum number of telescopes will be four, but observations could start with three telescopes or even with a "bare bones" system consisting of only two telescopes and one delay line. The best observing wavelength is the K band, since here the 1.8 m telescopes are fully coherent with a simple tip-tilt correction, most stars are unresolved with the VLTI baselines, and the isoplanatic patch is comparable to the field of view. In order to achieve the desired $10\,\mu$as precision, the program stars have to be observed for 1800 s per night. In order to reduce the effects of systematic errors, the total integration time will be split into three to six shorter observations spread over the course of the night. In the first step of the data reduction, the set of all differential delay measurements obtained during the night will be converted into difference positions for all target – reference star pairs. When 20 to 30 such observations have been acquired for each star, it will be possible to solve for the parallax and proper motion of the stars (relative to the respective references), and to look for the signature of planets in the residuals. Since giant planets around stars at 30 pc, say, will be detected on the 10 to 100 σ level, it will be possible to find multiple planets in a harmonic analysis of the astrometric data. The obvious problems that arise when the reference star turns out to be a member of a binary system can be avoided by using two independent references for each program star. With 300 observing nights spread over 10 years it will thus be possible to observe a sample of 200 stars.

5 Other Applications of Precise Narrow-angle Astrometry

While the implementation of a narrow-angle astrometric mode for the VLT Interferometer appears justified for the purpose of searching for extrasolar planets alone, it is clear that many other fields would benefit from this capability. For instance, it would be possible to obtain extremely precise parallaxes. Accuracies of 1% at 1 kpc (e.g. for Cepheids), and 10% at 10 kpc (e.g. for post-AGB stars) could be achieved. The main problem, namely the conversion of relative to absolute parallaxes, could perhaps be tackled by observing galactic stars projected towards the Magellanic Clouds, which provide reference stars with precisely known parallaxes. Another example are the masses of the individual components in single-lined spectroscopic binaries, which can be determined by measuring the interferometric orbit, and the parallax of the system. Narrow-angle astrometry could also be used to investigate the dynamics of globular clusters, and to determine the orbits of the stars in the Galactic Center region. The efforts required to tackle the additional complications, brought about by observations in crowded fields, might be rewarded by a definite measurement of the mass of the putative black hole in the center of our galaxy.

6 Conclusions

While several methods seem to be capable of detecting extrasolar planets, interferometric narrow-angle astrometry offers unique opportunities for the investigation of these exciting objects. The main advantages of this method are its outstanding sensitivity, which does not depend on the inclination of the orbit, and the capability to determine planetary masses. The VLT Interferometer has the potential to open the new field of study of the formation and evolution of planetary systems to European astronomers.

References

Angel, J.R.P. (1994), Nature 368, 203

Cochran, W.D., Hatzes, A.P. (1994), ApSS 212, 281

Colavita, M.M. (1994), A&A 283, 1027

Dekany, R., Angel, R., Hege, K., Wittman, D. (1994), ApSS 212, 299

ESO VLT Interferometry Panel (1989), The VLT Interferometer Implementation Plan, VLT Report No. 59b

Hummel, C.A., Mozurkewich, D., Elias, N.M., Quirrenbach, A., Buscher, D.F., Armstrong, J.T., Johnston, K.J., Simon, R.S., Hutter, D.J. (1994), AJ 108, 326

Lestrade, J.F., Jones, D.L., Preston, R.A., Phillips, R.B. (1994), ApSS 212, 251

Lindegren, L. (1980), A&A 89, 41

von der Lühe, O., Quirrenbach, A., Koehler, B. (1994), these proceedings

Quirrenbach, A., Mozurkewich, D., Buscher, D.F., Hummel, C.A., Armstrong, J.T. (1994), A&A 286, 1019

Shao, M., Colavita, M.M. (1992), A&A 262, 353

Part 2

Star Formation

Star Formation Studies at the VLT

Claude Bertout[1], Bo Reipurth[2], Fabien Malbet[3]

[1] Observatoire de Grenoble, Université Joseph Fourier, BP 53, 38041 Grenoble cedex 9, France
[2] European Southern Observatory, Casilla 19001, Santiago 19, Chile
[3] Jet Propulsion Laboratory, Pasadena, USA

Abstract. We present an overview of what can be done in the field of star formation using the VLT and its advanced instrumentation, and especially with the very high angular resolution provided by the VLT in the VLTI mode. Herbig-Haro objects and early stellar evolution, pre-main sequence binaries, circumstellar disks and surface properties of young stars are investigated. We show that the VLT and VISA will be not only very useful to better understand star formation processes, but indeed necessary in order to advance in certain subjects.

1 Introduction

Instead of trying to cover the entire field of star formation, we focus here on some areas where the gain in angular resolution brought about by the VLT and VLTI will bring decisive progress in our understanding of the physical processes that govern the formation of stars. We shall therefore discuss mainly the circumstellar environment of low-mass young stellar objects.

Over the last ten years, considerable progress about the nature of these objects has been achieved. Comprehensive surveys of molecular clouds uncovered several phases in the formation of stars, ranging from embedded protostars that are studied best at millimeter and submillimeter wavelengths to pre-main sequence, optically visible stars with various degrees of activity. Among these, one now distinguishes between: (a) weak-emission line T Tauri stars (WTTSs), which are magnetically active pre-main sequence stars with late spectral type that display strong, variable X-ray flux, flare activity, and large, cool spots; (b) classical T Tauri stars (CTTSs), which are thought to be WTTSs surrounded by accretion disks with accretion rates in the range 10^{-8} – 10^{-6} $M_\odot$/yr. The disk produces IR excess and interacts with the underlying star to produce the observed blue and UV radiative excess and to drive observed collimated jets and winds.

Some open questions currently being investigated are the following. Why are there two classes of young low-mass stars? What is the origin of magnetic field (fossil or dynamo-generated)? What is the nature of the angular momentum transport mecanism in the disk? What is the nature of star/disk interaction (boundary layer vs. accretion column)? What mechanism drives the jet and what is the nature of the observed wind/disk connection?

In addition to the two above mentioned types of young stars, there is a third. much more rare class, the FU Orionis stars, which are erupting CTTSs with apparent mass-accretion rates of up to 10^{-4} $M_\odot$/yr. FU Ori stars drive the optical jets discussed below as well as massive molecular outflows. The frequency of outbursts, their nature, and the nature of the jet driving engine are topics of active current research. It should, from the outset, be pointed out that star formation studies are among the most rapidly evolving subjects in contemporary astrophysics, and it is likely that some of the problems we outline here will have changed character by the time the VLT becomes operational.

2 Herbig-Haro Jets and Early Stellar Evolution

About 10 years ago it was recognized that some Herbig-Haro objects (which are nebulous patches commonly found in star-forming regions) take the form of highly collimated jets. In the intervening years such HH jets have been subject to intense study (for a recent review, see Reipurth & Heathcote 1993). Unlike the better known extragalactic jets, HH jets emit line radiation, they are so close that they are on the verge of being resolvable from the ground, and their kinematics can be explored through proper motion and radial velocity studies. Thus, it is possible to derive detailed physical properties for these objects, so that HH jets have emerged as Rosetta stones for the study of highly collimated flows and the earliest stages of stellar evolution.

In most cases HH jets are ejected from very young stars still deeply embedded in their parental clouds. When the extinction is not excessive one can sometimes observe two oppositely directed lobes, each with typical dimensions of 1 to 2 arcminutes, or roughly 0.2 pc. Such jets consist of a highly collimated chain of faint knots, with widths of $0.6''$ to $0.8''$ and separations of several arcseconds, terminating in a working surface where the flow rams into the ambient medium. The characteristic line spectra from HH flows have been successfully modelled as emission from shocks, and shock models mapping extensive grids of parameter-space have given important insights into the flow physics. Radial velocities and proper motions of HH jets yield typical space velocities of the order of a few hundred km/sec, suggesting dynamical ages of a thousand years or so.

It is commonly found that HH jets have multiple working surfaces along their flow axes. This is best understood in terms of episodic outbursts of the driving source, and HH jets have therefore been linked to FU Orionis eruptions (e.g. Reipurth 1989). Such events are interpreted as massive accretion episodes in the circumstellar disks of T Tauri stars. Although the precise formation mechanism of HH jets is still not known, they may be linked to the disposal of infalling high angular momentum material, probably through interaction with magnetic fields. As the jets stream away from the newborn stars, they interact with the ambient cloud material, and may actually drive the molecular flows commonly observed around young embedded stars (e.g. Raga & Cabrit 1993).

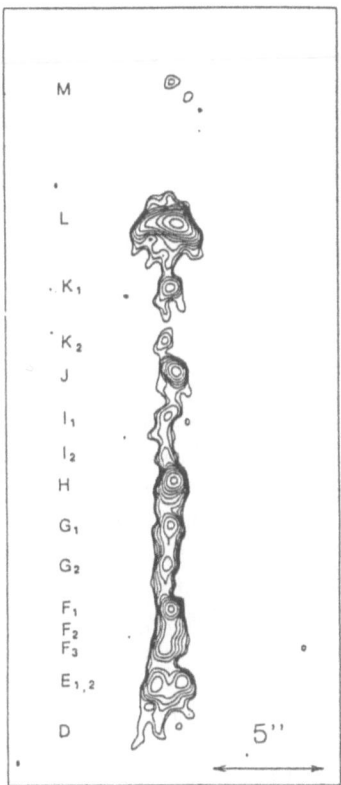

Fig. 1. A contour plot of the HH 111 jet, based on an Hα image obtained at the NTT in 0.7″ seeing, and deconvolved to a resolution of 0.39″. From Reipurth, Raga, Heathcote (1992).

2.1 High Resolution Imaging

HH complexes are often large (up to 10 arcmin or more) and they have a wealth of structure at the sub-arcsecond level. It would therefore be very attractive if the VLT was equipped with an imaging instrument like the Wide Field Direct Visual Camera (Wampler 1994), that fully exploits the imaging capabilities of an 8m telescope and takes advantage of the image deconvolution techniques which have become increasingly advanced in recent years. Many short exposures can be combined such that the poorest images are used to improve the S/N, while the best ones are defining the resolution of the final image (e.g. Lucy 1992). It is conceivable that resolutions of 0.1″ can eventually be achieved this way. HST images already obtained of HH complexes provide this resolution, but experience shows that even with 4-5 hour exposures, the 2.4m HST is limited to study only the very brightest HH jets.

High resolution and good S/N will allow the, as yet unresolved, question of the structure of jet knots to be addressed. If knots are due to time variability in the driving source, they should show the morphology of tiny bow shocks. Another question is whether the large diffuse structures we see in leading working surfaces are already fully resolved, or they have further sub-structure, an important issue for shock models. Further, it is known that working surfaces are divided into bow shocks and Mach disks, with different emission characteristics. It is not known if jet knots are similarly sub-divided, an issue also connected to their mechanism of formation. Finally, HH jets generally show large proper motions, around $1''$-$2''$ per decade. Hitherto, measured proper motions relate to bulk structures, and nothing is known about the detailed flow behaviour on scales that reveal the detailed interaction with the ambient medium.

2.2 2-Dimensional Spectroscopy

No high-resolution Fabry-Perot spectrometer is currently being considered for the VLT. But the area spectroscopy (ARGUS) mode of the Multi-Fibre Area Spectrograph (MFAS) allows at least 670 fibres with $0.2''$ or $0.7''$ spatial sampling within 5 or 18 square arcsec fields, with a resolving power of up to 30 000. This is ideally suited for studying the detailed kinematics of selected areas of HH flows. Firstly, accurate radial velocities are required, in combination with proper motions, to determine space motions of HH jets. Secondly, it is of great interest to search for a kinematic signature of the entrainment that takes place when a jet penetrates its environment, since entrainment is likely to have a major impact on the dynamics of the flow. Thirdly, a detailed kinematic study of HH jets can decide if the pronounced wiggling seen in many HH jets is due to precession of the source, flow along a curved tube, or meandering through an inhomogeneous environment, since each of these scenarios predicts a different position-velocity diagram.

2.3 Very High Resolution Spectroscopy

The best collimated HH jets lend themselves well to long slit spectroscopy. The UV-Visual Echelle Spectrograph (UVES) permits spectra with resolving powers of the order of 100 000. This resolution corresponds more or less to the thermal widths of the C, N, O, S etc. lines for a gas at 10000 K. With this resolution one would thus get "everything that there is to know" about the radial velocity structure of HH flows. This will allow a number of questions to be addressed: first, are flows in HH jets laminar or turbulent, or do they evolve from one mode into the other; second, models make specific predictions for the line widths and radial velocities of various emission lines in the recombination region behind a shock wave which can be tested by very high resolution spectroscopy; third, the origin of jets is not understood, and very high resolution spectroscopy of residual jets very close to T Tauri stars, as pioneered by Solf (1989), may give clues. Indeed, it is possible that vestiges of the jet phenomenon may be commonly observed in CTTSs if sufficently high spatial and spectral resolution is applied.

2.4   Infrared Imaging and Spectroscopy

Many HH jets have in recent years been found to emit strongly in molecular hydrogen and infrared [Fe II] lines. The Infrared Spectrometer and Array Camera (ISAAC), the High-Resolution Near-Infrared Camera (CONICA), the Mid Infrared Imager/Spectrometer (MIIS), and the VLT High-Resolution IR Echelle Spectrometer (CRIRES) will be important for this new and rapidly developing field. There are a number of current issues. First, the slower and weaker parts of the shocks, traced by molecular hydrogen emission, are expected to show different structures from the optical lines; when observed with sufficent spatial resolution, the structures in different emission lines can provide sensitive tests of theoretical models. Second, the youngest objects are most deeply embedded, and it is to be expected that a large population of infrared, optically obscured HH jets await discovery, as exemplified by the recent discovery of HH 212 (Zinnecker et al. 1995). These very young, infrared jets are likely to provide new insights into the origin of jets and their interaction with the ambient medium. Third, infrared lines permit determination of excitation temperatures, electron densities and extinction of HH flows (e.g. Gredel et al. 1992).

2.5   VLTI

Extended objects like HH jets will require very good coverage of the (u,v) plane. This might require observations spread out over months. However, with milli-arcsec resolution the large proper motions of jets become a problem: an HH jet would displace itself about one resolution element within 2 or 3 days. In its currently proposed configuration, VLTI is therefore unlikely to have a major impact on the study of HH jets.

3   Pre-Main Sequence Binaries

Most stars are members of binary systems. Thus, to understand the process of star formation we must understand the formation of binary stars. As a first step in this direction, a number of recent studies have focused on the properties of binaries which have not yet reached the main sequence (e.g. Reipurth & Zinnecker 1993, Leinert et al. 1993, Ghez et al. 1993). A major review has recently appeared by Mathieu (1994). A principal result deduced in these investigations is that binarity is higher among young stars than among the main sequence stars studied by Duquennoy & Mayor (1991). This could be because the distribution of separations of components is different for younger and older stars, even though the total binary frequency is constant, and that there is an excess only in the separation range sampled by the observations. Or perhaps essentially all stars are born in binary or multiple systems, which then subsequently are disrupted, possibly through close encounters with other stars. In either case, it is clear that young binaries undergo dynamical evolution towards the main sequence. It will be important in the coming years to do unbiased surveys for faint sub-arcsecond companions to young stars in the visible and the infrared, and the VLT will be able to produce outstanding results in this field. Speckle observations will push the resolution limit even further. But one should remember that, while for diffraction-limited speckle observations the resolution limit will improve going from a 3.6m to an 8m telescope, there is no similar gain in limiting magnitude.

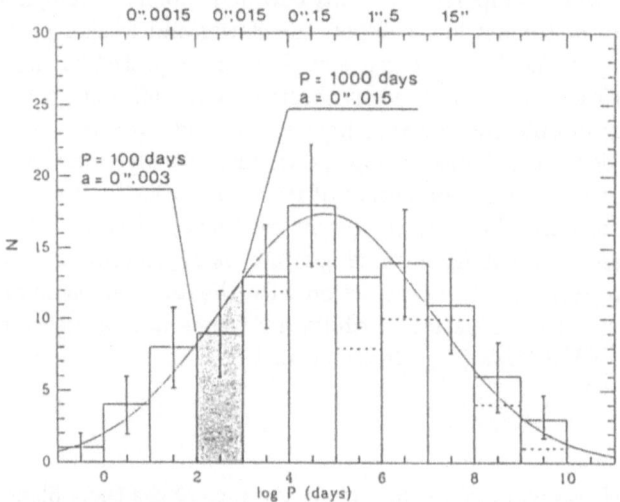

Fig. 2. The period distribution for nearby G dwarfs, from Duquennoy & Mayor (1991). The periods in days have been converted to semi-major axes in arcsec as described in the text. Practical observing limits with VLTI have been marked, the lower limit is $a = 0.003''$ ($P = 100$ days) and the upper limit is $P = 1000$ days ($a = 0.015''$). This range covers about 10% of all binaries in nearby (130 pc) star forming regions.

The really unique contributions to the study of pre-main sequence binaries will be made by the VLTI. The question is whether VLTI allows us to resolve the orbits of spectroscopic PMS binaries. If so, an orbit determination gives us the sum of the component masses, while the spectroscopic data give us the mass ratio (or the mass function if the system is only single-lined). Together the data would give us the first precise mass determinations of low-mass pre-main sequence stars, in much the same way it has been done for nearby low-mass binaries (e.g. Mariotti et al. 1990). This would for the first time allow a meaningful observational calibration of theoretical evolutionary tracks before the main sequence. If we assume observations to be done at λ 5500 Å with a VLTI baseline of 150 m, we derive $\lambda/D = 0.00076'' \approx 1$ milliarcsec. The distance of the nearest star forming complexes is about 130 pc, which we will adopt in the following, and let us assume that the binary components have $m_1 = m_2 = 0.5\ M_\odot$. We shall additionally assume that the period distribution of low-mass PMS binaries is not too different from the distribution of G-dwarfs determined by Duquennoy & Mayor (1991), an assumption that is not necessarily correct, but which is the best one can do before an actual determination is done. Figure 2 shows the G-dwarf period distribution, with periods converted into semi-major axes in arcseconds. If we assume that 3 milliarcsec (corresponding to a period of 100 days) is a lower limit for an orbit determination, and that systems with periods longer than 1000 days are difficult to observe in practice, then we find that about 10% of all PMS binaries have separations large enough for an orbit determination and periods short enough to be observable over a full orbital revolution. It is therefore of particular interest to find and study PMS binaries in the period interval from 100 to 1000 days.

However, it is important to recall that distances to star forming regions are not very precisely determined. Even if we could claim to know these distances to as well as 10%, this translates via Keplers law into a 30% uncertainty in the masses, since $[a(1+0.1)]^3 \approx a^3 + 0.3a^3 + \cdots$. It is therefore clear that an integral part of a PMS binary mass determination must be a *distance* determination, e.g. by measuring parallaxes via narrow-angle VLTI astrometry (see von der Lühe et al, these proceedings).

4 Circumstellar Disks

The presence of circumstellar disks around young stars is inferred from indirect clues rather than direct observations. The most tantalizing arguments for the presence of disks rely on *models* of (a) forbidden line profiles, (b) spectral energy distributions, and (c) polarization patterns.

Search for direct evidence of disks around young stellar objects is an active research area, and observations in various wavelength ranges uncovered several types of condensations and/or disk-like structures in the immediate vicinity of young stars. For example,

• Near-infrared adaptive optics observations at the ESO 3.6m telescope showed that a 400AU disk-like structure surrounds the FU Orionis binary s-tar Z CMa (Malbet et al. 1993).

• Millimeter and submillimeter range interferometry of the T Tauri star HL Tau show that the star is embedded in molecular condensations on various scales. Kinematic evidence for entrainment of molecular gas by the jet is apparent on the scale of a few arcseconds (Cabrit et al. 1994).

• Also in the millimeter range, the T Tauri binary star GG Tau was found to be surrounded by a circumbinary molecular ring (Dutrey et al. 1994).

These examples attest to the large variety of circumstellar environnments that are associated with young stars but give little evidence so far for the Keplerian accretion disks that are thought to be responsible for the FU Orionis outbursts and the strong activity of CTTSs. Clearly, a much higher spatial resolution than currently available is necessary to probe the structure of these disks and to test current ideas about the physical processes at work in these objects.

In order to compare expected accretion disk properties to expected performance of the VLT and VLTI, we computed synthetic images and visibilities of the *thermal* radiation emitted by a typical T Tauri disk model. Figures 3a to 3c display disk images and visibilities at 0.5, 2.2, and 10 μm and Table 1 summarizes the 2.2 μm performance at (a) one VLT telescope equipped with adaptive optics and used as an imaging device, and (b) the VISA interferometer with three 1.8m telescopes and up to 150m baseline. The visibility is a way to quickly see if an object is resolved with an interferometer and two or three different (u,v) plane configurations are enough to understand the global morphology of a disk (inclination, size,...).

Table 1. Observations of T Tauri Disks: VLT vs. VISA

	VLT with *maximum* adaptive optics correction (1 x 8m)	VISA (3x1.8m telescopes)
Technique	direct imaging (image deconvolution in principle not needed)	visibility measurement (model constraining); ultimately, image reconstruction
Field of view	30"-90"	2"
Resolution @ 2.2 μm (in mas)	60	5 (for 100m baseline)
Resolution in AU (d=150pc)	9	0.75
Limiting magnitude @ 2.2μm	13 (limited by AO ref. star)	8 - 12 (fringe tracker limited)
On source sensitivity	25	8 - 12

A first conclusion that can be drawn from comparing Figure 3 and Table 1 is the usefulness of the VISA interferometer for probing the physical properties of the inner accretion disk. Indeed, the 8m telescope resolution at 2.2μm with adaptive optics *full* correction is, optimistically, about 10 AU at the distance of the closest star-forming regions, while VISA allows sub-AU resolution at the same distance. Figure 3b demonstrates that a few interferometric measurements in the (u,v) plane will allow one to determine roughly the geometry of 2.2μm emission by simple visibility modeling.

A second obvious conclusion is the usefulness of 10μm measurements for studying the circumstellar environment by direct imaging. This is the result of two conspiring factors: (a) 8m telescopes are diffraction-limited at this wavelength; and (b) because of the low surface temperature of disks surrounding young stars, we see the entire disk at wavelengths close to or larger than 10μm. An additional bonus not taken into account here is the low extinction in the mid-IR, which allows one to study the dense star-forming cores. We thus wish to emphasize the strong need for modern mid-infrared arrays at the VLT if this facility is to be competitive in the field of star formation studies.

Such mid-IR detectors should also be made available for the VISA interferometer, as Table 2 demonstrates. There, we show that even in a very conservative estimate of limiting magnitude, which corresponds to a case of observations with no fringe tracker, the expected disk model flux is larger than VISA's limiting flux up to the near-IR. Given a sufficient coverage of the (u,v) plane, one could thus map the disk temperature structure and distinguish between the various

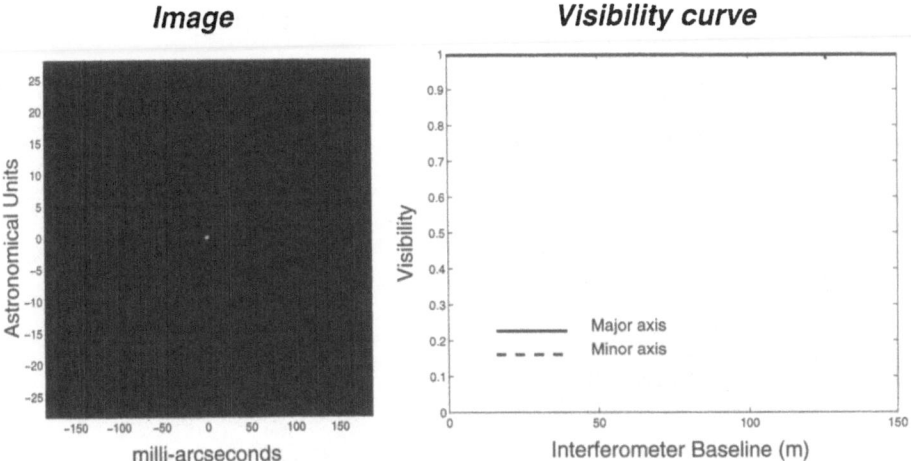

Fig. 3a. 0.5μm direct image (left) and visibility (right) of a typical T Tauri accretion disk. The visibility is constant and equal to 1 because the circumstellar environment is unresolved.

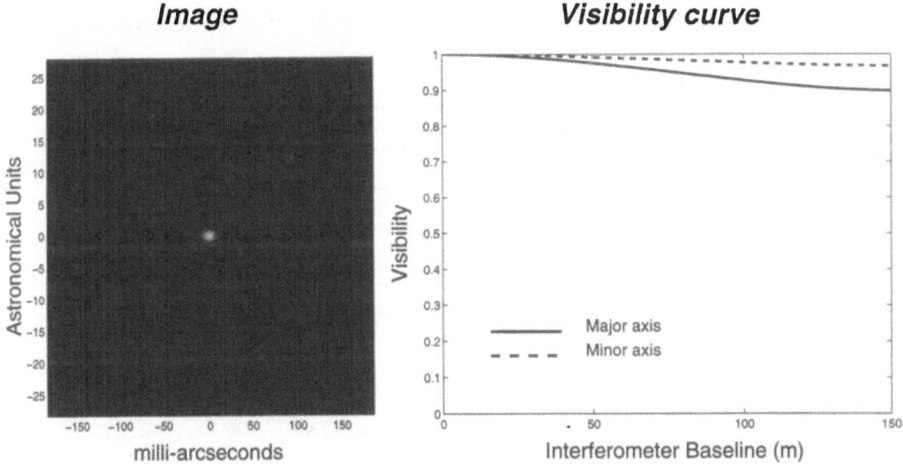

Fig. 3b. 2.2μm direct image (left) and visibility (right) of a typical T Tauri accretion disk. The decrease at long baseline shows the disk is resolved. A visibility accuracy of 1% will easily permit a visibility of 93% to be measured with 100-m baseline.

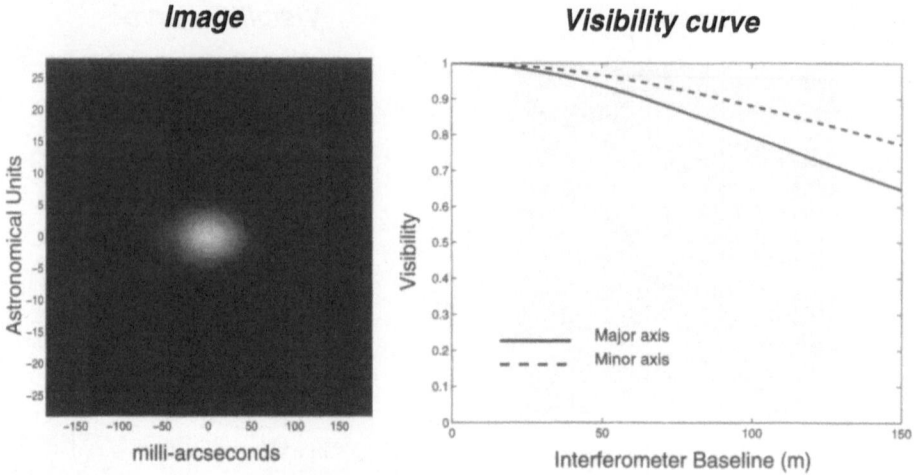

Fig. 3c. $10\mu m$ direct image (left) and visibility (right) of a typical T Tauri accretion disk. The visibility curve shows that the disk is easily resolved.

Table 2. Observations of T Tauri accretion disks with VISA: a conservative estimate (lim. mag. K=8)

Wavelength (μm)	VISA limiting flux (Jy)	Disk model flux (Jy)
2.2	0.8	3.8
5	0.4	4.8
10	3.2	4.5
20	11	3.8

Notes to Table 2. The following assumptions were made in deriving the above values.

- 3 x 1.8m telescopes
- Detector or background noise-limited case
- S/N=50
- Total optical efficiency 10%
- Visibility instrumental losses 30%
- Classical accretion disk model with mass-accretion rate of $10^{-6}\,M_\odot$/yr

physical processes that may be responsible for disk thermal emission (kinematic viscosity, magnetic viscosity, reprocessing of stellar light, etc).

In summary, we emphasize that any 8-10m telescope equipped with adaptive optics will allow for significant progress in understanding of circumstellar environments. However, a VISA-type interferometer is needed for qualitative breakthrough in studies of planetary system formation and evolution. The $10\mu m$ window provides the best resolution/flux compromise for studying disks around YSOs, and VISA (better science version) has the potential performance needed for detecting T Tauri and FU Orionis disks in the near-IR (strong deviations from axisymmetry are expected in the disk). One should note that the Palomar ASEPS-0 Testbed interferometer (prototype for the Keck interferometer), will soon be operational and is expected to reach K= 6-7 mag. Therefore, the so-called better science version is the minimum configuration that should be considered for a competitive VISA, even if its construction is delayed by one or two years.

5 Surface Properties of Young Stars

Progress on surface properties of bright stars has been very rapid in recent years, due to the conjunction of long-term synoptic observations and high-resolution spectroscopy. Young stars being quite faint, large collecting areas are necessary to accurately derive their surface properties. We focus in this section on two topics of current interest in which obvious progress can be achieved with the UVES spectrograph.

5.1 Magnetic Field Strength and Topology in Young Stars

On the basis of comparative studies with active magnetic main-sequence stars, one expects magnetic field strengths of 1-2 kG in T Tauri stars. Actual measurements are difficult: Zeeman broadening of photospheric lines is given by $\Delta\lambda = 5.7 \, 10^{-10} g \lambda^2 B$, where g is the effective Landi factor, λ is in Angstroms, B in kG, e.g., $\Delta\lambda$= 0.05Å for g=2.5, B=1kG, λ=6000Å. So, far, there is only one published measurement, which gives B = 1.0 ± 0.5kG for WTTS TAP 35 (Basri and Marcy 1990).

A current observing method for magnetic field strengths involves correlating the incident spectrum with numerical masks containing lines with similar Landi factors, thus giving high S/N correlation peaks which can be calibrated to give B (Queloz, Babel, Mayor, in preparation). Performance on existing 2m-telescope are R=40000, S/N=20, for a 1h integration time, allowing detection of B $\approx$ 2kG in a star with V=11.

Expected UVES performance is about the same but for V=16, so that T Tauri stars can be studied. Measurement of B in a large sample of WTTSs (V $\leq$ 14) with a time resolution of 1h will allow the study of: (a) the distribution of magnetic field on the stellar surface; (b) the origin of magnetic field (dynamo vs. fossil field) by looking for correlations between B and rotational period; (c) the evolution of B during the pms phase; (d) in conjunction with Doppler imaging (see below), the correlations between cold spots and chromospheric network.

Measurement of B in FU Orionis spectra, a difficult but potentially quite rewarding experiment, will allow one to study the distribution of magnetic field on the disk surface and to test the magnetodynamic origin of FU Ori winds and jets.

5.2 Doppler Imaging of Young Stars

For a V=11 star, i.e., for the brightest T Tauri stars, spectrograms with R=40000, S/N=100, obtained in 1h on an existing 2m-class telescope allow for Doppler-mapping of spots with diameter 15% of the star's diameter or larger, and a spot-to-photosphere temperature contrast larger than 800K (cf. Joncour et al. 1994).

A possible project with UVES will be to Doppler-image the stellar surface of relatively bright CTTSs to resolve either the small hot spots, expected if accretion is channeled by the magnetic field lines, or the (non-axisymmetric) boundary layer. With this technique, we will thus be able to study the nature of accretion onto T Tauri stars, a most timely and interesting problem in star formation research. Again, a word of caution is in order, since the HIRES spectrograph of the Keck telescope allows our Californian colleagues to solve this problem today.

Acknowledgements

We are grateful to Jérôme Bouvier, Christian Perrier, Alex Raga and Hans Zinnecker for valuable comments and discussions.

References

Basri, G., Marcy, G. (1990). Limits on the Magnetic Flux on a Pre-Main Sequence Star. In The Sun and Cool Stars: Activity, Magnetism, Dynamos (IAU Coll. No.130). Helsinki: Touminen.

Cabrit, S., Guilloteau, S., André, P., Bertout, C., Montmerle, T. (1994). Astron. Astrophys. submitted.

Duquennoy, A., Mayor, M. (1991): Astron. Astrophys. 248, 485

Dutrey, A., Guilloteau, S., Simon, M. (1994). Astron. Astrophys. 286, 149

Ghez, A.M., Neugebauer, G., Matthews, K. (1993): Astron.J. 106, 2005

Gredel, R., Reipurth, B., Heathcote, S. (1992): Astron. Astrophys. 266,439

Joncour, I., Bertout, C., Mńard, F. (1994). Astron. Astrophys. 285, L25-L28.

Leinert, Ch., Zinnecker, H., Weitzel, N., Christou, J., Ridgway, S.T., Jameson, R., Haas, M., Lenzen, R. (1993): Astron. Astrophys. 278, 129

Lucy, L.B. (1992): Astron.J. 104,1260

Malbet, F., Rigaut, F., Bertout, C., Léna, P. (1993). Astron. Astrophys. 271, L9

Mariotti, J.-M., Perrier, C., Duquennoy, A., Duhoux, P. (1990): Astron. Astrophys. 230, 77

Mathieu, R. (1994): Ann. Rev. Astron. Astrophys. 32, 465

Raga, A.C., Cabrit, S. (1993): Astron. Astrophys. 278,267

Reipurth, B. (1989): Nature 340, 42

Reipurth, B., Heathcote, S. (1993): in *Astrophysical Jets*, STScI Symp. Series vol.6, eds. D.Burgarella, M. Livio, C.P.O'Dea, p.35

Reipurth, B., Zinnecker, H. (1993): Astron. Astrophys. 278,81

Reipurth, B., Raga, A.C., Heathcote, S. (1992): Astrophys.J. 392,145

Solf, J. (1989): in ESO Workshop on *Low Mass Star Formation and Pre-Main Sequence Objects*, ed. Bo Reipurth, p.399

Wampler, E.J. (1994): in *Instruments for the ESO VLT*, ed. A.F.M. Moorwood, ESO, p.55

Zinnecker, H., McCaughrean, M., Rayner, J. (1995): in press

Circumstellar Disks with VLT/VLTI

Steven V. W. Beckwith

Max-Planck-Institut für Astronomie, Heidelberg, Germany

Abstract. Disk research in the era of the VLT will profit from the high angular resolution and high dynamic range made available from interferometry, adaptive optics and the sensitivity made possible by large collecting areas. Significant advances are likely to come from the unique aspects of the VLT: the VLT interferometer (VLTI) and thermal infrared observations with large, adaptively corrected apertures. High angular resolution will allow mapping of the inner regions of disks, revealing the presence of gaps cleared by orbiting planets and the surface density of small particles, either nascent (for young disks) or produced by the grinding of larger bodies in the older disks. High sensitivity at thermal infrared (10 & 20 μm) wavelengths will make it possible to search for disks around intermediate-aged and older stars and to study the surface brightness of particle emission within a few tenths AU of nearby stars.

1. Introduction

Circumstellar disks are believed to be a common feature of both young and main sequence stars in different stages of development (*e.g.* Beckwith & Sargent 1993; Backman & Paresce 1993). The disks probably account for many of the odd characteristics of young stars (Bertout 1989) and may well play a role in early stellar evolution as well as the formation of binary and multiple star systems (MacDonald & Clarke 1994). As such, circumstellar disks have become part of a paradigm in young stellar object research; there are many fairly straightforward observations proposed to check the details of this paradigm: mapping gas and dust distributions (Sargent & Beckwith 1991; Ohashi *et al.* 1994), correlations between disk properties and other environmental factors such as companion stars (Bouvier *et al.* 1993; Jensen, Mathieu, & Fuller 1994; Osterloh & Beckwith 1994), modelling of lines orginating above and below the disk plane (Appenzeller 1983; Appenzeller *et al.* 1984; Edwards *et al.* 1993), and study of the symbiosis between mass accretion through the disks and mass outflow in collimated winds (Cabrit *et al.* 1990), to name but a few.

There is little doubt that these checks will continue into the era of large telescopes. Indeed, the majority of research efforts in this area are likely to be tests of the standard paradigm. For the most part, the VLT will not be *required* for these checks, and it will, in any case, not be a unique instrument to extend extant research efforts; the Keck can already address most of the observations requiring large aperture; in a few years, the LBT and the twin Kecks will have good thermal infrared instruments, adaptive optics, and should tackle many of the ongoing programs with their enhanced capabilities. It would be quite a task to attempt a summary of all those programs, in any case, so I shall try to concentrate on those aspects of the VLT that bring unique capabilities to bear on the problems of disk research.

Perhaps the most interesting property of these disks is their resemblence to nascent planetary systems (Beckwith *et al.* 1990). The disks may hold the key to an understanding of our own origins and the propensity (or rarity) of planets around other stars. It is the understanding of planet formation which represents the next, obvious potential breakthrough in disk research. The VLT and VLTI will open up several means by which we might study planet formation in the disks; these are highlighted in what follows.

2. Background

We see disks from the youngest identifiable stages (protostars?) up to the main sequence, the former presumably before planets have formed and the latter after planets have already been created. There is limited information on the intermediate stages, the stages during which particles coagulate to create giant bodies. Some of this information will be filled in by two of the key projects on ISO which aim to detect thermal and far infrared emission from the particles. The VLT could also play a role via its greatly increased sensitivity to thermal infrared radiation.

The time scales of interest are listed in Table I. The protodisk and T Tauri phases last a few million years and are rather well studied. The main sequence phase is also well studied, but the issue of planet formation remains open. Disks around the youngest stars are fairly massive and almost certainly consist of particles accreted from the interstellar medium. By contrast, disks around main sequence stars are known to consist of short-lived particles, particles presumably generated by collisions of larger solid bodies near the stars (Aumann *et al.* 1984; Backman & Paresce 1993). There is already quite good *indirect* evidence for planets or asteroids around other stars, and the job of the VLT will be to use these indirect observations to quantify the distribution and production rate of small particles.

Table I. Characteristics of Disk Samples

Objects	Age(yr)	D(pc)	Star	Disk Properties
Early YSO	$< 10^5$	$\gtrsim 150$	Faint Vis. Bright IR	Large particle & gas mass, halos, bright NIR to FIR
T Tauri	10^5–10^7	$\gtrsim 150$	Bright Vis. Bright IR	Large particle & gas mass, halos?, gaps? bright IR
Cluster PMS	10^7–10^8	$\gtrsim 40$	Bright Vis.	Small particle mass, gaps?, planets?, bright FIR?
Main Seq.	$\gtrsim 10^8$	$\gtrsim 1$	Bright Vis.	Tiny particle mass, planets?, scattered Vis., faint FIR

There very few pre-main sequence stars closer than ~ 150 pc. There are many more well developed main sequence stars, such as Vega and β Pic, within about 20 pc. The available spatial resolution is dramatically better for the nearby, old disks than for the young ones in the planet-building stages. Nevertheless, to make great progress in understanding how disks develop and how planets were formed, we need to have astronomically interesting spatial resolution for objects which are about 150 parsecs away: interesting means a *spatial* resolution of 1 AU or better.

It is important to keep in mind that with the exception of the protostar (or embedded) phase the central stars of these disks are usually bright at optical and near-infrared wavelengths. That means that the large collecting area of the VLT is germaine only to a few kinds of observations: very high resolution spectroscopy or thermal infrared observations of very tenuous disks. However, the brightness of these stars is a great advantage for very high angular resolution imaging. Even without laser beacons, there is a large sample of stars of all ages for which full correction of the distorted wavefront using adaptive optics will be possible, since the stars themselves serve as wavefront references. In fact, the star/disk systems should be among the very best candidates to combine adaptive optics with multiple-telescope interferometry and use the VLT to its full capabilities.

Figure 1 indicates the principal components of radiation from a star disk system. The disk is a thick plane shown cross section, the disk photosphere falls off exponentially as an atmosphere, and a halo or nearly isotropic component surrounds the disk star system. This simplified diagram neglects the presence of a circumstellar wind or well collimated jet outflow; that is the subject of other talks in this Conference.

The disk itself probably extends to the stellar surface, approximately 0.01 AU, when it is first created. The outer parts of the disk extend to several hundred AU or more; in the images of β Pic, particles are seen to several thousand AU (Smith & Terrile 1984). These scales correspond to ~ 0.1 milliseconds (mas) to several seconds of arc, respectively, at 150 pc. At present, the best resolution achieved with large ground based telescopes at infrared wavelengths where these disks are prominent is several tenths of an arcsecond. To study disk structure in the younger and intermediate age disks, significantly better angular resolution will have to be achieved.

Figure 2 lays out schematically the typical temperatures, wavelengths of emitted radiation, and interesting characteristics at different distances from the star. This figure shows some of the potential science to come from imaging observations at different spatial resolutions. Of special interest are the scales at which thermal radiation peaks at $\sim 5\,\mu$m, since these could be resolved with the VLTI. Shorter wavelength interferometry will be productive if we can use the scattered light from the disk photospheres and halos to probe the circumstellar structure within an AU or so of the star, since the thermal emission at these wavelengths is below the resolution limit.

Main sequence stars are more than an order of magnitude closer than the young stars. The disk temperatures around stars such as Vega and β Pic are quite

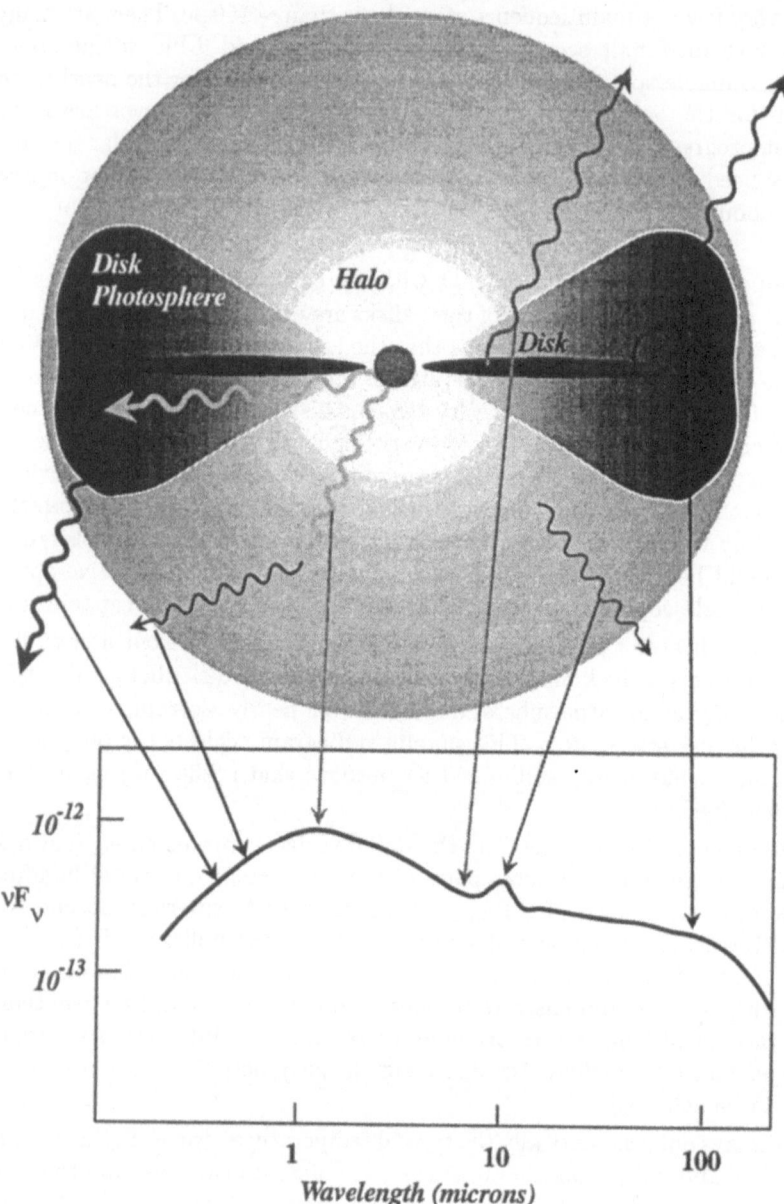

Fig. 1. A simplified diagram of a star/disk system indicating the different components of the radiation field: the star, an optically thick disk, a disk "photosphere" of exponentially decreasing opacity, and a tenuous halo. Thermal radiation and scattered light are both important at different wavelengths as shown. Any wind is ignored in this sketch.

similar to those of the "hot" disk shown in Figure 2. One can see immediately that thermal emission is easily resolved by adaptive optics and interferometry for some of the nearby stars, assuming high dynamic range can be achieved – the stars themselves are brighter than the disks. The structure of these disks is more amenable to observation owing to their proximity.

3. Strengths of the VLT/VLTI

The characteristics of the VLT/VLTI which will be new for the ESO community are:

1. Very high angular resolution ($\sim \frac{\lambda}{100\,\mathrm{m}}$) and good sensitivity through interferometry with the large telescopes (VLTI). This resolution should be unique in the world at least for a few years.
2. Very high dynamic range ($\gtrsim 1000$) coupled with high ($\sim \frac{\lambda}{8\,\mathrm{m}}$) angular resolution through the use of adaptive optics, especially in the thermal infrared ($\lambda \gtrsim 5\,\mu\mathrm{m}$). Both the Keck(s) and LBT could duplicate this capability even before the VLT is built, but it is not known if they will do so.
3. Very high photon-collecting capacity for high resolution spectroscopy and background-limited imaging. By the time the VLT is built, one expects approximately 5 other telescopes to have approximately the same capability: the two Kecks, the LBT, and the two Gemini telescopes. Furthermore, the upgraded MMT and the Magellan telescopes will have quite similar light gathering power for most observations.

The VLT interferometer in will be a unique instrument. Multiple telescope interferometry will not be possible with any other planned facility: phase closure requires three or more telescopes. The very large collecting area equivalent to a 16 m telescope is practically indistinguishable from the twin Keck telescopes (16 m vs 14 m equivalent apertures); many of the questions requiring large apertures can be addressed immediately with Keck; I assume most of them will be and will not discuss them further.

The interferometric resolution at $2\,\mu\mathrm{m}$ will be 4 mas, equivalent to 0.6 AU at the distance to the nearest young stars and close to the stellar radii of nearby main sequence stars. It is already well known that gaps develop in the inner disks out to a few tenths AU as some of these disks evolve (*e.g.*, Skrutskie *et al.* 1990). It is not yet known whether the gaps are the result of dynamical dynamic clearing by small planets or some other mechanism. The ability to resolve disk gaps and to understand disk structure with the resolution offered by the VLTI would be a marked advance in our ability to understand the early evolution of these disks. Even at $10\,\mu\mathrm{m}$ where much of the disk radiation is emitted, the resolution will be of order 3 AU at the distance to these young stars. That is a significant gain and allows one to tackle the question of how disk matter is distributed during the time of formation of the giant planets.

The prevailing theory is that Jupiter formed first in the primitive solar nebula and cleared a gap in the disk. The gap should have a width of order 1 AU at

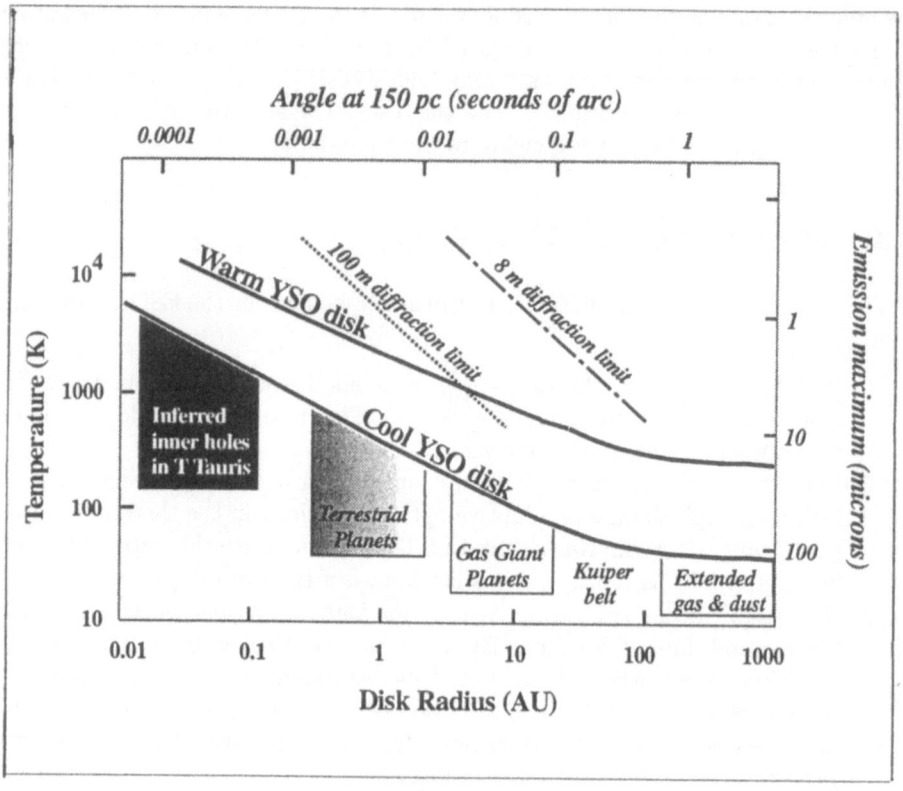

Fig. 2. The ordinate plots logarithmic disk radius vs distance from the central star; the left hand scale gives disk temperature and the right hand scale indicates the approximate wavelength of maximum emission at that temperature. The two heavy lines represent the hot (HL Tau) and cool (weak-line T Tauri) disks, respectively. The apparent angles subtended at a distance of 150 pc are shown at the top of the diagram. For nearby main sequence stars, the disk temperatures will be similar to the HL Tau disk, but the angular scales will be more than ten times larger, so even adaptive optics will have the ability to resolve some of these disks.

Jupiter's orbital radius of 5 AU. It could just be resolved in the thermal infrared by the VLTI. A resolution of 1 AU will occur for $\lambda \lesssim 5\,\mu$m, so it should be possible to directly search for such gaps in a large sample of young stars with disks in nearby star forming regions. Interferometric images of these disks would represent the first direct attack on our theory of the formation of the Solar System.

Around the more luminous stars (Herbig AeBe stars, FU Orionis stars), disk particles will be warm out to even larger distances than shown in Figure 2. Particles 1 AU or more from the star might be imaged directly at 10 and 20 μm;

Bertout presents simulations of the appearance of disks at various wavelengths elsewhere in these proceedings to demonstrate the need for high angular resolution at thermal infrared wavelengths.

Alternatively, the scattered light from particles and resonant gas transitions in the disk "photosphere" and halo may allow the high angular resolution at short wavelengths to observe the more extended parts of the disk[1]. Both approaches are likely to bear fruit. However, the thermal emission from the optically thick inner disk regions is understood much better at this time than the scattering envelopes, and it is difficult to make good estimates of the utility of the scattered light.

Our current prejudice is that the disk structure is understood *in the main*, save for the presence of gaps. The disk photospheres – perhaps *photoplanes* would be a more accurate word – are not yet understood, nor are the halos (*e.g.*, Natta 1993). Scattered stellar light may well provide the illumination needed to study the disk and halo structure without directly imaging the thermal radiation. At present, this is unknown, although there are hopeful signs on the basis of HST images that disks can be found which will be rather bright at short wavelengths, the proplyds, for example (O'Dell *et al.* 1994).

Adaptive optics will bring several advantages to disk research. The first is that by keeping the images small and concentrating the light, the dynamic range for observations will be greatly increased, particularly in the thermal infrared. Second, the sensitivity to thermal infrared radiation will be substantially better with adaptive optics, so much so that in fact observations with an adaptively corrected VLT telescope will compete with ISO at $10\,\mu$m

These advantages will be most pronounced for the study of nearby stars and the more evolved zodiacal type disks, such as one sees around β Pictoris and Vega. Currently, the images of short wavelength scattered light from β Pic indicate disk structure with density continuing to rise at small radii (Smith & Terrile 1987) and yet it is clear from the spectral energy distributions that the inner regions must be relatively free of small particles (Gillett 1986). It is impossible at the moment to image within radii of a few tens of AU (that is, within the size of the Solar System at the same distance as the star) owing to the lack of dynamic range even in very good coronographic images. We know now how one could make much better coronographs using apodising pupils and special masks. But, to work effectively with such coronographs, one has to take out the atmospheric phase errors that produce both large seeing disk and scattered light out to large distances. The sources themselves can be used for the wavefront sensing and correction ensuring that plenty of photons are available to correct to an extremely very high order (Angel 1994).

The expectation is that one should be able to observe the disk structure in the outer parts and look for gaps which might be created by dynamical clearing of

---

[1] Because disk temperatures should fall more slowly than the inverse distance from the star, the angular size of thermally emitting regions will grow faster than the diffraction limit for increasing wavelength, thus favoring long wavelengths for studies of *emission*.

large bodies especially to map the inner regions where we know that the particle density drops markedly. The particle density must drop owing to a lack of short wavelength thermal infrared emission. Direct detection of the density at small radii will require adaptive optics on a large telescope.

Lagage & Pantin (1994) made 10 μm images of β Pic using the TIMMI camera on the NNT. The images show a marked asymmetry in the scattered light. In addition to this asymmetry, a drop in intensity of light as one gets close as 20 AU to the star suggests there is a planet or other large body which clears out an inner hole. To confirm this result and to study the way in which such dynamical gap-clearing occurs, one needs to have extremely good angular resolution with very high dynamic range at thermal infrared wavelengths. Such will be provided by adaptive optics working on the VLT. The actual resolution, assuming diffraction-limited performance at 20 μm, corresponds to $\lesssim$5 AU at the distance to these stars. With very high dynamic range, it should be possible to map the distribution of particles right down into the place in which the gap is cleared and, therefore, to compare with dynamical models. The particles must be produced continuously by the grinding up of planets, asteroids or other large bodies in the outer part of the system. Good knowledge of the surface density of these particles should reveal the distribution of larger bodies that produce them. Adaptive optics should make it possible for first time to look for planets in a way that does not involve astrometry, radio velocity searches, and extremely long observing periods.

The spectral energy distributions of these stars indicate that, in extreme cases, one sees an excess above the photosphere at 10 μm (β Pic), but more often the excess appears first at 20 or 30 μm (Aumann et al. 1984). The excellent telluric transmission available at the Paranal site should allow observations at 20 μm to be made routinely as they are currently made at 10 μm from La Silla; and occasional observations at 34 μm will also be possible. With the very large collecting areas and clean configurations of these telescopes, the sensitivity of imaging observations will be superior to that from ISO. So superior will these be that searches for more candidates, especially at larger distances, could become an important long-term research program with the VLT. In addition, owing to the larger size of the VLT apertures, one will be able to image the radiation discovered in such searches and study the surface density of matter in the way needed to compare with theories of planetary formation and subsequent evolution (Cameron 1988). Should one be able to use interferometric capability as a follow-up to these surveys, the VLT will be unique in its ability to map the surface densities on scales corresponding to the size of planetary orbits, even in the inner terrestrial regions.

The key uncertainties in our study of the structure of disks have to do with whether large bodies form, whether these bodies clear out gaps through their dynamical interaction with the particles, and whether the surface density of the particles conforms to well-known predictions of different dynamical configurations: small planetary systems, grinding of larger bodies at large distances from the star, or gap clearing by winds or planets. These questions cannot be addressed with the current generation of telescopes, and many will be accessible for study only with an interferometer of high sensitivity. The VLTI should provide both the high spatial resolution and sensitivity. This will make the structure of disks in the infrared and optical bands subject to study for the first time, and will provide the link between Solar system observations long underway and astronomical observations, which are relatively new, but are needed to blend these two fields together.

References

Angel, J. R. P. 1994, *Nature*, **368**, 203.

Appenzeller. I. 1983, *Rev. Mexicana Astron. Astrophys.*, **7**:151–168.

Appenzeller, I., Jankovics, I., Östreicher, R. 1984, *Astron. Ap.*, **141**, 108.

Aumann, H. H. *et al.* 1984, *Ap. J. Letters*, **278**, L23–L27.

Backman, D. E. & Paresce, F. 1993, in *Protostars and Planets III*, ed. G. Levy and J. Lunine (Tucson:U. Arizona Press), p. 1253–1304.

Beckwith, S. V. W. and Sargent, A. I. 1993, in *Protostars and Planets III*, ed. G. Levy and J. Lunine (Tucson:U. Arizona Press), p. 521–541.

Beckwith, S. V. W., Sargent, A. I., Chini, R., and Güsten, R. 1990, *Astron. J.*, **99**, 924.

Bertout, C. 1989, *Ann. Rev. Astron. Ap.*, **27**, 351–395.

Bouvier, J., Cabrit, S., Fenandez, M., Martin, E. L., and Matthews, J. M. 1993, *Astron. Ap.*, **272**, 176.

Cabrit, S., Edwards, S., Strom, S. E., and Strom, K. M. 1990, *Ap. J.*, **354**, 687.

Cameron, A. G. W. 1988, *Ann. Rev. Astron. Ap.*, **26**, 441–472.

Edwards, S., Ray, T. P., and Mundt, R. 1993, in *Protostars and Planets III*, ed. G. Levy and J. Lunine (Tucson:U. Arizona Press), p. 567–602.

Gillett, F. C. 1986, in *Light on Dark Matter*, ed. F. P. Israel (Dordrecht: D. Reidel), 61–69.

Jensen, E. L. N., Mathieu, R. D., & Fuller, G. A. 1994, *Ap. J. Letters*, in press.

Kenyon, S. J., Hartmann, L., and Hewett, R. 1988, *Ap. J.*, **325**, 231–251.

Lagage, P. O. & Pantin, E. 1994, *Nature*, **369**, 628–630.

MacDonald, J. & Clarke, C. 1994, in *Disks & Outflows Around Young Stars*, ed. S. V. W. Beckwith, A. Natta, and J. Staude (Heidelberg:Springer-Verlag), in press.

Natta, A. 1993, *Ap. J.*, **412**, 761.

O'Dell, C. R., Wen, Z. and Hu, X. 1993,

Ohashi *et al.* 1994

Osterloh, M. & Beckwith, S. V. W. 1994, *Ap. J.*, in press.

Sargent, A. I. and Beckwith, S. V. W. 1991, *Ap. J. Lett.*, **382**, L31.

Shu, F. H., Ruden, S. P., Lada, C. J., and Lizano, S. 1991, *Ap. J. Lett.*, **370**, L31.

Skrutskie, M. F., Dutkevitch, D., Strom, S. E., Edwards, S., and Strom, K. M. 1990. *Astron. J.*, **99**, 1187.

Smith, B. A., and Terrile, R. 1984. *Science*, **226**, 1421.

Smith, B. Q. and Terrile, R. 1987, *Bull. Amer. Astron. Soc.*, **19**, 829.

Part 3

Stellar Astrophysics

Do We Need the VLT to Study Nearby Stellar Populations?

Bengt Gustafsson

Astronomical Observatory, Box 515, S-75120 Uppsala, Sweden

Abstract. The potential of, and the need for, the VLT in studies of the composition and history of stellar populations in the solar neighbourhood is illustrated and discussed. It is concluded that there are certain areas, of great significance for the understanding of stellar and Galactic evolution, where the contribution of the VLT will be of key significance.

1 Introduction: VLT for the Solar Neighbourhood?

The title of this presentation is not merely a rhetorical question, but has actually been asked by some colleagues. They are not necessarily uninterested in the nearby universe as such, but may ask the question nevertheless, out of pure innocence, as one might seriously question the use of the Space Shuttle for bird watching. What could be the motives for using such a far-sighted tool as the VLT to study the nearby stars, many of them accessible even to exploration by the naked eye?

There are several reasons for this. *One* is just the opposite to what was just stated: some of the stars in the solar neighbourhood are so absolutely faint that it has not been possible to study them in any detail with the presently largest telescopes. Examples of such stars are late-type M dwarfs, brown dwarfs and the cool white dwarfs. Other examples are the highly obscured stars, whether newly formed in dense dust clouds like the ρ Ophiuchi cloud, or wrapped in their own debris in late evolution on, or soon after, the asymptotic giant branch stage. *A second reason* may be that one wishes to explore some detail in the stellar spectrum at a particularly high spectral resolution and/or a very high S/N. Examples are the study of convection through measuring spectral line asymmetries and line shifts, Doppler imaging of spots on stellar surfaces, the measurement and mapping of stellar magnetic fields, or the study of very small light variations. *A third reason* may be the study of very rapid events, e.g. in compact objects. Thus, the accretion disks of X-ray binaries may be explored from light echoes of sudden bright events, provided that a time resolution better than about 1 s can be afforded, which certainly needs a large light-collection area. *A fourth reason* may be that very faint surroundings of nearby bright stars may need the VLT for their exploration: examples are the thin outer envelopes shed by nearby red giants, the accretion disks of T Tauri stars, or the very faint dwarf companions of bright supergiants or giants. *A fifth reason* for using the VLT is when high spatial resolution, in AU, is needed. That is of course simplest to acheive for nearby stars, although it may require the VLTI, i.e., the development of technical and financial resources that are presently not fully ascertained.

In this short presentation I shall mainly dwell on the first two reasons and the fourth reason. In particular, I shall leave the discussion of the (strong) motives for studying nearby pre-main-sequence stars, white dwarfs and compact binaries to others, and thus essentially concentrate on the good old "normal" stars. I shall explore the proposition that a considerable fraction of the VLT observing time should be used for investigating stars that were already studied by the Herschels, Friedrich Wilhelm August Argelander and Angelo Secci.

2   Solar-type Stars

The VLT can be used to determine parameters like metallicities and ages from *uvby* photometry for F and G type dwarfs as far away as the Magellanic Clouds (incidently, that is a project of considerable interest). But why would one use it for such stars in the solar neigbourhood? One reason would be to give such measures as those just mentioned an accurate calibration – the present uncertainty in this calibration mainly reflects our shortcomings in understanding of the stellar atmospheres. I shall take a somewhat different example to clarify this.

Our knowledge about the early evolution of the Galaxy, and even of the Universe as a whole, is, in important respects, based on the chemical analyses of halo dwarfs and subgiants. All these analyses are based on classical model photospheres. In these models the stars are assumed to be stratified in plane-parallel layers, and convection is treated in the very crude mixing-length approximation. It is, however, clear from the simulations of convection in solar-type dwarfs by Nordlund and Dravins (1990) that the real stellar atmospheres are highly inhomogeneous, with a structure very far from plane-parallel stratification. Thus, in a diagram where the distribution of temperature and pressure are plotted for a model of Procyon, a somewhat evolved Population I late F-type dwarf, one does not find a unique T-P relation at all; the structure seems to consist of two different almost unconnected regions in the the T-P diagram (Fig. 1). Note that temperatures around the effective temperature, so significant for the formation of the continuous flux in the plane-parallel approximation, are hardly represented at all! The convection effects are probably even more pronounced for the subdwarfs, since their atmospheres are more transparent, which makes the convective layers more visible from the outside. Also, the penetrating ultraviolet fluxes in the hot stellar granules enhance the non-LTE effects in these stars – there could be an interesting interplay between convection and non-LTE effects in these atmospheres which is not yet studied.

Why can we not model these inhomogeneous atmospheres of halo stars with simulations, similar to those produced by Nordlund and Dravins for Procyon? What made these simulations so persuasive for the Sun, and other Population I solar-type dwarfs, was the comparison with spectra of very high quality. Thus, the observed spectral-line asymmetries, represented as line bisectors, were repro-

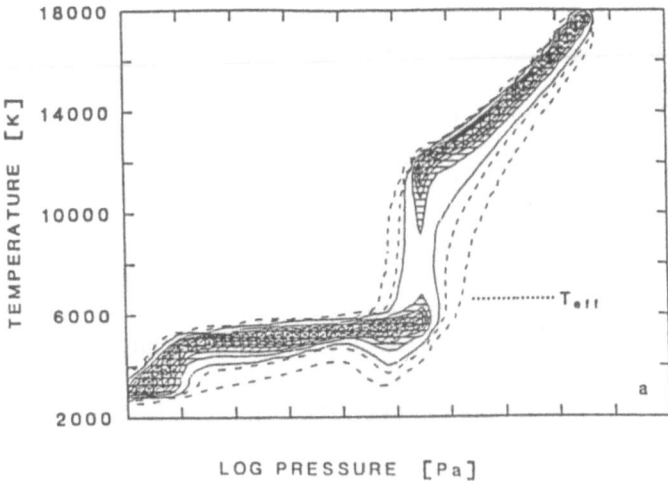

LOG PRESSURE [Pa]

Fig. 1. Statistical properties of a hydrodynamic simulation of the photosphere of Procyon, showing the distribution of temperatures at different gas pressures. The outermost contour contains 99 % of the T-P points in the the simulation, the next 95 %, while the solid ones correspond to 90, 70, 50, 30 and 10 %, respectively. The effective temperature (T_{eff} =6600 K) is marked and is obviously not very representative for the gas in the model. From Nordlund and Dravins (1990).

duced very nicely. Corresponding observations for metal-poor stars are totally missing and will be needed before one can trust the results of the convection simulations for Population II dwarfs. What is actually needed? Observations at a spectral resolution of at least 200,000 and S/N of at least 400 for Extreme Population II dwarfs and subgiants with [Fe/H] < −2.0. These stars have apparent magnitudes beyond V = 8, but spectra of satisfactory resolution will be obtainable with the planned VLT spectrometers for very high spectral resolutions. If this is not done, abundance determinations for these stars, with their far-reaching cosmological and galactic implications, will suffer from a lack of knowledge about the atmospheric structure as their basic source of error. It will, as a consequence of that, be difficult to argue that, e.g., the Li abundance, the B abundance or the abundances of O, Mg, Fe or Eu are known to better than 30 % or so. Some cosmologists urge us to reduce these errors.

A still more direct example of the need for observations of this type is offered by the recent attempts by Nissen, Lambert and Smith (Smith et al. 1993, Nissen et al. 1994) to measure the $^6Li/^7Li$ isotopic ratio for Population II dwarfs and subgiants. The measurements, which are very difficult, are based on measuring the shape of the LiI 6707 Å line. The authors used the McDonald 2.7 m spectrometer and the ESO 3.6 m with the CES spectrometer, at a resolution of 120,000 and a S/N of about 400. Although this resolution is not optimal, a probable detection of ^6Li is reported ($^6Li/^7Li$=0.05±0.02) for at least one star, HD 84937. As dicussed by Smith et al. (1993) and Steigman et al. (1993), the abundance of ^6Li in HD 84937 has about the value one expects from the measured Be and B abundances in metal-poor stars (Gilmore et al. 1992, Edvardsson et al. 1994) and the known cross sections for Cosmic Ray production of the light elements, provided that the depeletion of ^{6}Li is small, as predicted by standard stellar models. The corresponding depletion of ^{7}Li is negligible, and this therefore suggests that the ^{7}Li abundance measured in warm Population II dwarfs is close to the primoridal value and may be used to test Big Bang nucleosynthesis. Before such a far-reaching conclusion may be made with confidence, one needs to determine the lithium isotopic ratio in a large sample of Population II dwarfs and subgiants. However, only a handful, and not the most metal-poor ones, may be reached with present-day telescopes; this is a good example of a project where the VLT is needed to enable the necessary very high spectrum quality. Also, the Li profile is affected by the convective motions in the atmospheres, and these should be carefully measured and modelled in order to significantly improve the accuracy of the isotopic ratio.

Another example of a project which requires very high resolution and S/N is the dating of the Galaxy, using Thorium (with a half life of 14 Gyears) in solar-type stars as a chronometer. This method was devised by Butcher (1987) and later refined and explored by Morell et al. (1992) and François et al. (1993). The recent discovery by Sneden et al. (1994) of a very metal-poor giant (CS 22892-052, V=13.2, [Fe/H]=−3.1), with abundances of the heavy r-process elements enhanced by a factor of 100 relative to iron, means that the method can be applied to the very oldest stars in the Galactic halo, provided that the relative yields of the r-process elements were not abnormal. A basic problem with this method is, however, that the Th II line used, at 4019.13 Å, is a blend located in the wing of a much stronger Fe I line, which also carries other blending and partly unidentified lines. In order to reach safe results with this method, one has to secure better data, with fully resolved spectral lines (i.e., R $\gtrsim$ 150,000) and very high S/N ($\gtrsim$300), for a fair number of halo stars. In addition, efforts must be made to improve the atomic physics data for the relevant wavelength region, as well as again to study the asymmetries of the lines caused by convection.

With the UVES spectrometer at the VLT it will be possible to reach solar-type stars out to about 1 kpc for accurate abundance analyses (i.e. with R=40,000 and S/N=100 in one hour). Our recent study of 189 such stars in the immediate solar neighbourhood, based on observations with the ESO CAT and the McDonald 2.7 m telescopes, revealed a number of new and rather unexpected patterns in the relative abundances as a function of age and mean distance from the Galactic Centre (Edvardsson et al. 1993). For example a significant scatter (of about 0.15 dex) in overall metal abundances at given age and galactocentric distance is present, while the scatter in relative abundances, presumably arising from supernovae of different type, e.g. in the ratio of Mg to Fe, is negligible (< 0.05 dex). Rather drastic hypotheses, e.g., concerning clumpy infall of metal-poor gas during the disk evolution or triggering of star formation by Type Ia supernovae, or dominating production of iron-peak elements in SNe Type II, or very efficient heating of the Galactic disk (by massive black holes or infalling dwarf galaxies?) must be postulated. We also found a clear tendency for a ra-

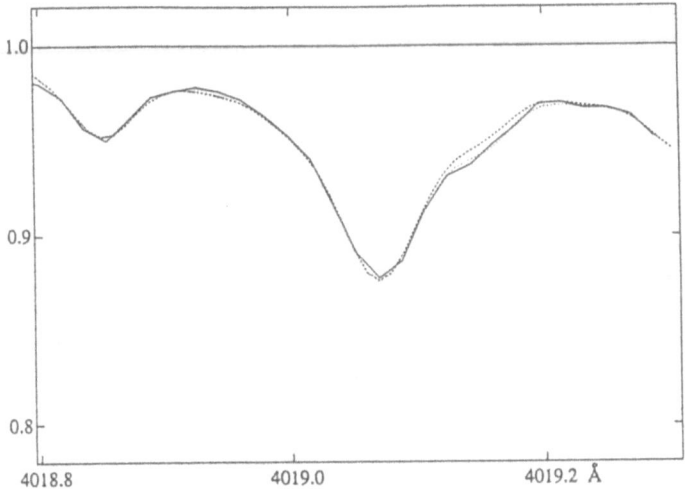

Fig. 2. The region around the Th II line at 4019.13 Å, in the spectrum of the Population II subgiant HR 8515, as compared with two model spectra (dashed and dotted) with a difference of 0.1 dex in the Thorium abundance. The Th line is located in the wing of the Fe I line 4019.043 Å, which is also affected by a blending Ni I line at 4019.067 Å and a probable unidentified line at 4019.10 Å (Morell et al. 1992). The wavelengths were given an offset of 0.025Å in order to match the stellar radial velocity. The figure was kindly prepared by O. Morell.

dial gradient in Mg, Si and Ca abundances relative to Fe for the oldest stars, indicating that star formation occurred more rapidly in the interior parts of our galaxy. Also a new group of stars with somewhat peculiar abundances of Na, Al and Mg was identified – with respect to this population the Sun seems iron rich.

All these results were obtained by studying stars at a distance within 100 pc, and the abundances at other galactocentric distances were found indirectly by measuring the stellar motions and calculating the Galactic orbits. However, the potential field in the Galaxy is not well known – in particular, little is known about its fine structure – and therefore our extrapolations to conditions at other distances from the centre of the Galaxy are uncertain. Moreover, we do not know how representative the immediate solar neighbourhood is even for Galactic fields at our distance from the Galactic Centre. So, it would be a very good idea to use the VLT for determining relative heavy-element abundances for solar-type stars of different ages at about 1 kpc distance from us in radial Galactic directions, and also in tangential directions, as well as about 1 kpc above the galactic plane. Such a grand study should, if possible, be complemented with measuring accurate proper motions for the programme stars – a task that could be suitable for the next generation astrometric satellite that is currently under discussion in the astrometry community.

Let me conclude this section by saying a few words about the use of the VLT for studying stellar oscillations, for which the success rate is thought to be greatest for stars with intensive convective motions, i.e. somewhat evolved solar-type stars (disregarding the non-radial oscillations found for white dwarfs). The large aperture of the VLT is a great advantage for such projects. However, if the oscillations are to be found and studied spectroscopically from Doppler shifts, the problem is to achieve the very high degree of wavelength stability needed, in spite of the considerable size of the spectrometers. Also, the need for very long observing runs, and for continuing campaigns with similar telescopes when the object sets below the horizon at Paranal, may seem to make the VLT less optimal for this type of project. However, the recent probable detection of oscillations of η Boo (G0IV) from Balmer-line strength variations with the Nordic 2.5 m telescope (NOT) (Kjeldsen et al. 1994) opens up the possibility for the VLT to play a significant role also in this very promising field of stellar astrophysics.

3   The Coolest Dwarfs

Olof Morell (1994) has just finished a thesis with detailed chemical analyses of about 20 K dwarfs, which is based on CAT/CES spectra. His study shows that data of high internal consistency and high accuracy may be obtained, as good as, or in certain respects even better than one may get for solar-type dwarfs. Some of these data are also complementary in terms of chemical elements and isotopes to what can be derived from F or G-type spectra. Moreover, any sample of K dwarfs is less affected by stellar evolution. Furthermore, one may advocate that effects on the chemical analysis of convection and non-LTE are probably less significant for these stars than for somewhat hotter dwarfs. In spite of the problems with line blending and extensive pressure-broadened line wings (which gives strong motives to carry out much of the study in the near-IR spectral region), one must therefore recommend further exploration of the possibility to use these stars systematically in studies of Galactic chemical evolution. In addition to this, the study of the properties of the low-mass end of the stellar mass distribution is worthwhile; e.g., we do not know whether the IMF is the same independently of the chemical composition of the pre-stellar gas. Such a question may be explored after further K dwarf analysis. However, in the close solar neigbourhood much of this continued work may be carried out without the VLT.

In the present perspective, however, the study of halo K dwarfs is relevant. In order to reach the most metal-poor K dwarfs with detailed chemical analyses, the VLT would be required. Not the least would that give CNO abundances with criteria complementary to those in current use for solar-type stars, which are somewhat debated. More experimental, but quite important, would be to attempt estimating the oxygen and carbon isotopic ratios for disk and halo late G and K dwarfs and subgiants by measuring lines of the CO 4.6 and 2.3 micron bands, with the planned VLT cryogenic echelle. Such projects would shed light on the significance of internal early mixing processes in the stars, as well as make it possible to trace the interstellar abundances of these isotopes as functions of age and galactocentric distance. With a required S/N of about 100, and a resolution of (at least) 80,000, one would be able to measure the isotopic ratios for a representative sample of about 100 dwarfs and subgiants in the solar neighbourhood, within less than 100 hours of observing time.

Another good reason for studying K dwarfs, and not least the younger rapidly rotating ones, is the possibility to map spots on their surfaces and their magnetic fields by Doppler imaging. As is further discussed in other contributions at this conference, such efforts require high spectral resolution (the slower rotation the higher), and, in particular, very high S/N. Projects of great interest that would require the VLT would be such studies in the nearby open clusters and star-forming regions.

The exploration of the chemical chemical composition and surface structures of M dwarfs is still in its infancy, but will improve very significantly within the next ten years, not only due to the erection of large telescopes but also the construction of cryogenic echelle spectrometers for the infrared. Several tens of thousands of M dwarfs will suddenly become accessible for detailed chemical analysis – at present less than 20 are available and hardly any analysis is truly reliable. Even some hundreds of the coolest M dwarfs, with masses around 0.1 solar masses, may soon be analysed, provided that the spectra are possible to model.

In spite of recent great advances in the modelling of M dwarf atmospheres (Kirkpatrick et al. 1993, Allard 1994, Brett and Plez 1993, cf. also Gustafsson and Jørgensen 1994), it is not clear how successful the theoretical modelling will be. The results of these efforts, if successful, will be of great significance for the understanding of the role of low mass objects in the evolution of galaxies. Of particular interest for the programmes at the VLT will be studies in the infrared at extra high resolution and very high S/N in mapping spots and magnetic fields in their atmospheres. A few such studies have been possible with existing telescopes, the sample of studied objects is, however, restricted to extreme cases namely very nearby, rapidly rotating stars with quite strong activity. In order to broaden these studies to more normal stars, higher resolution and S/N are needed, which is only possible with an increase in light collection area. These studies may well be necessary in order to put the chemical analyses on a reliable foundation, but are in particular of great interest in themselves and in order to improve our understanding of stellar activity and dynamos. Also these investigations should be systematically pushed towards the lowest mass objects in nearby clusters and star-formig regions. That will be possible with the VLT and the planned cryogenic echelle spectrometer. Another project for that intrument will be the study of convection, as a function of effective temperature and metallicity, in M dwarfs. Convective motions may reach high layers in these atmospheres as a result of H_2 dissociation and the study of them will be required for the understanding of other magneto-hydrodynamic phenomena; in addition to that, again the application of the results to improve stellar-atmosphere modelling, and thus to enable accurate chemical analyses, will be of fundamental significance. The convective velocities in M dwarfs are, however, expected to be relatively small, and a spectral resolution of at least 300,000 is needed for this project. The complex spectra, with numerous blends and broad damping wings, also make a very high S/N (considerably better than 100) necessary for these studies.

If the search for brown dwarfs is successful, the detected targets will certainly be interesting to observe with the VLT. It is not clear at which spectral resolution such efforts will and can be made. However, even at quite low resolution these spectra will contain interesting information, not only on the physical properties of the objects, but also on the basic physical processes in their atmospheres, such as the collision-induced absorption of H_2 which, strangely enough, may cause the flux distribution to peak more and more towards the blue when the object cools, at least for metal-poor objects (cf. Saumon et al. 1994). These opacities are still uncertain but remind us that simple order-of-magnitude estimates on what fluxes to expect, and in which wavelength regions, may be totally in error. If brown dwarfs are found, attempts to determine their chemical composition will be interesting.

4 Red Giants

The red giants may be reached for rather detailed analysis in the infrared over almost all the Galaxy with the VLT. Even at a resolution of the cryogenic echelle of at least 100,000 one may study the upper parts of the giant branches of ω Cen and 47 Tuc, and thus carry out analyses of the carbon and oxygen isotopic ratios. Among studies of very great interest in this area are detailed and systematic studies of stars in the Galactic Bulge.

However, for the relatively nearby red giants a number of tasks at very high spectral and spatial resolution will be of great interest, related to the lack of adequate knowledge about the dynamics of the photospheres and the mass loss mechanisms. For example, it is not known on which scale(s) convection occurs in them, and this lack of knowledge makes detailed numerical simulation more difficult and questionable, in particular if, in real stars, several widely different scales are of significance. Also, the coupling between convection and pulsations is a novel and interesting field to explore, as well as the detailed dynamics and structure of the circumstellar envelopes.

The study of convection in nearby giants is an obvious task for the VLTI and VISA, with the aim of imaging granular patterns on the stellar surfaces. One should note, however, that one probably wants to observe not only the largest structures, which may even be on the order of the stellar diameter, cf. Wilson et al. (1992), but reach across the full efficient power spectrum of the granulation. This is critical as a guidance for numerical simulations and may require at least 10,000 resolution elements per stellar disk. The interferometric studies could be combined with measurements of spectral line asymmetries and variations, measured again at very high resolution and S/N (typically 200,000 and 400, respectively). The monitoring of these spectral line-shape variations for pulsating giants of different types is another obvious task for the VLT. One of the most promising projects of this type is detailed spectroscopy of the vibration-rotation lines of CO at very high spectral resolution (about 200,000) for a set of stars, more or less variable, with different fundamental parameters, in attempts to empirically trace the structure of the outer atmosphere and its variations. Tsuji (1986) and more recently Wiedemann et al. (1994) have clearly demonstrated the potential of such studies, that will hopefully, in the end, lead to much better understanding of mass loss from red giants.

Another field of investigation is the study of circumstellar shells of red giants in scattered star light in resonance lines. A small number of shells have been traced and mapped in this way (see Plez and Lambert 1994, and references cited therein). The discovery that, e.g., almost all carbon stars have molecular CO shells, about 1/10 of them thin, detached from the star and with clumpy structures (Olofsson et al. 1990, 1992, and unpublished research), makes it tempting to try to trace these shells also at optical and near IR wavelengths to obtain velocity and intensity maps at higher spatial resolution. We have discovered three of these shells in the K I 7699 Å line (Eriksson et al. 1994) but have not succeeded in mapping their structure in any detail, due to lack of photons. The most extended shells such as that of S Sct (larger than one arc minute accross) are the far most difficult ones to image in the scattering lines. They certainly present an interesting challenge for detailed exploration with the VLT.

5   Final Recommendations

Astronomical knowledge is, in many respects, a house of cards. A famous example of this is the astronomical distance scale, but it also holds true for much of our astrophysical understanding. If an upper level in this house is loaded with great amounts of new information, this also motivates ambitious attempts to strengthen the lower floors and the foundations. Otherwise, the house may get unbalanced - even if it does not fall down it may look more impressive than useful, and it is hardly possible to continue the construction and reach the sky. I thus recommend that a considerable fraction of the time at the VLT be used to strengthen our understanding of the nearby stars, which consequently means of all stars, and of all galaxies in the universe.

May I finally just add that some important parts of the projects mentioned above might be possible to carry out at smaller telescopes, e.g. the ESO 3.6 m telescope, if it were equipped with two new instruments: one state-of-the-art visual spectrometer, with a resolution of at least 200,000, and one cryogenic echelle with properties similar to that proposed for the VLT. The impact of these two instruments would indeed be very great if they could be constructed within the next few years - and they would diminish the pressure on the VLT and improve the quality of the projects finally carried out there, as well. If ESO cannot undertake this additional task in its present situation I propose that the ESO member states, and interested groups in them, be invited to assist in, or undertake, the construction of these two instruments.

Acknowledgements. A number of ideas in the present paper are due to Poul Erik Nissen, who is also thanked for valuable comments on the manuscript. Kjell Eriksson also contributed comments and assistance in preparing it.

References

Allard, F.: 1994, In P. Thejll and U.G. Jørgensen (eds.), Poster Session Proceedings of IAU Coll. 146, Copenhagen University, p. 1

Brett, J.M., Plez, B.: 1993, Proc. AS Australia, 10, 250

Butcher, H.R.: 1987, Nature, 328, 127

Edvardsson, B., Andersen, J., Gustafsson, B., Lambert, D.L., Nissen, P.E., Tomkin, J.: 1993, A&A, 275, 101

Edvardsson, B., Gustafsson, B., Johansson, S.G., Kiselman, D., Lambert, D.L., Nissen, P.E., Gilmore, G.: 1994, A&A, in press

Eriksson, K., Gustafsson, B., Olander, N., Olofsson, H.: 1994, in preparation

François, P., Spite, M., Spite, F.: 1993, A&A, 274, 821

Gilmore, G., Gustafsson, B., Edvardsson, B., Nissen, P.E.: 1992, Nature, 357, 379

Gustafsson B., Jørgensen, U.G.: 1994, A&A Rev, in press

Kirkpatrick, J.D., Kelly, D., Rieke, G.H., Liebert, J., Allard, F., Wehrse, R.: 1993, ApJ, 402, 643

Kjeldsen, H., Bedding, T.R., Viskum, M., Frandsen, S.: 1994, in preparation

Morell, O.: 1994, PhD Thesis, Uppsala University

Morell, O., Källander, D., Butcher, H.R.: 1992, A&A, 259, 543

Nissen, P.E., Lambert, D.L., Smith, V.V.: 1994, The ESO Messenger, 76, 36

Nordlund, Å; Dravins, D.: 1990, A&A, 228, 115

Olofsson, H., Carlström, U., Eriksson, K., Gustafsson, B., Willson, L.A., 1990, A&A, 230, L13

Olofsson, H., Carlström, U., Eriksson, K., Gustafsson, B., 1992, A&A, 253, L17

Plez, B., Lambert, D.L.: 1994, ApJ Letters, 425, L101

Saumon, D., Bergen, P., Lunine, J.I., Hubbard, W.B., Burrows, A.: 1994, ApJ, 424, 333

Smith, V.V., Lambert, D.L., Nissen, P.E.: 1993, ApJ, 408, 262

Sneden, C., Preston, G.W., McWilliam, A., Searle, L.: 1994, preprint

Steigman, G., Fields, B.D., Olive, K.A., Schramm, D.N., Walker, T.P.: 1993, ApJ, 415, L35

Tsuji, T.: 1986, ARA&A, 24, 89

Wiedemann, G., Ayres, T.R., Jennings, D.E., Saar, S.H.: 1994, ApJ, 423, 806

Wilson, R.W., Baldwin, J.E., Buscher, D F., Warner, P.J.: 1992, MNRAS, 257, 369

Nucleosynthesis in the First Galactic Stars

J. Andersen[1], B. Nordström[1], T.C. Beers[2], R. Cayrel[3], F. Spite[3], M. Spite[3], P.E. Nissen[4] and B. Barbuy[5]

[1] Copenhagen University, Denmark
[2] Michigan State University, U.S.A.
[3] Observatoire de Paris, France
[4] Aarhus University, Denmark
[5] Sao Paulo Observatory, Brazil

Abstract. Our aim is to study the detailed chemical composition of the first galactic stars, which contain an important record of the products of nucleosynthesis in the Big Bang and the earliest stellar generations. Stars with [Fe/H]$\leq$-3 are being identified from candidates in the surveys by Beers et al. (1985, 1992). Determination of the detailed elemental and isotopic composition of these faint stars (V$\leq$ 15) will require the VLT and UVES.

1 Introduction

The very first stars formed in our Galaxy hold important clues to its origin and to the processes of galaxy formation in general. The detailed elemental and isotopic abundances of these stars are our most significant data regarding:(i) test of Big-Bang nucleosynthesis from the abundances of the light elements Li, Be, and B; (ii) evolution of CNO abundances in the earliest stellar generations; (iii) the relative importance of r and s-process nucleosynthesis in primordial stars; and (iv) mixing and dredge-up processes in extremely metal-deficient stars. For the metal-rich young disk, Edvardsson et al. (1993) recently demonstrated how a wealth of new information can result from careful analysis of high-resolution, high S/N spectra of a carefully selected sample of stars. There is every reason to expect similarly accurate data on very metal-poor stars to also yield rich rewards.

The first Galactic stars should survive today as extremely metal-poor halo objects ($\leq$0.1% of the Sun's heavy-element content). Such stars are extremely rare in the solar neighbourhood and were long thought not to exist at all. Even amongst stellar samples with the most extreme kinematical characteristics, i.e. high proper- motion stars in retrograde Galactic orbits, stars with [Fe/H]$\leq$-3 are still almost non- existent (see Fig. 1). Nonetheless, significant numbers of extremely metal-poor stars are in fact turning up in the objective-prism surveys of Beers et al. (1985, 1992; BPS). We have initiated a programme to identify and obtain the necessary additional data in for a sizeable sample of these stars in preparation for a detailed spectroscopic study with the VLT and UVES.

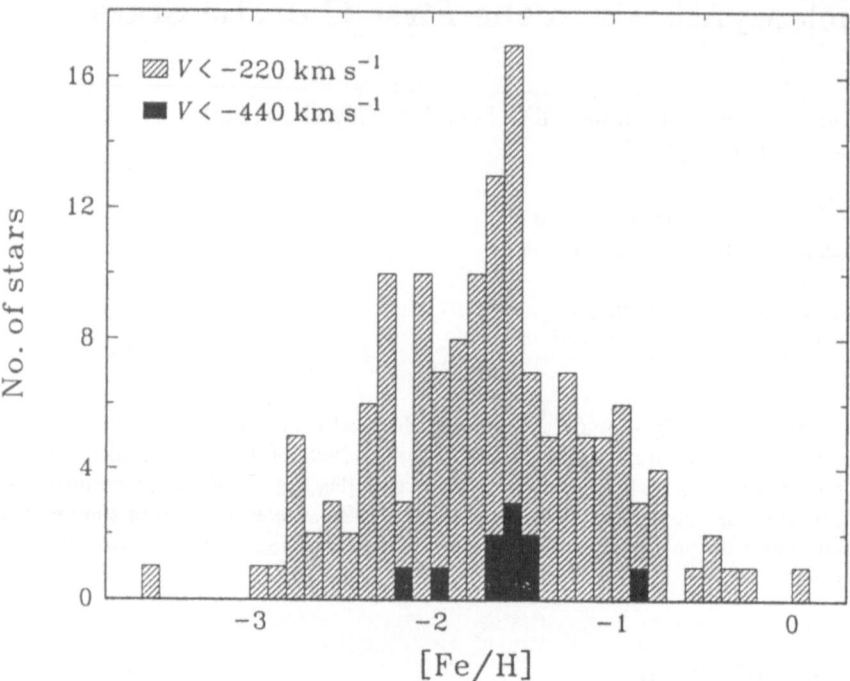

Fig. 1. Metallicity distribution for halo stars with retrograde orbits in the proper-motion sample of Carney et al. (1990); stars with the most extreme orbits are shown in black.

2 Preparatory Observations

From a total of ~8000 candidate stars in the BPS surveys, ~4000 are observable in the southern hemisphere. In order to explore the multi- dimensional parameter space of this as yet virtually unknown population, a final sample of at least ~200 stars will be needed (cf. e.g. Edvardsson et al. 1993). Our current ESO program, supplemented with northern-hemisphere observations from Observatoire de Haute-Provence and coordinated with similar efforts by Norris and Ryan from Australia in the south, aims at identifying and further characterising the sample of stars with $Z/Z_\odot \lesssim 10^{-3}$, i.e. below even the most metal-poor globular clusters.

The investigation is laid out as follows:

1. Intermediate-resolution spectroscopy ($\Delta\lambda \approx 1$Å) for the most promising BPS candidates to provide an estimate of the metallicity (to ±0.2 dex) and radial velocity (to ±10 km s^{-1}).

2. Strömgren uvbyβ and Cousins R-I photometry of the same stars to provide estimates of the fundamental parameters (T$_{eff}$, log g) and reddening.

3. CORAVEL radial-velocity observations to improve the accuracy and detect possible binaries in the sample.

4. Higher-resolution spectroscopy with the ESO 3.6-m telescope and CASPEC to explore the detailed abundance patterns of the brightest metal-poor stars.

3   Initial Results

Through August 1994, some 275 stars have been observed with the ESO 1.52-m telescope and the Boller and Chivens spectrograph, and most of the complementary photometry has also been obtained. From these observations, we have identified some 30 stars with metallicity below [Fe/H] = -3, so these stars are turning up in roughly the expected numbers. One star is near [Fe/H] = -4, as would be expected if the number of stars per metallicity bin on a linear scale is constant (see Fig. 2). As the sample continues to increase, we may expect to detect a few stars at or even below [Fe/H] = -5, an as yet unexplored domain, especially if we can extend the sample to V~15.

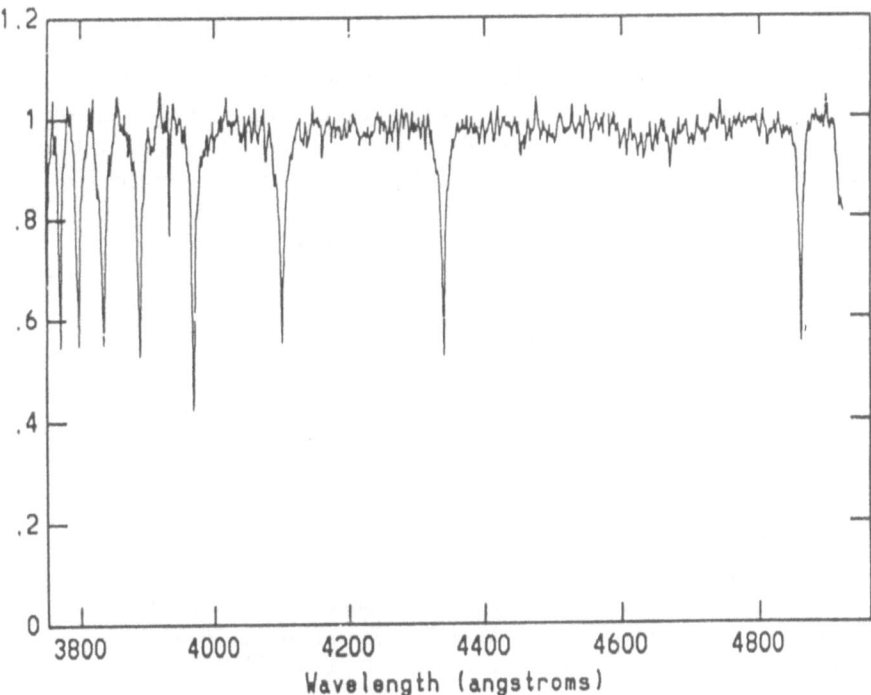

Fig. 2. Medium-resolution spectrum from La Silla of an extremely metal-poor star from our current programme. Using the equivalent widths of Hδ and Ca II K as temperature and metallicity indicators, respectively, we estimate [Fe/H]$\approx$-3.9.

The bulk of the sample must await detailed spectroscopic analysis with the future generation of large telescopes. However, exploratory work on some of the brighter stars has already provided tantalizing glimpses of a nucleosynthesis history quite unlike that of previously well-studied disk and halo samples, which have shown remarkably homogeneous proportions of the elements heavier than hydrogen (Edvardsson et al. 1993, Nissen et al. 1994a).

Especially, the remarkable star CS 22892-052 studied by Sneden et al. (1994) shows large overabundances of the r-process elements, increasing with atomic number. This pattern strongly suggests that we are witnessing the result of a localized nucleosynthesis event in the early Galactic halo, perhaps a single low-mass Type II supernova. Clearly, we are beginning to look back to a time when our Galaxy evolved in ways quite different from our familiar models.

4 The Ultimate Rewards

The most metal-deficient stars should tell us not only about what went on in the early Galaxy, but also what existed even before that stage. Hence, careful spectroscopic abundance determinations of the light elements and their isotopes (i.e. ^{6}Li, ^{7}Li, Be and B) play a key role in constraining models of Big-Bang nucleosynthesis (Gilmore et al. 1992, Boesgaard & King 1993, Nissen et al. 1994b). As discussed by the latter authors, ground-based work on these species requires very high spectral resolution (R≈100,000) and S/N ratio (∼400 or more). Coupled with the need, outlined above, to reach V≈15 in order to study the extremely metal-deficient stars, the bulk of the most interesting results in this area will certainly require the full power of the VLT and its UVES spectrograph.

References

Beers T.C, Preston G.W., Shectman S.A., 1985, AJ 90, 2089
Beers T.C, Preston G.W., Shectman S.A., 1992, AJ 103, 1987
Boesgaard A.M., King J.R., 1993, AJ 106, 2309
Carney B.W., Latham D.W., Laird J.B., 1990, AJ 99, 572
Edvardsson B., Andersen J., Gustafsson B., Lambert D.L., Nissen P.E., Tomkin J., 1993, A&A 275, 101
Gilmore G., Gustafsson B., Edvardsson B., Nissen P.E., 1992, Nature 357, 379
Nissen P.E., Gustafsson B., Edvardsson B., Gilmore G., 1994a, A&A 285, 440
Nissen P.E., Lambert D.L., Smith V.V., 1994b, The Messenger 76, 36
Sneden C., Preston G.W., McWilliam A., Searle L., 1994, ApJ 431, L27

Elemental and Isotopic Abundances in Metal-Poor Stars

Pierre Magain

Institut d'Astrophysique, Université de Liège, B-4000 Liège, Belgium

Abstract. Typical observing programs aimed at studying the early chemical evolution of the Galaxy are presented: the determination of relative abundances and cosmic scatter, the analysis of the most extreme metal-poor stars, and the study of the isotopic abundances. It is shown how the VLT could contribute to these studies. The instrumental requirements are considered, and it is concluded that not only a large telescope but also a very high resolution spectrograph is needed if significant progress is to be made in this field.

1 Introduction

The chemical composition of the atmospheres of "normal" unevolved stars is thought to be essentially identical to that of the interstellar gas out of which these stars formed. Thus, the determination of the chemical composition of stars of different ages allows one to trace back the history of the chemical enrichment of the Galaxy, and to better understand the stellar nucleosynthetic processes.

Metal-poor halo stars have a particular importance in this respect since they are the oldest stars in the Galaxy. Their study thus allows the exploration of the very first stages of the chemical evolution of the Galaxy.

The VLT should be very useful for their study since these stars are much less numerous in the solar neighborhood than more metal-rich disk stars. In order to get a significant sample, one thus has to go to larger distances, and thus to fainter objects. This is especially true for the most extreme metal-poor stars, whose space density is particularly low.

In the following, we present a number of typical questions which may be addressed, and discuss the impact of the VLT on our ability to tackle them. We restrict our discussion to what we consider are non trivial aspects. Obvious points do not need discussion, and will not be considered. (A typical example of such obvious programs is the multi-fiber spectroscopy of globular cluster stars.)

We start by the discussion of the most classical aspect of the abundance analyses, namely the determination of the relative abundances as a function of metallicity. To this very classical theme, we add a new dimension in which the VLT might bring significant impact, i.e. the study of the cosmic scatter in these relative abundance ratios.

The second aspect we address is the study of the most extreme metal-poor stars, objects which are particularly interesting and for which a large telescope is clearly required. However, we wish to make a few remarks, showing that, while very promising, their study will not be obvious if we wish to extract meaningful information.

Finally, we discuss the determination of the isotopic abundances, a subject which is still in its infancy. We show how information on the isotopic ratios may shed new light on some problems and discuss the instrumental requirements for such isotopic abundances to be measurable.

Note that throughout this paper, we do not address the question of the very light elements, such as Li, Be or B, whose behaviour deviates significantly from that of the heavier elements, and which are considered in another paper (Andersen et al., these proceedings).

2 Relative Abundances and Cosmic Scatter

2.1 Relative Abundances

In a typical abundance analysis, medium or, preferably, high resolution spectra are used to obtain equivalent widths (EWs) of the spectral lines. These EWs constitute the basic observational data. They are interpreted with the help of a model atmosphere (whose parameters: effective temperature, surface gravity and chemical abundances, are obtained through photometric or spectroscopic measurements). An estimate of the elemental abundance can thus be deduced from each line of the element considered.

It is generally considered good practice to get the equivalent width of as many lines as possible, in order to check for consistency of the derived abundances, and to reduce the uncertainty on the mean value.

This is obviously correct, *provided that the observation of a larger spectral range is not made at the expense of the accuracy of the data.*

Indeed, the accuracy of the data can be easily degraded in several ways.

First, with CCDs as detectors, and their generally square shape, the observation of a large spectral range at high resolution requires the use of cross-dispersed Echelle gratings. One thus has several orders on the detector, these orders being generally curved and tilted, and the orientation of the slit image even sometimes changing with position on the detector. There is also a significant level of interorder light. All this renders the proper reduction of the data relatively tricky, resulting in the inclusion of what is sometimes referred to as "reduction noise".

Secondly, if some elements, like Fe, show plenty of spectral lines, many others just have a few lines, some of which might not be adequate for abundance determinations because of improper strength, of blends, or problems with the continuum determination, etc. The inclusion of more lines might thus result in a degradation of the accuracy of the results.

These problems are illustrated in Fig. 1, which compares the determination of the same relative abundance, i.e. the abundance of zirconium relative to yttrium, in similar samples of stars. The analyses of Luck and Bond (1985) and Gilroy et al. (1988) were based on cross-dispersed Echelle spectra, the first with an electronic camera, the second with a CCD as detector. Despite the improvement in spectral resolution and signal-to-noise ratio (S/N), the scatter is nearly as

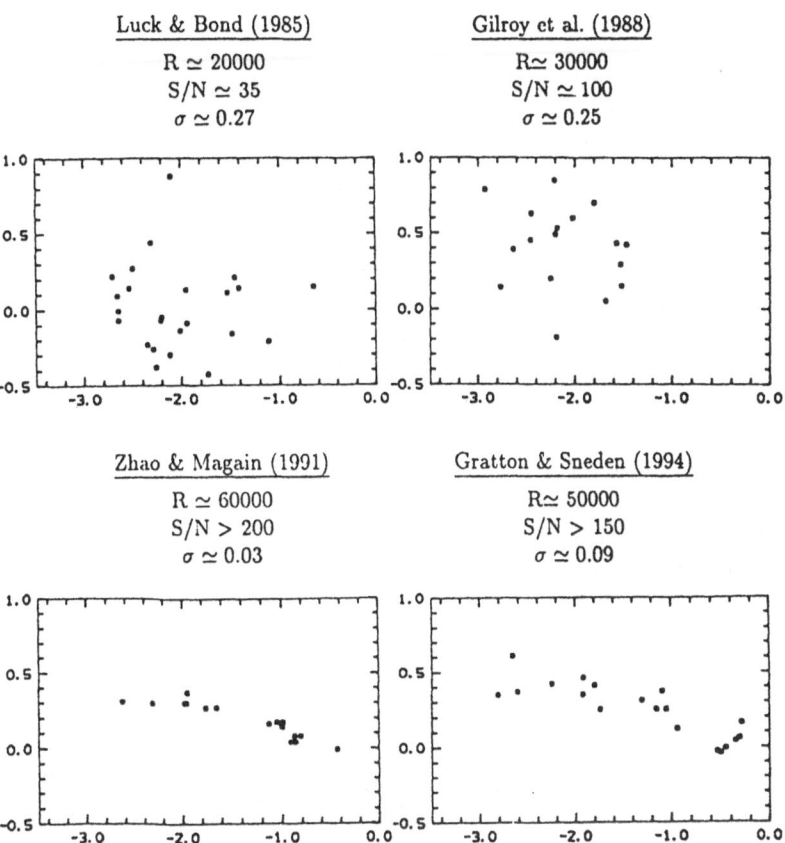

Fig. 1. Plots of [Zr/Y] as a function of [Fe/H] in metal-poor stars, as determined in different investigations. The resolving power and the S/N are indicated, as well as the scatter σ in the resulting relative abundances. See text for further details

large in the second analysis. The analysis of Zhao and Magain (1991) was based on higher resolution, higher S/N spectra obtained with the CAT/CES at La Silla. In this case, only a short spectral range is observed. Three lines of yttrium and a *single* line of zirconium were used, the other lines of this element being considered as unsuitable for a proper abundance determination. Despite the use of such a small number of lines, the scatter is reduced by a factor 10! Finally, Gratton and Sneden (1994) used the same instrument as Zhao and Magain, but their spectra have a slightly lower S/N and include more lines. The scatter is not as small as in Zhao and Magain (1991), probably because of the inclusion of lines which are not optimal.

So, since not only Zr, but also several other elements have a very limited number of suitable spectral lines, one has to be able to get high accuracy data for *individual lines*. It is often not possible, nor desirable, to include many lines in order to improve the precision of the abundance determinations.

2.2 Cosmic Scatter

The accuracy presently achievable for the brightest metal-poor stars, and which will be within reach of the VLT for a significant sample of stars, opens new interesting possibilities. In fact, when the precision on the relative abundances gets below the 10% level, as in the work of Zhao and Magain (1991), the cosmic scatter on many abundance ratios can be reliably measured.

This gives a new important constraint on the nucleosynthesis and galactic evolution models. Let us, as an example, show what can be deduced from the same Y and Zr analysis (Zhao and Magain 1991). We have seen that the observed scatter in the Zr/Y ratio is very small, and compatible with the observational uncertainties. As far as we can say, these two elements are perfectly correlated. Such a nice correlation disappears when Y or Zr is compared to Fe or Cr. Here, the scatter amounts to 0.11 dex (25%). On the other hand, Y and Zr are better correlated with Ti, with a scatter of 0.06 dex (14%).

Since Fe and Cr are thought to be predominantly produced by Type I supernovae, while Ti is a typical massive star (Type II supernova) nucleosynthesis product, the observed correlation suggests that such massive stars played a significant role in the early production of Y and Zr. Such a conclusion could not have been drawn from the examination of the relative abundances alone.

2.3 Instrumental Requirements

If useful results are to be obtained from the analysis of the chemical abundances in metal-poor stars, the instrumental requirements are the following. We need:
(1) very good photon statistics;
(2) to avoid line blending;
(3) an accurate continuum determination and;
(4) an accurate zero-point determination.

Thus, obviously, both high resolution and high S/N are required. An absolute minimum is a resolving power of at least 50000 and a S/N of at least 200.

3 The Most Extreme Metal-Poor Stars

The most extreme metal-poor stars are especially fascinating, as, e.g., their surface abundances should be very close to the primordial ones. However, their space density is so low that very few of them are presently within reach of a detailed chemical analysis.

Present estimates suggest that less than 10 dwarf stars with [Fe/H] < −4 are brighter than 13th visual magnitude (Nissen 1992). To build a significant sample (∼ 100) of such stars, one has to go down to 15th magnitude. A large telescope is therefore required to get high resolution, high S/N spectra of such objects. Their study is thus an obvious project for the VLT.

However, it has to be recalled that these stars are not only very rare, they also have extremely weak lines. Thus, high resolution and high S/N spectra are mandatory if the spectral lines are to be detected and measured with the required accuracy.

A typical cross-dispersed Echelle spectrograph (like UVES) works at a resolving power $R \sim 40000$ and $S/N \sim 100$. Diffuse light, reduction noise, etc., make it quite hard to reach a significantly higher S/N. With such an instrument, the one sigma uncertainty on EWs is of the order of 1.3 mÅ. Thus, only lines having an EW larger than 13 mÅ can be measured with a 10% precision.

In a typical, solar-type, extremely metal-poor star with abundances reduced by a factor 10^5 relative to solar, the only elements which would be accessible with such a instrument are Ca, Ti, Fe, Ni, assuming that the entire spectral range from the UV atmospheric cutoff to the near IR is accessible. One can hardly promise that great scientific achievements would result from the determination of the abundances of these 4 elements.

On the other hand, a very high resolution spectrograph with an efficient design would allow a $S/N \sim 500$ to be reached without any difficulty. The increase in resolving power and in S/N would mean that weaker lines would be measurable. With $R = 10^5$ and $S/N = 500$, the one sigma error bar on the EWs is 0.12 mÅ, meaning that an EW of 1.2 mÅ would be measured with a 10% precision.

Considering the same typical star and the same spectral range as above, the abundances of 15 elements would be available: Na, Mg, Al, Si, Ca, Sc, Ti, Cr, Mn, Fe, Co, Ni, Cu, Sr, Ba. Such a sample would certainly allow interesting discoveries.

It is thus obvious, if meaningful results are to be obtained at all, that one will have to pay the price. Of course, the exposure time needed to reach a the high R and high S/N required is much higher than for a typical observation with an UVES-like instrument. This is, however, unavoidable.

4 Isotopic Ratios

The determination of the isotopic abundances in metal-poor stars opens entirely new prospects in studies of the chemical evolution of the Galaxy. We shall illustrate this statement by considering what can be learned from a determination of the odd-to-even isotopic ratio of barium. We begin by showing what can be done with present instruments and, then, consider what the VLT could bring to that subject, and what are the instrumental requirements. These requirements also apply to the determination of other isotopic abundances, such as those of lithium.

4.1   The Case of Barium

The elements heavier than the iron peak are synthesized by neutron capture processes. Two processes are traditionally considered, depending on the magnitude of the neutron flux. Both of these processes generally contribute, to various degrees, to the synthesis of the heavy elements.

The s-process (slow) occurs when the neutron flux is low. It is thought to operate mainly in the atmospheres of asymptotic giant branch (AGB) stars. Its timescale is relatively long and, in addition, it requires seed (iron-peak) nuclei to be present in the star before operation of the nucleosynthetic processes. Thus, although the s-process dominates the production of heavy nuclei at the present time, its products are not expected to be present in significant amounts in the atmospheres of the oldest stars.

On the other hand, the r-process (rapid) requires high neutron fluxes, such as can be encountered in explosive situations. Type II supernovae are one of the most likely sites for its operation. In that case, the r-process can operate on seed nuclei synthesized by the progenitor star itself. This fact, together with the short lifetime of the progenitor, implies a much shorter timescale for the synthesis of r-process products than for the s-process ones.

It has been suggested by Truran (1981) that elements such as barium which, in solar-system material, are mostly synthesized by the s-process, are in fact r-process products in very-metal-poor stars. Subsequent spectroscopic analyses (Gilroy et al. 1988 and references therein), seemed to support that hypothesis on the basis of a fair agreement of the observed abundance distribution with the predictions of r-process models.

However, the abundance distributions provide only an indirect test for comparison with very approximate models, and one would like to have a much more direct way of testing Truran's hypothesis.

Such a test is provided by the isotopes of barium. In solar-system material, barium is mainly represented by five stable isotopes, with mass numbers 134 to 138. While the s-process dominates the production of the even isotopes, the r-process contributes significantly to the synthesis of the odd isotopes. The determination of the odd-to-even isotopic ratio thus allows measurement of the relative contributions of these two processes.

The isotopic shifts of the spectral lines being completely negligible for elements as heavy as barium, the spectral lines of the different isotopes cannot be separated, and another method must be used. This method is based on the hyperfine structure (HFS) splitting, which affects only the odd isotopes.

The fraction of odd isotopes is thus directly related to the width of spectral lines affected by HFS. High resolution, high S/N observations should allow the line broadening and, thus, the relative importance of the two processes to be determined.

Such an analysis has been carried out for the resonance line of BaII at 4554Å in the classical metal-poor star HD140283. Figure 2 shows the observed line profile, compared to the model calculations with 0, 50 and 100% of odd isotopes. It is clearly seen that the observations favour a low value. The best fit value is

8(±6)%, in agreement with a pure *s*-process synthesis (for which models predict 11(±2)%), and in complete disagreement with the hypothesis of an enhanced *r*-process contribution. This more direct test thus contradicts the previous expectations.

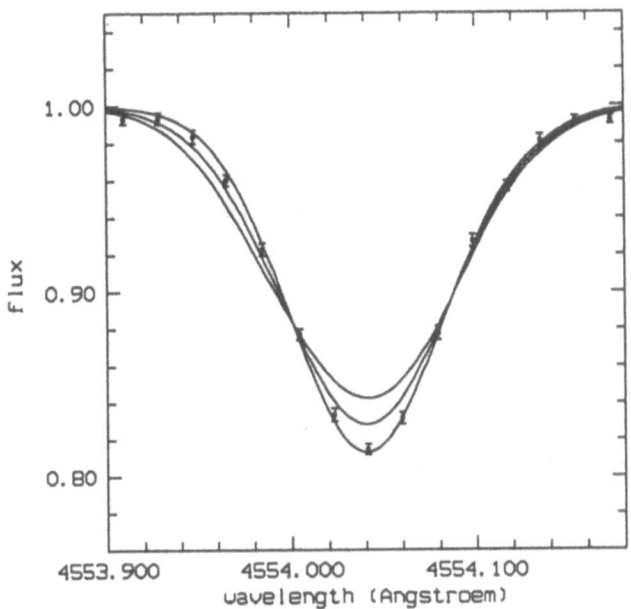

Fig. 2. The observed profile of the 4554Å resonance line of BaII (dots with error bars) is compared with synthetic profiles computed with 100, 50 and 0% of odd isotopes. The deepest, narrowest profile which fits the observations corresponds to 0% of odd isotopes

4.2 Impact of the VLT

The preceding example shows how the inclusion of a new dimension — here, isotopic abundances — may shed entirely new light on a problem.

However, this kind of observational programme is not easy to conduct with the facilities presently available. Indeed, both high resolution and high S/N are mandatory, and halo stars are rather faint.

The work on HD140283 was carried out with the CES fed by the 1.4m CAT. The resolving power of 10^5 was slightly too low for our purpose, so that special observing and reduction techniques were used to increase the resolution. In particular, 30 different spectra were obtained, with slightly different centerings in wavelength, so that the resulting spectrum is much better sampled than an individual spectrum. The details may be found in Magain (1994).

The total exposure time amounts to 20 hours, for a star of $m_V = 7.2$. This star is, in fact, the brightest one suitable for this program. Very few are brighter then $m_V = 9$ (maybe two more). In order to secure a significant sample, one would have to push the magnitude limit to $m_v \sim 11$, so that a few dozens of stars might be available.

The instrumental requirements would be the following: $R \sim 2 \cdot 10^5$, $S/N \sim 500$. Using a lower resolution would not be wise as: (1) the increased S/N would not compensate the loss of discrimination caused by the lower resolution, so that one would have to use longer exposures and (2) the result would become much more sensitive to uncertainties in the instrumental profile.

From the above discussion, it is clear that a large telescope equipped with a very high resolution spectrograph is mandatory if the observing time is to be kept within reasonable limits.

5 Conclusions

We have seen that the VLT can have a significant impact on our knowledge of the early chemical evolution on the Galaxy. However, it is clear that a large telescope, if necessary, is not sufficient in this respect. It has to be equipped with a very high resolution spectrograph, able to reach a resolving power of $2 \cdot 10^5$, and a S/N of 500.

References

Gilroy, K.K., Sneden, C., Pilachowski, C.A., Cowan, J.J.: ApJ **327** 298 (1988)
Gratton, R.G., Sneden, C.: A&A, **287** 937 (1994)
Luck, R.E., Bond, H.E.: ApJ **292** 559 (1985)
Magain, P.: A&A submitted (1994)
Nissen, P.E.: in Proc. of the ESO Workshop on High Resolution Spectroscopy with the VLT, M.-H. Ulrich, ed., ESO Conference and Workshop Proceedings No. 40, p. 49 (1992)
Truran, J.W.: A&A **97** 391 (1981)
Zhao, G., Magain, P.: A&A **244** 425 (1991)

Stellar Surface Structure:
Doppler Imaging with the VLT-UVES

Klaus G. Strassmeier

Institut für Astronomie, Universität Wien, A-1180 Wien, Austria
strassmeier@astro.ast.univie.ac.at

Abstract. I present a feasibility study for Doppler imaging with the VLT unit telescope and UVES and introduce a future key program for mapping open cluster stars.

1   Scientific Objectives

The idea that cool starspots, hot plages, and huge coronal loops exist on stars with convective envelopes is now reasonably well accepted. While the hotter surface features require wavelengths not accessible from the ground, cool starspots have their largest impact in the optical region of the electromagnetic spectrum. Typical spot sizes are a factor of hundred larger than the largest sunspot, sometimes even covering the rotational pole of the star. Such configurations cause some serious problems for hydrostatic equilibrium in the stellar envelope, e.g., where and how dissipates the energy that is blocked by such huge starspots? Since we do not even know the variability timescales or lifetimes of these spots as a function of *stellar age*, or their detailed morphology, we propose to map a series of stars in open clusters of well-known ages. This has the unique advantage that the approximate evolutionary status of the stars to be mapped is already known. Of course, because cluster stars are faint, this can only be done with a *very large telescope* equipped with an efficient high-resolution spectrograph.

2   Seeing a Star's Surface

As of mid 1994 only 15 late-type stars have a published surface map! This small number of stars is the result from 11 years of work of various groups from all over the world and shines some light on the stringent observational requirements !

This section outlines the Doppler imaging technique, §3 summarizes the observational requirements and §4 presents a possible key program for the VLT.

2.1 The Solution of the Direct Problem

A rotationally broadened spectral line profile as a function of rotational phase φ can be written as

$$R_{\text{obs}}(\lambda, \varphi) = \frac{\iint I_c(M, X(M)) R_{\text{loc}}(M, X(M), \lambda + \Delta\lambda_D(M, \varphi)) \cos\theta \, dM}{\iint I_c(M, X(M)) \cos\theta \, dM} \quad (1)$$

where I_c is the continuum intensity, R_{loc} the local line profile, M the position on the stellar surface, $\Delta\lambda_D$ the Doppler shift of element dM at a particular rotation phase, and $X(M)$ the unknown surface parameter that is solved for. In our case $X(M) \equiv T_{\text{eff}}(M)$. The direct problem then is to alter the assumed initial $T_{\text{eff}}(M)$ until a satisfactory fit to the observations is found.

2.2 The Inverse Problem

The inverse problem amounts to finding the surface parameter $X(M)$ from the observed profiles such that

$$D_\lambda(\varphi) = \frac{1}{n_\varphi n_\lambda} \sum_{\varphi=1}^{n_\varphi} \sum_{\lambda=1}^{n_\lambda} g(\varphi, \lambda) [R_{\text{obs}}(\varphi, \lambda) - R_{\text{th}}(\varphi, \lambda)]^2 \leq \sigma^2. \quad (2)$$

Since this is an ill-posed problem we introduce an additional criterion – a regularization functional $r(X(M))$ – and minimize

$$E(X(M)) = D(X(M)) + \Lambda \, r(X(M)). \quad (3)$$

The choice of $r(X(M))$ is related to the choice of X and two regularization functionals are currently used for temperature mapping,

$$\text{Maximum Entropy}: \quad r(X(M)) = \iint_M X(M) \log(X(M)) dM \quad (4)$$

$$\text{and} \quad \text{Thikonov}: \quad r(X(M)) = \iint_M |\nabla X(M)|^2 \, dM. \quad (5)$$

Note that the "physics" go into the computation of the local line profiles and the continuum intensities as well as the model atmospheres from which the profiles are computed (see, e.g., Piskunov & Rice 1993, Strassmeier et al. 1993).

2.3 Mapping the Magnetic Field

From the solar analogy we know that starspots are regions of strong magnetic fields and cause different levels of polarization for different parts on the solar/stellar surface. An *optical fiber polarimeter* in combination with UVES could provide for high resolution spectroscopy in circularly polarized light and allow magnetic surface mapping of late-type stars. Extracting spatial information from all four Stokes parameters, however, requires extremely high quality spectra (Semel et al. 1993). Because the Zeeman splitting depends on $\lambda^2 H$ one could additionally choose magnetically sensitive lines in the infrared wavelength region. The proposed *cryogenic echelle spectrometer* (Wiedemann et al. 1994) would be the instrument for such a task.

2.4 Continuum Variations

Stars with cool spots show significant light variations in bands as U, B, V, R, and I (e.g. Rodonó & Cutispoto 1992). These continuum variations are used as an additional constraint for the line profile solution. Thus, good Doppler images need *simultaneous multi-color photometry.*

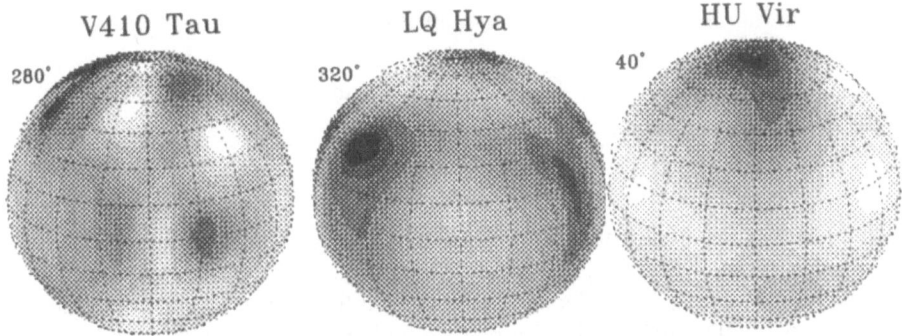

V410 Tau **LQ Hya** **HU Vir**

280° 320° 40°

Fig. 1. A crude age sequence of stellar surface images. Shown are, from left to right, the K4 pre-main sequence star V410 Tau, the K2 dwarf LQ Hya (age of the Pleiades), and the evolved K0 subgiant HU Vir. The apparent V brightness for these stars is 11.2, 7.8, and 8.7 mag, respectively. The VLT could push the brightness limit to 15th magnitude at twice the spectral resolution! The grey scale represents the temperature difference from the "unspotted" photosphere (darker = cooler).

3 Observational Requirements

Two stellar parameters – brightness and rotational broadening – dictate the instrumental requirements for Doppler imaging. The former sets a limit for the achievable S/N ratio and the latter determines the size of a resolution element on the stellar surface. In detail, there are five observational criteria to be considered:

3.1 Ratio of Local Line Broadening to Rotational Broadening

This ratio is fixed for a given star and should be as large as possible. It limits the resolution on the stellar surface even at infinite instrumental resolution,

$$\frac{r_{\text{spot}}}{R_{\text{star}}} \geq \frac{FWHM_{\text{star}}^{\text{local line}}}{FWHM_{\text{star}}^{\text{Doppler}}} \tag{6}$$

where r is the spot radius (for an assumed circular feature) in units of the stellar disk radius R.

Remark: Only stars with $v \sin i \geq 20$ km s^{-1} are usable for Doppler imaging!

3.2 Noise and Photon Statistics

Since the spot shape on the stellar surface is determined by the shape of the bump in the line profile, we need the highest possible S/N. Previous observations showed the existence of line profile variability on the sub-percent level with respect to the continuum.

If we would like to detect a 1 % spot-bump amplitude (R_{bump}) at a level that is n times the uncertainty of a profile point ($\sigma_{profile}$), then:

$$R_{bump} \geq n\,\sigma_{profile} \text{ and} \tag{7}$$

$$\sigma_{profile} = f\left(\Delta t, \dot{N}_\lambda, N_{RON}, \nu_{sampling}, 1 - R_{line}, \text{ff}, \text{sky}...\right) \tag{8}$$

where Δt is the integration time, $\dot{N}_\lambda$ the stellar photon flux, N_{RON} the read-out-noise of the CCD, $\nu_{sampling}$ the sampling frequency of pixels through the line profile width, $1 - R_{line}$ the depth of the spectral line, ff the "stability" of the flat field of the CCD, and sky the sky background contribution. Obviously we need a signal-to-noise ratio of ≈ 300 for a 3σ detection. Order extraction from large format echelle spectra must be done extremely carefully to avoid systematic line profile errors due to, e.g., a slight misalignement between dispersion direction and CCD pixel rows (Hensberge 1994).

Remark: The desired S/N ratio per wavelength bin is at least 300:1!

3.3 Instrumental Resolution

The best resolution on the stellar surface is achieved when

$$FWHM_{instr} \leq FWHM_{line} \tag{9}$$

where $FWHM_{line}$ is the intrinsic (local) line profile. Of course, even for stars that do not rotate, the observed line widths are significantly broadened. For example, a solar-type star with $T_{eff} = 5000\,\text{K}$ has a thermal line width of 1.2 km s^{-1}, a microturbulence of ≈ 2 km s^{-1}, and a macroturbulence of ≈ 4 km s^{-1}. While the latter two mechanism can be deconvolved from the observations the thermal width remains as the intrinsic barrier. If we now consider the instrumental resolutions usually applied for Doppler-imaging work, i.e. $\lambda/\Delta\lambda \approx 35\,000 - 100\,000$ or ≈ 8–3 km s^{-1}, we see that, even when $\lambda/\Delta\lambda$ is 100 000, one gives away resolution on the stellar surface by approximately a factor of two. Nevertheless, reliable images can be reconstructed even when only few spectral resolution elements are available across the line.

Remark One should aim at $\lambda/\Delta\lambda \approx 200\,000$ for stars with small $v\sin i$ and 100 000 for stars with moderate to large $v\sin i$!

3.4 "Bump" Smearing

During a CCD integration the spots keep moving through the profile and consequently a spot bump will be eventually smeared out. We can easily show that the smearing width, δ in % of the full width of the line profile, is four times the integration time divided by the stellar rotation period. With

$$\left(\frac{d\lambda}{dt}\right)_{\text{spot}} = \frac{2\pi}{P_{\text{rot}}} \frac{\lambda_0 v \sin i}{c} \quad \text{and} \quad \Delta\lambda_{\text{rot}} \simeq \frac{\pi}{2} \frac{\lambda_0 v \sin i}{c} \tag{10}$$

we get

$$\delta(bump) = \frac{\dot{\lambda}\Delta t}{\Delta\lambda_{\text{rot}}} \simeq 4 \frac{\Delta t}{P_{\text{rot}}}. \tag{11}$$

As an example consider the RS CVn binary EI Eri with $P_{\text{rot}} = 1.945$ d and $v \sin i = 50$ km s^{-1}. If the integration time $\Delta t = 1^h$ then $\delta(bump)$ is $\approx 8\%$ or $\approx 10°$ on the stellar surface.

Remark: Integration times must be as short as possible. Aim at $\Delta t \leq 0.01 P_{\text{rot}}$.

3.5 Data Phasing

Doppler imaging requires six to eight line profiles well distributed over a rotation cycle to recover the full stellar surface. If these profiles are taken during different rotation cycles one usually phases the data with the stellar rotation period. However, broad-band lightcurves of spotted stars tell us that there can be changes of the spot distribution with time scales as short as one stellar rotation. Although this might not be true for every star it must be assured that the light curve shape did not change during the spectroscopic observations!

Remark: Ideally, the observing time interval should not exceed the length of one stellar rotation period!

4 The VLT UVES Approach: Doppler Imaging of Open Cluster Stars

To establish an observational sequence of magnetic dynamo activity from the pre-main sequence phase up to the AGB we propose to spatially resolve the surface of stars in several *open clusters of well-known ages*. Figure 2 shows two color-magnitude diagrams representative for a young and an old cluster while Fig. 3 shows the decay of chromospheric activity with stellar age. There is little or no knowledge about the age dependence of *photospheric* activity so far and more and better observations are clearly needed.

Knowing the surface geometry of magnetic regions as a function of stellar age, and thus also of stellar rotation, one likely will be able to incorporate the Sun's magnetic surface activity into the broader picture of stellar magnetism and its evolution throughout the H-R diagram. Thus also better understand the

92

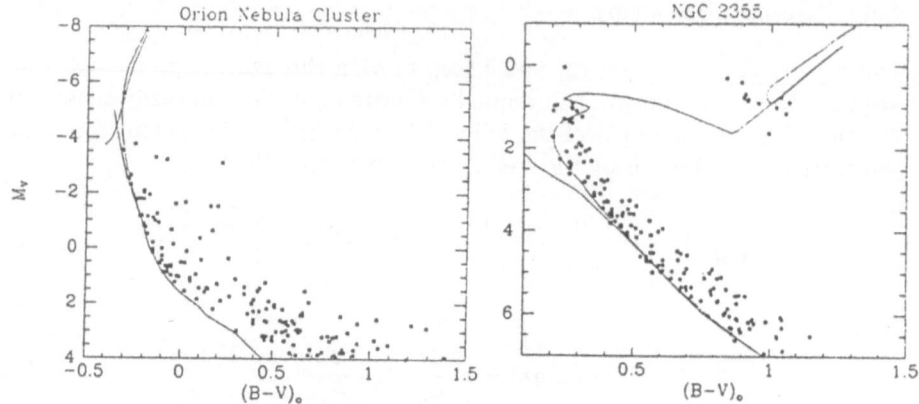

Fig. 2. Color-magnitude diagrams for two candidate clusters. The left panel shows stars from the Orion Nebula cluster at age $4\,10^6$ years and the right panel shows stars from NGC 2355 at age $9.5\,10^8$ years (taken from Meynet et al. 1993).

SOLAR-TYPE STARS

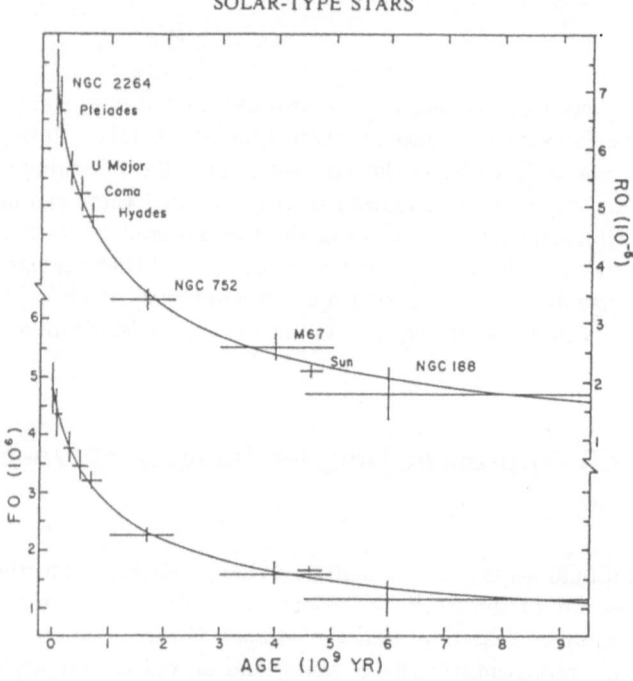

Fig. 3. The age dependence of chromospheric activity of stars with $B - V = 0.60$ in several open clusters and the Sun. $R0$ is the ratio of chromospheric flux to bolometric flux and $F0$ is the absolute chromospheric flux (taken from Barry et al. 1987).

Table 1. Expected limiting magnitudes for Doppler imaging with UVES

$\lambda/\Delta\lambda = 120,000$		
S/N ≈ 1000	$\Delta t \leq 20$ min	V≈8 mag
	$\Delta t = 60$ min	V≈9 mag
S/N ≈ 300	$\Delta t \leq 20$ min	V≈11 mag
	$\Delta t = 60$ min	V≈12 mag
S/N ≈ 100	$\Delta t \leq 20$ min	V≈13 mag
	$\Delta t = 60$ min	V≈14 mag
S/N ≈ 70	$\Delta t = 90$ min	V≈15 mag

Sun itself. *However, stars in open clusters are relatively faint for high-resolution spectroscopy and require the light collecting power of the VLT.*

Table 1 shows the UVES performance for Doppler imaging according to the ESO VLT instrument book (Moorwood et al. 1994). Listed is the approximate limiting magnitude in V as a function of S/N ratio per wavelength bin and typical integration times as required for Doppler imaging. We see that, in principle, a 15th magnitude star of moderately long period could be mapped. This is four (!) magnitudes fainter than what has been achieved so far with a 4-m class telescope.

Acknowledgements. Work on stellar activity at Vienna University is supported by FWF grant P-8942 and ÖAW grant OWF-40 to KGS.

References

Barry D. C., Cromwell R. H., Hege E. K., 1987, ApJ 315, 264
Hensberge H., 1994, priv. communication
Meynet G., Mermilliod J.-C., Maeder A., 1993, A&AS 98, 477
Moorwood A. F. M. (ed.), 1994, *Instruments for the ESO VLT*, ESO Instrumentation Department
Piskunov N. E., Rice J. B., 1993, PASP 105, 1415
Rodonó M., Cutispoto G., 1992, A&AS 95, 55
Semel M., Donati J.-F., Rees D. E., 1993, A&A 278, 231
Strassmeier K. G., Rice J. B., Wehlau W. H., et al., 1993, A&A 268, 671
Wiedemann G., Delabre B., Moorwood A. F. M., 1994, *High-resolution infrared spectrometer for the VLT*, ESO STC Working Group

Observation of Surface Activity on Cool Giants with the VLT Interferometer

O. von der Lühe[1], M. Schüssler[2], S.K. Solanki[3], P. Caligari[2]

[1] European Southern Observatory, Garching, Germany
[2] Kiepenheuer-Institut für Sonnenphysik, Freiburg, Germany
[3] Institut für Astronomie, Zürich, Switzerland

Abstract. Surface activity can reveal crucial information on the processes which generate magnetic fields in the interior of cool giants. We discuss how the interferometric mode of VLT can be used to observe active regions directly on near-by stars, and we compare this approach with Doppler imaging.

1 Scientific Goals

Rapidly rotating, cool stars with extended outer convection zones typically show signs of violent magnetic activity such as starspots, activity cycles, flares, hot chromospheres, formation of coronae and generation of fast stellar winds. However, detailed observations of activity within the photosphere are presently only possible for the Sun. Indirect methods like Doppler or Zeeman imaging indicate the presence of huge magnetic starspots on a number of stars, many of which are cool giants with deep outer convection zones.

The generation of magnetic fields by dynamos is a fundamental astrophysical process which influences the structure and the physical characeristics of many objects – like stars, accretion disks, and even galaxies. However, the underlying magneto-hydrodynamic mechanisms are still not well understood. Investigations of the dynamics of magnetic structures on stellar surfaces provide crucial insight, because they form the connection between the observable surface fields and the generally observationally inaccessible dynamo regions within or below the outer convection zones of cool stars.

The temporal variation of the distribution in latitude and longitude of magnetic flux emergence can be used as a diagnostic for propagating dynamo waves, whose properties reveal important information about the inner working of the stellar dynamo process. If magnetic flux predominantly emerges at low latitudes, magnetic buoyancy must dominate the Coriolis force due to rotation, which provides a lower threshold for the initial field strength in the dynamo region. On the other hand, very rapid rotation of a star leads to a dominance of the Coriolis force on the dynamics of erupting flux tubes which form starspots; as a consequence, the flux tubes are forced to move parallel to the axis of rotation and magnetic flux gathers around the poles of the star (Schüssler & Solanki, 1992). Magnetic braking of rotation depends on the latitude distribution of the field: braking is quite inefficient if the fields are concentrated at the poles while the reverse is true for fields close to the equator. This effect may cause two populations of stars: initially rapidly rotating stars suffer little braking; the rates of initially slow rotators with fields emerging closer to the equator will diminish rapidly. In addition, the dynamo efficiency varies with the rotation rate.

The longitudinal distribution of magnetic field would reveal the azimuthal wave number of the dynamo if active latitudes exist. For close binaries, one can determine the effects of tidal forces on the erupting flux and possibly their influence on the dynamo process itself.

Direct imaging of the magnetic (polarimetric) and intensity structure of stellar surfaces with extremely high angular resolution is needed in order to unambiguously derive the necessary information for a sufficiently large sample of stars (Solanki, 1991). Detailed theoretical models and numerical MHD simulation tools are available for comparison with observations (Schüssler et al., 1994; Caligari et al., 1994).

The interferometric mode of the VLT offers an angular resolution of better than one milli-arcsec in the red spectral range and in the near infra-red. This will be sufficient to cover a significant sample, mostly red giants, with a few to a few dozen pixels. Highly resolved photometric and polarimetric observations obtained with the VLTI will, therefore, contribute to improve the constraints on current and future models of stellar activity. Although the sources are relatively bright, observations will require the combined mode including the 8m Unit Telescopes in order to achieve a sufficiently high signal within narrow spectral bands for high resolution polarimetry.

2 Interferometric Observations

The following types of observations are needed:

a. *direct broadband imaging in the visible and near IR* to study positions and morphology of active regions like starspots and their temporal evolution.

b. *spectroscopy* with a resolution of some $\frac{\lambda}{\Delta\lambda} \approx 40000$ to study the temperature distribution and possibly to infer magnetic field strength.

c. *spectro–Polarimarey* with a resolution of some $\frac{\lambda}{\Delta\lambda} \approx 40000$ to measure field strength and polarity.

The prime observable of an astronomical interferometer is the complex Fourier transform ("complex visibility") of the source's intensity distribution which is encoded in the contrast of the interference fringes. Any pair of telescopes in an interferometric array measures the Fourier component with an angular frequency $s = \hat{B}/\lambda$, where $\hat{B}$ denotes the projected baseline (vector of separation between the two telescopes projected onto a plane which is perpendicular to the observing direction) and λ is the mean observing wavelength. Angular frequencies are measured in units "line pairs per radian". Several baselines measured simultaneously with an array of more than two telescopes, as well as the changes of the projected baselines during the observation in the course of the night (Earth rotational synthesis), generally permit mapping a fair fraction of the Fourier plane. A map of the source is generated by inversion of the data with deconvolution techniques quite similar to those used in Radio interferometry. The angular resolution is determined by the largest projected baseline and may be anisotropic.

Since close giants have angular diameters of a few dozen milli-arcsec at best, only the interferometric mode of VLT will be able to resolve these stars with a useful number of elements. A comprehensive description of VLTI has been given by Beckers (1991), the most recent update of VLTI's present status can be found in von der Lühe et al. (1994). The basic approach will be to obtain visibility measurements in broadband light for baselines up to 128 m with the Unit Telescopes of the VLT and up to 200 m with Auxiliary Telescopes for direct broadband imaging, and spectrally resolved or spectro-polarimetric visiblity observations, from which a limited number of "spectroasterograms" in the vicinity of interesting lines will be generated. The better the angular resolution, the better the discrimination of spectral features between active and quiet stellar surface regions will be.

To illustrate the nature of these measurements, Figure 1 shows two artificial stellar surface maps, one without spots, the other one with three active regions (left column). Both stars are limb darkened by a linear law with a coefficient of 0.9. The center column shows the logarithms of the corresponding Fourier transforms, and the right column shows the corresponding Fourier phases. It is seen that the presence of starspots becomes manifest in deviations of the Fourier magnitudes and phases from the "Airy pattern" shown for the unspotted star, and in a structured phase in the rings of order two and higher (the information in the first ring mainly encodes the location of the image centroid). The maximum visibility magnitude in the rings of order two and higher varies between $4 \cdot 10^{-2}$, and a few times 10^{-3}, so they must be measured with a precision of order 10^{-3}.

Figure 2 shows the simulated Fourier modulus (left) and the phase (right) for a K giant with 20 solar radii at a distance of 59 pc which as it be observed with all four Unit Telescopes of the VLT at 650 nm. The declination of the source is 29 degrees north (!), observing is done for zenith distances up to 60 degrees. The dark patches in Figure 2 represent missing data due to the limited Fourier plane coverage of the available baselines. Although there are substantial holes, a lot of detailed structure can be measured in the Fourier transform. This type of data does not lend itself to simple Fourier inversion and the best approach for analysis still needs to be established. Fitting of parameters which describe the limb darkening coefficient, the number, the positions, and the diameters of a few active regions in a model intensity distribution is probably a good start. Spectrally resolved data (i. e., Fourier maps for several wavelengths across suitable spectral lines) which exploit spectral signatures of active regions will provide additional constraints and therefore improve tremendously the fidelity of stellar maps thus generated.

3 Comparison with Doppler Imaging

Doppler imaging (Vogt and Penrod, 1983) and its derivates, like Doppler-Zeeman imaging (Donati et al., 1992), has been used in the recent past to detect activity on a small sample of stars (see also the contribution by Strassmeier in these proceedings). The comparison of this technique with interferometry quickly reveals that both apply to complementary samples of objects, rather than being competitors. Aside from the obvious circumstance that each technique can, for some sources, be used to confirm the findings of the other, the requirements on the object are somewhat different. Interferometric imaging is diffraction limited by the longest baseline, whereas there is no *a priori* such limitation for Doppler imaging. Only the closest stars, within about 100 pc, are therefore candidates for interferometric imaging.

However, Doppler imaging requires a value for $v \cdot \sin i$ within the range of $25 \ \mathrm{kms}^{-1} \cdots 100 \ \mathrm{kms}^{-1}$, and is therefore limited to rapid rotators. Also, the inclination angle i of the stellar rotation axis should be in the range of 30 to 60 degrees for generating two-dimensional surface maps. Interferometry does not have these limitations. For mapping fast rotators, observations probably need to be limited to one to a few hours worth of data unless a clever inversion technique would take the stellar rotation into account. The interferometer would essentially be operated in "snapshot mode" which severely limits the capability to generate structurally complex maps. Slow rotators can be mapped easier because the apparent structure changes less rapidly, and more data can be combined into a single map. Eventually, observations from several nights and with different configurations of Auxiliary Telescopes can be integrated into a single map.

Doppler imaging requires spectral resolution of some $4 \cdot 10^4$ with high SNR, which limits the technique to bright stars. Spectral lines used for Doppler imaging also need to be blend-free, which also limits spectral types which are suitable. Interferometry can operate with fairly broad spectral ranges which are limited only by fringe contrast loss due to spectral dispersion. Spectral resolution is only needed for spectral imaging.

4 Conclusions

Interferometry with the VLT can provide direct and unique information about surface activity on stars. Because of the limited angular resolution of at best 0.5 milli-arcsec, the sample of resolvable stars is limited to nearby (< 100 pc) giants and subgiants. Candidate stars are nearby, so they are bright *per se* and therefore are ideal for on-source referencing adaptive optics and on-source fringe tracking. The requirement to measure fringe visibilities with a precision of 10^{-3} appears to be difficult; however, recent measurements with a much smaller (5 cm) interferometer already demonstrated precisions of order 10^{-2} with 25 nm spectral bands on bright stars (Quirrenbach et al., 1994). Visibility measurements with sufficient precision in broad spectral bands should not be difficult to achieve with the VLT Sub-array (VISA) which consists of only the re-configurable Auxiliary

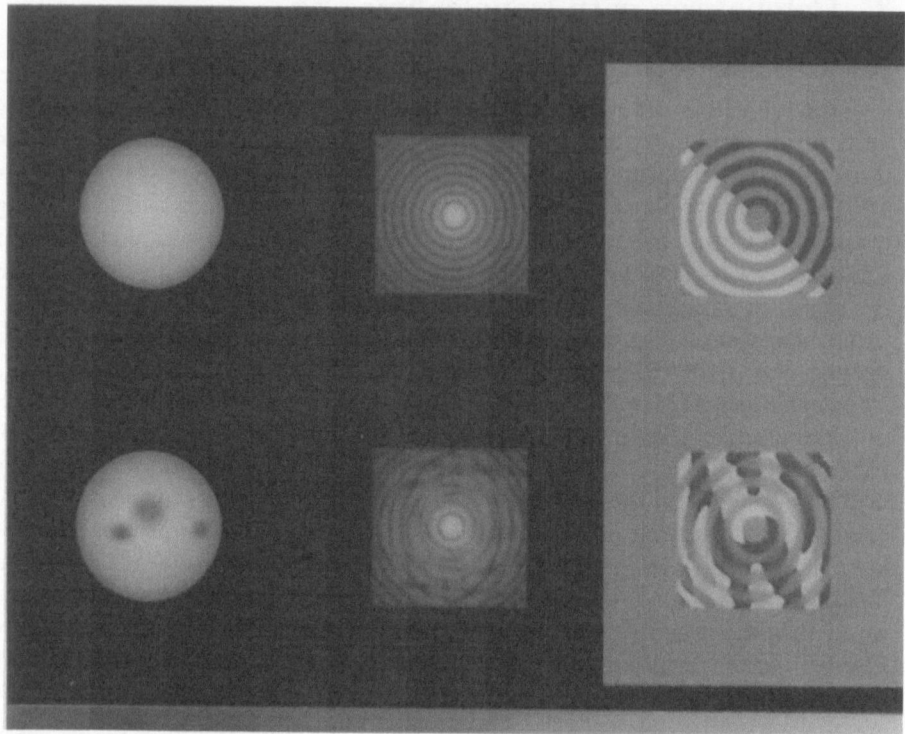

Fig. 1. Left: two limb-darkened stellar surfaces, one showing starspots, center: logarithm of the modulus (3 decades) of the intensity Fourier transforms, right: phase of the transforms.

Telescope array. High resolution spectroscopy and polarimetry will, however, require the Unit Telescopes combined coherently.

Several (more than 3) simultaneous baselines and VISA's rapid reconfiguration capability are needed if observations are limited to periods of a few hours for a single map, due to fast internal changes in the source structure due to rapid rotation, for instance. This capability will permit comparing the results of interferometry with those from Doppler imaging of the same sources. Otherwise, these two techniques are complementary and address different classes of objects.

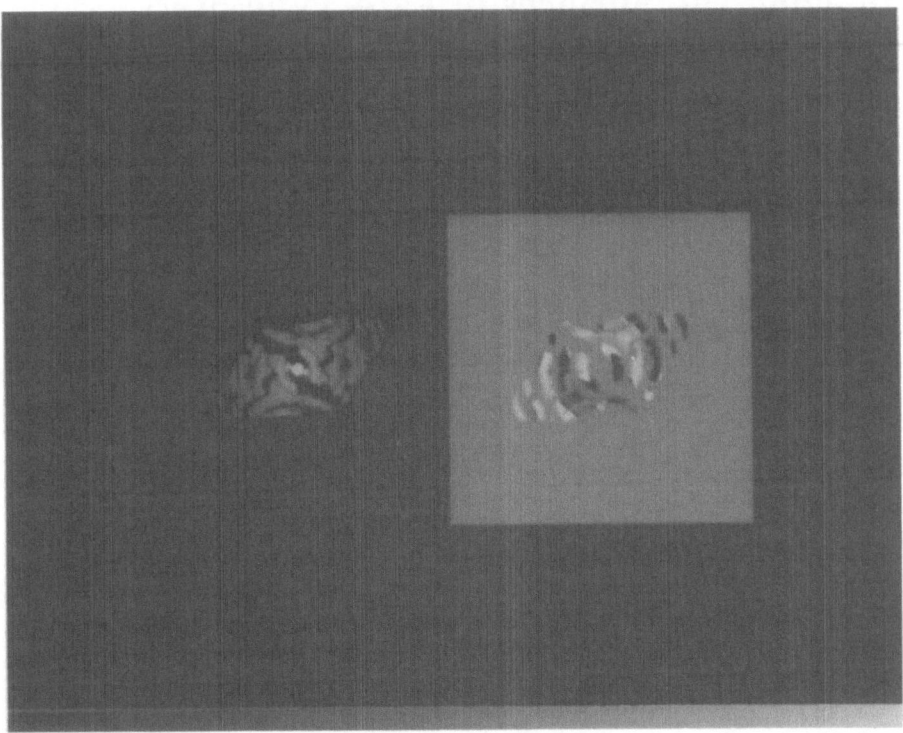

Fig. 2. Fourier modulus (left) and phase (right) data of a limb-darkened stellar surface showing starspots, as observed with VLTI at 650 nm. The apparent stellar diameter is 3.1 milli-arcsec.

References

Beckers, J. M. (1991): J. Optics (Paris) 22, 73–83.

Caligari, P., Moreno-Insertis, F., Schüssler, M. (1994): Astrophys. J., submitted

Donati, J.-F., Brown, S. F., Semel, M., Rees, D. E., Dempsey, R. C., Matthews, J. M., Henry, G. W., and Hall, D. S. (1992): Astron. Astrophys. 265, 682–700.

Quirrenbach, A., Mozurkewich, D., Buscher, D. F., Hummel, C. A., and Armstrong, J. T. (1994): Astron. Astrophys. 286, 1019-1027.

Schüssler, M., Solanki, S. K. (1993): Astron. Astrophys. 264, L13–L16.

Schüssler, M., Caligari, P., Ferriz-Mas, A., Moreno-Insertis, F. (1994), Astron. Astrophys. 281, L69–L72.

Solanki, S. K. (1991): in *Cool Stars, Stellar Systems and the Sun VII,"* J. Bookbinder, M. Giampapa (Eds.), PASP Conference Series, 211-222.

Vogt, S. S., Penrod, G. D. (1983): PASP 95, pp. 565–576.

von der Lühe, O., Ferrand, D., Koehler, B., Zhu, N., and Reinheimer, Th. (1994): in *Amplitude and Intensity Spatial Interferometry II,* J. B. Breckinridge (Ed.), SPIE Proc. Vol. 2200, pp. 168–179.

Imaging the Surfaces of Late-Type Stars

T. R. Bedding, O. von der Lühe and A. A. Zijlstra

European Southern Observatory, Karl-Schwarzschild-Str. 2
D-85748 Garching, Germany

Abstract. We discuss the prospects for resolving the disks of red-giants and Mira variables with the VLT and the VLTI. We also describe infrared aperture-masking observations with the NTT, from which we have measured angular diameters of 50 mas for R Dor (M8 III) and 44 mas for W Hya (Mira variable).

1 Scientific Aims

Optical interferometry is beginning to show that red supergiants are not simply featureless spheres. The best studied case is α Ori. Evidence for time-varying hotspots on the surface of this star has been found by a group at Cambridge, using the technique of aperture masking with the William Herschel Telescope (Buscher et al. 1990; Wilson et al. 1992). These hotspots are probably convective cells. Indeed, Schwarzschild (1975) suggested that the dominant convective elements of red giants and supergiants are so large that only a small number of them occupy the surface of the star at any time. This model explains the irregular variability in brightness and polarization seen in these stars. An alternative model, which cannot yet be ruled out, is that we are seeing non-radial oscillations.

Two other supergiants have apparent sizes large enough to allow imaging with a 4-m telescope. Of these, α Her has also been reported by the Cambridge group to have a hotspot (Tuthill et al. 1994a), while α Sco has so far failed to exhibit such features (Wilson et al. 1992; Bedding et al. 1994).

Red giants show brightness fluctuations and polarization characteristics similar to those of the supergiants, so there is good reason to expect them also to have large-scale surface features. No direct evidence for hotspots on giant stars has been reported; their smaller diameters demands higher angular resolution than has so far been achieved.

The subclass of red giants known as Mira variables show regular brightness fluctuations that are due to pulsations. The main interest currently lies in understanding the mechanism of pulsation and the details of the shock front formation. This is linked to mass loss, another poorly understood aspect of these stars. Several independent observations of o Cet have shown an elongated atmosphere, which is probably associated with the pulsation of the star (Karovska et al. 1991; Haniff et al. 1992; Wilson et al. 1992; Quirrenbach et al. 1992). Another Mira variable, R Cas has recently been found by the Cambridge group to have an even stronger ellipticity than o Cet (Tuthill et al. 1994a). If these ellipticities are found to be common among Miras, mass-loss models will have to be substantially revised.

Angular-diameters, although less glamorous to observe than asymmetries, are nonetheless important. Measuring the angular diameter of a star is the only direct way to determine its effective temperature, and the temperature scale for late-type giants is poorly known. For Miras, diameter measurements will also give important information on the pulsation mechanism. In the case of R Leo, Tuthill et al. (1994b) have combined a single-epoch angular-diameter measurement with an accurate distance to rule out fundamental-mode pulsation. As they point out, the forthcoming availability of Hipparcos parallaxes offers the chance to extend this work to a dozen or more Miras.

We now discuss the technical requirements for making high-resolution observations of a sample of red giants, supergiants and Mira variables.

2 Aperture Masking

Most of the results described above were obtained by masking the apertures of 4-metre class telescopes. Modifying the telescope pupil with an aperture mask brings benefits which, at least for some observations, outweigh the loss of photons (see Haniff 1994 and Bedding et al. 1993 for recent reviews). The main reason for this is that each r_0-sized patch on the telescope pupil has an unknown wavefront error which must be compensated before high-resolution information can be extracted. Therefore, although each patch on the pupil adds photons to the image, it also introduces an extra unknown. This is why the optimum strategy is not always to use as much of the pupil as possible. If we use a pupil mask containing a small number of r_0-sized holes, we measure only a limited set of visibility amplitudes and closure phases, but with much higher signal-to-noise ratios.

Another big advantage of using an aperture mask with sub-apertures of size $\sim r_0$ is that the observations are much less affected by changes in the seeing. In practice, accurate calibration of the seeing is frequently the limiting factor in interferometry and is the cause of a substantial credibility problem. An aperture mask, by producing data with higher signal-to-noise and better calibration, helps significantly in this regard. In addition, the mask does not taper down the weights of the high spatial frequencies and so allows one to achieve better angular resolution than with an unmasked pupil.

We do not wish to give the impression that masking a telescope is always the best strategy. However, for observations of bright barely-resolved stars, aperture masking has proved to be a powerful technique.

3 Observations with the NTT

We used the NTT for aperture-masking observations in August 1993, using the SHARP infrared camera (Eckart et al. 1991). Observations in the near-infrared (NIR) have two advantages: (i) cool stars emit most of their flux in this spectral region; and (ii) atmospheric conditions are better than in the visible, making interferometry much easier. The drawback of NIR observations is the loss of angular resolution: $\theta \propto \lambda/\text{baseline}$. The NTT's compact alt-azimuth design allowed us to place an aperture mask on the baffle in front of Mirror 3, where the converging $f/11$ beam has a diameter of 45 cm. Further details of the experimental setup are given in Bedding et al. (1993).

A big disadvantage of using a non-redundant mask with a small number of holes is the poor coverage of spatial frequencies. A good compromise in the photon-rich infrared regime is an annular mask (Haniff & Buscher 1992). This has full spatial frequency coverage while being minimally redundant: roughly speaking, each baseline is measured twice. Furthermore, a thin annulus largely retains the other advantages mentioned above, namely accurate calibration and enhanced resolution. So far this method has only been tested on binary stars (Haniff et al. 1989).

We made observations of several stars in the near infrared using both an annular mask and a 7-hole non-redundant mask. Our main target was R Doradus. This is an M8 giant and is the brightest, and therefore closest, star with such a late spectral type (Wing 1971). We found it to have an angular diameter of about 50 mas at a wavelength of $1.5\,\mu$m (no correction has been made for limb darkening). By comparison, the infrared ($2.2\,\mu$m) angular diameter of α Ori is only 44 mas (Dyck et al. 1992). This means that R Dor replaces α Ori as the star with the largest apparent size apart from the Sun. Indeed, Wing (1971) already predicted this result.

We also observed W Hydrae, which is a small-amplitude Mira with spectral type about M8 (Wing 1971). We found this star to have a diameter of about 44 mas, less than R Dor, but still impressively large. We see no evidence for asymmetries in either star, but higher resolution is clearly needed. These stars will make excellent targets for the VLT.

4 Observations with a VLT Unit Telescope

The 3.5-m aperture of the NTT is barely able to resolve these stars, let alone make an image. Observing with a VLT unit telescope will increase the telescope diameter by a factor of 2.3, a significant improvement. A further gain will come from dedicated instruments: CONICA for the NIR and VHARC for the visible. CONICA has been approved and is now being designed, while VHARC is still in the "definition" phase. The following technical requirements are relevant to observations of giant stars.

Observations: Changes in α Ori and Miras probably occur on timescales from weeks to months. Frequent observations would be desirable and would probably require less than one hour per star per epoch. Once an observing procedure has been established, observations in service mode should be possible.

Exposure times: Fringe contrast and accuracy are both improved by having short exposures. The shortest exposure time foreseen for CONICA is 30 ms, which should be adequate, although even shorter times might be preferred for bright stars.

Filters: For CONICA to achieve the best possible resolution, it should include a filter at the short-wavelength limit. We suggest $\lambda = 1.0 \, \mu m$, $\Delta \lambda = 0.1 \, \mu m$.

Masks: Placing a full-sized mask over the primary mirror of a large telescope is difficult and dangerous. The best solution is to place a small mask at a pupil image inside the instrument (this was not possible with SHARP on the NTT). For CONICA, annular masks and two-dimensional non-redundant masks should be included. An annular mask is not a good choice in the visible regime, where photon noise generally dominates over atmospheric noise. Here, Buscher & Haniff (1993) advocate the use of a long slit, although this has not yet been confirmed by observations. VHARC should include masks with slits and also with non-redundant arrays of holes.

Calibrator stars: The transfer function of the telescope plus atmosphere must be measured by observing a star of known angular size (preferably unresolved). This applies to observations both with and without an aperture mask. The calibrator star should be close to the object of interest (preferably within a few degrees) and of comparable brightness. This is a problem in the NIR because the objects of interest (α Ori, R Dor, o Cet, W Hya) are the brightest infrared stars in the sky. This follows, of course, from the fact that they are the largest stars in the sky and means that the calibrator star will generally be fainter and/or bluer than the object. For our NTT observations, the brightness ratio between R Dor and its calibrator (γ Ret) was about 20:1, leading us to use a neutral density filter for R Dor to obtain comparable count rates for the two stars. CONICA and VHARC should include one or two ND filters. Future development of phase-diversity imaging (Gonsalvez & Childlaw 1979) might eliminate the need to observe a calibrator, at the expense of a more complicated dual-channel instrument.

5 Observations with the VLTI

The VLTI will provide infrared interferometry over long baselines, allowing observations of many more objects that would be unresolved with a unit telescope. The VISA mode, which is a reconfigurable array of three 1.8 m auxiliary telescopes, is well suited to making regular angular diameter measurements of a sample of Miras throughout their pulsation cycles. Observations at minimum are much easier in the IR than in the visible because variability is much less. With these measurements, it should be possible to identify the oscillation modes and investigate the pulsation mechanism. Finally, we note that recent 11 μm observations of dust shells around late-type stars by Danchi et al. (1994) have shown that a mid-infrared capability for VISA would be extremely valuable.

References

Bedding, T.R., von der Lühe, O., Zijlstra, A.A., Eckart, A., & Tacconi-Garman, L.E., 1993, Messenger 74, 2

Bedding, T.R., Robertson, J.G., & Marson, R.G., 1994, A&A, 290, 340

Buscher, D.F., & Haniff, C.A., 1993, J. Opt. Soc. Am. A 10, 1882

Buscher, D.F., Haniff, C.A., Baldwin, J.E., & Warner, P.J., 1990, MNRAS 245, 7P

Danchi, W.C., Bester, M., Degiacomi, C.G., Greenhill, L.J., & Townes, C.H., 1994, AJ 107, 1469

Dyck, H.M., Benson, J.A., Ridgway, S.T., & Dixon, D.J., 1992, AJ 104, 1982

Eckart, A., Hofmann, R., Duhoux, P., Genzel, R., & Drapatz, S., 1991, Messenger 65, 1

Gonsalvez, R.A. & Childlaw, R., 1979, In: Tescher, A.G. (ed.), *Proc. Soc. Photo-Opt. Instrum. Eng., 207*, 32

Haniff, C.A., 1994, In: Robertson, J.G., & Tango, W.J. (eds.), *Proc. IAU Symp. 158, Very High Angular Resolution Imaging*, p. 317, Kluwer, Dordrecht

Haniff, C.A., & Buscher, D.F., 1992, J. Opt. Soc. Am. A 9, 203

Haniff, C.A., Buscher, D.F., Christou, J.C., & Ridgway, S.T., 1989, MNRAS 241, 51P

Haniff, C.A., Ghez, A.M., Gorham, P.W., Kulkarni, S.R., Matthews, K., & Neugebauer, G., 1992, AJ 103, 1662

Karovska, M., Nisenson, P., & Papaliolios, C., 1991, ApJ 374, L51

Quirrenbach, A., Mozurkewich, D., Armstrong, J.T., Johnston, K.J., Colavita, M.M., & Shao, M., 1992, A&A 259, L19

Schwarzschild, M., 1975, ApJ 195, 137

Tuthill, P.G., Haniff, C.A., & Baldwin, J.E., 1994a, In: Robertson, J.G., & Tango, W.J. (eds.), *Proc. IAU Symp. 158, Very High Angular Resolution Imaging*, p. 395

Tuthill, P.G., Haniff, C.A., Baldwin, J.E., & Feast, M.W., 1994b, MNRAS 266, 745

Wilson, R.W., Baldwin, J.E., Buscher, D.F., & Warner, P.J., 1992, MNRAS 257, 369

Wing, R.F., 1971, PASP 83, 301

A Sharper View on (Post-) AGB Evolution

L.B.F.M. Waters[12]

[1] Astronomical Institute 'Anton Pannekoek' University of Amsterdam
[2] SRON Laboratory for Space Research Groningen

Abstract. The VLT, with its planned instrumentation for high resolution spectroscopy and imaging, opens exiting new possibilities to investigate the physics of mass loss of stars on the Asymptotic Giant Branch (AGB). Studies on the brightest AGB stars in the Solar neighbourhood with 4-meter class telescopes already indicate the potential of high resolution near-IR and mid-IR spectroscopy. The VLT will allow such studies to be carried out systematically, giving a much better view on the mass loss process in different physical conditions. The high resolution imaging possibilities, especially in the infrared, allow direct mapping of the surface of red giants and supergiants, allow pulsation mode identification and the study of grain formation processes in the outflow.

1 Introduction

The evolution of stars on the Asymptotic Giant Branch (AGB) and beyond is an area in stellar evolution that is still far from being understood. Low- and intermediate mass stars (with masses between 0.8 and 6-8 $M_\odot$) pass through this phase at the end of their life, and expel large amounts of material via a slow, dense and dusty stellar wind (for a review see Habing 1990). Mass loss rates range between 10^{-7} and 10^{-4} $M_\odot$ yr^{-1}, and outflow velocities are typically 5 to 30 km s^{-1}. The dust which forms in the outflow obscures the star at optical wavelengths and causes most of the energy to be emitted in the infrared. Many molecules can be found in the photosphere and stellar wind, producing a very rich spectrum at infrared wavelengths. It is clear that infrared spectroscopy of AGB and post-AGB stars contains information that is essential for a better understanding of AGB evolution and the mass loss process. The unique imaging capabilities of the VLT at optical and infrared wavelengths will allow a view of the surface of these very extended atmospheres. The physical processes in the outer atmospheres of these stars can thus be studied directly: limb darkening, convection, and the onset of the stellar wind are a few areas that will benefit from these new possibilities.

In this review we will describe some examples of fundamental questions in AGB and post-AGB evolution that can be tackled with the VLT or VLT interferometer. We conclude with some recommendations for the VLT instrumentation presently being studied or being designed.

2 The Mass Loss Process

Despite considerable efforts, AGB evolution still presents us with fundamental questions that need to be resolved. During AGB evolution, surface mass loss dominates the course of events. This is because the rate at which material is removed from the envelope through mass loss exceeds the nuclear burning rate by several orders of magnitude. Therefore it is essential that we understand the mass loss process and its dependence on stellar parameters. Although progress in models for AGB mass loss has been made (e.g. Wood 1979; Bowen 1988; Bowen & Willson 1991; Pijpers & Habing 1989; Fleischer et al. 1992; Dominik et al. 1993), we are still far from having a reliable theory that describes how mass loss in AGB stars depends on mass, luminosity and chemical composition. In the model of Bowen (1988), shock waves are generated in the outer atmosphere of the star, which results in a large increase of the scale height. The density of the atmosphere at distances where dust can form is increased substantially, and radiation pressure on the dust then pushes the material to infinity. This mechanism can produce the correct order of magnitude mass loss rates and expansion velocities, but is still rather ad hoc.

In particular the large variations in the mass loss rate, especially obvious in C-rich AGB stars (e.g. Willems & de Jong 1986; Olofsson et al. 1992) are still far from understood. Such variations are, to a lesser extent, also seen in M giants and S stars (Zijlstra et al. 1992) and seem to be a common property of AGB stars. It has been suggested that the variations in mass loss are related to so-called thermal pulses, during which the energy is produced in a He-burning shell, rather than the H-burning shell. These thermal pulses may cause changes in the surface temperature and gravity, which directly affect the mass loss process (Vassiliadis & Wood 1993).

2.1 Imaging

The VLT interferometer is capable of imaging the stellar surface of nearby AGB stars and red supergiants, allowing the study of surface phenomena such as convection and starspots. The importance of direct imaging of AGB stars has recently been demonstrated by Tuthill et al. (1994), who obtained images of R Leo at 833 and 903 nm. Their angular diameter in combination with the known parallax of the star result in a stellar radius of 495 $R_\odot$. Such a large radius excludes fundamental mode pulsation in R Leo, given a pulsation period of 314 d (Tuthill et al. 1994), and suggests first or higher overtone as the pulsational mode.

Wilson et al. (1992) obtained images of α Ori, and o Ceti at wavelengths of 546 and 710 nm with a spatial resolution between 20 and 30 mas. The images of o Ceti clearly show asymmetries, suggesting that the star is not round but elongated, possibly as a result of stellar pulsations. If such deviations from spherical symmetry can be linked to mass loss, this would mean a major break-through in our understanding of the mass loss process. For α Ori clear variability in the surface brightness distributon was observed over a period of two years. These variations may be due to the appearance and disappearance of large convection cells.

Table 1. Some interesting scales for AGB and Post-AGB stars

R_* ($R_\odot$)	L_* ($L_\odot$)	T_{eff}	D (pc)	scale	ang. size (marcsec)
		AGB star			
300	3000	2500	500	R_*	2.8
				extended atm.	2.8-28
1000	35000	2500	500	R_*	9
				extended atm.	9-90
		post-AGB star			
50	3000	6000	500	R_*	0.5
				pagb wind	0.5-5
				AGB remnant	1500
				circum-sys-	
				tem disc	10-100

Our understanding of AGB evolution is seriously hampered by the unknown distance to individual AGB stars. This makes it impossible to reliably derive their luminosity and place them accurately in an HR diagram. The VLT interferometer in principle can be used to measure the distance to individual AGB stars. This can be done by simultaneously measuring the radial velocity curve of a pulsating star as well as the change in radius using direct imaging of the stellar surface. The radial velocity curve gives the displacement of the stellar surface in linear dimension, while the measurement of the change in stellar angular diameter gives the same quantity in units that depend on the distance. From this the distance can be derived. Such a measurement could be performed in service mode, since the objects are bright and integration times short, and since the pulsation periods are typically several 100 days.

In Table 1 we list some interesting scales in AGB and post-AGB stars that would become accessible with the VLT interferometer.

2.2 Spectroscopy

High resolution near-IR and mid-IR spectroscopy of AGB stars offers the possibility to investigate the chemical and dynamical structure of the outer atmospheres of AGB stars. As the gas moves away from the photosphere, it will cool and more and more complex molecules will be able to form. It is expected that the atmospheres will show stratification in the appearance of certain molecules (e.g.

Cherchneff & Glassgold 1993). Since the atmosphere is also slowly accelerating in this region of the outflow, we can study how this acceleration occurs by measuring the absorption lines and line shapes of different kinds of molecules. The inner part of the AGB wind, in which the physical processes that determine the outflow are active, is not observable with millimeter or centimeter observations of molecules. This is because at these wavelengths only rotational transitions are found that probe the outer regions of the outflow, in which the terminal velocity has been reached. At large distance from the star, the interstellar UV radiation field dissociates molecules. For instance, in a C-rich chemistry H_2C_2 will form C_2, which can be studied using optical and near-IR spectroscopy.

Table 2. Spectral resolution required for (post)-AGB stars

AGB	Species	v_{exp} (km/s)	Spectr. Res.	λ range (μm)
photosphere	molecules	-	10^5	0.5 - 10
wind	molecules	10	10^5	0.5 - 10
wind	dust	-	100-1000	1 - 20
Post-AGB	Species	v_{exp} (km/s)	Spectr. Res.	λ range (μm)
photosphere	atoms	-	$5\ 10^4$	0.3 - 5
pagb wind	atoms,ions	10-100	$2\ 10^4$	0.3 - 1
AGB remnant	dust	10	100-1000	1 - 20
AGB remnant	molecules	10	$3\ 10^4$-10^5	1 - 20

Up to now such studies have been restricted to only the brightest AGB stars, but the potential has been clearly demonstrated. For instance, the 10 μm spectrum of IRC+10 216 (Keady & Ridgway 1993) shows absorption from SiO, CS, C_2H_2, HCN, NH_3 and SiH_4. Analysis of the line profiles suggests that CS is depleted from the gas phase at distances larger than 100-1000 R_*, possibly as a result of condensation onto dust grains. Such observations provide strong constraints on the dust formation and grain growth process, which is intimately linked to the mass loss process. Bakker et al. (1995) used the Utrecht Echelle Spectrograph (UES) attached to the 4.2m William Herschel Telescope (WHT) to detect the Phillips C_2 band in the outflow of IRC+10 216 with an expansion velocity of 17.9 km s^{-1}, indicating that the C_2 is located where the flow has reached its terminal velocity.

The infrared energy distribution of AGB and post-AGB stars is characterized by thermal emission from dust. As was mentioned earlier, this dust plays an important role in the dynamics and energy balance of the stellar wind. Knowledge of the chemical composition of circumstellar dust and its emissivity are essential for a better understanding of the dust formation process, and for the determination of dust mass loss rates. Dust mass loss rates derived from the IRAS 12-100 μm emission from circumstellar dust are widely used (e.g. van der Veen & Olofsson 1990). Study of the dust properties therefore is important to improve the accuracy of the dust mass loss rates. Many dust features can be found in the atmospheric windows, such as the family of infrared unidentified features (3.3, 3.4, 3.5, 7.7, 11.2 μm), the unidentified 21 μm feature, the Silicate features (9.7, 18μm) and the SiC emission at 11.3 μm. The VLT can be used to obtain low resolution spectra of AGB stars in the Galactic Centre and in the Magellanic Clouds (see also the contribution of Käufl et al., these proceedings) in the near-IR and mid-IR in order to determine the chemical composition of the star. In nearby AGB stars, the spatial distribution of dust features can be studied by using low-resolution spectral imaging in the 10 and 20 μm window. Possible changes in the properties of the dust as the material moves away from the star can be measured. The spectral resolution required for dust features is typically 100-1000.

In Table 2 we list some requirements for the spectral resolution of VLT instruments for the study of AGB and post-AGB winds.

3   Post-AGB Evolution

High spatial and spectral resolution observations of post-AGB stars are necessary to resolve some basic questions concerning the way stars evolve off the AGB and (may) form a Planetary Nebula (PN). We mention two examples here: high resolution optical and infrared spectroscopy of heavily obscured post-AGB stars, and mid-IR imaging of AGB remnants.

3.1   Spectroscopy

The optical spectra of post-AGB stars contain valuable information concerning the nature of the central star *and* of the AGB remnant (see Figures 1 and 2). The photospheric absorption lines can be used to determine T_{eff}, log g and surface abundances reliably. The enhancement of S process elements and of C, N and O can be measured. It is important to measure these quantities for stars that have left the AGB very recently, because this allows detailed comparison with evolutionary calculations. Young post-AGB stars tend to suffer from substantial circumstellar reddening from the surrounding dust shell, which makes them very faint. HD 56126 is one of the brighter post-AGB stars with V =8.8, but many objects are found with V>14. High quality spectroscopy requires a large collecting area such as provided by the VLT.

An exciting new area of research is the study of molecular aborption and emission from the AGB remnant in the optical spectra of post-AGB stars. Figure 1 shows the C_2 Phillips band in HD 56126 (Bakker et al. 1995) taken with the WHT. The C_2 absorption is due to gas near the inner radius of the AGB remnant, and allows us to study the molecular component of the material ejected

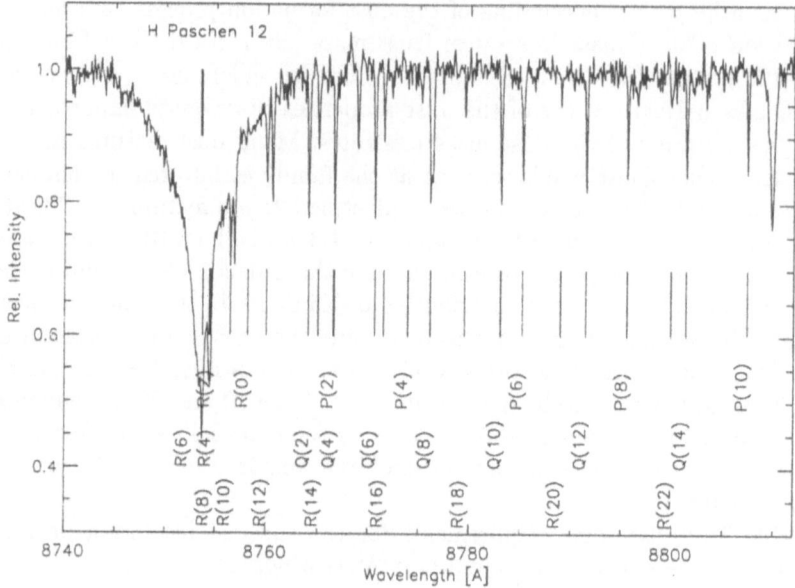

Fig. 1. High resolution red spectrum, taken with the 4.2m WHT of HD 56126 showing the photospheric P12 line and the C_2 Phillips band in absorption

at the very end of the AGB. This is very difficult using other observing techniques: the millimeter emission lines probe material at a larger distance from the star than C_2. Using a simple model for the excitation of C_2, Bakker et al. (1995) derive a very high mass loss rate of $2.8\ 10^{-4}\ M_\odot\ yr^{-1}$, which is more than a factor 10 higher than the mass loss rate derived from CO millimetre observations (Omont et al. 1993), but agrees well with the dust mass loss rate (Bakker et al. 1995). This may indicate that the AGB mass loss increased substantially over a period of typically 10^4 yrs, shortly before the star left the AGB. The high spectral resolution allows an accurate measurement of the C_2 expansion velocity, which was found to be slightly smaller than that seen in CO (8.8 vs 10 km/s).

In Figure 2 we show the 4.6 μm spectrum of the F supergiant IRC+10 420, taken with UKIRT and CGS4 (Oudmaijer 1995). The spectrum, with a resolution of about 16,000, shows the CO ro-vibrational fundamental transitions as wel as the Pfβ line. Here we see two mass loss episodes: the CO emission is formed in a molecular outflow which probably stopped only very recently, while the Pfβ emission is formed in partly ionized gas very near the star. An analysis of the optical and IR spectrum of this star suggests that the geometry of the two outflows is quite different. The P Cygni profiles observed in the CO lines suggest that they are formed in a roughly spherically symmetric outflow, while the Hydrogen emission lines indicate a strongly non-spherical geometry. IRC+10 420 is in a very rapid phase of evolution, and may be a post-red supergiant rather than a post-AGB star (Jones et al. 1993; Oudmaijer et al. 1994).

3.2 Imaging

The shaping of Planetary Nebulae (PNe) is closely linked to the morphology of the AGB wind. Numerical calculations of Mellema et al. (1991) show that, in order to reproduce the observed asymmetry in young PNe, substantial asymmetries need to be present in the AGB wind. Evidence for non-spherical mass loss in AGB stars has been scarce so far. In a recent paper, Plez & Lambert (1994) show that in four AGB stars the distribution of light scattered in the KI resonance line is quite asymmetric. The asymmetry in the AGB wind can be determined by mid-IR imaging of the AGB remnant observed in post-AGB stars. Several 10 μm images, taken with 3 meter class telescopes, have been published already (e.g. Skinner et al. 1993), showing substantial deviations from spherical symmetry in AGB remnants, *before* a fast wind has had the opportunity to shape the nebula. This implies that an asymmetry in the mass loss process must develop at some point near the end of the AGB. Figure 3 shows a TIMMI 10 μm image of HD 56126. The object is resolved and the peak intensity is offset from the center, again suggesting an asymmetric distribution of dust. A similar image taken by Meixner et al. (1994) also shows asymmetry. An important point to realize is that the combination of images at different wavelengths constrains the dust emissivity *and* the density distribution of dust shells much more strongly than an image at only one wavelength. The importance of the 20 μm window should therefore be stressed. It is also clear that the much better spatial resolution of the VLT, and the VLT interferometer, will spectacularly improve the image quality over what is possible at present!

4 Conclusions

The VLT and VLT interferometer open new and exciting possibilities for AGB and post-AGB research. The planned VLT instrument development matches the requirements for the proposed research well. However it is important to emphasize the impact of the VLT interferometer (see Table 1), which has indeed the potential to revolutionize research in this area. Also the importance of high resolution, high S/N spectroscopy should be mentioned. Specifically near-IR and mid-IR spectrographs with a resolution of 30.000 to 100.000 are required to carry out some of the projects mentioned above (Table 2). At lower spectral resolution the study of dust features is important, and we would like to emphasize the 20 μm window for spectral imaging of thermal emission from dust. The VLT at Paranal should be a very good site for 20 μm observations.

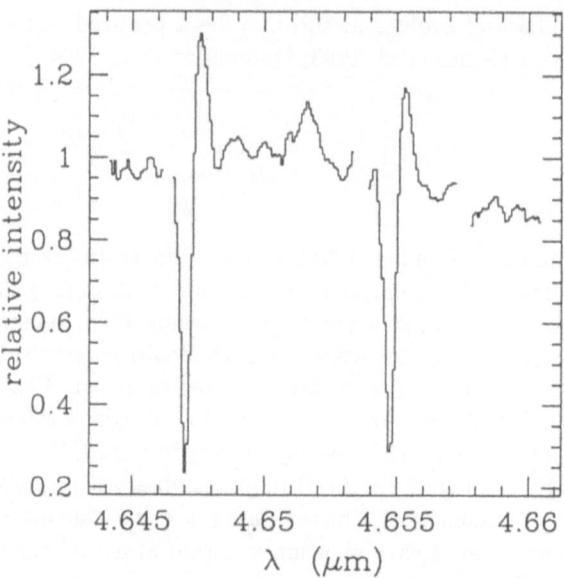

Fig. 2. High resolution infrared spectrum of IRC+10 420 taken with UKIRT (CGS4), showing the fundamental CO ro-vibrational transition (with a P Cygni profile) as well as the Pfβ line (to higher wavelength). Some parts of the spectrum are missing due to poor sky cancellation

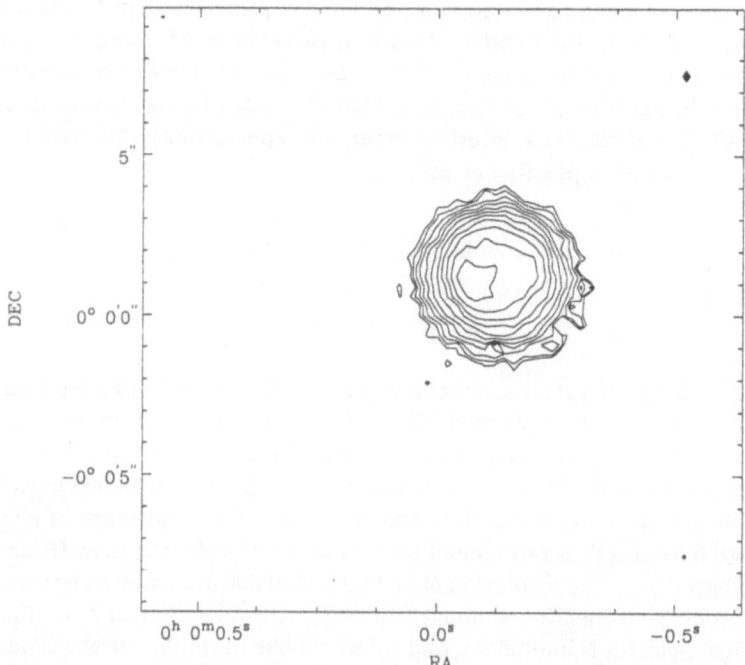

Fig. 3. 10 μm image of HD 56126 taken with TIMMI at the 3.6m ESO telescope. Notice that the source is resolved and is not spherically symmetric

Acknowledgements

I thank Eric Bakker and René Oudmaijer for providing the spectra of HD 56126 and IRC+10 420, and Ulli Käufl for invaluable help in obtaining the TIMMI image of HD 56126. This paper was written while the author was financially supported by the Royal Dutch Academy of Arts and Sciences.

References

Bakker, E.J., Lamers, H.J.G.L.M., Waters, L.B.F.M., Schoenmaker, T. (1995): in proc. of the Edinburgh conference *Circumstellar Matter*, Ed. G. Watt (in press)

Bowen, G.H. (1988): ApJ **411**, 323

Bowen, G.H., Willson, L.A. (1991): ApJ **375**, L53

Cherchneff, I., Glassgold, A.E. (1993): ApJ **419**, L41

Dominik, C., Sedlmayr, E., Gail, H.-P. (1993): A&A **277**, 578

Fleischer, A.J., Gauger, A., Sedlmayr, E. (1992): A&A **266**, 321

Habing, H.J. (1990): in *From Mira's to Planetary Nebulae: Which Path for Stellar Evolution?*, eds. M.O. Mennessier and A. Omont, Editions Frontières, p. 16

Jones T.J. *et al* (1993): ApJ **411**, 323

Keady, J.J., Ridgway, S.T. (1993): ApJ **406** 199

Meixner, M., Graham, J.R., Skinner, C.J., Hawkins, G.W., Keto, E., Arens, J.F., Jernigan, J.G. (1994): Exper. Astr. **3**, 53

Mellema, G., Eulderink, F., Icke, V. (1991): A&A **252**, 718

Plez, B., Lambert, D.L. (1994): ApJ **425**, L101

Olofsson, H., Carlström, U., Eriksson, K., Gustafsson, B. (1992): A&A **253**, L17

Omont, A., Loup, C., Forveille, T., te Lintel Hekkert, P., Habing, H.J., Sivagnanam, P. (1993): A&A **267**, 515

Oudmaijer, R.D., Geballe, T.R., Waters, L.B.F.M., Sahu, K.S. (1994): A&A 281, L33

Oudmaijer, R.D. (1995): in proc. of the Edinburgh conference *Circumstellar Matter*, Ed. G. Watt (in press)

Pijpers, F.P., Habing, H.J. (1989): A&A **215**, 334

Skinner, C.J., Meixner, M.M., Hawkins, G.W., Keto, E., Jernigan, J.G., Arens, J.F. (1994): ApJ **423**, L135

Tuthill, P.G., Haniff, C.A., Baldwin, J.E., Feast, M.W. (1994): MNRAS **266** 745

Vassiliadis, Wood, P.R. (1993): ApJ **413**, 641

van der Veen, W.E.C.J., Olofsson, H. (1990): in *From Mira's to Planetary Nebulae: Which Path for Stellar Evolution?*, eds. M.O. Mennessier and A. Omont, Editions Frontières, p. 139

Willems, F.J., de Jong, T. (1986): ApJ **309**, L39

Wilson, R.W., Baldwin, J.E., Buscher, D.F., Warner, P.J. (1992): MNRAS **257** 369

Wood, P.R. (1979): ApJ **227**, 220

Zijlstra, A.A., Loup, C., Waters, L.B.F.M., de Jong, T. (1992): A&A **265**, **L5**

The Asymptotic Giant Branch in the LMC, SMC and the Galactic Center

H. U. Käufl[1], L. B. F. M. Waters[2],
G. Wiedemann[1], A. A. Zijlstra[1],

[1] ESO, D-85748 Garching b. München, Fed. Rep. of Germany
[2] Laboratory for Space Research Groningen, PO Box 800, NL-9700 Groningen, The Netherlands

Abstract. Mass loss on the AGB is extremely important for galactic chemical evolution but is poorly understood. A data set of a coherently selected sample of AGB stars is required to resolve this issue. The VLT with its suite of infrared instruments allows the Large Magellanic Cloud, the Small Magellanic Cloud and the Galactic Center to be completely surveyed for AGB stars. The survey could be done using only morning and evening twilight, which is of no use to other programs. The limiting flux for this survey is expected to be 2-3 orders of magnitude fainter than that of the IRAS 12.5 μm channel. The survey will result in the first complete catalog of AGB stars at a known distance. The objects would then be studied more in depth, employing high spatial resolution IR imaging and medium to high resolution spectroscopy with a special emphasis on rotational-vibrational molecular transitions. This program could lead to an understanding of this important but as yet enigmatic phase of stellar evolution, especially the physics of mass loss and the thermonuclear evolution of AGB stellar cores.

1 Scientific Rationale

Less massive stars in the red giant phase, after their life-time on the main sequence, are powered by H-shell or He-core burning. Further exhaustion of nuclear fuel leads to He-shell burning while forming a degenerate core. This phase of stellar evolution is referred to as the **A**symptotic **G**iant **B**ranch (AGB). All stars up to $\approx 8 M_\odot$ follow this track and return up to 85% of their mass to the interstellar medium (see e.g. Weidemann, 1993 or Koester, these proceedings). While $\int \dot{M}(t) dt$, is well established, the mass loss $\dot{M}(t)$ is neither understood theoretically nor strongly constrained observationally (see e.g. the review by Lafon and Berruyer, 1991). Studies of AGB-stars and the associated mass loss today are problematic. Objects from the IRAS data base have been investigated mostly photometrically. Van der Veen and Habing, 1988, e.g. classify AGB-objects phenomenologically according to their location in the IRAS color-color diagram and postulate an evolutionary track which, however, must remain purely hypothetical. Generally the distance, and consequently the absolute luminosity (certainly a fundamental property), are uncertain. For galactic objects it is often difficult to distinguish circumstellar from interstellar reddening and extinction. Radio surveys for OH stars deliver a database which may be complete. A precise knowledge of their distances, however, remains restricted to a few cases while requiring a prohibitive amount of observing time at radio interferometers.

Most objects are obscured by vast amounts of gas and dust. Detailed studies of the envelope by high resolution IR spectroscopy are today restricted to few exceptionally bright objects, due to the lack of large enough telescopes with suitable instrumentation. These objects represent a rather arbitrary sample, again with still mostly unknown distances. It is doubtful if general conclusions about AGB evolution can be derived from data on exceptionally bright objects. Spatially resolved AGB stars (see e.g. Waters in these proceedings) are by themselves extremely interesting objects, but the above reservations with respect to general conclusions most likely also hold here.

Understanding the thermonuclear evolution of AGB stars is based largely on theoretical models and conjectures, only marginally constrained by observation.

2 A Survey for AGB Stars with the VLT

2.1 The Survey and its Completeness

From the set of variable Galactic OH stars with good photometry and known (phase-lag) distances given by Herman and Habing, 1985, we estimated the apparent magnitudes of AGB objects in the L-band ($\lambda \approx 3.6\mu m$) and the N-band ($\lambda \approx 10\mu m$) at 75 kpc. Comparison with sensitivity estimates for the VLT infrared instruments (see e.g. Moorwood, 1992 or Käufl and Delabre, 1992) yields:

   a. at $\lambda \approx 3.6\mu m$: for a field of 4 arcmin^2 a complete survey for most AGB-stars at 75 kpc (at the 10σ level) needs an exposure time of 4 minutes;

   b. at $\lambda \approx 10\mu m$: for a field of 0.5 arcmin^2 a complete survey for the reddest AGB-stars at 75 kpc (at the 10σ level) needs an exposure time of 30 sec.

The surveys at 3.6 and 10μm are complementary. Each one will pick up those AGB-stars missed by the other. For comparison, the limiting flux for pointsources (10σ) at $\lambda \approx 10\mu m$ will be 6 mJy, i.e. the survey will be nearly 3 orders of magnitude fainter than the IRAS point-source catalogue.

Observations at $\lambda \geq 3.6\mu m$ are dominated by the thermal emission of sky and telescope. They can be performed even in daytime and clearly during twilight, i.e. in conditions which are otherwise of no use. Assuming that 20 minutes morning and evening twilight were available on 300 days per year, more than 1 deg^2 per year could be surveyed in the LMC and SMC. Depending on the observing overhead times, the survey will proceed 5-10 times faster for the Galactic Center.

2.2 The Identification and Creation of the Catalog

After detection, the objects need to be positively identified as AGB-stars. This is best done by photometry from 500nm to 24μm. In addition the fields need to be monitored repeatedly (6-8 weeks intervals), since most AGB-objects are variable. The chemical composition of the dust shell will be determined by low resolution spectroscopy, especially at $\lambda \approx 10\mu m$. If carried out in twilight time only, the survey would proceed somewhat slower than given above (because of sharing of twilight periods with the follow-up observations).

At the end of say a period of three years a catalogue of AGB stars will be available containing the following information:

a. spectral distribution, and absolute bolometric luminosity for the LMC and SMC objects, (relative for the objects in the Galactic Center region);
b. variability and phase as a function of wavelength;
c. chemical composition of dust shell;

In addition a general deep catalog of infrared point sources at $\lambda \approx 3.5 \mu m$ and $\lambda \approx 10 \mu m$ will be available, which by itself may also be quite interesting.

2.3 A Distance Law for AGB Stars

After a few years, several thousand AGB stars will be available and almost certainly will allow a distance law, e.g. by relating luminosity, color and variability, to be propounded. This might allow the merging of the observations of the LMC and SMC objects with the Galactic Center ones. If this fails then the distance to the Galactic Center objects could be constrained by other methods, e.g. statistical parallaxes, measurement of stellar diameters (e.g. by using lunar occultations in the infrared), derived from morphology for the cases of spatially resolvable objects, or constrained by association.

Establishing the distance to as many galactic objects as possible, and thus merging the data, is of paramount importance for two reasons:

a. at 75kpc, high resolution spectroscopy is restricted to the very brightest objects (contrary to galactic AGB stars). In order to analyze the spectra of galactic objects with radiative transfer calculations, the absolute luminosities should be available;
b. assessing what differences exist between AGB evolution for the LMC, the SMC and our galaxy (e.g. caused by the difference in metallicities).

3 The Case of High Resolution Spectroscopy at $\lambda \approx 10 \mu m$

Absorption lines from rotational-vibrational transitions of circumstellar molecules originate from a confined zone above the stellar photosphere. Close to the star the molecules are either unstable or the stellar atmosphere is not yet optically thin. In the stellar atmosphere, the population of the molecular rotational vibrational states is a function of radiation, pressure and temperature (n.b. typically $T_{vib} = \frac{h*\nu}{k_B} \approx 200\text{-}500\text{K}$). A strategic selection of the quantum numbers, and hence transitions, allows for altitude-resolved observations in the stellar atmospheres. Thus these spectra provide unique information about the physical and chemical conditions of the region where dust forms and where the outflowing material is being accelerated. This is complementary to radio observations which are usually restricted to pure rotational transitions of molecules in the vibrational ground state, i.e. they give only information about the outermost region of the envelopes where dust is already formed and the outflow is already accelerated to its terminal velocity. Molecules with suitable transitions in the $10\mu m$ window are e.g. SiO, SiS, SiC, NH_3, PH_3, CH_4, SiH_4, C_2H_2, C_2H_4 or C_2H_6. The isotopic shift in most molecular transitions is large enough to allow for an easy discrimination of bands belonging to different isotopes. SiO for example comes in 9 different isotopic configurations ranging from $^{28}Si^{16}O$ to $^{30}Si^{18}O$. Thus precise isotopic ratios can be derived. For an example see Schrey et al., 1986, who show that for planetary atmospheres, spectroscopy achieves a precision which matches insitu mass spectroscopy. These ratios can be compared to codes describing AGB thermonuclear evolution. Contrary to observations at shorter wavelengths, spectra at $\lambda \approx 10\mu m$ are least contaminated by circumstellar dust absorption and reemisson.

The maximum sensitivity will be achieved when the spectral resolution matches the intrinsic linewidth, i.e. a resolution $\lambda/\Delta\lambda$ approaching 50-100 000 would be best suited. Once the general structure of the spectra is understood, objects less luminous, especially again those in the Magellanic Clouds, could be observed also at lower resolution. Käufl and Stanghellini, 1992, give an example of molecular absorption spectroscopy and analysis of the data for the case of low resolution spectroscopy.

4   The VLT Specific Advantage

Obviously, observation of the LMC, SMC and Galactic Center needs a telescope in the southern hemisphere. For the VLT at $\lambda \approx 10\mu m$ the signal-to-noise ratio will be 4-times that achievable at a 4m class telescope, and the survey will proceed 16 times faster. Since it is also important to search for morphological structure in AGB objects, the spatial resolution of the VLT, together with adaptive optics in the near-infrared, are required. The diffraction limit of the VLT at $\lambda \approx 3\mu m$ is 0.1arcsec which corresponds to 6000AU at the LMC or $\approx$1000AU at the Galactic Center.

Particularly beneficial, if not essential, for this project is the operational concept of the VLT, allowing for a quasi-simultaneous access to different multimode instruments. The advantages are:

a. the survey can be carried out without affecting the normal astronomical observations of the VLT;

b. distributing the observing over 300 days per year 20–40 minutes each, is better suited for the project than 8–15 nights completely scheduled;

c. clearly the survey will be best carried out in a quasi-automatic mode or by remote observation. The VLT is planned to have the required infrastructure to accommodate such observing;

d. the reduction of the survey data will benefit substantially from the catalog handling tools which are planned to be available at the VLT.

5 Conclusions

The VLT with its presently planned suite of multi-mode instruments will allow a complete survey for AGB-stars in the LMC, SMC and the Galactic Center to be carried out. Areas of the order of one square-degree per year can be observed using twilight time only, which is of little or no use for other projects. Based on the survey data a catalog of AGB stars with well known distances can be created. The primary scientific questions which can then be addressed are manifold:

a. to produce a distance scale for AGB stars;
b. to understand the enigmatic but important mass loss process;
c. to constrain models for the thermo-nuclear evolution of AGB-stars;
d. to get a sample of "missing links" between the AGB and Planetary Nebulae;
e. to constrain models on the early evolution of white dwarfs.

References

Herman, J., Habing, H.J. (1985): Phys. Rep. **124** 255

Käufl H.U., Delabre, B. (1992): in Proc. ESO conference on Progress in Telescope and Instrumentation Technologies, ed. by M.-H. Ulrich (European Southern Observatory) pp. 597-600

Käufl, H.U., Stanghellini, L. (1993): in Proc. 2nd ESO/CTIO Workshop on Mass Loss on the AGB and Beyond, ed. by H.E. Schwartz (European Southern Observatory) pp. 189-193

Lafon, J.-J.P., Berruyer, N. (1991): Astron. Astrophys. Rev. **2** 249

Moorwood A.F.M. (1992): in Proc. ESO conference on Progress in Telescope and Instrumentation Technologies, ed. by M.-H. Ulrich (European Southern Observatory) pp. 567-576

van der Veen, W.E.C.J., Habing, H.J. (1988): Astron. Astrophys. **194** 125

Schrey, H., Rothermel, H., Käufl, H.U., Drapatz, S. (1986): Astron. Astrophys. **155** 200

Weidemann, V. (1993): in Proc. 2nd ESO/CTIO Workshop on Mass Loss on the AGB and Beyond, ed. by H.E. Schwartz (European Southern Observatory) pp. 55-59

Prospects for Circumstellar Physics. Observations and Models

Jean-Pierre J. Lafon[1]

[1] **Groupement de Recherches "Milieux Circumstellaires"**,   Observatoire de Paris-Meudon, DASGAL/URA CNRS D0335
92195 Meudon Cedex, France

Abstract. In addition to information derived from spectrophotometric measurements of unresolved objects, observations at much increased angular resolution can provide constraints on new types of models of circumstellar media. Models of circumstellar envelopes of young or evolved stars are now quickly evolving because of developments in mathematics, new techniques and increased power of computers. The models will benefit greatly from the facilities of the VLT and the VLTI.

1   Circumstellar Media

It is now well established that, at both ends of a star's life, when mass loss and/or accretion are important, the circumstellar environment has a considerable importance for the evolution of both the internal structure of stars and the interstellar medium. The trajectory of a star in the H-R diagram is highly sensitive to the rates at which matter, momentum and energy flow from the former to the latter, or conversely. These rates are governed by several physical phenomena efficient in circumstellar envelopes, particularly in their inner layers at small scale sizes. This means that a better knowledge of the close circumstellar environment is crucial for a better understanding of the evolution of stars and galaxies.

The overall structure of circumstellar envelopes of evolved stars can be divided roughly into four zones with fairly different scale sizes. The first one is a thin layer, at the basis of the envelope, in which physics is not clearly different from that of the interior of the star; it is usually represented by some time-dependent surface through which the star fixes the boundary conditions; in accreting objects it is a boundary layer. In the next layer (within a few stellar radii), several mechanisms linked to the instability of the medium produce momentum and energy: Alfvén waves, acoustic waves, shock waves, thermal pressure gradients, etc. Many scale sizes related to strong gradients characterize these processes: wavelengths, dissipation lengths, sizes of inhomogeneities, etc. In the third zone (within a few hundred stellar radii), the grains condense and grow, and the wind is accelerated by radiation pressure on grains. Finally the largest zone is the outer one in which chemical processes on the gas and the grains (sputtering and surface chemistry), in a less perturbed outflow, are increasingly important.

In hot stars, the main driving mechanisms are thermal pressure gradients and radiative pressure through absorption in lines, throughout the envelope.

In young objects strong asymmetries appear with fast polar flows, which can become jets in very young objects.

Qualitative and quantitative improvements of models describing the physical processes or the overall structure of circumstellar media have been obtained during the last decade. However, the problem is that there is still a significant gap between what can be modelled accurately, i.e. with accurate description of the physics involved, and what can be observed precisely, i.e. with optimized spectral or angular resolution. In circumstellar envelopes, matter and radiation are frequently out of any kind of equilibrium, and still out of the state in which they have been usually reproduced in the laboratory, or modelled under normal conditions.

Thus, there is a strong need for new models under little-explored conditions. It is also necessary to derive from these new models quantitative data more suitable for comparison with observed quantities.

2 Physics and Observations of Circumstellar Media

There is still a controversy concerning the applicability of models with spherical symmetry, compared to 2D or 3D models. In fact, we think that the problem is ill-posed if only the condition of fitting the results to some given observations is considered. Indeed, as already mentioned, there is no model able to be compared to all available data, and what can be derived with reasonable accuracy from a model depends strongly on the quality of the basic data and also on the consistency of the model. There is no need to describe accurately radiative transfer, or hydrodynamics or thermodynamics, or any other aspect of physics if all the effects to which the expected quantity is sensitive are not described with the same degree of accuracy. Crude descriptions are only allowed for parameters or phenomena to which the expected data are weakly sensitive, but this is not always possible because fundamental physical phenomena efficient in the model may not be understood well enough.

A first class of possible models are more or less crude models ("crude models" for simplicity) which can provide some information on large scale sizes, even when they are oversimplified provided that they are based on good input physics, depend on a small number of free parameters, and can be fitted to a large number of observations. Most of these models ignore the conditions in the inner layers of the envelopes, where small scale phenomena initiate mass loss (or accretion), or they use phenomenological descriptions of the dominant phenomena using adequate parameters (such as "some radius at which dust grains condense" in evolved stars, or some "exponent characterizing the optical depth of driving lines" in the expression of the driving force for a wind from a hot star, and so on ...). These models are not self consistent, but always semi-empirical with more or less detailed physics. Systematic analyses of this kind mainly give information on the nature of materials in the envelopes and their optical properties depending on the stages of star evolution, and sometimes also on mass loss (or accretion) rates which are precisely sensitive parameters for the evolution of the subphotospheric structure (Le Bertre, 1988a, b).

Then, different strategies can be and are effectively developed for breakthroughs in the field of circumstellar envelopes. One of them consists of many surveys, as complete as possible with the available techniques, performed or in progress, at various wavelengths. They provide a lot of data mostly used to find out evolutionary sequences in diagrams more or less equivalent to the H-R diagram, or temporal sequences of similar objects. "Crude models" in the above mentioned sense could be used as suitable tools for analysing survey data from a theoretical point of view.

A second class of more elaborate models is made of semi-empirical models. Whereas "crude" models are in general based on the assumption of spherical symmetry, semi-empirical models often relax this assumption. They generally describe some properties of the circumstellar media using less detailed assumptions for other properties; models of this type have recently been developed (Lafon, these proceedings). In the case that the dynamics of the grains and the gas surrounding an evolved star is ignored, then the density map can be a basic free parameter (reduced to a small number of free parameters using realistic assumptions including axisymmetry). Energy exchange and radiative transfer problems are solved carefully in a self-consistent way, so that the model provides both images and spectra. However, while spectra can be directly compared to observations, Fourier transforms of the images are necessary to generate visibility curves, the only data which can be compared to available observational data. There are also in this class simplified self consistent models with spherical symmetry based on the assumption of dust driven mass loss (Winters, 1994).

A third class of models includes full self-consistency between the involved phenomena, but the number of these phenomena is reduced and approximations are introduced in order to make the problem tractable and the results understandable. This class of models is illustrated by the model of hot (Be) star (Lafon, these proceedings). Such models produce spectra and images, but can be built only for a still more limited number of objects. This approach corresponds to a specific strategy for breakthroughs in this field: looking at details in selected objects.

A fundamental problem in all the models is the determination of a reliable set of boundary conditions at both ends of the circumstellar envelope. Whatever the assumptions concerning symmetry, when there is no accretion and unless interaction between the outer envelope and the interstellar medium is concerned, most of the models assume that the envelope "extends" to infinity with no radiation coming from outside. In case of accretion, the external medium is assumed uniform and with specified equilibrium physical conditions. The problem is much more complicated for the inner layers of the envelope because it is difficult to define some inner "limiting surface"; indeed, this can be clearly done in general only for each process separately, which means that the inner boundary is not the same for radiation at different wavelengths, for dusty (or non-dusty) winds, for winds driven by Alfvén waves or acoustic waves, or shock waves (Lafon and Berruyer,

1991), etc. Finally, one has always to define some limiting surface (usually called false or true "photosphere"), which can be submitted to time-dependent deformations, on (or through) which the characteristics of the phenomena involved in exchanges of matter, momentum and radiation are known. Since, in turn, what happens deeply under the photospheres depends on what happens above it, a good knowledge of the outer layers of the star/inner layers of the envelope is crucial for good modelling.

One type of strategy to deal with such problems is to build up better and better models of each relevant physical mechanisms and to mix some of them with some weighting factors. However, this is a long-term strategy (several decades have been necessary to reach the present knowledge of shock waves and no global model is still available). In any case, this strategy can be helped significantly by systematic observations of the media: the identification of the efficient mechanisms is certainly more easy when chemical, physical and thermodynamical conditions are better known. At least, this may suggest laboratory experiments able to reproduce specific mechanisms.

Systematic observations at 3.6 μm and 10 (and 20?) μm of AGB envelopes and Proto-Planetary Nebulae with ISAAC and the proposed MIIS spectrometer respectively, as suggested by Kaüfl et al (these proceedings), would be very useful to check the photosphere/wind interface by finely tuning the altitude at which the elements are present. Some elements peak where dust forms and this occurs at small scales close to the star surface, so that 1D-models with (spherical or even plane) symmetry would probably be adequate for comparison with observations.

Finally, conditions for qualitative and quantitative improvements of models of circumstellar media have been gathered, or are about to be gathered, to provide an exceptional opportunity to increase significantly our knowledge. This is due both to the developments of spectroscopic facilities, as already mentioned above, and also the developments of interferometric (or other) facilities for High Angular Resolution, especially in the range of resolution in which small scale sizes, typical of small scale phenomena, can or soon will be resolved (0.1 to 0.001 arcsec).

Indeed, in this domain, the VLTI is would be a formidable facility, even with the sub array VISA, because the priority for detailed analyses of circumstellar objects relies on a few tens of bright objects, which should be observed at regular intervals of time. Of course, if possible, especially for circumstellar envelopes, phase closure and full imaging would be substantial improvements, compared to visibility curves obtained with a few telescopes (but with a sufficient coverage of the phase plane).

Apart from the VLTI, the design of VHARC on the VLT is adequate for investigation of envelopes and mass ejection of Miras, red giants and supergiants, Luminuous Blue Variables and Be stars. Moreover, it would be complementary to VISA for acquisition of visibility points at small frequencies, which are necessary for a good interpretation of visibility curves.

3 Conclusions

Theory and modelling must take new facilities into account. Several different models are in competition, with strong controversies: High Spectral and High Angular Resolution measurements can provide crucial tests. Adding such constraints to the classical constraints leads to a new insights into physics of circumstellar matter and, thence concerning stellar evolution.

References

Lafon, J.-P. J., Berruyer, N. (1991): "Mass Loss Mechanisms in Evolved Stars", Astron. Astrophys. Rev., **2**, pp. 249-289

Le Bertre, T. (1988 a): Astron. Astrophys., **190**, pp. 79-86

Le Bertre, T. (1988 b): Astron. Astrophys., **203**, pp. 85-98

Winters, J.M. (1994): "Internal Structure and Optical Appearance of Circumstellar Dust Shells around Cool Carbon Stars", Thesis, Technischen Universität, Berlin

Compact Stars
In and Out of Binaries

Frank Verbunt

Astronomical Institute, Postbox 80.000, 3508 TA Utrecht, the Netherlands

Abstract. The Very Large Telescope will allow us to determine more orbital periods and component masses of cataclysmic variables, low-mass X-ray binaries and high-mass X-ray binaries. These are required to put evolutionary models for these systems on a quantitative footing. The masses of neutron stars in low-mass X-ray binaries will set the best limits on the equation of state at high densities. Fundamentally new measurements that become possible with the Very Large Telescope include the detection and low-resolution spectroscopy of nearby single neutron stars, and the structure of the accretion disks in low-mass X-ray binaries. The most useful instruments for this research are a low-resolution infrared spectrograph, and a fast multicolour photometer.

1 Introduction

Compact stars are difficult to study in the optical, due to their small surface area and accordingly low flux. With the advent of Very Large Telescopes, equipped with state-of-the-art detectors, many opportunities for the study of such stars open up. In this review I give some examples for the study of single neutron stars (Section 2), of white dwarfs and neutron stars or black holes accreting matter from companion stars (Sections 3, 4), and of white dwarfs accompanying neutron stars (Section 5). The required instrumentation is summarized in Section 6.

2 Single Neutron Stars

When neutron stars were first conceived, e.g. by Baade and Zwicky (1934), it was thought that the radius of about 10 km would preclude any detection in the optical. A neutron star consists of a degenerate, isothermal core with an atmosphere on top. The core cools mainly by emitting neutrinos, but some energy leaks away via the atmosphere. The cooling time scale of a newly born neutron star is about 10^6yr (e.g. Shapiro and Teukolsky 1983). A rapidly rotating neutron star with a strong magnetic field can be a radio pulsar. The details of the pulsar mechanism still elude us, but presumably charged particles, say electrons, are pulled from the neutron star by the electromagnetic force of the rotating magnetic field and accelerated. If the acceleration is sufficiently high, interaction with the magnetic field causes the electron to emit a γ-ray photon, which interacts with the field to produce an electron–positron pair. These particles are accelerated in turn, produce γ-photons, etc., and a whole cascade is produced of electrons accelerated away from the neutron star, and positrons accelerated towards the neutron star surface. The latter may heat this surface for as long as the radio pulsar is active, i.e. on the order of 10^7yr (Bhattacharya et al. 1992). In some cases visual radiation emitted by the accelerated electrons can be seen: the Crab pulsar is an example.

Recent studies of radio pulsars show that there may well be active radio pulsars at distances less than 100 pc from the Sun (Tauris et al. 1994). The γ-ray source Geminga is a single neutron star at a distance of possibly only 38 pc (Bertsch et al. 1992). As pointed out to me by Jan van Paradijs, the optical surface fluxes emitted by such neutron stars may well be in the range of the Very Large Telescope. For example, a sphere with a radius of 10 km, and emitting a blackbody spectrum at a temperature of 5×10^5K, will have an apparent visual magnitude of 21 at a distance of 10 pc, and of 26 at a distance of 100 pc. At this high temperature, the colour is at the Rayleigh-Jeans limit $B - V = -0.31$.

Obtaining colours of neutron stars will help us in understanding their atmospheres at a very rough level; obtaining spectra will provide gravitational redshifts – much in excess of expected spatial velocities – and therefore a relation between the mass and radius of the neutron stars, and thus of the equation of state at neutron-star densities.

3   Accreting White Dwarfs: Cataclysmic Variables

In cataclysmic variables a white dwarf accretes matter from a low-mass companion star, thought to be on or close to the main-sequence. The accretion often occurs via a disk, which dominates the ultraviolet and optical spectrum when the accretion rate is not too low, as shown in Figure 1. The distance between white dwarf and red dwarf is typically on the order of a solar radius, which implies an orbital period on the order of hours. A typical distance to Earth is of order 100 pc. At the moment some 400 cataclysmic variables are known.

The white dwarf must have been at the center of a giant, once, which implies that the size of the binary orbit must have been much bigger in the past. Indeed, cataclysmic variables are thought to evolve out of Algol-type progenitors via a spiral-in process, during which the low-mass star is engulfed by the rapidly expanding envelope of its more massive companion. The orbital motion is braked by friction, until the energy released is enough to expel the giant's envelope, leaving its core in close orbit around the main-sequence companion. Subsequent loss of angular momentum brings the low-mass star into contact with its Roche lobe, and mass transfer to the core, cooled now into a white dwarf, ensues (Webbink 1979).

The study of cataclysmic variables has two major goals. The first of these is to measure the masses of the components and the orbital periods. This enables us to compare the properties of the assumed progenitors with those of the cataclysmic variables, and thus to find out how much mass and angular momentum is lost from the binary during the spiral-in process. The second goal is to understand the details of the mass transfer process, and therewith hopefully the underlying mechanism for it.

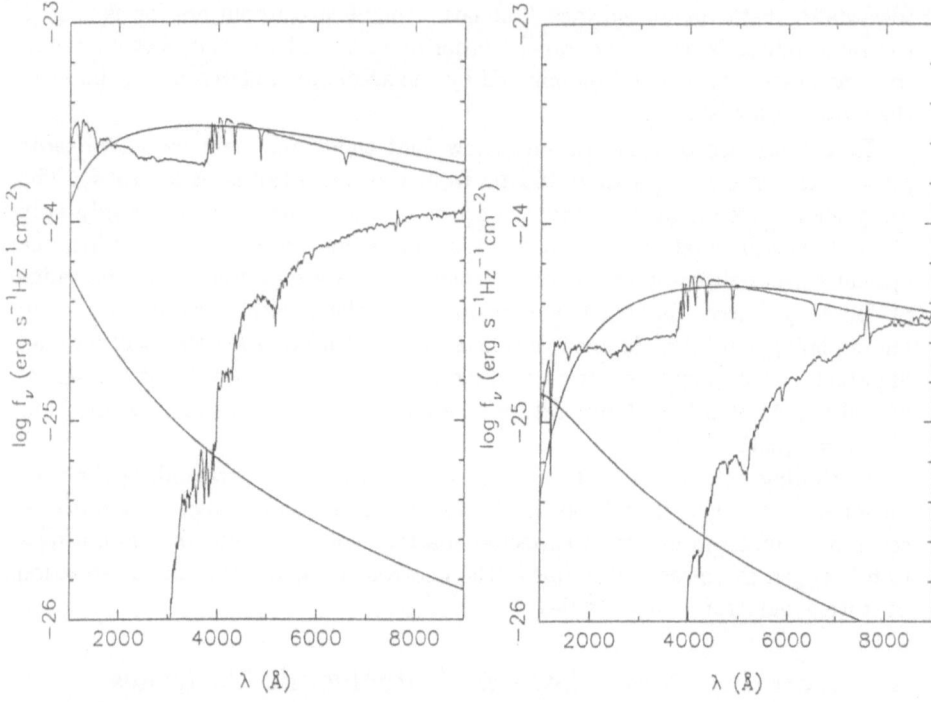

Fig. 1. Components of the spectrum of a cataclysmic variable. Left frame: a K4 V star is assumed to transfer $4 \times 10^{-9} M_\odot/\mathrm{yr}$ to a $0.6\,M_\odot$ white dwarf, half of the accretion luminosity is assumed to be emitted by a standard accretion disk, and half by an area equal to 10% of the white-dwarf surface, the boundary layer. The two top lines show the spectrum emitted by the accretion disk, as calculated by summing black bodies (smooth line) or model-atmosphere stellar spectra (jagged line). The stellar spectrum of the K4 V star (lower jagged line) peaks in the infrared, the spectrum from the boundary layer (lower smooth line) peaks in the EUV/X-ray. The fluxes are given for a distance of 100 pc. In the ultraviolet-optical range the disk spectrum dominates (after Bath and Pringle 1985). Right frame: a similar plot for an M0 star transferring $4 \times 10^{-10} M_\odot/\mathrm{yr}$ to a $0.6\,M_\odot$ white dwarf.

3.1 Mass Determinations

Mass determinations in cataclysmic variables, as in all binaries, are best done by obtaining radial velocity curves for both stars in an eclipsing system. The number of systems for which this can be done currently is rather limited, and in addition beset by problems, due mainly to the fact that the optical and ultraviolet light is dominated by the accretion disk, which makes it difficult to obtain radial velocities for the white dwarf and for the companion star. For the white dwarf velocity, one often takes the velocity of the emission lines in the accretion disk; this has to be done with care, however, as the disk emission is not symmetric, and its measured velocity therefore does not directly reflect the motion of the white dwarf at its center. With good signal-to-noise, the asymmetries can be studied, understood, and one then may correct for them (e.g. Marsh et al. 1990). For the red dwarf velocity, one may go to the infrared, where this star contributes more significantly to the total flux (see Figure 1). In addition it is useful to observe the binary at times of low accretion rates, when the contribution of the disk is minimal. For these studies, the VLT will bring better signal to noise, and therefore allow more systems to be measured accurately. However, the limitation to eclipsing systems and the inherent problems of line asymmetries, may limit the actual progress.

An indirect, but still useful way to determine the mass of the red dwarf is to make a spectral identification. This can be done in the infrared by observing the ratio of the TiO bands at 7165 and 7665 Å (Wade and Horne 1988). In addition to the spectral type, and via this, the mass of the red dwarf, this method provides a spectroscopic parallax to the binary. This is important not only for the determination of the spatial density of these types of systems, but also for their accretion luminosities, and thus mass accretion rates. The VLT can contribute very significantly here by determining distances to a large number of fainter, and presumably more distant, cataclysmic variables, thus providing a badly needed improvement of the currently rather meager statistics.

3.2 The Accretion Disk

The study of the mass transfer in most cataclysmic variables entails a detailed study of the accretion disk. In eclipsing systems, the red star gradually covers and uncovers the accretion disk, and thus provides a one-dimensional scan of the disk. Under suitable assumptions, the distribution of the light over the accretion disk may be derived, and by doing this in various colours, the temperature distribution over the disk surface, for comparison with theoretical models for the radial structure of accretion disks. Even though accretion disks are ubiquitous, they can be studied in most detail in eclipsing cataclysmic variables. In modern studies, the eclipse maps are made in spectrophotometry, providing information about continuum and line fluxes simultaneously as a function of location in the accretion disk (Rutten et al. 1993). Low-resolution spectroscopy allows one to use the velocity information present in the lines to determine the loci in the disk which contribute to the line radiation. A study of time-dependent disks will show how mass is transported through the disk, and thus contribute to our understanding of the viscous processes that cause the mass transport (Rutten et al. 1992). The VLT will contribute to this field in two ways. First, it will allow a large number of systems to be studied. More important however, in my view, is the possibility that the VLT will allow much more detailed maps of the disk line emission to be made, thus possibly allowing measurement, via the Zeeman line broadening, of the strengths of the magnetic fields present in the accretion disk (Horne, private communication).

In systems with a relatively low accretion rate, the eclipse of the white dwarf can be separated from the accretion disk, and compared with the eclipse of a (semi-)sphere, to determine the limb-darkening of a white dwarf. The best current observations provide only a marginal result (Bunk 1993), but the VLT should allow this measurement to be done.

3.3   The Supersoft X-ray Sources

A new class of accreting white dwarfs has been identified with ROSAT, in which white dwarfs accrete matter which immediately undergoes thermonuclear fusion on the white-dwarf surface. Such binaries are recognizable by their supersoft (i.e. emitting only at energies less than ~ 0.5 keV) X-ray spectra. These binaries are very interesting as possible progenitors for type Ia supernovae, and as possible progenitors for binaries in which the neutron star is formed via accretion-induced collapse of a white dwarf (Van den Heuvel et al. 1992). Because of the havoc wrought on the X-ray spectrum by interstellar absorption, the supersoft sources are hard to find in our own galaxy, and most systems now identified are in the Magellanic Clouds and in M31 (e.g. Hasinger 1994). At such large distances, their optical spectra are difficult to study. However, the larger area of the VLT, as well as improved instrumentation, will allow time-dependent spectroscopic studies to be made, in which we can learn the nature of the star that donates the mass to the white dwarf, and the size of the orbit. Both these parameters are required for us to place the supersoft sources in an evolutionary scheme of binaries.

4   Accreting Neutron Stars or Black Holes: X-ray Binaries

Neutron stars and black holes can be discovered via the X-rays they emit when they accrete mass from a companion star. The mass donors turn out to be either massive O or B stars with $M > 10 M_\odot$ or low-mass stars with $M < 1.0 M_\odot$. In the first category, the high-mass X-ray binaries, the optical flux is dominated by the light of the O or B mass donor, with a small contamination by the accretion disk. The orbital periods of these systems range from a day to more than a thousand days (see Figure 2). In the second category, the low-mass X-ray binaries, the optical flux is dominated by optical emission from the accretion disk, caused by reprocessing of X-rays that impinge on its surface. The orbital periods of these binaries range from only 685 s (!) to several hundred days. The main goals of the study of X-ray binaries are determinations of: a) their orbital periods, which reflect the prior evolution of the binary; b) the masses of the neutron stars or black holes, because the maximum possible mass of a neutron star is set by the properties of the equation of state at very high densities; c) the structure of their accretion disks.

4.1 Orbital Periods

Orbital periods are obtained either from photometry, via eclipses or via ellipsoidal variations due to the tidal deformation of the mass donor star, or from radial velocity studies (e.g. Van Paradijs and McClintock 1995). Photometric detections of orbital periods works mainly for close orbits: in wide orbits the ellipsoidal variations are negligible, and the probability of an eclipse is small. Radial velocities also are largest in close orbits; to measure the velocity of the compact star one often uses disk emission lines, with similar problems as in cataclysmic variables. With infrared detectors it is now becoming possible to study the O or B stars in high-mass X-ray binaries in other spiral arms, which were hitherto inaccessible due to interstellar absorption (see Kudritzki 1995). Low-mass X-ray binaries near the center of our Galaxy, where they are concentrated, can also be studied, mainly photometrically, in the infrared (Miller et al. 1993).

The contribution of the Very Large Telescope will be to allow spectroscopy in the infrared for the high- and low-mass X-ray binaries near our Galactic Center, and in the visual and infrared for X-ray binaries in the Magellanic Clouds. A study of hard X-ray sources near the Galactic Center is required to adress the puzzling observation that most high-mass X-ray binaries are relatively close to the Sun, even though the X-ray observations are suffiently sensitive to detect similar sources throughout our Galaxy (Verbunt 1988).

4.2 Mass of the Neutron Star or Black Hole

Radial velocities of an O or B star companion to a neutron stars are on the order of tens of km/s in a close orbit, and those for a massive companion to a black hole and for a low-mass companion to a neutron star of the order of a hundred km/s. In high-mass binaries, a good signal-to-noise is required to determine the presence of blends and correct for them (Penrod and Vogt 1985). The number of high-mass X-ray binaries in which the velocity of the O or B star can be measured is limited, as most high-mass systems are wide. In low-mass X-ray binaries, the donor is only detectable in a few wider systems, with evolved donors, and in transients in their quiescent stage, when the X-ray source is off. Surprisingly, many of the transients turn out to contain black holes. The velocity of the compact star is derived from the disk emission lines, and as in cataclysmic variables this method is suspect.

The Very Large Telescope will help enormously, by providing the much better signal to noise required for the correct separation of the donor absorption lines from disk emission, and in allowing better data to be obtained for the often very faint counterparts of transients in quiescence. It will also allow more reliable data to be obtained from the X-ray binaries in the Magellanic Clouds. Mass determinations for the neutron stars in low-mass X-ray binaries are very interesting as theory predicts that such neutron stars may have accreted a sizable fraction of their mass – unlike their counterparts in massive binaries (Verbunt 1993) – so that measuring their mass provides stringent constraints simultaneously on evolutionary scenarios, and on the maximum possible mass for a neutron star and, via this, on the equation of state.

4.3 Disk Structure

From the X-ray observations it appears that the disks in low-mass X-ray binaries are much thicker than those in cataclysmic variables (Mason 1986). Studying this in some detail is important for understanding other, less directly accessible systems with accretion disks: for example, the interpretation of observations of active galactic nuclei hinges on assumptions about the structure of accretion disks in them. In many low-mass X-ray binaries, the X-ray luminosity undergoes sudden bursts, lasting some ten seconds, during which the X-ray flux rises rapidly by a factor up to a hundred, and then declines again. As most of the optical flux is due to reprocessing of X-rays in the disk, the variations of the optical flux concurrent with the X-ray burst allow us to map the form of the disk by using the different light-travel times from the neutron star to Earth via various places in the disk. A first study has been done for 4U1735-44, and demonstrated that some of the reprocessing must take place in the companion star (Bunk 1993). As this analysis is essentially a deconvolution process, high signal to noise is required, such as may be provided by a Very Large Telescope with a multicolour fast photometer. Apart from constraining the disk size and form, it will also constrain the size and form of the donor star, and thus (via the theory of Roche-lobes) the mass of the donor. Experience shows that the required simultaneity between groundbased optical and satellite X-ray observations is not easily achieved.

5 Recycled Pulsars with White Dwarf Companions

During the last decade a new class of radio pulsars has been discovered, the recycled radio pulsars, thought to have descended from X-ray binaries, and characterized by relatively short pulse periods and weak magnetic fields of the neutron star. Many of these recycled pulsars are accompanied by a low-mass white dwarf in a circular orbit. The temperature of the white dwarf provides a handle on its age, and thus on the age of the necessarily older neutron star. The first result of this method is to show that the magnetic field of the neutron star cannot decay on a time scale as short as previously thought (Kulkarni 1986); and it is now accepted that the magnetic field of neutron stars may not decay at all (Verbunt 1994). Temperature determinations of a larger sample of white dwarf companions will allow us to compare the age distribution of the white dwarfs with the distribution of the characteristic ages $\tau_c \equiv 0.5P/\dot{P}$ of their neutron star companions. Abundance analysis of their spectrum will tell us whether the white dwarfs are formed from the cores of originally intermediately massive stars (around 5 $M_\odot$, say) or of low-mass stars, and at what evolutionary stage. With a Very Large Telescope it may even be possible to determine the surface gravity of the white dwarf, and thus via atmosphere models, its mass. (Distances can be determined from the radio observations.)

A radio pulsar, as a rapidly rotating magnet, loses energy in Poynting flux and in the form of highly-energetic particles, at a rate set by the loss of rotational energy: $\dot{E} = I\Omega\dot{\Omega}$, where I and Ω are the moment of inertia and the angular rotation velocity of the neutron star. The directly observed radio luminosity of a radio pulsar corresponds only to a minute fraction (on the order of 10^{-7}) of the overall energy loss. A radio pulsar can irradiate a close companion, and thus strongly affect its structure, to the extent of causing serious mass loss and damage to it. The observation of such a companion and of the ionization of the mass blown from it provides direct information on the form (radiation or particles) of the bulk energy loss of a radio pulsar (Kulkarni et al. 1992).

6 Summary

A larger surface area and new detectors on the Very Large Telescope will enable us to do more of the same, and intrinsically new work. In the first category fall the determinations of more orbital periods and component masses in cataclysmic variables, low-mass X-ray binaries and high-mass X-ray binaries. An increased number of reliable distances and component masses for these systems is badly needed to put the evolutionary models that relate these binaries to their progenitors, on a quantitative footing. Masses of neutron stars in low-mass X-ray binaries or of the recycled pulsar will provide the best tests for the equation of state at nuclear densities, by determining the maximum allowed mass of a neutron star. In the second category fall the measurements of optical flux from the neutron star surface, in an ideal case even an optical spectrum; the determination via eclipse and velocity mapping of details on accretion disks in cataclysmic variables, possibly including the determination of the magnetic field strength of small structures, and the first maps of disks in low-mass X-ray binaries; and the determination of the limb darkening of a white dwarf by measuring its eclipse. The instrumental requirements are set by the time resolution that one needs (see Figure 2) and by the magnitudes of the target objects.

The absolute magnitude of the O or B star in a high-mass X-ray binary is $M_V \sim -5$. Thus, near the Galactic Center and ignoring absorption its apparent magnitude is about 10, in the Magellanic Clouds about 13.5. To determine the orbital period in these systems, infrared spectroscopy at high-resolution (10 km/s) at these magnitudes must be done at good signal to noise with exposure times on the order of hours. Low-mass X-ray binaries now studied have apparent magnitudes in the range 15-20; in the Magellanic Clouds the magnitudes of such systems and of the optical counterparts of the supersoft X-ray sources are around 20-23. Period determinations of these systems are probably best done

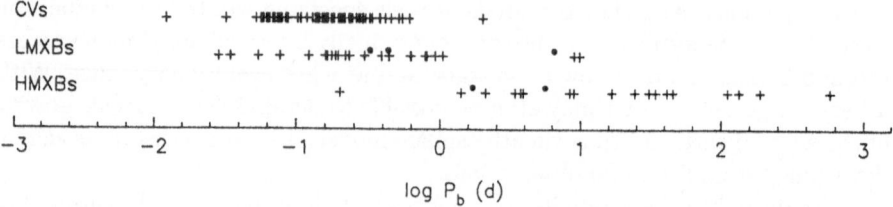

Fig. 2. Orbital period distributions of cataclysmic variables (CVs), low-mass X-ray binaries (LMXBs) and high-mass X-ray binaries (HMXBs). Each symbol indicates a system, the black-hole binaries are indicated •. The time resolution required to resolve the orbit is about $0.05P_b$, i.e. an hour for high-mass X-ray binaries, and five minutes for cataclysmic variables and low-mass X-ray binaries. The time resolution required to resolve structures on the order of $0.01R_\odot$ in an accretion disk, or to resolve the eclipse of the white dwarf, is $0.05 \times 0.01R_\odot/(\pi a)$ where a is the semi-major axis of the orbit; this is about 1/300 the times required to solve the orbit, for $a = R_\odot$.

with spectroscopy, and require good spectra in the optical (for the Magellanic Clouds) or infrared (near the Galactic Center) obtained at low resolution in about 15 minutes at magnitudes 20-23. Determining the disk structure is feasible via eclipse mapping and via the observation of a reprocessed X-ray burst, with a multicolour photometer that has a time resolution of a second or less at magnitudes down to ∼ 20 (see Barwig 1995).

Studying detail in cataclysmic variable disks requires low resolution spectroscopy at time resolution of the order of 1 s for systems with apparent magnitudes around 10. Finally, a 21-26 magnitude neutron star may be identified unambiguously by detecting its pulse period with a photometer with a time resolution going down to 0.01 s. At the bright end of this range a low-resolution spectrum could allow determination of the gravitational redshift.

References

Baade, W., Zwicky, F. (1934): Phys. Rev. 45, 138.

Barwig, H. (1995): these proceedings.

Bath, G.T., Pringle, J.E. (1985): in *Interacting Binary Stars*, eds. J.E. Pringle and R.A. Wade, p.129 (C.U.P., Cambridge).

Bertsch, D.L., Brazier, K.T.S., Fichtel, C.E., Hartman, R.C., Hunter, S.D., Kanbach, G., Kniffen, D.A.,Kwok, P.W., Lin, Y.C., Mattox, J.R., Mayer-Hasselwander, H.A., v. Montigy, C., Michelson, P.F., Nolan, P.L., Pinkau, K., Rothermel, H., Schneid, E.J., Sommer, M., Sreekumar, P., Thompson, D.J. (1992): Nature 357, 306.

Bhattacharya, D., Wijers, R.A.M.J., Hartman, J.W., Verbunt, F. (1992): Astron. Astroph. 254, 198.

Bunk, W.H. (1993): PhD Thesis, Ludwig Maximilian Universität München.

Hasinger, G. (1994): in *The evolution of X-ray binaries*, eds. S.S. Holt and C.S. Day, p.611 (American Institute of Physics, New York)

Kudritzki, R. (1995): these proceedings

Kulkarni, S.R. (1986): Astroph. J. (Letters) 306, 85.

Kulkarni, S.R., Phinney, E.S., Evans, C.R., Hasinger, G. (1992): Nature 359, 300.

Marsh, T.R., Horne, K., Schlegel, E.M., Honeycutt, K., Kaitschuk, R.H. (1990): Astroph. J. 364, 637.

Mason, K.O. (1986): in *Physics of accretion onto compact objects*, eds. K.O. Mason, M.G. Watson and N.E. White, p. 29 (Springer Verlag, Berlin).

Miller, B.W., Margon, B., Burton, M.G. (1993): Astron.J. 106, 28.

Penrod, G.D., Vogt, S.S. 1985, Astroph. J. 229, 653.

Rutten, R.G.M., Kuulkers, E., Vogt, N., Van Paradijs, J. (1992): Astron. Astroph. 265, 159.

Rutten, R.G.M., Dhillon, V.S., Horne, K., Kuulkers, E., Van Paradijs, J. (1993): Nature 362, 518.

Shapiro, S.L., Teukoslsky, S.A. (1983): *Black Holes, White Dwarfs and Neutron Stars, The Physics of Compact Objects* (Wiley Interscience, New York).

Tauris, T.M., Nicastro, L., Johnston, S., Manchester, R.N., Bailes, M., Lyne, A.G., Glowacki, J., Lorimer, D.A., D'Amico, N. (1994): Astroph. J. (Letters): 428, L53.

van den Heuvel, E.P.J. Bhattacharya, D., Nomoto, K., Rappaport, S.A. (1992): Astron. Astroph. 262, 97.

van Paradijs, J., McClintock, J.E. (1995): in: *X-ray binaries*, eds. W.H.G. Lewin, J. van Paradijs and E.P.J. van den Heuvel, in press (C.U.P.)

Verbunt, F. (1988): in: *The physics of neutron stars and black holes*, ed. Y. Tanaka, p.159 (Universal Academy Press, Tokyo).

Verbunt, F. (1993): Ann. Rev. Astron. Astroph. 31, 93.

Verbunt, F. (1994): in *The evolution of X-ray binaries*, eds. S.S. Holt and C.S. Day, p.351 (American Institute of Physics, New York)

Wade, R.A., Horne, K. (1988): Astroph. J. 324, 441.

Webbink, R.F. (1979): in: White dwarfs and variable degenerate stars, IAU Coll. 53, eds. H.M. Van Riper and V. Weidemann, p. 426 (University of Rochester).

VLT High-Speed Spectro-Photometry: A Powerful Tool for Exploring Compact Stellar Objects and Related Phenomena

H. Barwig, K.-H. Mantel

Universitaets-Sternwarte Muenchen, FRG

Abstract. In the parameter domain of short timescales, we expect exciting answers to various astrophysical problems related to the understanding of compact objects and phenomena in their environments. Several of these investigations will not only benefit from the light collecting power of the VLT, but are only possible using telescopes of the 8-m class. Examples are given for investigation of close interacting binary systems, a field of astrophysics where application of VLT high-speed spectro-photometry will increase our knowledge dramatically. Proposed observing programs include: eclipse mapping of accretion disks with different spectral resolutions; Doppler tomography of accretion flows; exploration of flickering and search for associated light echoes. In order to measure such short timescale phenomena, the high-speed multi-object spectro-photometer FRISPI has been proposed for the VLT.

1 Introduction

Among the many instruments proposed for the VLT, observational facilities for adequately exploring compact stellar objects and related phenomena like white dwarfs, neutron stars and black holes are missing. These targets predominate the observational parameter domain of short time scales due to their small geometrical extent. Furthermore, in their environment very high energy densities and enormous magnetic fields are present. Maximum radiation is released at short wavelengths (UV - X-rays), which makes these objects appear relatively faint in the optical. Alhough of fairly low absolute visual luminosity, these objects play an important role in several fields of astrophysics.

Adequate observations require the light collecting power of 8-m class telescopes as well as high-speed spectro-photometric instruments with integration times in the range between milliseconds and several seconds. Relevant observing programs (time resolution in brackets) include: investigations of cataclysmic variables (0.1s - 1 min); search for millisecond-pulsars in globular clusters (0.1-10 ms; phase-resolved spectro-photometry of pulsars (0.1-10 ms); magnetic field tomography of eclipsing AM Her stars (0.1s); light-echo mapping of X-ray binaries (0.1s); as well as search for optical counterparts of gamma-ray bursters (0.1-10s). In the following, examples are given for high-speed photometric observations of accretion phenomena in close binaries.

2   Eclipse Mapping of Accretion Disks in Close Binaries

2.1   Broad Band Observations

Among compact targets, close binary systems e.g. cataclysmic variables (CV) are of enormous scientific interest. CVs are interacting binaries with a white dwarf primary and a red dwarf secondary which fills its Roche lobe. Orbital periods range between 80 minutes and several hours. A gas stream flowing through the inner Lagrangian point transfers matter from the secondary to an accretion disk around the primary. The transfer of angular momentum outwards allows matter to move inward towards the white dwarf, but the physical mechanism of this accretion-disk viscosity is not yet understood. An important group among the CV systems are the dwarf novae (DN) which show quasi-periodic outbursts. The process of mass accretion in a disk has turned out to be of outstanding significance for many astrophysical processes. It is likewise found in disks of protostars and AGNs and is also relevant, e.g., for the evolution of galaxies and planetary systems as well. CVs provide ideal laboratories for studies of mass accretion mechanisms and for testing related theories, since the stellar components (including the hot spot) in these laboratories are fairly well known through observation of their spectra and their photometric signatures. Under these "calibrated" conditions, the disk itself can be studied. Most convenient for these investigations are DNs with high orbital inclination where the eclipse by the secondary provides scans across the accretion disk during quiescence and outburst. Time and spectrally resolved observations of eclipse phenomena using large telescopes like the VLT, combined with high speed spectro-photometric instruments, provide the input data for the famous eclipse-mapping method first invented by Horne (1985).

As an example, first results (Bobinger 1994) from the application of the eclipse-mapping technique to photometric data obtained in the V-band of a double eclipsing DN are presented in Fig. 1. It shows two-dimensional disk maps reconstructed from one-dimensional eclipse light curves which had been observed with a 2 m telescope during 4 different nights. Spatial brightness structures of the accretion disk are clearly visible at the micro-arcsecond level. Simulations of eclipses in CV systems ,assuming a steady state disk with two artifically shaped hot spots, clearly prove that the quality of the eclipse-mapping reconstruction strongly depends on the S/N of the input data. Increasing the S/N by averaging over several orbital periods has the disadvantage of smearing out spatial structures. In order to get information on the actual processes of accretion and to achieve spatial resolution of sub-microarcsecond structures, single eclipse events have to be analysed, a task which cannot be performed without the light-collecting power of an 8 m telescope.

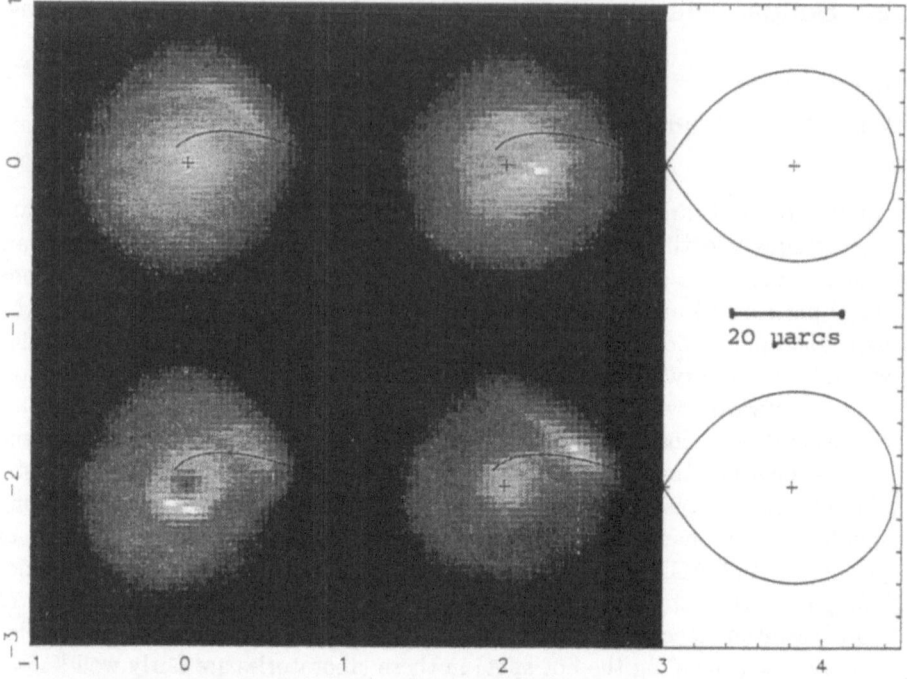

Fig. 1. Eclipse maps of an accretion disk reconstructed from four different eclipse light curves of a DN. On the right side, the Roche lobe filling secondary is indicated to show the relative system dimensions.

2.2 Narrow-Band Observations

The need for large telescope apertures becomes even more stringent for applications involving the spectrally-resolved eclipse-mapping technique, a straightforward extension of the original maximum-entropy eclipse-mapping method. From many narrow-band images, low resolution spectra from isolated parts of the accretion disk can be synthesized (Rutten et al. 1994). From comparison with theoretical disk spectra, the physical conditions in the accretion disk are derived. Also the spectra of the hot spot and the secondary, respectively, can be extracted. This technique requires accurate spectro-photometric data of high S/N and sufficient phase resolution to slice up the accretion disk. Our proposed spectro-photometer FRISPI mounted at the VLT provides s uitable instrumentation for these investigations (for details see Section 5).

3 Doppler Tomography and Spectral Eclipse Mapping

Doppler tomography, developed by Marsh & Horne (1988), provides another powerful tool for exploring accretion phenomena. It allows one to map the e-mission regions associated with the accretion flows in CVs. For this purpose the two-dimensional data consisting of the velocity profiles as a function of binary phase are used. A combination of the Doppler tomography method with the spec-tral eclipse-mapping technique allows the velocity fields in accretion disks to be studied and hence the processes relevant for understanding the outburst mecha-nism. Phase-resolved high-resolution spectro-photometry of high S/N, covering single orbital revolutions, cannot be investigated successfully with anything less than an 8-m class telescope.

4 Investigation of Flickering

Flickering seems to be quite a common signature of the accretion process and therefore could probably be related to the viscosity mechanism. This associa-tion also seems most important for studying the pre-outburst phase of CVs. The flickering timescales range from a few seconds, possibly indicative of Keplerian motion near the white dwarf surface, to a fraction of the orbital period. Ob-servations of the phase-dependent disappearance and reappearance of flickering activity observed in eclipsing CV systems indicate that this phenomenon is not exclusively confined to the hot spot region. In order to localize the site of ori-gin, the eclipse-mapping technique can be applied to data which show the rms of the residuals obtained by subtracting a smoothed light curve from the original e-clipse light curves averaged over many orbital cycles (Horne and Stiening, 1985). Knowing the site of flickering, high-speed spectro-photometry of the individual flickering events could provide information on the flickering process. Further-more, one may search for light echoes using cross-correlation techniques, a first step towards a possible echo mapping of CV systems. The spectrally resolved photometry of single flickering events with integration times in the sub-second range can only be achieved using 8-m class telescopes.

5 Required Instrumentation

For observational programs in high-speed astrophysics like those mentioned above, the multi-object spectro-photometer **FRISPI** (*Fast Recording Imager and Spectro-Photometric Instrument*) has been proposed for the VLT (Fig. 2). A prototype (MEKASPEK) has already been developed at the Universitäts-Sternwarte München (Mantel et al. 1989, 1993). It is characterized by fiber-input channels that can be directed by remote control to different targets in the focal plane of the telescope. Their light signals pass a spectrograph equipped with a two-dimensional photon-counting (PCA or MAMA) detector providing a maximum time resolution in the sub-millisecond range. For high-speed field

FRISPI
the Fast Recording Imager and Spectro–Photometric Instrument

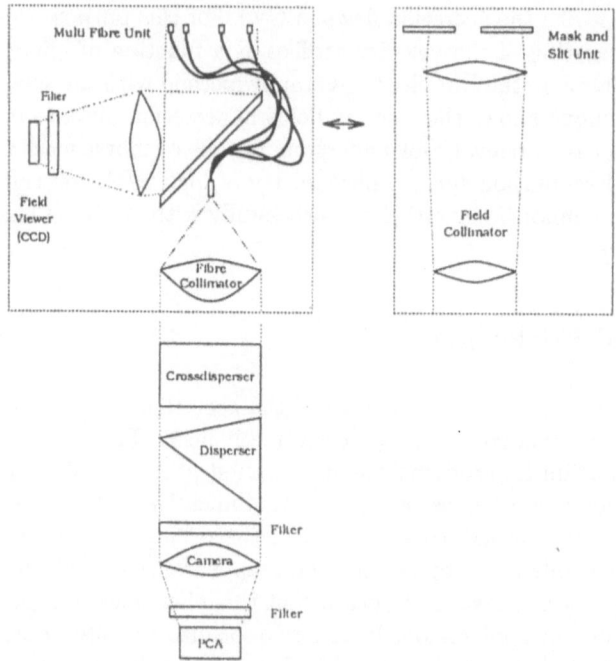

Fig. 2. Layout of the high-speed spectro-photometer proposed for the VLT by H. Barwig, S. Kiesewetter and K.-H. Mantel, USM

photometry, the fiber unit can be exchanged by a direct-imaging device allowing a field of view of some arc-minutes to be monitored. (For details see: Moorwood 1994.)

References

Bobinger, A. (1994), private communication

Horne, K. (1985), Mon. Not. R. Soc. **213** 129

Horne, K., Stiening, R.F. (1985), Mon. Not. R. Soc. **216** 933

Mantel, K.-H., Barwig, H., Kiesewetter, S. (1993) in Proc. of IAU Coll. **136**, p.172

Mantel, K.-H., Barwig, H. (1994), in NATO ASI Series, eds. C. Sterken and M. de Groot, p.329

Marsh, T.R. and Horne, K. (1988), Mon. Not. R. Soc. **235** 269

Moorwood, A.F.M. (1994): *Instruments for the ESO VLT*, pp. 46-49

Rutten, R.G.M., Dhillon, V.S., Horne, K., Kuulkers, E. (1994), A&A **283** 441

Observational Astrophysics on Milli-, Micro-, and Nanosecond Timescales

D.Dravins, L.Lindegren, & E.Mezey

Lund Observatory, Box 43, S-221 00 Lund, Sweden

Abstract. Instrumentation and observing methods are being developed for a program in optical high-speed astrophysics, an exploratory project entering the domains of milli-, micro-, and nanosecond variability. Current studies include accretion flows around compact objects, stellar scintillation, and astronomical quantum optics. To study such rapid phenomena is not possible everywhere in the spectrum (e.g., X-ray studies are constrained by the photon count rates feasible with current spacecraft). The ground-based optical is a promising region, for which we have constructed a dedicated instrument, *QVANTOS* ('Quantum-Optical Spectrometer'). Its first version was used on La Palma to study atmospheric scintillation on timescales between 100 milli- and 100 nanoseconds. For very high time resolution, light curves are of little use, and statistical functions of variability have to be measured. The noise in such functions decreases dramatically with increased light collecting power, making very large telescopes much more sensitive for the study of rapid variability than 2-4m class ones.

1. The Challenge of High-Speed Astrophysics

The present project is part of an exploratory program in high-speed astrophysics, entering the previously unexplored domains of milli-, micro-, and nanosecond variability. The goal is to explore the possible very rapid variability of astrophysical objects, e.g. accretion systems around compact objects in the Galaxy. The environments of such objects are promising laboratories to search for very rapid phenomena: the geometrical extent can be very small, the energy density very high, the magnetic fields enormous, and a series of phenomena, ranging from magneto-hydrodynamic turbulence to stimulated synchrotron radiation might well occur. Some processes may occur over scales of only kilometers or less, and there is no immediate hope for their spatial imaging. Insights can instead be gained through studies of their small-scale instabilities, such as hydrodynamic oscillations or magneto-hydrodynamic flares. These events could be observable in the time domain, on timescales of seconds, milli-, or even microseconds. Phenomena which might be found on these subsecond scales include:

* Atmospheric intensity scintillation of stars on the shortest timescales

* Plasma instabilities and the fine structure in accretion flows onto white dwarfs and neutron stars

* Small-scale [magneto-]hydrodynamic instabilities in accretion disks

* Radial oscillations in white dwarfs ($\simeq$ 100-1000 ms), and non-radial ones in neutron stars ($\lesssim$ 100 μs)

* Optical emission from millisecond pulsars ($\simeq$ 1-10 ms)

* Fine structure in the emission ('photon showers') from various objects

* Photo-hydrodynamic turbulence ('photon bubbles') in very luminous stars

* Stimulated emission, e.g. synchrotron radiation, from magnetic objects ('cosmic free-electron laser')

* Non-equilibrium photon statistics (i.e. non-Bose-Einstein distributions)

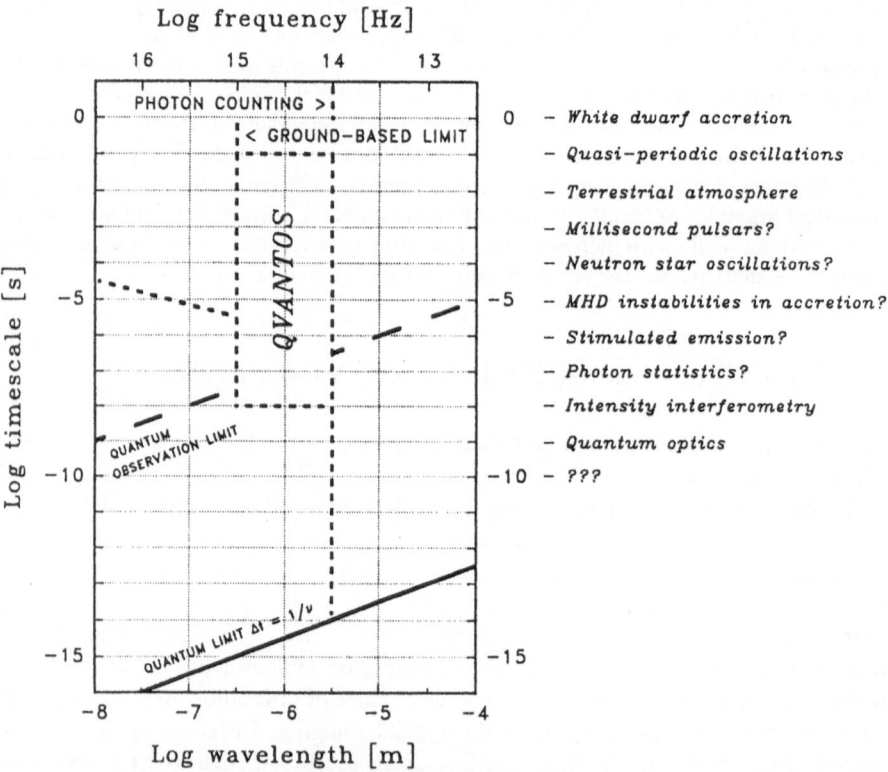

Fig. 1. Parameter domains of astrophysics, subdividing electromagnetic radiation with respect to timescale of study. This diagram shows some regions of shorter timescales around the optical, with expected astrophysical processes marked. The domain for the QVANTOS instrument is indicated.

2. Parameter Domains of Astrophysics

The science of astronomy can be subdivided into parameter domains with respect to electromagnetic wavelength, and the timescale of study. Classical astronomy, for example, was largely confined to wavelengths accessible from the ground, and timescales between about 0.1 seconds and 10 years. The aim of the present project is to push the time resolution to the highest possible values, and to ultimately reach the region where quantum statistical effects of radiation may be detectable. Increasing the temporal resolution to microseconds, one should encounter successively more rapid events, on timescales such as those theoretically expected for magnetic instabilities in accretion systems, or for non-radial oscillations is neutron stars. However, we are not aware of any macroscopic processes in astrophysical sources that are expected to be observable in the nanosecond domain. Such resolutions, however, lead into the microscopic realm of quantum optics, and the quantum-mechanical statistics of photon counts.

2.1. Advantages of Observing in the Optical

Rapid astrophysical events are generally expected near compact sources such as white dwarfs, neutron stars or presumed black holes. There are quasi-periodic oscillations, flashes, pulsars, and other phenomena. For best detection and visibility, X-rays could appear to be most attractive, since they often originate in high-temperature regions quite close to the compact object. Nevertheless, the optical region may in practice be the best for the detailed study of the most rapid phenomena. The reason is that the number of photons that can be detected per second (and especially per millisecond!) is often much greater from the optical parts of the sources (as observed with large telescopes), than that from their X-ray parts, observed with current space instruments. Foreseeable X-ray satellites will not be able to collect more than perhaps a thousand X-ray photons per second, even from quite bright objects. While this will be adequate to explore many phenomena on subsecond timescales (quasi-periodic oscillations, pulsars, etc.), it is probably not adequate in searches for very rapid fluctuations.

For comparison, optical light curves of some accretion sources showing periodicity on the scale of seconds can be quite prominent, when recorded with telescopes in the 1.5 m class (e.g. Larsson 1985; Imamura et al. 1990). Using an 8-meter telescope, and a detector efficiency improved by a factor 3, implies some 100 times more photons collected. Detailed light curves could then be seen if their periodicity was merely tens of milliseconds (with count rates on the order of a million optical photons per second). Another advantage of the optical is the feasibility to ultimately detect quantum phenomena. Some quantum effects are easier to observe in radio (e.g. the bunching of photons appears as 'wave noise'), but a limit is set by the difficulties of photon counting at wavelengths beyond the infrared. While this does not hinder studies of rapid variability as such, it does preclude some studies of quantum effects in radiation. The most promising domain thus seems to be the ground-based optical, for which we have constructed a dedicated instrument: QVANTOS ('Quantum-Optical Spectrometer').

3. QVANTOS: An Optimized Observing Instrument

A number of criteria can be defined for optimizing an instrument for high-speed astrophysics:

*   *Data glut:* 1 ms resolution means 3.6 million points an hour, and 100 Mb of data in three nights. However, 1 μs gives 100 Gb, and a hypothetical plot of the light curve (at 300 *dpi* laser printer resolution) would emerge at a real-time speed of $\simeq$ 100 meters per second! Thus, a need for real-time data analysis, and a data reduction to manageable statistical functions only.

*   *Faint sources:* To study variability also on timescales shorter than typical intervals between successive photons, a statistical analysis of their arrival times is required to test for deviations from randomness.

*   *The terrestrial atmosphere* causes rapid fluctuations of the source intensity. The study of astrophysical fluctuations requires a correspondingly accurate measurement and correction for atmospheric effects. This may require the simultaneous observation of a calibration star, and perhaps other measures (such as a possibility for apodizing the telescope entrance aperture).

*   *Time resolution:* The highest resolution that is meaningful is set by quantum-optical properties of light, e.g. the bunching of photons in time. Such properties are fully developed on times equal to the inverse frequency bandwidth ($\simeq 10^{-14}$ s for white light), but may be detectable also on the much longer (nanosecond) timescales.

*   *Efficient detectors:* Rapid photon counting has in the past been feasible only with photocathode detectors As discussed further below, the signal-to-noise ratios in measured statistical functions improve very rapidly with improved detector efficiency. Therefore, silicon detectors (avalanche photodiodes) are currently being tested for the new QVANTOS Mark II version.

The Mark I version of QVANTOS was used on La Palma to test instrumentation and observation methods. At the same time, it was used to explore what challenges in understanding the terrestrial atmosphere must be met, before astrophysical variability on such timescales can be convincingly demonstrated to exist. For more details on these instrumentation efforts, see Dravins et al. (1994).

Previous experiments by other groups in observational high-speed astrophysics illustrate that meeting all requirements is non-trivial. The pioneering *MANIA* experiment at the Russian 6-meter telescope (e.g. Beskin et al. 1982), recently used also in Argentina (Beskin et al. 1994), has limitations in the maximum photon count rates that can be processed. Recent global observational campaigns, using a network of large telescopes in searches for stellar oscillations, were limited by atmospheric intensity scintillation. Instruments in space avoid the terrestrial atmosphere: the High Speed Photometer on the Hubble Space Telescope was a major effort (Bless 1982), but had limits on the quantities of data that could be stored onboard. Our instrument also has limits in the sense that the full data stream is not saved, only certain statistical functions.

4. The Role of Very Large Telescopes

For very high time resolution, classical light curves are of little use. Measurements will thus have to be of autocorrelations, power spectra, or other statistical properties of the arriving photon stream. All such statistical functions depend on a power of the average intensity that is higher than 1. For example, an autocorrelation (obtained by multiplying the intensity signal by itself, shifted by a time lag) is proportional to the square of the intensity. Due to this dependence, very large telescopes are *much more sensitive* for the detection of rapid variability than ordinary-sized ones. A search for e.g. magneto-hydrodynamic instabilities in accretion disks around supposed black holes, using autocorrelation techniques, will benefit a factor $(8.2/3.6)^4 \simeq 27$ if using one 8.2 meter telescope instead of a 3.6 m one, rather than the ratio $(8.2/3.6)^2 \simeq 5$ that is valid for the intensity. For other measures, e.g. those of the fourth-order moments of the photon distribution, the signal will increase as the fourth power of the intensity, making a full Very Large Telescope with four 8-meter units some 185,000 times more sensitive than a 3.6 m one (implying that one night of observing on the full VLT gives the same signal as 500 years of integration with a 3.6 m!! – Figure 2).

	Telescope diameter	Intensity $< I >$	Second-order intensity correlation $< I^2 >$	Fourth-order photon statistics $< I^4 >$
	3.6 m	1	1	1
	8.2 m	5	27	720
	4 * 8.2 m	21	430	185,000

Fig. 2. Comparisons between the observed signal of source intensity (I), its square and fourth powers, for telescopes of different size. The signal for classical quantities increases with the intensity I; the signal in power spectra and similar functions of relevance for variability searches, as I^2; and that of four-photon correlations as I^4, as relevant for quantum statistics studies. The advent of very large telescopes greatly increases the potential for high-speed astrophysics.

Acknowledgements

The high-speed astrophysics project is supported by the Swedish Natural Science Research Council and the Swedish Council for Planning and Coordination of Research.

References

Beskin, G.M., Neizvestnyi, S.I., Pimonov, A.A., Plakhotnichenko, V.L., Shvartsman, V.F.: 1982, in C.M.Humphries, ed. *Instrumentation for Astronomy with Large Optical Telescopes*, IAU coll. **67**, Reidel, p. 181

Beskin, G., Neizvestny, S., Plokhotnichenko, V., Zhuravkov, A., Benvenuto, O.G., Feinstein, C., Méndez, M.: 1994, A&A **289**, 141

Bless, R.C.: 1982, in D.N.B.Hall, ed. *The Space Telescope Observatory*, NASA **CP-2244**, p. 106

Dravins, D., Hagerbo, H.O., Lindegren, L., Mezey, E., Nilsson, B.: 1994, in D.L.Crawford & E.R.Craine, eds. *Instrumentation in Astronomy VIII*, SPIE proc. vol. **2198**, p. 289

Imamura, J.N., Kristian, J., Middleditch, J., Steiman-Cameron, T.Y.: 1990, ApJ **365**, 31

Larsson, S.: 1985, A&A **145**, L1

Problems and Prospects in White Dwarf Stars

Detlev Koester

Institut für Astronomie und Astrophysik, Universität Kiel,
D-24098 Kiel, Germany

Abstract. We present a number of open problems related to white dwarf research, where we expect significant progress to be made once the VLT and its instrumentation become available. These questions are intrinsic to white dwarf studies: chemical composition of surface layers and the determining physical processes; gravitational redshift and mass radius-relation; rotation; and magnetic fields. White dwarfs can, however, also be helpful to study questions of broader interest; as examples we discuss the critical mass for white dwarf formation vs. supernovae, or magnetic field decay in neutron stars.

1 Introduction

White dwarfs are intrinsically faint stars, with only few objects brighter than V = 12, and some of the brightest, e.g. Sirius B or Procyon B, posing additional observational problems due to the close vicinity of very bright stars. For many interesting subclasses, the brightest members are much fainter than that. Although the theoretical basis for the understanding of the interior structure and mass-radius relation has been available for 60 years, progress on the observational side, with detection of highly unusual surface abundances, extreme magnetic fields, etc. has only become possible with the advent of large telescopes and modern detectors during the last 15 years. On the other hand, the spectral lines in white dwarfs are usually very broad due to the high atmospheric pressures. High spectral resolution is therefore not always necessary, and some problems can be attacked using moderate resolutions of 5 to 20 Å. Exceptions do exist, however, as will be shown below.

2 Surface Abundance Patterns

White dwarf spectral types can be generally divided into two major groups, hydrogen-rich and helium-rich atmospheres. With few exceptions one element is extremely dominant and all others constitute only traces, usually with abundances much below the solar values. The standard explanation for these facts is gravitational separation (Schatzman 1949): in the high gravitational fields the heavier elements sink down and the lightest ones present (H or He) remain at the top. A recent review on the evolution of surface abundances is Koester (1987).

The diffusion velocities are so large that, at the low temperature end of e.g. the helium-rich sequence, no metals are expected, and indeed, the spectral class

DC shows a featureless optical spectrum (He-rich atmosphere). However, in the same temperature range the DZ white dwarfs are found, defined by the presence of traces of metals: typically Ca in the optical and Mg, Fe, Si in the UV. The current explanation for these abundances is in terms of an accretion/diffusion scenario: matter is accreted out of the interstellar matter, and then diffusion on a timescale of about 10^6 years leads to an exponential decline of heavy element surface abundances. Hydrogen, however, the lightest element, is always present in interstellar matter, but remains at the top of the atmosphere, and should therefore be detectable in these objects! The detection and the determination of accurate hydrogen to heavy element abundance ratios has remained elusive except for very few cases, which tend to put the accretion/diffusion scenario into difficulties, because the hydrogen abundance is smaller than expected. As an example, indicating the observational difficulties, we mention the case of L745-46A, where Koester (1987) reported the detection of a 100 mÅ Hα line. L745-46A is the brightest member of the DZ class (V = 13) and was observed at 0.2 Å resolution with CASPEC. Most other members of this class are more than two magnitudes fainter, and the resolution and signal-to-noise necessary to detect such faint lines are beyond the capabilities of current ESO instruments.

3 Hα Line Cores in Faint Objects

Another class of programs with similar requirements is the observation of Hα in hydrogen-rich (DA) white dwarfs. In the effective temperature range of 8000 to 30000 K, this line has wings extending to more than 150 Å from the line center, and at medium resolution a very broad and flat line core, making precise wavelength determinations almost impossible. At sufficiently high resolution (0.1 - 0.2 Å) and S/N ($\approx$ 100), however, a very sharp core becomes visible, which is explained by non-LTE effects (Koester and Herrero 1988). The highly precise wavelength determination with this sharp core opens up a number of previously unexpected possibilities (e.g. Koester and Herrero 1988): determination of gravitational redshifts, if a binary or common proper motion companion is present to determine the true space velocity; rotation velocities from the broadening of the sharp core, and even sensitive upper limits on magnetic fields from the absence of Zeeman splitting. All these projects are currently limited to a few objects with magnitudes brighter than 14.5. Extending the limit to about 16 with VLT will give us a large sample of rotational velocities and allow a study of the implications for the transport of angular momentum during stellar evolution from the main sequence to a white dwarf.

4 Magnetic Fields in White Dwarfs

About 3% of white dwarfs show magnetic fields, which are discovered either through the Zeeman splitting of spectral lines, circular polarization measurements, or unusual spectral features that are not easily identified. Field strengths range between a few to 1000 MG (e.g. Schmidt 1989). Independent of the astronomical importance of understanding the origin of these fields and their connection with the progenitors of white dwarfs, these objects offer the chance to observe the behavior of atoms (mostly hydrogen) in fields that cannot be achieved in the terrestrial laboratory. These observations can be compared with theoretical computations that are extremely demanding on computing power and therefore became possible only very recently.

Most magnetic white dwarfs are very slow rotators, although there are a few exceptions with the shortest period currently being about 100 min. The change of viewing angle with time offers the most promising opportunity to determine the field strength and field configuration through a comparison with theoretical models for magnetic white dwarfs: the observation of time-resolved spectroscopy and polarimetry. To my knowledge ESO currently has no significant polarimetry capabilities. The current limit at the 5m Palomar telescope with an exposure time of one hour is about 16th magnitude.

We give one recent example, to compare these capabilities with the requirements of current research. A very exciting magnetic white dwarf (HE1211-1707) was detected recently in the Hamburg Quasar Survey. It is very hot, probably one of the hottest magnetic white dwarfs known. It also shows spectral variations on time scales of 20 min, which, if caused by rotation, would point to a rotation period of about one to two hours. Time resolved spectroscopy of this 17th magnitude object may marginally be possible with EFOSC at the 3.6m telescope. However, time resolved spectropolarimetry, with resolution of 2 - 5 Å, high S/N and polarization errors less than 1% will become only possible with FORS1 or a similar instrument at the VLT.

5 Stellar Evolution: Critical Mass for White Dwarf Formation

Our current understanding of the final phases of stellar evolution implies that low-mass stars end up as white dwarfs, while massive stars lead to a supernova with the possible creation of a neutron star or black hole. The question of the limiting mass between the two regimes (or perhaps range of masses, if additional parameters such as rotation play a role) is of great importance for an understanding of stellar evolution, the physics of mass loss, the mass budget of the Galaxy, and the evolution of stellar populations in the Galaxy.

Theoretical stellar evolution calculations using empirically calibrated mass loss formulae is one possible method of study (see e.g. Van der Veen 1989). A much more direct method, however, is the study of white dwarfs in young open clusters. An early attempt was made by Sweeney (1974); an important breakthrough was the study by Romanishin and Angel (1980). This work was followed up with similar work by Koester and Reimers; most of the earlier work is summarized in Reimers and Koester (1988), and a more recent paper is Koester and Reimers (1993).

The basis of this method is very simple: white dwarf candidates are identified as faint blue objects in the field of young Galactic clusters; spectroscopy is then used to identify them as white dwarfs, confirm cluster membership, and determine parameters (Teff, log g, mass, cooling age, etc.). The presence of white dwarfs in a cluster with a turnoff mass of 5 $M_\odot$ indicates, that stars of at least 5 $M_\odot$ on the main sequence still end as white dwarfs. A more refined analysis using cooling ages and evolutionary ages of progenitor stars can then be used in favorable cases to establish one point in the relation between initial (main sequence) and final (remnant) mass.

The currently best result for the critical mass is 8 $M_\odot$, with a possible range from 6 to 11 (Reimers and Koester 1988). In order to confirm this result and increase the accuracy, the next step in this program would be the observation of the rich clusters NGC6067, NGC6664, NGC3766, and M25, with turnoff masses between 4.7 and 6.0 $M_\odot$. The expected V magnitudes of white dwarfs in these clusters range from 21.3 to 23.7, leading to the following requirements: resolution 6 - 10 Å, S/N $\geq$ 10, and $V_{lim} \approx 23.5$. This will require the VLT and an instrument like FORS1.

6 Decay of Magnetic Fields in Neutron Stars

In this last example observations of white dwarfs are again used as a tool for the solution of an astrophysical problem. There has been a considerable debate over the last years on whether the magnetic fields in neutron stars decay through ohmic decay on time scales of a few million years or not. Theoretical calculations seem to indicate at most a very small and slow decay (Sang and Chanmugam 1987), whereas observational evidence from population statistics suggest a fairly rapid decay as suggested by Gunn and Ostriker (1970).

Several pulsars are members of binary systems, where the secondary is known or suspected to be a white dwarf. Both competing scenarios for the explanation of these systems — evolution of the neutron star from a massive progenitor through a supernova stage, or accretion-induced collapse of a white dwarf — predict that the present white dwarf member is younger than the neutron star. If it is possible to determine the cooling age of the white dwarf, this gives immediately a lower limit to the neutron star age, and, with the known magnetic field strength, an indication of the life time of the magnetic field. In the first such application, Kulkarni (1986) concluded for PSR0655+64 that the cooling age of the white dwarf was $5\,10^8$ years, whereas the field strength still is about 10^{10} G. More recent examples for this technique are Koester et al. (1992) and Danziger et al. (1993). In the system of PSR0820+02 (Koester et al. 1992) the white dwarf is at V = 22.8 and spectroscopy was impossible in this case. With the VLT, these systems and a few more are within reach of real quantitative spectral analysis.

7 Conclusions

There are a number of open questions and problems regarding white dwarfs directly, or involving white dwarfs as tools, that are likely to make significant progress with the use of the VLT. The instrumental requirements divide typically into two classes:

i. Low resolution (5 - 20 Å) spectroscopy of very faint (V = 23) objects with low to intermediate S/N ($\approx$ 10 - 20)

ii. 0.1 - 0.2 Å resolution of objects with V $\leq$ 17 with high S/N ($\approx$ 50 - 100)

The VLT instruments of the first generation that cover these requirements and will certainly be used most heavily in white dwarf research are FORS and UVES.

References

Gunn, J.E., Ostriker, J.P. (1970), Astrophys.J. 160, 979

Koester, D. (1987), IAU Coll. 95, p. 329

Koester, D., Herrero, A. (1988), Astrophys.J. 332, 910

Koester, D., Chanmugam, G., Reimers, D. (1992), Astrophys.J. 395, L107

Koester, D., Reimers, D. (1993), Astron. Astrophys. 275, 479

Kulkarni, S.R. (1986), Astrophys.J. 306, L85

Reimers, D., Koester, D. (1988), The Messenger 54, 47

Romanishin, W., Angel, J.R.P. (1980), Astrophys.J. 235, 992

Sang, Y., Chanmugam, G. (1987), Astrophys.J. 323, L61

Schatzman, E. (1949), Publ. Köbenhavns Obs. No. 149

Schmidt, G.D. (1989), IAU Coll. 114, p.305

Sweeney, M.A. (1974), Astron. Astrophys. 49, 375

Mass Determination of Very Low Mass Stars

A. Duquennoy[1], J.-M. Mariotti[2], M. Mayor[1], C. Perrier[3]

[1] Observatoire de Genève, CH–1290 Sauverny, Switzerland
[2] Observatoire de Paris, DESPA, F–92195 Meudon, France
[3] Observatoire de Grenoble - LA, BP 53X, F–38041 Grenoble, France

Abstract. We discuss the precision achieved with the present observing techniques for the mass determination of low and very low mass stars. We show the gain that will be achieved with high precision radial velocity measurements combined with diffraction–limited imaging with 8-m class telescopes and the VLTI.

1 Introduction: Scientific Objectives

Despite the fact that 90 % of the stars of our Galaxy are less massive than the Sun, our knowledge of low mass and very low mass (VLM) stars is much poorer than that of their more massive counterparts. This is of course due to the lower luminosity of the former, and to their lower surface temperature which shifts the wavelength of maximum emission towards near infrared wavelengths, a domain for which high-performance astronomical instrumentation has been available for only two decades. As a consequence, several questions involving VLM stars (we will use here the widely accepted definition $M < 0.3 M_\odot$) are still under debate. As examples, we can list the following three questions:

    a. What is the lower end of the mass spectrum produced by star formation processes?

Although the present day mass function seems well defined down to $0.5 M_\odot$, considerable discrepancies subsist for the lower range $M < 0.2 M_\odot$ (see, e.g., Kroupa et al., 1990). In particular, the mass spectrum and even the existence of sub-stellar objects (brown dwarfs) are still fully open questions.

    b. What is the mass–luminosity relation at the red end of the Main Sequence? The most comprehensive work on this subject (Henry and McCarthy, 1993) illustrates the recent advances in this domain provided by new observational techniques. However, it is still not possible to infer the influence of parameters like age or metallicity from the luminosity of a star close to the Hydrogen-burning limit.

    c. How are binaries formed, how many, down to which masses? The large collection of data provided, during the last decade, by radial velocity surveys with precisions down to a few hundreds of m/s, allows us to establish statistically significant laws on binary parameters such as the distributions of periods, eccentricities, mass ratios,... (Duquennoy and Mayor, 1991). This may constitute a fundamental tool for investigating such questions as:

        1 What is the predominant scenario of binary system formation: fragmentation, fission, capture, accretion,...?

2 Is there a dependence of the binary formation process with M_2?

3 Is the binary rate decreasing with M_1?

2 Stellar Mass Determination

2.1 Method

The most direct method of measuring stellar masses is to combine a set of radial velocity (RV) measurements with a set of angular separations and position angles measurements (ρ, θ) of a binary. Usually, the RV data are superior to determine the period P, epoch of periastron passage, eccentricity,..., while the (ρ, θ) data yield in addition the inclination of the orbit and its orientation on the sky, the former parameter being important for mass determination. However the most appropriate approach must be based on a combination of the overall set of data and the extraction of the orbital parameters by a general least squares fit (e.g., Morbey, 1975).

It should be noted that if the spectroscopic binary is an SB1 (i.e., for all phases, lines of the secondary are never resolved), an external estimation of the parallax of the system is required in order to link the absolute scale of the RV data with the angular scale of the (ρ, θ) data.

2.2 Observing Program

Our current observing program focuses on nearby Main Sequence binaries with no mass transfer history, to insure that the masses are unchanged since the formation process. Our targets are cooler than F7V in order to obtain the required RV precision and have an expected separation larger than or equal to the diffraction limit of a 4 m-class telescope at $2.2\,\mu m$, i.e., $a_{min} \geq 0.12$". Finally, they all exhibit spectroscopic $f(M)$ indicating a minimum companion mass below $0.3\,M_\odot$.

The RV data predominantly come from CORAVEL, an instrument of the 1 m Swiss telescope at Observatoire de Haute-Provence. These observations now span 15 years. The 'imagery' data have been obtained on various 4 m-class telescopes (ESO 3.6 m, KPNO 4 m, CFHT 3.6 m) with infrared diffraction-limited imaging techniques: speckle interferometry (1–D and 2–D) and Adaptive Optics.

2.3 Results

Our program covers about 60 nearby Red Dwarfs with $\delta < 30$ deg. Presently, about one fourth of the sources has been fully resolved at least for one observation, another fourth has been marginally resolved, the rest being still unresolved. Since our preliminary results (Mariotti et al., 1990; Duquennoy et al., 1992; Perrier et al., 1992), F. Beck (Obs. Grenoble) has adapted a code from A. Tokovinine (1992), allowing to combine successfully the RV and the (ρ, θ) data. Publication of the orbital data and masses for about ten binaries is in preparation.

2.4   Uncertainty on the Derived Masses

The intersection of accessible regions in a plane $(P, M_2/M_1)$ between spectro-scopic and imaging techniques remains small for nearby objects containing VLM companions and makes both techniques work at the limit of their present perfor-mances (Fig. 1). Also, precision on parallaxes is presently a problem, although there are good hopes that this will be solved shortly thanks to HIPPARCOS.

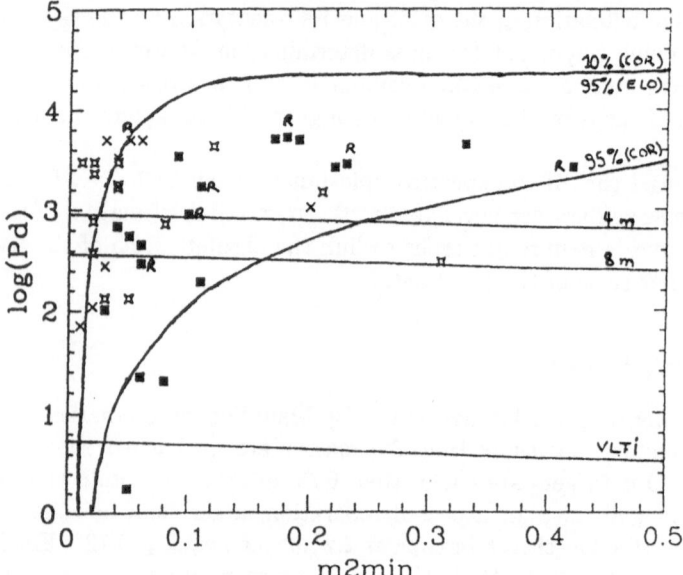

Fig. 1. Distribution of our ESO 3.6 m program sources in the $m_{2,min} - P$ (period in days) plane. The symbols denote the quality of the spectroscopic orbit: good (full squares), poor (stars) or doubtful (crosses). The 'R' indicates a source resolved by IR imaging. The 95% and 10% limits for probability of detection by CORAVEL are drawn, as well as the limit set by the diffraction limit of a 4-m telescope, an 8-m telescope and the VLTI for a source at 0.09" parallax (median value of our sample).

Indeed, for our present results, the final uncertainties on the derived masses are about equally shared between the RV, the imaging, and the parallax uncer-tainties. An Improvment in the precision can only be achieved by increasing the number of observations of each source, or by using new instruments providing ultra-accurate measurements

3   New Targets for the VLT

3.1   RV Background and Future

Since 1991, we are conducting a new survey of 250 M dwarfs at La Silla with CORAVEL. The objectives are to determine the distribution of orbital elements among M dwarf binaries in order to compare them with those of G and K dwarfs, and to detect new VLM companions, possibly including brown dwarf candidates. Present estimations show that the on-going RV surveys of M dwarfs will provide about 30 VLM companion candidates with $M_{2,min} \leq 0.15\,M_{\odot}$.

In the meantime, a new series of RV instrumentation is being developed and tested: ELODIE, mounted since 1993 at the 193 cm telescope of the Observatoire de Haute-Provence, and its twin CORALIE, for the 1.2 m Swiss telescope in La Silla, have been built by a collaboration between OHP, Marseille and Geneva. They use fiber-fed Echelle spectrographs optimized for RV measurements with a numerical cross-correlation technique. Preliminary results obtained with ELODIE show a precision in RV determination of $\sim 15\,\mathrm{m/s}$ for a 20 min exposure on an $m_V = 8.5$ star. To illustrate the gain provided by this 20-fold increase in RV precision, one can refer to Fig. 1 where the 10% detection probability line for CORAVEL would correspond to a 95% detection probability for ELODIE, opening access to the detection of VLM companions in most parts of the plane of the figure.

3.2   The VLT

The 8-m telescopes of the VLT will of course provide a very substantial gain for high angular resolution observations of red dwarf binaries: the resolution will reach ~ 0.06" at $2.2\,\mu m$ with CONICA, but the use of coronographic masks in conjunction with Adaptive Optics will also allow detection of extremely faint companions a few Airy disks away from bright primaries (Beuzit and Lagrange, 1994). It is foreseen that this gain will be sufficient to resolve most of the yet still unresolved pairs.

Nevertheless, the real break-through in precision will be provided by the VLTI/VISA. Here, the crucial fact is not so much that the increase in angular resolution will allow us to resolve a few more binaries, but that it will provide separation measurements of the previously observed objects with a much better precision.

The gains in mass determination achieved by the introduction of these new observing techniques and more accurate parallaxes are detailed in Table 1.

Table 1. Expected precision on stellar masses for various configurations (relative uncertainties, in %). Masses above 0.08 $M_\odot$ correspond to *actual* observations, while the sub-stellar cases are for *estimations* with the following assumptions: binary with $M_1 = 0.7\,M_\odot$, $\pi = 0.06"$, P = 3 yrs, e = 0.5; $N_{mes}(\text{RV}) = 20$, $N_{mes}(\rho, \theta) = 4$.

Case			$M_2(M_\odot)$	0.03	0.07 est.	0.11	0.21 obs.	0.33	Precision limited by
CORAVEL	+ 4 m	+	'old' parallax	106	87	38	15	8	
"	+ 8 m	+	"	77	59				
"	+ "	+	Hipparcos	63	44				
"	+ VISA	+	"	36	17				CORAVEL
ELODIE	+ 8 m	+	Hipparcos	32	31				8 m
"	+ VIMA	+	"	5	4				
"	+ VISA	+	"	4	3				

4 Conclusions

Combining the gain in RV determination, provided by the currently implemented digital correlation technique, and the gain in parallax precision yielded by e.g. HIPPARCOS, with the gain in diffraction-limited imaging provided by the VLT (Adaptive Optics on the 8-m and VISA), it will be possible to determine masses of VLM stars at the sub-percent level. Noticeably, the mass of a hypothetical brown dwarf companion of 0.03 $M_\odot$ could be estimated with $\sim 5\%$ accuracy, establishing undoubtedly its sub-stellar nature.

References

Beuzit, J.-L., and Lagrange A.-M. (1994): in, IAP (Paris) Coll. "Circumstellar dust disks and planet formation"
Duquennoy A. and Mayor M. (1991): *Astron. Astrophys.* **248**, 485
Duquennoy A., et al. (1992): in IAU Coll. # 135, Atlanta
Henry T.J. and McCarthy D.W. (1993): *A. J.* **106**, 773
Kroupa P., Tout C.A., Gilmore G. (1990): *M.N.R.A.S.* **244**, 76
Mariotti J.-M. et al. (1990): *Astron. Astrophys.* **230**, 77
Morbey C.L. (1975): *P.A.S.P.* **87**, 689
Perrier C., et al. (1992): in IAU Coll. # 135, Atlanta
Tokovinine A. (1992): in IAU Coll. # 135, Atlanta

Evolution of Massive Stars and the VLT

Corinne Charbonnel[1], André Maeder[2]

[1] Observatoire Midi-Pyrénées, 14, av. E.Belin, 31400 Toulouse, France
[2] Observatoire de Genève, CH-1290 Sauverny, Switzerland

Abstract. Massive star evolution is a key ingredient to understand the nearby and distant Universe. It has enabled us to understand the relationships between the various kinds of massive stars (O-stars, blue and red supergiants, WR stars), their properties and chemistries, their relative numbers in our Galaxy and in galaxies of other metallicities Z, which is a necessary step to make models of photometric, spectral and chemical evolution of galaxies. However, current theoretical models of massive stars still contain non negligible uncertainties, and still fail in explaining some fundamental observations. In this discussion, we point out the need for further theoretical and observational studies of massive stars. We show how FUEGOS on the VLT will allow a decisive and unique improvement in our understanding of the evolution of massive stars in galaxies with different metallicities.

1   Scientific Interests

Massive stars play a major role in the evolution of galaxies thanks to their aptitude to emit large amounts of mass, radiation and momentum. These highly energetic objects are important sources of mechanical power through their winds and supernovae explosions; they are involved in the release of star formation processes. They strongly contribute to the UV radiation and to the far-infrared luminosities through the heating of dust. They are the main contributors to nucleosynthesis and chemical evolution through the enrichment by their ejecta, either by supernovae or by stellar winds.

Conversely, observations of massive stars in distant parts of our Galaxy and in galaxies of the Local Group reveal the high dependence of their properties and evolution on the properties and chemical composition of their cocoon environments. Large differences exist in massive star population statistics among galaxies, suggesting that the initial metallicity plays an important role (Maeder 1991). As a consequence, we must proceed to careful tests of the current stellar models by observing massive stars in nearby galaxies before interpreting integrated spectra of more distant galaxies.

However, even if sophisticated stellar models exist, disagreements still persist between real and model stars. These failures are mainly due to the fact that the current input physics is to a large extent arbitrary. We need new observational constraints on the basic, but still uncertain physical ingredients, which govern the evolution of massive stars, i.e. the mass loss through the stellar winds, the mixing processes in the stellar interior, and the metallicity effects.

2 Mass Loss by Stellar Winds

Evaporation by stellar winds is one of the main characteristics of massive star evolution. However, in spite of the crucial role winds play in all model predictions (Maeder 1991, Schaerer & Maeder 1992), mass loss rates are not known with a high precision. Stellar wind models have been developed by several groups (Abbott 1982, Pauldrach et al. 1986, Owocki et al. 1988, Kudritzki et al. 1987, Schaerer & Scmutz 1994), but until very recently the observed mass loss rates were generally larger than predicted (Lamers & Leitherer 1993). This explains why in all evolutionary models empirical mass loss rates are still used.

The latest developments of the radiation driven wind theory seem to be in better agreement with observations. Nevertheless, hardly any observations at non-solar metallicity are available, and we have to rely on theoretical predictions for the mass loss dependance on metallicity. This clearly shows that we need new observations of stellar mass loss rates in different chemical environments and for a large stellar population.

Spectroscopy is absolutely necessary to measure the strength of the winds. The most useful range is in the blue, but the infrared and the visible also contain crucial lines ($H\alpha$ emission). The existing excellent non-LTE atmosphere models allow interpretation of these data, if the spectra are obtained with a sufficient resolution, around 10^4. Observations are needed in a significant sample in various parts of the HR diagram, and for objects at different metallicity, both in different parts of our Galaxy and in galaxies of the Local Group. For this purpose, FUEGOS and the VLT are resolutely essential.

3 Mixing Processes in Massive Stars

The weakness of the theoretical description of the mixing processes in stellar interiors leads to major uncertainties in massive star models. Several disageements between real and model stars would probably fade away if the physics of the non-convective mechanisms that mix the stars was better understood. In particular, some chemical anomalies observed in different points of the HR diagram could be explained by rotation-induced mixing.

One of the main problems which seems to be related to rotation concerns the He and N enrichments observed in fast rotating OB stars. Among the main sequence objects that present normal He and N abundances, the fast rotators generally show He and N enhancements (Herrero et al. 1992), that are not predicted by standard models with no extra-mixing at solar metallicity. By using the formalism proposed by Zahn (1983) for rotation-induced mixing, several authors have shown that considerable He and N increase can be obtained on the main sequence in the case of fast rotation (Maeder 1982, 1987, Langer 1991). However, even if they are encouraging, these results are still preliminary, since the physics of rotation-induced mixing is not completely understood.

In order to go further in the understanding of the influence of rotation on massive star evolution, the determination of the abundance of some key elements like He, C and N in massive stars is required. A fine spectral classification of stars first selected by colours will first allow the locatation of stars in a theoretical HR diagram. Then high resolution spectroscopy will be necessary to determine the abundances of the interesting elements from their rather faint lines. A spectral resolution of 10^4 with a S/N ratio of 100 is required. For fast rotators however, a lower resolution will be sufficient, thanks to the line broadening. We need an extensive and detailed spectroscopic study both for main-sequence and evolved objects in our Galaxy and in galaxies of the Local Group. Since this kind of analysis has to be performed for several hundred of stars, a multi-object spectrograph is needed. FUEGOS is the appropriate instrument.

4 Massive Stars in Galaxies of the Local Group. Metallicity Effects

Before studying massive stars throughout the Universe, we need to understand and neatly describe their structure and evolution in galaxies of the Local Group. This is necessary in order to disentangle the various effects of evolution, metallicity, IMF and SFR. In particular, this will enable us to examine how metallicity influences massive star evolution, a problem which cannot be solved by observing stars only in our Galaxy.

Massive star evolution is highly different at different metallicities. Among the stellar properties that show very large gradients in the Milky Way and very large differences from one galaxy to another, the number ratios WR/O, WC/WR and WC/WN are extreme examples, as shown in Table 1. The number ratio WR/O is observed to vary by one order of magnitude from M31 to the SMC. Some WR subtypes are totally missing in some galaxies (there are no late WC stars in the LMC and SMC, while there are many in the Milky Way), even if very young star populations are present.

The differences of WR populations in galaxies with active star formation was first discovered by Smith (1968, 1988). Maeder et al. (1980) suggested that the observed differences could be due to the dependence of mass loss rates on the local metallicity. As we have said before, evaporation by stellar winds is the dominant factor for massive star evolution. If there was no relation between mass loss rates and metallicity, massive star evolution would be about the same everywhere. Models of stellar winds have suggested the existence of a relation of the form $\dot{M} \propto Z^\alpha$, with α=0.5-0.7 (Kudritzki et al. 1987). Metallicity effects enter massive star evolution through this dependence. The observational trends described in Table 1 are generally explained by larger winds which lead to earlier formation of WR stars at higher metallicity (Maeder & Meynet 1994).

However, this is may not be the end of the story. Indeed, the role of rotation in the WR formation scenario is still not clear, and it is urgently necessary to extend the statistical sample in our Galaxy and in galaxies of the Local Group. The statistical nature of this study requires the observation of a large number of stars forming unbiased subsamples. For a complete study of the HR diagram, several hundred of stars should be observed, mainly in each Magellanic Cloud.

This major program would require an impossible length of time with a single-object spectrograph. FUEGOS on a VLT 8m telescope is perfectly adapted, and even unique for this purpose.

In conclusion, FUEGOS with the VLT will allow a decisive improvement in our understanding of the evolution of massive stars in galaxies with different metallicities.

Table 1. Observed number ratios WR/O, WC/WR and WC/WN in galaxies of various average metallicities Z. SN means solar neighbourhood, rings of different galactocentric distances being indicated (Table 6 from Maeder & Meynet 1994)

GALAXY	Z	WR/O	WC/WR	WC/WN
M31	0.035	0.24	0.44	0.79
MILKY WAY				
SN 6 – 7.5 kpc	0.029	0.205	0.55	1.22
SN 7.5 – 9 kpc	0.020	0.104	0.48	0.92
SN 9.5 – 11 kpc	0.013	0.033	0.33	0.49
M 33	0.013	0.06	0.52	1.08
LMC	0.006	0.04	0.20	0.26
NGC 6822	0.005	0.02	–	–
SMC	0.002	0.017	0.11	0.13
IC 1613	0.002	0.02	–	–

References

Abbott D.C. (1982) : Ap. J. 259, 282-301

Herrero A., Kudritzki R.P., Vilchez J.M., Kunze D., Butler K., Haser S. (1992) : A&A 261, 209-34

Kudritzki R.P., Pauldrach A., Puls J. (1987) : A&A 173, 293-98

Lamers HJGLM, Leitherer C. (1993) : ApJ 472, 771-791

Langer N. (1991) : IAU Symp. 143 on *WR stars and iterrelations with other massive stars in galaxies*, Eds. Van der Hucht, Hidayat, Kluver, 431

Maeder A., Lequeux J., Azzopardi M. (1980) : A&A 90, L17

Maeder A. (1982) : A&A 105, 149

Maeder A. (1987) : A&A 173, 247

Maeder A. (1991) : A&A 242, 93

Maeder A., Meynet G. (1994) : A&A 287, 803

Owocki S.P., Castor J.I., Rybicki G.B., (1988) : Ap.J. 335, 914-30

Pauldrach A., Puls J., Kudritzki R.P. (1986) : A&A 164, 86-100

Schaerer D., Maeder A. (1992) : A&A 263, 129

Schaerer D., Schmutz W. (1994) : A&A 288, 231

Smith L.F. (1968) : MNRAS 141, 317

Smith L.F. (1988) : ApJ 327, 128

Zahn J.P. (1983) : *Astrophysical Processes in Upper Main Sequence Stars*, 13th Saas-Fee Course, Eds. Hauck & Maeder

Part 4

Stellar Clusters

Globular Clusters with the VLT

F. Fusi Pecci[1], C. Cacciari[1], F.R. Ferraro[1], R. Gratton[2], L. Origlia[3]

[1] Osservatorio Astronomico, Via Zamboni 33, I-40126 Bologna, Italy
[2] Osservatorio Astronomico, Vicolo dell'Osservatorio 5, Padova, Italy
[3] Osservatorio Astronomico, Strada Osservatorio 20, Pino Torinese, Italy

Abstract. We present a review of the most important scientific areas and topics where Globular Clusters can play a significant rôle, with particular emphasis on cosmology, galactic formation and evolution, and stellar evolution and dynamics. The impact of the VLT and the comparison with other instruments are discussed.

1 Introduction

Globular clusters (GC) are the best example of a "simple" stellar population, *i.e.* stars with the same age and chemical composition (with a few exceptions) whose only varying parameter is mass. Therefore, they are ideal laboratories to study stellar astrophysical problems such as: the evolution of Population II stars; phenomena related to the environment (e.g. internal dynamics, binary formation and evolution, star interactions such as captures and mergers, X-ray sources, pulsars, etc.); the formation, evolution, and dynamics of the parent galaxies. Hence they provide constraints on some important cosmological quantities like the age of the universe, the distance scale, the primordial helium content, etc.

Given the extent of the subject and the connections to many fields, we shall only touch upon some topics of major scientific interest and present some e-valuations of the most efficient telescope/instrument combination quoting for simplicity four prototype classes (4m-class / 8m-class / HST / ISO).

2 Cosmological Tests

Cosmological models can be tested using two main aspects of GC properties, namely absolute ages which are related to the age of the universe, and dark matter content which is related to the baryonic matter in the universe.

2.1 Absolute Ages

In the Colour-Magnitude (or HR) diagram the optimal *clock* is the Main Sequence (MS) turn-off (TO), whose magnitude and colours (i.e. luminosity and temperature) provide the cluster age using the theoretical isochrone with the same chemical composition. Since stellar clocks are intrinsically based on stellar models, the *verification* of the validity of these models is both *complementary* and *necessary* to any dating procedure.

A few rough estimates may be useful before proceeding. Assuming an absolute age of 15 Gyr, an uncertainty of ±1 Gyr is given formally by an error of 0.07 mag in the TO luminosity, or 0.30 dex in the metallicity [m/H], or 0.03 dex in the helium content Y. The errors currently obtained on TO absolute luminosity (~0.2 mag), metallicity (~ 0.2 dex), and helium abundance (~ 0.02) lead to an error on the absolute age not smaller than 3 Gyr.

Therefore the problems, from the observational point of view, are:

⊕ *clock reading:*

• Derive accurate TO apparent magnitudes and colours from the observations. The required photometric accuracy (~ 0.01-0.02 mag for the individual stars at the TO) can be obtained if the observations reach at least 2-3 mag below the TO (V ~ 20-27, depending on the cluster distance). This can already be done for the Milky Way GCs with good CCD equipment and 4m-class telescopes on sufficiently well populated external zones of the clusters. Outside the Galaxy, HST can presently meet these requirements as far as the Fornax dSph galaxy (the TO region in M31 is already inaccessible to HST), whereas VLT with Adaptive Optics will be able to reach the TO stars as far as M31 (V ~ 28-30), although with a somewhat lower accuracy.

⊕ *clock running:*

• Test the evolutionary models (assumptions, input physics, approximations, etc.), (See Sect. 4), and calibrate the clock (mixing length, overshooting, etc.).

⊕ *accurate chemical abundances*

• Overall metallicity [m/H], and the relative abundances of elements such as Fe,C,N,O, etc. must be known with an accuracy of at least 0.1 dex (and about ten times better for the helium abundance Y). High resolution spectroscopic observations of RGB stars can be obtained with 4m-class telescopes, but the "truly unevolved" MS stars are much fainter (V > 17) and require the use of an 8m-class telescope and spectrographs such as MFAS and UVES.

⊕ *accurate distances*

This relies upon the use of various types of stellar "standard candles", for example:

• The RR Lyrae variables. Absolute magnitudes can be obtained using the Baade-Wesselink method (feasible with 4m-class telescopes).

• The Horizontal Branch (HB) luminosity level, calibrated on GCs in M31. The HST is presently the only suitable instrument.

• The RGB tip. Useful data can be secured also with small-medium size telescopes.

• The field subdwarfs with known accurate parallaxes (*e.g.* from Hipparcos) and metallicity determinations. All the complementary photometric and spectroscopic observations can be done with small-medium size telescopes.

• The White Dwarfs (WDs). This is a typical HST program. The VLT can be used profitably on certain aspects of this program (e.g. spectroscopy), and also to search for WD candidates in the more external regions.

♣ *GCs could also be "standard candles"*

If the GC Luminosity Function (LF) is constant or a known function of the parent galaxy morphological type and mass, the peak of the LF may be used as "standard candle" (see Harris 1991, for a review). This requires very deep imaging over large fields. One could determine the galaxy distances up to a distance modulus (m-M)> 35. With increasing distances, a transition from 4m-class telescopes to HST and 8m-class telescopes is necessary. In particular, VLT plus FORS, WFDVC, VHARC, MFAS, NIRMOS are fully suitable both for searching the candidate clusters, and for yielding medium-low resolution spectroscopy necessary to test the GC nature and membership and to secure chemical abundance determinations.

2.2 Very Low Mass Stars, Brown Dwarfs and Dark Matter

The stars at the low mass end of the Initial Mass Function are among the most popular candidates for baryonic dark matter. These degenerate dwarfs are commonly divided into two sub-groups, *i.e.* the very low mass stars (VLM) above the hydrogen-burning limit ($\sim 0.08M_\odot$), and the brown dwarfs (BD) with M < $0.08M_\odot$. Though the low metallicity makes the VLM and BD brighter than their solar neighborhood and Galactic disk counterparts, they still are fainter than V $\sim$ 25-27, and securing statistically complete and uncontaminated LF's is extremely difficult with the available tools.

According to the WFDVC specifications this imager at the VLT would be capable of providing a spatial resolution of $\sim$ 0.1 arcsec FWHM (almost comparable with HST), a sufficiently large field of view and 10 times the HST collecting area. Assuming the limiting magnitude I$\sim$ 28.5 (for S/N $\sim$ 2 in 10 hours) one could for instance extend the LF obtained in NGC 6397 by Paresce *et al.* (1994) by at least two magnitudes.

On the other hand, VLM's and BD's have high density and cool atmospheres ($T_{eff} < 3000^\circ K$) dominated by H_2 molecules, hence they emit predominantly in the red and IR bands. Both ISAAC and NIRMOS would thus be most suitable for both detection and low resolution spectroscopy of very faint candidates. For example, NIRMOS can provide unique LF's as faint as J=25 and H=24 with S/N=10 in 1 hour with pixel scale 0.3 arcsec, and also low-resolution spectroscopy (R=200) for candidates as faint as J=22 and H=21 with S/N=10 in 4 hours.

3 Tests of Galactic Formation and Evolution

To distinguish between different models of Galactic formation and chemical evolution (see Majewski 1993 for references) one should know: (a) "relative" ages to better than 0.5 Gyr; (b) global metal abundance and abundance ratios to better than 0.1 dex; (c) primordial helium abundance and the amount of helium dredged-up during subsequent evolution, to better than 0.01 dex; (d) individual cluster radial velocities to better than 10 Km/s; and (e) individual cluster proper motions as accurate as possible to derive the cluster orbits. Items (b) and (c) are discussed in Sect. 2.1.

3.1 Age Spread and the HB "Second Parameter" Problem

In a *resolved cluster* the age can be estimated by measuring the magnitude difference ΔV_{HB}^{TO} between the TO and the Horizontal Branch (HB) (see Buonanno *et al.* 1989), or the colour difference $\Delta(B-V)_{RGB}^{TO}$ between the TO and the RGB (VandenBerg *et al.* 1990). To achieve the error in age of ± 0.5 Gyr, the errors in magnitudes and colours must be smaller than 0.03-0.04 mag and 0.01 mag, respectively. The chemical abundances must be known at the level of accuracy requested for the absolute age determination. These requirements can be met using (V,B-V) or (V,V-I) photometry and spectroscopy, both currently feasible with 4m-class telescopes for most of the galactic halo and Magellanic Cloud GCs. In the bulge, however, IR (JK) observations are needed because of the very large extinction in that region of the Galaxy, and 8m-class telescopes are necessary to obtain accurate magnitudes at least 2-3 mag fainter than the TO (K(TO) $\geq$ 16) and properly define the unevolved MS. Moreover, since the required spatial resolution is very high, Adaptive Optics techniques are crucial. The scheduled VLT instruments ISAAC and CONICA, equipped with large format arrays and AO facility, can satisfy these requirements, allowing deep and both high (for the central regions) and medium (for the external zones) spatial resolution imaging in various near-IR filters.

In *non-resolved clusters*, ages can be estimated from integrated colours and spectroscopic indices, after proper calibration on the local template clusters, provided no "second parameter effect" is present (see below). FORS and MFAS on VLT are the ideal instruments for this purpose, thanks to the size of the field of view, the number of slits or fibers, and the spectral resolution.

The use of VLT opens a new window to this type of study as deep imaging and spectroscopy can be extended into the IR, where one can find many interesting absorption stellar features, due to atomic (neutral metals such as Si, Mg, Al, etc.) and molecular (CO, OH, H_2O, CN, etc.) species, which are very sensitive to the variation of the fundamental stellar parameters of cool stars. Presently, only the ~ 20 brightest clusters both in our Galaxy and in the Magellanic Clouds are observable in the IR at medium-low resolution with a 4m telescope, having K< 16 mag per square arcsec. Instruments like ISAAC and NIRMOS are crucial, especially to observe GC systems in the Local Group galaxies.

Another possible approach to deal with relative characteristics between G-GCs is to study their HB morphologies, in particular the so-called "HB second parameter effect", *i.e.* the occurrence of extremely-blue or red HBs in clusters having the same (intermediate) metallicity. Age is presently a favourite second parameter candidate (see for references Zinn 1993, but note also Fusi Pecci *et al.* 1993).

HBs are sufficiently bright as far as $d \sim 150$ Kpc (which includes the Magellanic Clouds and the Fornax dSph galaxy) to allow accurate photometry with

4-m class telescopes (apart from the highly crowded central regions which require the use of HST). Useful spectroscopy of individual HB stars in the MCs and Fornax, however, cannot be done without 8m-class telescopes and instruments like FORS, MFAS, NIRMOS etc., to yield abundances and velocities as accurate as those presently obtained for the Galactic GCs.

For the GCs in M31 and beyond, the use of HST or of VLT with highly sophisticated Adaptive Optics devices is indispensable to detect individual HB stars because of the faintness of the stars and the spatial resolution necessary to resolve them.

3.2 Dynamics of the GC System and the Parent-Galaxy Total Mass

For the Galactic GCs, radial velocities can be obtained with the highest accuracy from the analysis of a large number of individual cluster stars, and the use of 4m-class telescopes is adequate to this purpose. Obtaining space motions and orbital shapes is however the crucial item. According to the latest estimates (see Tinney, this volume) important progress can be made using CCD-devices and 8m-class telescopes, as accurate proper motions could be obtained using a relatively short baseline (5-10 years).

Spectroscopic observations of extragalactic cluster candidates require typical integration times of several hours per cluster in order to reach a S/N sufficiently good for metallicity estimates and for radial velocities with $\sigma \sim 50$ km s^{-1}, or possibly better. In addition, since clusters beyond ~ 2 Mpc can hardly be distinguished from foreground stars and background galaxies, the observing efficiency will be considerably lowered by the contamination from spurious objects. The use of Multi Object Spectroscopy on VLT is therefore essential to make these programmes feasible and efficient.

4 Stellar Evolution Tests

4.1 Some Spectroscopic Tests of Stellar Evolution

The original surface composition of stars is altered during their evolution by various mechanisms. Therefore the study of abundances of stars in different stages of their evolution, but likely with the same original composition, provides basic and sensitive tests of stellar evolution.

$\oplus$ *Diffusion and Lithium abundance in main-sequence and turn-off stars*

Diffusion on the MS may affect the abundance of Li for stars in the TO region of GCs. Spite *et al.* (1984), found a constant Li content among a large group of unevolved metal-poor old stars, that is likely to be the Big Bang signature. However Li is manufactured and destroyed by various processes which depend on details of the internal stellar structure on the MS and TO. Stars in GCs are thus very important, since their physical parameters are known with much better accuracy than for field stars.

The Li abundance for a TO star in NGC 6397 (the closest cluster, V(TO) $\sim$ 16) has been obtained recently by Molaro and Pasquini (1994) using EMMI at the ESO-NTT. However, these observations (R= 28,000, 4 × 90min) are clearly at the limit of a 4m-class telescope possibilities, and a systematic study of stars of different luminosity in various clusters can only be done using an 8m-class telescope. The UVES at VLT is very well suited for these studies.

⊕ *Mixing and environment*

Dredge up of CNO-processed material is expected to occur in GC stars at the base of the RGB (first dredge up), and possibly during the latest stages of AGB evolution (third dredge up). Theoretical predictions are generally rather well satisfied for most field metal-poor stars, while stars in GCs show a far more complex picture (Smith 1987).

Several candidates have been proposed as responsible for the detected peculiarities: environmental effects, enhanced core rotation (perhaps due to close encounters between protostars), pollution by (possibly temporary) companions or even by other cluster members, and/or some still unknown mechanism (mixing?) at work during evolution. The dependence of these mechanisms on evolutionary phases and cluster dynamical/structural parameters is different, and a systematic study of large samples of stars at different luminosities and positions in several clusters is decisive. While specific observations for a few bright giants may be carried out with a 4-m class telescope, a fiber instrument like MFAS (both in the Medusa and Argus mode) at an 8m-class telescope is required for a systematic and complete study of fainter stars.

⊕ *Mass loss and intracluster matter*

Stellar evolution models predict a mass loss of $\sim$0.2 $M_\odot$ prior to the HB phase and $\sim$0.1 $M_\odot$ on the asymptotic giant branch (AGB) (*e.g.* Fusi Pecci and Renzini 1976). This mass loss is required to explain the observed HB morphologies in GCs and the lack of any significant population of AGB stars brighter than the RGB tip.

Our knowledge of stellar mass loss is still very poor. The basic problems are: (a) direct observations of mass loss rates; (b) dependence of mass loss rate on abuncances (if any); (c) detection of intracluster matter, residual of stellar mass loss; (d) mechanism(s) causing stellar mass loss; (e) evolutionary feed-back on star evolution.

With VLT plus UVES one could extend to GC stars the type of studies carried out by Reimers (1975) on field stars, based on the detailed analysis of high resolution Ca H+K line profiles. Useful data could also be obtained using MIIS on the brightest GC stars for instance at 9.7, 11.5 and 18.0 μm, typical features already observed in bright nearby objects with high mass loss rates.

Concerning intracluster matter, a typical GC population of $\sim 10^3$ post-TO stars is expected to release a few 10^2 $M_\odot$ of stellar matter during the $\sim 10^8$ *yr* periods between each cluster passage through the galactic plane. This matter should be present if no "cleaning" mechanism is at work.

The intracluster gas could be in the form of atomic (neutral or ionized) hydrogen and/or molecular H_2 and CO, but all searches carried out so far have been unsuccessful but not conclusive. On the contrary, some evidence of cold dust was found in the intracluster medium of GGCs.

In many clusters IRAS point sources have also been detected within their tidal field (integrated fluxes in a 30" aperture between 0.2 and 1 Jy, Lynch and Rossano 1990). A few of them are extended at 12 μ, with typical sizes of 2-4 *arcmin*, and this could be a significant indication of the presence of cold dust in the intracluster medium, as also suggested by multicolour polarization observations.

ISO observations can provide a deep survey (0.1-1 mJ) of the cluster central regions with low spatial resolution and small field of view, due to the small format arrays. The use of an 8m-class telescope and large format arrays would allow to make almost as deep surveys (down to $\sim$1 mJy per square arcsec) on a much larger field of view (a few arcminutes) and with high spatial resolution (0.1"/pixel and an Airy disc of 0.6" at 10 μ). A complete mapping of the emission regions would thus be possible, even for quite extended and low surface brightness areas. Note that the high spatial resolution, available only with the VLT, will allow a distinction to be made between diffuse emission (dust) and point sources (very cool stars, e.g. brown dwarfs or carbon stars). An instrument like MIIS could satisfy these requirements.

⊕ Gravitational settling, diffusion and rotation for HB stars

It is known that the combined action of gravitational settling and outward diffusion by radiation pressure, observed in Pop-I B8-F2 stars, may change significantly the surface abundances of He and metals in the outer radiative envelope of hot HB stars, since the typical timescale of these phenomena (10^8 yr) is close to the HB lifetime. On the other hand, diffusion might be inhibited by rotation.

Observations of lines of He and heavier elements might then provide the age of stars on the HB, and then be used to distinguish between ZAHB and evolved stars once additional information on rotational velocities is available (see below). With proper modelling, one could get information on the direction of evolution off and along the HB. Earlier explorative observations of He lines have been done by Crocker and Rood (1988); however, an 8m-class instrument is required for extensive observations of different elements as quite high spectral resolutions (UVES with R> 40,000) are necessary and the stars are faint (V $\sim$ 14-20).

Peterson (1983) found a high frequency of larger-than-normal rotational velocities in BHB stars of M13. While the relation between surface and core rotation is not clear yet, these relatively high rotational velocities could help explain the very blue HB of M13, a typical example of the "second parameter" phenomenon. A confirmation of Peterson's result would thus be of paramount importance as one could eventually get direct hints on both rotation and mass loss. However, these observations are very difficult and uncertain as the blue HB stars are faint and reliable rotational velocities require high resolution spectra with high S/N ratios. The use of an 8m-telescope with a MOS capability is thus highly desirable.

4.2 Binaries

It is common belief that any object which does not fit into the "standard evolution theory" of normal stars could somehow be related to a binary system. Though it seems unlikely that binarity is responsible for so many different types of stars, it may be interesting at least to mention several categories of "unusual" cluster members: blue stragglers, subdwarfs O and B, cataclysimic variables, dwarf Cepheids, extreme-blue HB stars, novae, Ba-stars, CH-stars, UV-bright objects, "naked" or "nude" very blue stars, X-ray sources, millisecond pulsars.

To study these variegated classes of objects different techniques have been successfully used in the optical, X-ray, and radio bands. Several interesting programmes could easily be carried on with the VLT, for the sake of example we list here a few of them:

1. Surveys to detect radial velocity variables ($\sigma \sim 1$ km s^{-1}) with MFAS.
2. Surveys to detect photometric variables (e.g. eclipsing binaries) and to study their periods using both MFAS and FRISPI. In particular, the study of variability could be focused on blue stragglers as about 25% of them have been found to vary (Stryker 1993). The properties of spectrophotometric binaries could also be used to derive distances.
3. Spectroscopy of cataclysmic variables and novae. UVES high resolution spectra in the UV region can be used to assess membership and study in detail their properties. Since these stars are faint (V> 18 even in the closest clusters) VLT is absolutely necessary.
4. Detection of a "second" MS. Binary systems formed by approximately equal mass components are expected to produce a second "binary MS" parallel to that of single stars. The use of spectra obtained with MFAS may allow the detection of radial velocity variations in the candidates found via very deep imaging.
5. Blue Straggler stars (BSS) can be formed via several mechanisms mostly involving the interaction and merging of stars in binary systems (see for ref. Fusi Pecci *et al.* 1992). During their post-MS evolution, they are expected to end up on the red extreme of the HB due to their large mass. The merging process is expected to produce some peculiarities. In particular, if the system had not been able to lose a substantial fraction of the angular momentum acquired with the merging, a large rotational velocity is expected. The presence of a large convective envelope would then probably cause a strong dynamo effect, and hence a rather strong activity (analogous to that observed in FK Com objects). These effects should be detectable with appropriate spectroscopic observations. Finally, He, CNO (mainly O isotopic ratios) and Li abundance anomalies might be present, even though the theoretical background is not well defined at present (Pritchet and Glaspey, 1991). Adequate observations with a 4m-class telescope are feasible only for a few very bright BSS descendants in some clusters, an 8-m telescope is required for most BSS-progeny, as well as for the direct observation of BSS themselves. The need of a statistically significant sample of stars requires the

use of a fiber-fed, optical and IR medium-high resolution spectrograph (the determination of the O isotopic ratio can best be done with high resolution observations of the CO bands in the K wavelength region) .

6. Study of X-ray sources and millisecond Pulsars. This topic is so wide and the possible observations so many that we simply mention it. In this respect, especially spectrophotometry with MFAS and FRISPI are crucial to both identify and study the optical counterparts.

5 Cluster Internal Dynamics

GCs are ideal sites to test dynamical models of stellar systems, since they are relatively simple structures where large samples of individual objects can be observed. These programmes require high precision measures of radial velocities (error < 1km s^{-1}) for a large sample of stars in various cluster regions.

On a 4m-class telescope these observations are limited to a few bright giants, whereas VLT + MFAS are very well suited for these programs, the Argus mode being useful for the cluster core and the Medusa mode for the outer regions.

VLT in its interferometric configuration is the *only* instrument capable of observing the very central regions (a few fractions of a parsec) with the maximum possible spatial resolution. If the nominal resolution of about 0.004 arcsec could be reached, one could really make an incredible step forward even compared to the best results obtainable from HST in its highest spatial resolution configuration ($\sim$ 0.02 arcsec).

6 Conclusions

It is quite evident even from our schematic and incomplete review of GC studies that the VLT will be an extremely important tool for yielding a better insight into most of the current hot problems. A fruitful complementarity exists between the results uniquely achievable from the space (with HST, ISO, etc.) and those one can better obtain from the ground with the VLT and its many detectors. In this respect, it is important to note that (i) there are crucial observing programmes which require not only the use of the already planned VLT instruments (*i.e.* FORS, ISAAC, CONICA, UVES, and MFAS) but also of some instruments presently under study or just proposed like MIIS, NIRMOS, FRISPI, and WFD-VC; and (ii) several extremely important issues in the study of GC problems can best (or only) be addressed in the IR wavelength range. Since neither HST nor ISO for different reasons are able to carry out the necessary observations, IR instruments for the VLT (especially in the near and intermediate IR) provide a unique opportunity and should have a very high priority in the selection, construction and commissioning.

References

Buonanno R., Corsi C.E., Fusi Pecci F. 1989, A&A, 216, 80

Crocker D.A., Rood R.T. 1988, in Globular Cluster Systems in Galaxies, IAU Symp. No. 126, eds. J.E. Grindlay and A.G.D. Philip, Kluwer, Dordrecht, p. 509

Fusi Pecci F., Renzini A. 1976, A&A, 46, 447

Fusi Pecci F., Ferraro F.R., Corsi C.E., Cacciari C., Buonanno R. 1992, AJ, 104, 1831

Fusi Pecci F., Ferraro F.R., Bellazzini M., Djorgovski S.G., Piotto G., Buonanno R. 1993, AJ, 105, 1145

Harris W.E. 1991, ARA&A, 29, 543

Lynch D.K., Rossano G.S. 1990, AJ, 100, 719

Majewski S.R. 1993, ARA&A, 31, 575

Molaro P., Pasquini L., 1994, A&A, 281, L77

Paresce F., De Marchi G., Romaniello M., 1994, preprint

Peterson R.C. 1983, ApJ, 275, 737

Pritchet C.J., Glaspey J.W. 1991, ApJ, 373, 105

Reimers D., 1975, Mem. Soc. R. Sci. Liege, 6(8), 369

Renzini A. 1985, Astronomy Express, 1, 127

Smith G.H. 1987, PASP, 99, 67

Spite M., Maillard J.P., Spite F. 1984, A&A, 141, 56

Stryker L.L. 1993, PASP, 105, 1081

VandenBerg D.A., Bolte M., Stetson P.B. 1990, AJ, 100, 445

Zinn R.J. 1993, in The Globular Cluster-Galaxy Connection, eds. G.H. Smith and J.P. Brodie, ASP Conf. Ser. No. 48, p. 38

Detailed Analysis of Stars in the Galactic Bulge with the VLT

Roger Cayrel[1] and Poul. E. Nissen[2]

[1] Observatoire de Paris, F-75014 Paris, France
[2] Intitute of Physics and Astronomy, Aarhus University, DK-8000 Aarhus C, Denmark

Abstract. It is shown that the VLT, in combination with the planned Multifibre Area Spectrograph, can bring a bulk of new basic information on the Galactic Bulge, unobtainable in a reasonable time with 4m class telescopes. Such information would be of great interest for extragalactic astrophysics too, because of the deep connection between bulges, starbursts and the QSO phenomenon.

1 Introduction

In a sense the stellar population of the bulge is at the same time one of very high astrophysical significance, and one of the least known in our Galaxy.

It is of high astrophysical significance, because it is a unique population, not represented in the solar neighbourhood, and which has undoubtedly played a central role in the early evolution of the Galaxy. It is also important because it is known from the observation of high redshift objects that bulges are the sites of the most energetic phenomena known in the Universe, QSOs, and the search for a QSO remnant at the center of our own Galaxy, could be usefully complemented by the study of the expected effects of the occurrence of an active QSO on the chemical and dynamical history of the bulge. The bulge is also the place in our Galaxy where astration has gone further that anywhere else, apparently in a rather short time, and very early in the galactic history. This problem is of course linked to the assumed major starburst preceeding the formation of a QSO showing metallicities similar to, or higher than, solar in the broad band region (Hamann and Ferlan 1993).

The stellar population of the bulge is the least known because of its distance (not negotiable) and because of the substantial amount of obscuration by interstellar absorption, miraculously reduced to 1.5 mag in Baade's window at Galactic coordinates ($l = 1^o$, $b = -4^o$). The scientific committee of the Multifibre Area Spectrograph has listed the observation of stars in the bulge at a spectral resolution of 30,000 and a signal/noise ratio of 75-100 as a very high scientific priority. The aim is to obtain, by detailed analysis, the abundance of the elements for a sufficient number of stars in order to understand the chemical history of the bulge. There has been no detailed analysis made on stars of the bulge, but a study at medium resolution (Rich 1988, 1990) allowed a metallicity histogramme for the bulge to be obtained and the effective yield in the bulge to be determined. Very recently McWilliam and Rich (1994) have succeeded in performing the first detailed analysis of 12 giant stars in Baade's window at CTIO, but with a spectral resolution of 17,000 and a S/N ratio of 30 to 70. Samples of these spectra have been presented by Rich at this Workshop. So although the groundbreaking work on elemental abundances in the bulge has begun, there is still a great gain to be realized with the VLT-MFAS combination, the VLT collecting area permitting an increase in the spectral resolution and the S/N ratio to the desired value. MFAS will allow the study, in a single exposure, of about 80 objects, enabling the study of the requisite large number of objects necessary to understand a population with a large range of metallicities.

2 Extant Problems of the Galactic Bulge

2.1 The Metallicity Distribution

The last analysis by McWilliam and Rich does not contain enough objects to derive a metallicity histogramme for the Galactic Bulge. However, it shows that the former metallicity histogramme (Rich 1990), in perfect agreement with the "simple model" prediction for the bulge, considered as a closed system, for a high yield ($\simeq 2Z_\odot$) is to be revised towards a lower value of the yield, probably below $Z_\odot$. There is also a need to extend the sample towards lower metallicities, in particular below $0.1Z_\odot$, where too few objects are present. The limits of the agreement with the simple model have also to be checked, various authors having advocated important exchanges of matter between the bulge and the rest of the Galaxy (inflow from the halo, Wyse and Gilmore (1992), accretion of small systems, Quinn et al. (1993), 'fountain effects', Arimoto and Yoshii (1987).

2.1 Nucleosynthesis

One of the interesting questions is to find out if element abundance ratios are solar or not, in the bulge. The first results of McWilliam and Rich (1994) do not allow a simple and clear-cut conclusion. There is some evidence that a few α -elements are enriched with respect to iron, but also that some others are not. The problem is difficult, in particular because the temperature scale of metal-rich stars is still uncertain. It has not yet been possible to study the dependence of those features on global metallicity. A deficiency of enrichment by SN Ia with respect to enrichment by SN II seems probable, in comparison with the situation for the disk population.

2.3 The Time Scale of Star Formation in the Bulge

Linked with the delayed enrichment in iron produced by type Ia supernovae, comes the problem of the age distribution of stars in the bulge. It is known from the pioneering work of Blanco et al. (1984) that carbon-stars are excessively rare in Baade's window, putting a severe constraint on the age of evolved stars in the bulge. From the extant works, it appears that no stellar formation nor significant accretion of external systems has occurred in the last 5Gyr (Rich, this workshop).

3 The Connection with Bulges of other Galaxies

The bulges of other galaxies are often easier to observe than the bulge of our own Galaxy, the interstellar absorption being not as lethal as for the Galactic Bulge. However, because of the much larger distance, only studies of the integrated light are practicable. There is a large body of interesting data on the stellar population of galactic nuclei (Bica 1988, Pastoriza et al. 1994, and references therein), obtained by low resolution spectroscopy, interpreted by decomposition of the spectrum over a base of star cluster spectra, and spanning a wide range of age and metallicities (Bica & Alloin 1986). These works show that there is a tremendous range of age distributions in galactic nuclei, ranging from an exclusively old population, to 60% of the light (at 5870 Å) coming from stars younger than 10^8 years, not speaking of active nuclei. Replacing the study of our own bulge in the frame of the study of bulges in general, raises the problem of understanding, why some bulges have stopped forming stars very early (gas blown off by activity?) and why some others have had a rather continuous star formation rate.

Understanding the nucleosynthesis in the Galactic Bulge is of course a key element in the game.

4 Proposed Observations

In order to derive reliable elemental abundances we consider that it is necessary to work with a resolution:

$$R = 30,000$$

and a signal to noise ratio:

$$S/N = 100$$

Two spectral domains are needed to cover elements coming from the major nucleosynthesis processes (iron peak, several α elements, r and s elements). We are proposing the two intervals:

Table 1

$\lambda\lambda$	Elements
6257 – 6481	O, Mg, Ca, Ti, Cr, Fe, Ni, Y, La
6642 – 6829	Li, Al, Si, Fe, Ni, Y, La, Eu

The spectral resolution is just the one supplied by MFAS.

The desired S/N requires exposures of about 6 hours each (24 hours with a 4m telescope !), which must be broken into at least three exposures for cosmic ray cleaning. The two domains are in the 6th order of the echelle grating.

One of the most delicate problems is the subtraction of the background in such a crowded field. We are proposing to use the following pattern for background subtraction:

Fig. 1. Pattern for sky subtraction

The setting of the 80 fibres is first done on the targets, then three exposures are done at the three positions indicated by the dotted circles. The exposure time is shorter on the background positions, the optimal value being determined by the average ratio of the fluxes coming from the targets and from the background in each fibre.

A companion programme, still in the status of feasibility study, is the observation of *subgiants* in the bulge, photometrically and spectroscopically (unavoidably at a lower resolution) in order to obtain simultaneously the age, the metallicity and the α elements/iron ratio, in many bulge stars.

5   Conclusions

In a recent IAU Symposium on Galactic Bulges (No. 153, 1993) Ivan King in an introductory paper on the Galactic Bulge, made the statement:

"While I am on the topic of instrumental developments, there is another one very important for this field. Until recently, nearly all spectra were taken one at a time. But now, nearly every major observatory has, or is developing, a multi-object spectrograph. *It will be of tremendous value to studies of the Galactic Bulge to be able to get radial velocities, spectral classes, and chemical abundances in great numbers ...*"

Our conclusion is that the VLT, combined with the 80 fibre MFAS instrument, is exactly the combination that can achieve such a programme.

References

Arimoto N. and Yoshii Y. (1987) A&A **173**, 23

Bica E. (1988) A&A **195**, 76

Bica E. and Alloin D. (1986) A&A **162**, 21

Blanco, V.M., McCarthy, M.F. and Blanco, B.M. (1984) A.J. **89**, 636

Hamann F. and Ferland G.J. (1993) Ap.J. **418**, 11

King I. (1993) in IAU Symposium 153 *Galactic Bulges*, eds. H. Dejonghe and H.J. Habing, Kluwer Academic Publishers, Dordrecht, p.3

Pastoriza M.G., Bica E., Maia M, Bonnato Ch; and Dottori H. (1994) Ap.J. **432**, 128

Quinn P.J., Hernquist L., and Fullagar D.P. (1993) Ap.J. **403**, 74

Rich R.M. (1988) A.J. **95**, 828

Rich R.M. (1990) Ap.J. **362**, 604

McWilliam A. and Rich R.M. (1994) Ap.J.Suppl. **91**, 749

Wyse R.G.F. and Gilmore G. (1992) A.J. **104** , 144

Deep
Star Counts and Cosmological Backgrounds: A Powerful New Model for the Point Source Sky

Martin Cohen[1],[2]

[1] Jamieson Science and Engineering, Inc., 5321 Scotts Valley Drive, Suite 204, Scotts Valley, CA 95066, U.S.A.
[2] Radio Astronomy Laboratory, 601 Campbell Hall, University of California, Berkeley, CA 94720, U.S.A.

Abstract. I report further development of "SKY", the model for the point source sky developed by Wainscoat et al. (1992) and recently considerably enhanced by Cohen (1994a). SKY predicts cumulative or differential source counts, and integrated surface brightness of the sky due to smeared point sources, in any direction, both for many "hardwired" far-UV, optical, and infrared filters and, in a customised mode, for any filter lying entirely within the 2.0-35.0 μm range. SKY realistically represents the Galaxy using disk, spiral arms, local spurs, Gould's Belt, molecular ring, bulge, halo and a representation of the IR extragalactic sky. Science applications are: interpreting deep optical and near-IR star counts in terms of the interplay between different geometric components (e.g. halo:disk ratios at high latitudes; disk, arms, and ring at low latitudes) and seeking cosmological IR background radiation. Practical issues are establishing confusion limits due to high surface density of sources. Emphasis will be on the comparison between predicted source counts and theoretical performance for relevant instruments intended for the VLT.

1 An Introduction to the "SKY" Model

Several models of the sky have been generated to interpret rocket and balloon observations of the surface brightness of the near-IR (e.g. Ishida & Mikami 1982) and far-UV sky (e.g. Henry 1977) or, less frequently, to predict star counts (e.g. Kawara et al. 1982; Garwood & Jones 1987; Eaton et al. 1984; Ruelas-Mayorga 1991). But their geometry has always been very restricted (usually only disk and halo, or disk and bulge) and their origins have been rooted in the optical. Mamon and Soneira's (1982) work already shows how difficult it is to extend optically-based models even to 2μm. The SKY model offers substantial advantages over all these other models in: its geometric realism: the rich detail of its sky with respect to types of source incorporated; the inclusion of nonstellar sources; its extremely broad wavelength coverage; and its diverse output products.

In addition to the fidelity of its point source sky, SKY's real power is epitomized by the breadth and flexibility of its products. Simply stated, it predicts the number of sources to be seen in a given direction at a certain magnitude and wavelength. It can then be adjusted to give a better representation. In this way we build up a physically realistic and empirical model driven by observations. To provide diagnostics of any discrepancies and to guide such "adjustments", SKY produces any or all of the following outputs requested, as a function of apparent magnitude (or in-band flux, equivalently): (1) total differential or cumulative source counts, broken down into the separate contributions of disk, halo, arms, ring, bulge, etc.; (2) total cumulative surface brightness due to the smearing of all point sources in an area, as well as the distinct differential contributions of the geometrical components; (3) the breakdown of each geometric component's source counts among the 87 types of galactic source and 4 types of galaxy, in a user-selected "magnitude window"; (4) a color histogram; and (5) all of the above either for a default 1 sq. deg. patch of sky centered on the demand direction (l,b), or in a finite user-designated area.

The basic model was originally designed to predict mid-infrared point source counts (Wainscoat et al. 1992: hereafter WCVWS). However, the guiding philosophy was not to produce a model specific to the infrared but rather one that stressed geometric realism and accuracy of physical attributes of stars and non-stellar objects over a very broad range of wavelengths. SKY operates with a number of broadband "hardwired" filters (1400, 1565, 1660A; standard Johnson B,V,J,H,K; IRAS [12] and [25]; and the Japanese narrowband balloon bandpass at $2.4\mu m$). It can also handle customized filters because it draws upon an embedded complete library of continuous infrared spectra for all categories of source. The initial library used by WCVWS was based on the IRAS Low Resolution Spectrometer (7.7- 22.7 μm). Cohen (1993) extended this spectrally continuous library to the 2.0-35.0 μm range. This provides the flexibility to make predictions for the present and next generations of infrared satellites and ground-based IR array cameras.

The newest version of SKY (Cohen 1994a; Cohen et al. 1994) incorporates six fully detailed Galactic components: an exponential disk; a Bahcall-type bulge; a de Vaucouleurs halo; a 4-armed spiral with 2 local spurs; Gould's Belt; and a circular molecular ring. Each geometrical element may contain up to 87 source categories based upon detailed analyses of the content of the IRAS sky (Walker et al. 1989). These comprise 33 "normal" stellar types; 42 types of AGB star, both oxygen- and carbon-rich; 6 types of object that are distinct from others only by their mid-infrared high luminosity; and 6 types of exotica including HII regions, planetaries and reflection nebulae. Every source category has its own set of absolute magnitudes in the hardwired pass bands; its dispersion in M12; its own individual scale heights and volume density in the local solar neighborhood. In addition, some sources may be absent from some geometrical components (the arms and ring are made rich in high-mass stars; the halo deficient in these). SKY also accommodates external galaxies, through their IRAS luminosity functions, but only for wavelengths beyond 5 μm, where spatial-spectral variations are negligible. These many attributes, all drawn from the copious astronomical literature, are entirely equivalent to the customary luminosity functions. But this explicit categorization facilitates augmentation of the content of the model sky by adding other populations, or updating of physical parameters as these become better known from independent studies.

2 Scientific Questions Relevant to the VLT

2.1 Galactic Structure

The broad scientific issues for galactic structure are perhaps obvious. One might simply remark that at high latitudes one can preferentially select for topics such as the ratio of halo:disk counts as a function of wavelength. The question of exactly how far out of the galactic plane the sun lies is an interesting one because, without an accurate representation of this displacement, no model can avoid introducing a specious north-south asymmetry into its predictions of point source counts. Indeed, any search for near-infrared cosmological background radiation is utterly dependent on highly accurate subtraction of both the contributions of sunlight scattered by zodiacal particles and of smeared foreground point sources. Interpretation of DIRBE data already mandates these corrections to the total observed diffuse fluxes (Cohen 1994b) and observations from ISO and the VLT will also necessitate the best quantitative models for the point source sky. Consequently, one needs to have established as precisely as possible the correct relative contributions of halo and disk, and the solar location with respect to the plane, before attempting the quest for cosmic infrared backgrounds. These are both dominantly high latitude questions.

By contrast, on low and intermediate lines-of-sight, one can study the detailed interplay of disk, halo, bulge, arms, and molecular ring as functions of galactic direction and wavelength. Issues are, e.g., how barlike our bulge is; what happens in the anticentre direction (where multi-wavelength constraints from deep optical, IRAS, and near-infrared already appear in conflict (Cohen 1994b); what the true distribution of reddening along the plane is. All are amenable to study by interpretation of observed star counts. SKY offers a profound advantage over other models in these investigations because it offers a self-consistent treatment of the far-ultraviolet, optical, and mid-infrared point source skies. Further, SKY can make predictions customized for VLT instruments' passbands and site (by including the influence of the atmosphere on bandpass profiles).

2.2 The Search for Exotic Populations

Finally, having established these geometrical matters, one can readily seek exotic populations such as brown dwarfs (BDs) with SKY. At thresholds of K=20-24, brown dwarfs might be detectable (if any exist in abundance), but recognition of their presence will require that the halo:disk ratio and the solar displacement are already extremely well-characterized. We note that SKY's dependence on an explicit table of categories of source, rather than on a luminosity function, lends itself readily to the addition of new populations. One can determine from recent literature the important attributes of BDs that need to be incorporated in SKY (scale height; to which geometric component of the Galaxy they belong, e.g. halo alone; absolute magnitudes, colors, and local space densities); compare SKY's predictions with and without BDs in order to determine which combination of galactic direction, apparent magnitude, and wavelength is most sensitive to the presence of BDs; and thereby investigate whether any substantive limits can be placed on BD space densities using deep VLT (e.g. ISAAC and FORS) counts. This is a very different approach from that typical in BD studies where the emphasis is traditionally to hide as much mass as possible in an unobservable form. Using SKY and VLT data I believe one could empirically place meaningful quantitative constraints on the BD distribution and population by direct comparison with star counts.

3 Practical Questions Relevant to the VLT

Given the tremendous sensitivity of the VLT and its proposed instruments, it is perhaps more useful to consider not what *can*, but rather what *cannot*, be observed. By that I mean, do high source densities limit VLT sensitivity in specific galactic directions? In other words, when is an instrument or passband compromised in terms of its achieved performance because too many sources fall on one pixel, however small? Then there are specific questions about near-infrared adaptive optics such as determining the surface density of natural faint guide stars over the entire sky. SKY can help with these too. To illustrate the value of SKY in this context, I show some examples of predictions for deep star counts in the most benign (as regards confusion) direction, namely the north galactic pole (NGP: the SGP is very similar), and the most troublesome, namely towards the Galactic Centre. Figure 1 is relevant to ISAAC and customised for L'; Figure 2 is for N with the proposed Mid-Infrared Imager/Spectrometer (former "MIIS", now MIDIRIS). Each figure shows a pair of predictions for the cumulative source counts in 1 sq. deg. in the directions indicated on the plots. Total count is the solid line; the grey (dotted) line refers to the disk; long dashes to the arms (including Gould's Belt); long dash dot to the halo; short dashes to the bulge; short dash dot to the ring; long dash short dash to the extragalactic sky at wavelengths beyond 5 μm. Such plots can be rendered for differential counts too. For all these predictions I have placed the sun 15 pc north of the galactic plane.

These results are summarized in Table 1 where I present for the different VLT instruments the passband, technical limit, source density for the onset of "confusion" (sources per sq. deg.), the cumulative logN, and the corresponding magnitude limit for the Galactic Centre line-of-sight. This table was constructed for a VLT image quality [FWHM] of 0.26"; tracking of 0.1"; median seeing (0.8"); diffraction; and is averaged over all proposed pixel sizes. A uniform criterion for confusion was adopted, of 1 source per 25 *beams*. Only at N does the atmospherically defined performance accord with the confusion limit; at other wavelengths confusion is the stronger limit.

Table 1. Source confusion towards the Galactic Centre

INST.	BAND	TECHNICAL LIMIT	CONFUSED DENSITY (sources/sq deg)	LOG N (CUMUL)	CORRESP. MAG. LT.
FORS	B	27.5	8.10e05	5.91	26.1
FORS	V	26.5	8.10e05	5.91	23.6
ISAAC	K	23	7.97e05	5.90	12.6
ISAAC	L'	18.3	6.91e05	5.84	12.3
ISAAC	M	?	5.84e05	5.77	12.1
MIIS	N	12.1	5.00e05	5.70	12.1
MIIS	Q	8	2.31e05	5.36	11.5

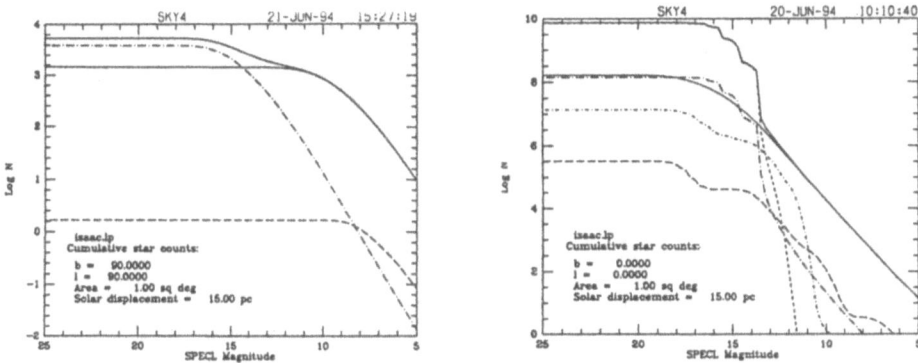

Fig. 1. NGP and Galactic Centre cumulative counts for ISAAC at L'

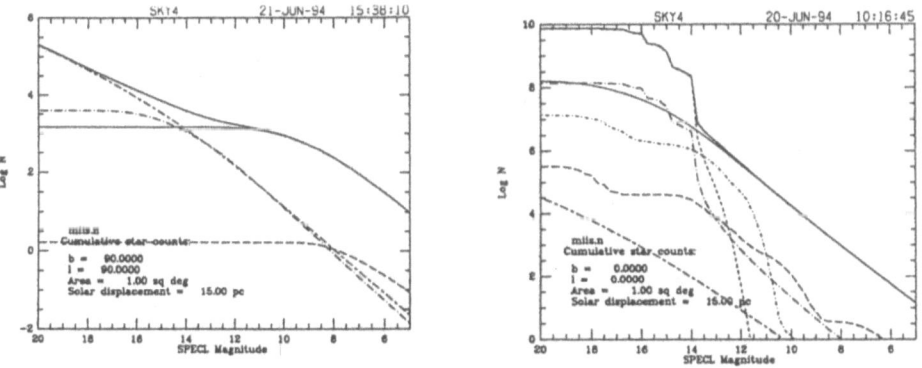

Fig. 2. NGP and Galactic Centre cumulative counts for MIIS at N

References

Cohen, M. (1993): Astron. J. **105** 1860

Cohen, M. (1994a): Astron. J. **107** 582

Cohen, M. (1994b): Proc. Les Houches Workshop "Science with Astronomical Near Infrared Sky Surveys", Ed. N. Epchtein, Kluwer Academic Press, in press

Cohen, M., Sasseen, T., Bowyer, S. (1994): Astrophys. J., **427** 848

Eaton, N., Adams, D. J., Giles, A. B. (1984): Mon. Not. Roy. Astron. Soc., **208** 241

Garwood, R., Jones, T. J., (1987): Publ. Astron. Soc. Pacific **99** 453

Henry, R. C. (1977): Astrophys. J. Suppl, **33** 451

Ishida, K., Mikami, T. (1982): Publ. Astron. Soc. Japan **34** 89

Kawara, K., Kozasa, T., Sato, S., Kobayashi, Y., Okuda, H., Jugaku, J. (1982): Publ. Astron. Soc. Japan **34** 389

Maihara, T., Oda, N., Sugiyama, T., Okuda, H. (1978): Publ. Astron. Soc. Japan **30** 1

Mamon, G. A., Soneira, R. M. (1982): Astrophys. J. **255** 181

Ruelas-Mayorga, A. (1991): Revista Mex. Astron. Astrof. **22** 27

Schwartz, D. E. (1989): Astron. J. **98** 2163

Wainscoat, R. J., Cohen, M., Volk, K., Walker, H. J.,

Walker, H. J., Cohen, M., Volk, K., Wainscoat, R. J., Schwartz, D. E. (1992): Astrophys. J. Suppl.**83** 111

Part 5

Interstellar Medium

High Resolution Spectroscopy of the Interstellar Medium

Roger Ferlet

Institut d'Astrophysique de Paris, CNRS 98 bis Bd. Arago, 75014 Paris, France

Abstract. It is demonstrated that high resolving power and high signal-to-noise ratio absorption line spectroscopy has led to most of the recent advances in the field of circumstellar, interstellar and intergalactic media. An ultra-high resolution mode would very greatly enhance the scientific return of the VLT in these fields of cosmogonical and cosmological significance.

1 Introduction

Some years ago, M. Harwit showed that most cosmic discoveries were connected with the occurence of a significant technical step or the opening of a new window on the Universe. Without question, high resolution spectroscopy associated with very large collecting area will be such a new major tool to attack vital astrophysical problems. Beside the serendipitous, there are several areas of research to be investigated with a high resolution VLT spectrographic capability. The VLT should not miss the opportunity to venture into an almost virgin Universe and make discoveries.

Here, we are talking about resolving power $R = \lambda/\Delta\lambda$ of at least 1 to 1.5 $\times$ 10^5, simply because many important pending problems cannot be tackled with lower resolution. Many of these are reviewed in the proceedings of the ESO Workshop on High Resolution Spectroscopy with the VLT held in Garching in 1992. We will briefly give here only a few exciting examples of potential achievements in the interstellar medium field through high resolution absorption spectroscopy at the VLT.

2 Technical Considerations

Measurements of equivalent widths of interstellar (IS) lines are indeed quite independent of the resolution. However, it is well known that they can yield very large errors on column densities when multiple components are present on the line of sight and when lines are saturated (on the flat part of the curve of growth). The former situation appears very general, even for the short sight lines in the solar neighbourhood. The latter one is less common, at least for part of the IS lines in the visible spectrum. But both cases will become severe limitations when probing the diffuse IS medium beyond what is currently reachable in the Galactic plane with available telescopes.

Sufficient resolution is required to resolve the velocity structure of the lines of sight. This is the only way to gain information on individual clouds, on average separated by about 2 km s^{-1}, that is now badly needed instead of integrated properties. It is true that IS resonance lines are much more numerous in the UV domain, and the R $\sim$ 10^5 observing mode of the Hubble Space Telescope (HST) represents indeed a very significant step in the field. However, by using visible data at R = 5 $\times$ 10^5 of the readily accessible D lines of Na I, Welty, Hobbs and York (1991) have clearly demonstrated that information is already partly lost in HST data. This is why higher resolution ground–based spectroscopy will still be extremely useful for interpreting UV data.

Amongst physical parameters, temperature is directly accessible only through the intrinsic width of the lines. For purely thermal broadening, the width increases with decreasing mass of the studied ions, and the lightest element having an observable resonance line in the visible is lithium (the lighter elements hydrogen and deuterium show up in the currently unacessible far UV, except Ly alpha observable with the HST but which is always heavily saturated and blended, and helium in the EUVE satellite domain but with a very poor resolution). Nevertheless, even through the Li I doublet, IS gas at 1000 K gives a full width at half maximum of 2.57 km s^{-1}, requiring already R $>$ 1.2 $\times$ 10^5. Higher resolution is mandatory in order to derive true b-values from cooler IS gas, even though careful profile fitting analysis of several lines may help to artificially increase the resolution by a factor of perhaps 2 or so, depending on the signal-to-noise ratio. Indeed, the lithium lines are so faint that a significant number of temperature measurements will be more easily done by using the two main ions observable from the ground: Na I and Ca II. For these, even the warm phase of the interstellar medium will require R $>$ 10^5. Moreover, using lines of different ions in the same velocity components will also allow determination of any turbulent velocity, another important and poorly known parameter for the dynamics of the interstellar medium.

In the case of the Na I – D lines, one has also to consider the two hyperfine structures (hfs) components of the ground level which give rise to a velocity splitting of 1.05 km s^{-1}, in a strength ratio of 5/3 for the optically thin limit. This value exceeds the thermal line width at T $\approx$ 80 K by a factor of almost 3. With a resolution of 0.5 km s^{-1} (R = 6 $\times$ 10^5), Welty, Hobbs and Kulkarni (1994) were therefore able to partially resolve a thermally broadened pair of hfs components formed in a cloud with T = 500 K, thereby immediately revealing a conspicuous signature of cold gas. In a sample of 276 individual clouds, they found that many have T $<$ 160 K and V$_{turb}$ $<$ 0.24 km s^{-1}, demonstrably subsonic internal turbulent motions. However, even at their resolution, the effects of instrumental broadening is already clear, implying that the typical temperature of most neutral IS gas is probably T $\lesssim$ 80 K.

3   Astrophysical considerations

3.1   Interstellar medium

Beside very high spectral resolution, studies of the diffuse interstellar medium also require very high signal-to-noise ratios. A typical example concerns the very local IS medium. Its kinematical complexity has been recognized by Ferlet et al. (1986) through the discovery of three Ca II absorption components toward Altaïr, a star located at only 5 pc from the Sun, and confirmed by Lallement et al. (1990) through a Ca II – k survey toward nearby stars. Prior to these observations with S/N ratio ∼ 1000, this local medium remained undetected from the ground. Furthermore, the study of the Na I/Ca II ratio is an indicator of the ionization state in the very local medium (Bertin et al., 1993). Recently, the long standing problem of the connection between these absorption measurements and the parameters of the interstellar wind flowing through the interplanetary medium, as derived from in situ backscattering data, has been resolved: we have definitively identified the cloud in which the solar system is embedded (Bertin et al., 1993; Lallement et al., 1994).

The determination of the IS abundances of light elements, in particular beryllium and lithium which are only observable in the visible range around 313.0 and 670.7 nm respectively, are of extreme importance. The knowledge of their present-day abundances, and even more uniquely of the ratios Be/Li and ^{7}Li/^{6}Li, is a key-milestone to severely constrain cosmological scenarios and galactic chemical evolution models (see the Proceedings of the ESO/EIPC Workshop on Light Elements Abundances, Elba, May 1994). This kind of observation needs incredibly high S/N ratios. For instance, IS ^{6}Li has been recently detected for the first time, towards ρ Oph (Lemoine et al., 1993), thanks to a S/N ratio ∼ 2700 per pixel, implying a limiting detectable equivalent widh of 0.0043 nm (3 σ). This was achieved after 13 hours of integration time with the CES at R = 10^5 fed by the ESO 3.6 m telescope via a 35 m fiber link from the Cassegrain focus, the total Li I equivalent width being ∼ 0.22 pm. More recently, we derived the ^{7}Li/^{6}Li ratio towards ζ Oph (Lemoine, Ferlet and Vidal–Madjar, 1994), still using CES + 3.6 m data but this time with a S/N ratio ∼ 7000 per pixel (which, by the way, seems to represent an intrinsic limit set by the possibility to effectively correct for interference fringes remnants at a level of ∼ 10^{-4} rms on the CCD) ! Towards both stars, using a sophisticated profile fitting analysis, we detect at least two IS absorption components. The totally unexpected result is that the ^{7}Li/^{6}Li ratio is found in both sight lines to be strikingly very low: ∼ 2 in one cloud; while for the other cloud, it is in much better agreement with the meteoritic value (12.3), which is thought to be representative of the early solar system. As a matter of fact, these values cannot as yet be explained by the most recent and elaborate evolutionary models. A very elegant explanation of the lowest values has been put forward by Cassé, Lehoucq and Vangioni–Flam (1994), which involves a SNII exploding inside an IS cloud and accelerating α and ^{16}O nuclei to interact with the ISM. However, the well studied ζ Oph sight line is well known to be complex, and observations at much higher resolution are needed to fully resolve the velocity structure and check its influence on the derived ratios. They are underway at the AAT in Australia, at $R = 6 \times 10^5$. Even with a 4 m class telescope, these studies are restricted to the very few stars on the whole sky which are bright enough, but still shining through a dense cloud to enhance the expected Li column density. They should be done in many individual clouds in order to statistically establish the significance of the first results, to directly test the cosmic ray spallation production of Li and derive the truly representative value for the ISM; it has to be noted that due to severe financial problems, the CES + 3.6 m instrumental configuration is not any more available at La Silla ! The gain from one VLT unit should even allow possible variations in the supernovae rate through IS clouds in other Galactic spiral arms to be detected.

Measurements in the visible range of molecules like CH, CH^+, CO^+ or C_2 are also of prime importance for interstellar chemistry; they can also yield the carbon isotopic ratio $^{12}C/^{13}C$ and the temperature of the dark clouds (see e.g. Hobbs, 1973; Crane et al., 1991; Vladilo et al., 1993). Several examples given in Black and Van Dishoeck (1988) show the absolute necessity for the highest possible resolution ($R > 4 \times 10^5$), if reliable evaluations are to be done. Interstellar abundance studies need now to be performed towards more and more distant targets, in very different environments or where galactic gradients become sensitive, or in other spiral arms. In several hours of integration time with the VLT, the most abundant species might be detectable in the Magellanic Clouds, which would give access to less processed extragalactic material, to be compared to the more extreme metallicity towards the Galactic Center. Equivalent widths of several pm in neutral IS sodium around 590 nm, good tracer of H I, are already gathered at $R = 10^5$ towards the very brightest Magellanic supergiants with the 3.6 m telescope, but Ca II at 393 nm, providing information on ionization, is just barely feasible at that resolution (Molaro et al. 1993; Vladilo et al., 1993). The use of an 8 m telescope would allow a much more complete mapping of the Magellanic Clouds, yielding in particular their still poorly known depth structure. One should also mentioned the study of the Galactic gaseous halo and the well known high velocity clouds discovered long ago in H I 21 cm, in order for instance to determine their distances and metallicities (see e.g. Meyer and Roth, 1991, who have shown that one of these clouds, observed towards the supernova 1991 T and the quasar 3C273, which are located within about 1° the sky, is made of already processed material).

3.2 Extragalactic

Deuterium is another light element of crucial importance for both cosmology and chemical evolution of galaxies. It is directly observable only in the far UV through the Lyman lines series in absorption towards early type stars, or towards nearby cool stars against the Ly α stellar emission line. With the help of ground–based data at higher resolution, whenever available, the present-day IS value of the D/H ratio is still controversial, since variations around 2×10^{-5} might exist in the very local ISM (see e.g. Ferlet, 1992). New determinations are underway with HST, but restricted to very short sight lines since only the very strong Ly

α line is observable and very low column densities are required to resolve the D
I line from the H I one. Knowing the present IS deuterium abundance, one has
to go back to the primordial value – which is of major cosmological significance
since it is highly sensitive to the universal baryon density – by applying astration
models which are still very uncertain. Therefore, a better approach is to estimate
D/H in "more priomordial" matter, as in the so–called Lyman alpha Forest seen
towards high redshift quasars. Very recently, such an important measurement
has been performed by two groups towards the same QSO, one at the Keck
Telescope (Songaila et al., 1994) and the other at Kitt Peak (Carswell et al.,
1994). They both have detected an absorption feature that they claimed to be
deuterium at z = 3.32 yielding a D/H ratio of $\sim 2 \times 10^{-4}$. If it is confirmed,
this result has dramatic consequences: the present evolutionary models which
suggest a relatively low astration of deuterium (a factor 2-4) are totally wrong
and most of the dark matter in the Universe is non baryonic ! However, an H I
feature mimicing deuterium cannot be ruled out.

Another exciting example is the measurement of the cosmic background radiation temperature, T(CBR), at high redshifts. Besides black body microwave
observations, IS lines from the lowest rotational levels of the CN molecule around
387.5 nm have been used for many years to derive the precise CBR temperature
at the present epoch (e.g. Palazzi et al., 1992). However, this is the only point
determined up to now. According to the Big Bang theory, T(CBR) depends in
a simple way upon the redshift, and its determination at different z would really be a new and extremely strong observational argument in favour of the Big
Bang. A convenient radiometer for directly measuring the background radiation
at earlier epochs would be neutral carbon. The ground state of C I splits into
three levels separated by energies of the same order as kT(CBR) at z > 0.5, for
which C I lines show up in the visible range. At such epochs, contrary to the
galactic ISM where collisions are much more efficient, C I should be appreciably
excited by the CBR. Until recently, because of the lack of resolution available
even for the brightest QSOs to resolve the C I lines splitting, only an upper
limit of 16 K in one absorption system at z = 1.776 was assigned by Meyer et al.
(1986). Similarly using lines from the ground state and the fine structure split
state of C II, Songaila et al. (1994) derived T(CBR) < 13.5 K, to be compared
with the prediction of 10.66 K at the redshift 2.909 of the observed system.

Obviously, the VLT will make these D/H analysis and T(CBR) determinations in QSOs' absorption systems not only more easy but also more numerous,
at different z over a wide range in metallicities, in order to be able to discriminate between theoretical models of galactic chemical evolution and place tight
constraints on the Big Bang.

3.3 Planetary systems

The study of protoplanetary systems is another field of research we would like to briefly mention here. The β Pictoris disk is the prototype of this new area which will become more and more important in the near future with the discovery of other systems. As for the previous fields, this requires high spectral resolution at high S/N ratios, but also high temporal resolution. In the case of β Pic, our extensive UV and visual circumstellar absorption data base is presently interpreted as the signature of cometary-like bodies evaporating when grazing and falling onto the star, perhaps through perturbations of a huge reservoir of small bodies by a planet orbiting β Pic (see e.g. Ferlet et al., 1993; Deleuil et. al., 1993; Vidal–Madjar et al., 1994; and references therein). Also predicted by model calculations (see e.g. Beust et al., 1990), such events last a few hours and have to be spectroscopically followed in velocity and equivalent width simultaneously in different lines (like the Ca II doublet), or as close as possible in time (see the Proceedings of the IAP Meeting on Circumstellar Dust Disks and Planet Formation, Paris, July 1994). Recent observations by Crawford et al. (1994) at ultra-high resolution ($R \geq 9 \times 10^5$ at AAT) have demonstrated that, as stated in the introduction, any significant instrumental gain allows the detection of previously unobserved changes.

The search for extra-solar system giant planets through extremely precise radial velocity measurements, is also now entering the discovery phase, since a ± 5 m s^{-1} sensitivity has been reached for very bright stars by two techniques: Fabry–Perot and gas absorption cell. It has been vigorously advanced as part of the first stage of the TOPS Program of NASA: Toward Other Planetary Systems. For instance, Jupiter induces an amplitude variation of 13 m s^{-1} in the radial velocity of the Sun over a period of 11.9 years. The limits for such long term follow-up of very many stars are more imposed by the apparent brightness of the observed stars. This is why TOPS is intended to be performed in partnership with the Keck Observatory.

4 Conclusions

It is obvious to claim that the jump from 4 m to 8 m (and a fortiori from 2 m) class telescopes will very much increase the number of observable targets, so that entirely new results will undoubtedly come up. The type of scientific questions we have discussed here are only tractable with VLT instruments with a resolving power of greater than 1 to 2×10^5. But an ultra-high resolution mode (around 6×10^5) would be an open window on a new Universe; an enormous amount of information on physical processes is awaiting for that mode, as it was already pointed out in the ESO Workshop on High Resolution Spectroscopy with the VLT, and as briefly shown here. It should be noted also that a small spectral range, say few nm, is often sufficient. Once again, it would be a mistake not to implement these spectral modes at the VLT while they are already in use at a few 4 m class telescopes and are planned, or will be operational, on several 8 m class telescopes before the VLT.

References

H. Beust, A.M. Lagrange–Henri, A. Vidal–Madjar, R. Ferlet: Astron. Astrophys., **236**, 202 (1990)

P. Bertin, R. Lallement, R. Ferlet, A. Vidal–Madjar: Astron. Astrophys., **278**, 549 (1993)

P. Bertin, R. Lallement, R. Ferlet, A. Vidal–Madjar: J. Geophys. Res., **98**, 15193 (1993)

J.H. Black, E. Van Dishoeck, E.: Astrophys. J., **331**, 986 (1988)

R.F. Carswell, M. Rauch, R.J. Weymann, A. Cooke, J.K. Webb: Monthly Notices Roy. Astron. Soc., **268**, L1 (1994)

M. Cassé, R. Lehoucq, E. Vangioni–Flam: Nature, , in press (1994)

P. Crane, D. Hegyi, D. Lambert: Astrophys. J., **378**, 181 (1991)

I.A. Crawford, J. Spyromilio, M.J. Barlow, F. Diego, A.M. Lagrange: Monthly Notices Roy. Astron. Soc., **266**, L65 (1994)

M. Deleuil, et al.: Astron. Astrophys., **267**, 187 (1993)

R. Ferlet: in *Astrochemistry of Cosmic Phenomena*, IAU Symp. No. 150, Dordrecht: Reidel, p. 85 (1992)

R. Ferlet, R. Lallement, A. Vidal–Madjar: Astron. Astrophys., **163**, 204 (1986)

R. Ferlet, et al.: Astron. Astrophys., **267**, 137 (1993)

L.M. Hobbs: Astrophys. J., **181**, 795 (1973)

R. Lallement, R. Ferlet, A. Vidal–Madjar, C. Gry: in *Physics of the Outer Heliosphere*, Adv. Sp. Res., 37 (1990)

R. Lallement, P. Bertin, R. Ferlet, A. Vidal–Madjar, J.L. Bertaux: Astron. Astrophys., **286**, 898 (1994)

M. Lemoine, R. Ferlet, A. Vidal–Madjar, C. Emerich, P. Bertin: Astron. Astrophys., **269**, 469 (1993)

M. Lemoine, R. Ferlet, A. Vidal–Madjar: Astron. Astrophys., , in press (1994)

D. Meyer, J. Black, F. Chaffee, C. Foltz, D. York, D.: Astrophys. J. Letters, **308**, L37 (1986)

P. Molaro, G. Vladilo, S. Monai, S. D'Odorico, R. Ferlet, A. Vidal-Madjar, M. Dennefeld: Astron. Astrophys., **274**, 505 (1993)

E. Palazzi, N. Mandolesi, P. Crane: Astrophys. J., **398**, 53 (1992)

A. Vidal–Madjar, et al.: Astron. Astrophys., , in press (1994)

G. Vladilo, M. Centurion, C. Cassola: Astron. Astrophys., **273**, 239 (1993)

G. Vladilo, P. Molaro, S. Monai, S. D'Odorico, R. Ferlet, A. Vidal-Madjar, M. Dennefeld: Astron. Astrophys., **274**, 37 (1993)

A. Songaila, L.L. Cowie, C.J. Hogan, M. Rugers: Nature, **368**, 599 (1994)

D.E. Welty, L.M. Hobbs, V.P. Kulkarni: Astrophys. J., , in press (1994)

D.E. Welty, L.M. Hobbs, D.G. York: Astrophys. J. Suppl., **75**, 425 (1991)

The Interstellar Medium in Galaxies

J.M. van der Hulst[1]

[1] Kapteyn Astronomical Institute, Postbus 800, NL-9700 AV Groningen, The Netherlands

Abstract. The prospects for research of the Interstellar Medium (ISM) in galaxies with the VLT are discussed. The focus is on measuring the physical conditions and structure of the ISM, and on probing the conditions in regions of circumnuclear star formation which are heavily obscured and only well accessible in the infrared region of the spectrum.

1 Introduction

The Interstellar Medium (ISM) in galaxies consists of several phases which are thought to coexist in pressure equilibrium. The original idea goes back to the seminal paper of Field, Goldsmith and Habing (1969) who considered the energy balance in the ISM assuming ionization and heating by cosmic rays. Since then much attention has been given to further investigating the physics of the ISM and the picture has been much more refined. An overview can be found in Shull (1987). The most detailed studies of the structure, the kinematics and the physical properties of the ISM, concern the Milky Way and other members of the Local Group such as the Magellanic Clouds, M 33 and M 31. A good overview of the status of all aspects of such studies can be found in IAU Symposium 144 (ed. Bloemen 1991).

Studies of the ISM in galaxies in the optical and the near and mid IR thus far have concentrated mostly on spectroscopy of H II regions and Supernova Remnants for determining abundances (Vila Costas and Edmunds 1992, and references therein), emission line imaging (mostly Hα, [O III], and [S II]; Rand et al. 1990, Dettmar 1990, Walterbos and Braun 1992, 1995, Hunter and Gallagher 1992) and studies of the dust using IRAS and ground based imaging (Telesco 1988, and references therein))

Large optical facilities at excellent sites such as the VLT will greatly improve studies of the ISM. Larger sensitivity will enable studies of faint structures in the ISM and make distant galaxies more accessible. Studies in the IR will furthermore be enhanced tremendously because in this spectral region the benefit from the larger collecting area is tremendous as simple tip-tilt corrections to the incoming wave fronts will produce diffraction limited operation of the telescope.

This overview wil be divided into three areas of interest for future studies of the ISM with the VLT: the structure and physical condition of the diffuse ionized gas in galaxies; the properties of the ISM around star forming regions of galaxies; and the properties of the ISM in the nuclear regions of galaxies.

2 The Diffuse Ionized Gas (DIG) in Galaxies

2.1 The Structure of the DIG

Emission line imaging of the ISM in nearby galaxies such as M 33 (Courtès et al. 1988), M 31 (Walterbos and Braun 1992), the Magellanic Clouds (Davies et al. 1986), NGC 891 (Rand et al. 1990, Dettmar 1990), NGC 4631 and NGC 4565 (Rand et al. 1992) and several Irregular galaxies (Hunter and Gallagher 1990, 1992, Hunter et al. 1993), has shown that the structure of the ionized gas is very filamentary and pervasive. For a recent review see Walterbos (1991) and Dettmar (1992). In addition to the obvious H II regions in galaxies there is ionized gas throughout the disk of galaxies at levels of emission measure a few pc cm^{-6}. This component has a low average density ($n_e \sim 0.2$ cm^{-2}) and a temperature of $< 10,000$ K. Yet it contributes more to the ionized gas mass of a galaxy than the H II regions (Reynolds 1989).

In edge-on galaxies such as NGC 891 there are clear indications for ionized gas up to a few kpc above the plane with structure suggesting the presence of chimneys venting gas from the disk out into the halo, e.g. such as expected from the combined action of stellar winds and supernova explosions in regions of massive star formation (Norman and Ikeuchi 1989). Figure 1 shows an Hα image of NGC 891 (Rand et al. 1990). Observing programs to search for these features in other galaxies are underway, but the main limitation is sensitivity. For the same reason most imaging has been done in Hα only, other emission lines such as [S II] and [O III] often being too weak. Yet imaging in different emission lines makes it possible to map out the distribution of the excitation of the ionized gas. This provides clues as to whether the gas is photo-ionized by a dilute interstellar radiation field as proposed by Mathis (1986) or whether it is shock ionized in which case one expects [S II]/H$\alpha > 0.6$ (Shull and McKee 1979). In the Galaxy (Reynolds 1988) and M 31 (Walterbos and Braun 1995) values are found between 0.3 and 0.5, suggestive of photo-ionization.

With the VLT and an imaging capability such as FORS one can improve considerably with respect to the present studies. By observing more galaxies of different types, in different environments and with different massive star formation rates it will be possible to link the structure and physical state of the DIG to other fundamental properties such as galaxy type and star formation rate. One expects for example a thicker ionized gas layer in galaxies with higher star formation rates.

2.2 The Excitation of the DIG

Little is known about the excitation of the DIG. Based on imaging and spectroscopy of a few galaxies, including the Milky Way, the impression is that the [S II]/Hα ratio increases from about 0.2 in and around H II regions to a constant value of about 0.5 in the genuine DIG (Walterbos and Braun 1992, Rand et al. 1990, Keppel et al. 1991). In the edge-on galaxy NGC 891, Keppel et al. (1991)

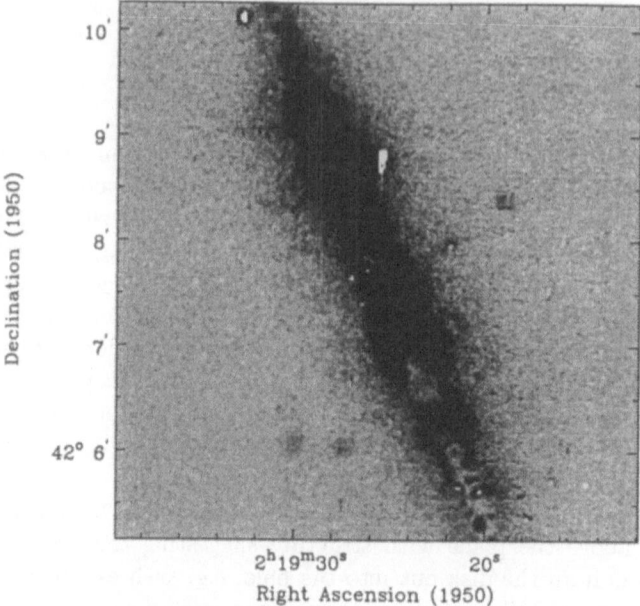

Fig. 1. The distribution of Hα emission in the edge-on galaxy NGC 891 (Rand et al. 1990). North is at the top and East is to the left. The faintest features have an emission measure of a few pc cm^{-6}.

find that the [N II]/Hα ratio increases by about a factor 2.5 with increasing distance from the midplane. The available data is insufficient to firmly distinguish between photo-ionization (Mathis 1986) or shock ionization (Sivan et al. 1986).

The low emission measures of the DIG imply that for spectroscopic studies with sufficient sensitivity to detect major emission lines such as [O III] 5007, 4959 and 4363Å, [O II] 3727Å, [N II], [S II], the Balmer lines, and weaker lines such as [O I] 6300Å, a large telecope and sensitive spectrograph are required. The combination of the VLT and FORS are ideal for obtaining spectroscopy over the whole spectrum of the DIG in nearby galaxies. Spectra of the quality one now obtains routinely for bright H II regions in nearby galaxies, can then be obtained for the DIG in nearby galaxies. This will make a proper analysis of the excitation mechanism and energy input into the DIG possible. It is of particular interest to measure the change in conditions as a function of distance to the sites of massive star formation as outlines by the H II regions and OB associations in a galaxy. The most detailed of such studies can be done in the Magellanic Clouds.

In addition to probing the excitation distribution, one can also examine the Balmer decrement to trace changes in dust content of the galaxy. This does, however, have the complication that the Balmer decrement depends on what fraction of the obscuring and scattering dust is mixed in with the tenuous ionized gas.

3 Extragalactic H II regions

Much is known about the properties of H II regions in nearby galaxies. Though not a particular component of the ISM, H II region studies are relevant for our understanding of the characteristics of the ISM. H II regions trace the distribution of massive star formation in a galaxy, i.e. the distribution of the major sources of energy input in the ISM and provide one of the few ways to measure elemental abundances of the ISM. Most of our knowledge of the metallicity of the ISM in galaxies has been derived from spectroscopic studies of the H II regions. For a review of this subject see Shields (1991).

3.1 Imaging

Extensive studies of the distribution of H II regions in galaxies have been made for about 100 or so nearby galaxies. Much of the pioneering work in this field is that by Kennicutt, Hodge and collaborators and includes studies of the luminosity function of H II regions in galaxies (Kennicutt et al. 1989, Hodge and Kennicutt 1983a, 1983b). The latter are particularly interesting and show that the structural properties of the H II regions change with galaxy type: the most luminous H II regions are found in the later type galaxies and also the number of H II regions per unit blue luninosity inceases toward later Hubble types. The luminosity funtion reflects this as a slight flattening toward later Hubble types. A marked difference has been found between the arm and interarm regions (Kennicutt et al. 1989, Rand 1992). The spiral arms contain a significantly higher fraction of bright H II regions.

Similar studies of more distant galaxies, including galaxies in high (cluster) and low (void) density environments, and low surface brightness galaxies require higher sensitivity *and* spatial resolution. In particular the low surface brightness galaxies are of interest, as they appear to be a major fraction of the general galaxy population (McGaugh 1994, Ferguson and McGaugh 1994). An example of such a low surface brightness galaxy is given in Figure 2, which shows an Hα and an R image of F568-3 obtained with the 2.5 Isaac Newton Telescope at La Palma. Is is obvious that studies of the H II region population in such galaxies to similar quality of the work on nearby bright spirals requires an instrument like the VLT and FORS. As such galaxies probably have a star formation history quite different from the well studied high surface brightness spirals (de Blok et al. 1995, McGaugh 1992) it is of extreme interest to study the characteristics of the H II regions in these rather unevolved objects.

An important improvement over existing facilities are the capabilities in the near- and mid-IR, partly because of an improvement in instrumentation, partly because of the increase in collecting area. The IR emission lines from H II regions and their vicinity are faint and only accessible in very nearby galaxies such as the Magellanic Clouds. Interesting transitions are the Hydrogen recombination lines such as Brα, Brγ and Paβ, fine structure lines such as [Ne II], [Ar III] and [S IV] from the ionized gas, and the rotational lines of warm H$_2$ (e.g. S(0) 1$-$0, S(1) 1$-$0, S(1) 2-1) and several PAH features from the surrounding warm molecular

F568−3

Hα + continuum

Hα

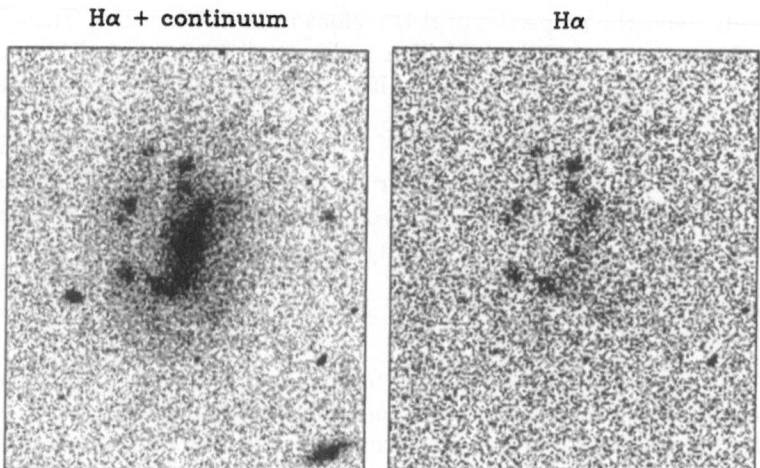

Fig. 2. Hα + continuum image (left panel) and Hα image (right panel) of the low surface brightness galaxy F568-3

gas and dust. Imaging in these lines has the great advantage that effects of extinction do not play a significant role anymore and that one can image the distribution of ionized and warm molecular gas unambiguously. Imaging in e.g. the [Ne II], [Ar III] and [S IV] lines provides an excellent means of mapping out the distribution of excitation. Figure 3 demonstrates this clearly. This figure shows a [Ne II] and [S IV] image of the high excitation H II region N160A in the Large Magellanic Cloud. The general structure corresponds well with the known distribution of Hα emission except for the source in the north-west which does not show up in optical continuum or emission line images (Heydari-Malayeri and Testor 1986) but is visible in [Ne II] and in the mid-IR continuum. This apparently is an embedded, low excitation object, only discovered in the IR.

3.2 Spectroscopy

Spectroscopy of H II regions in nearby galaxies has led to the general picture that galaxies have radial abundance gradients (Shields 1990, and references therein, Vila-Costas and Edmunds 1992 and references therein, Dinerstein 1990), and that the central abundance correlates with the mass of the galaxy and the morphological type, the higher central abundance occurring in the earlier type galaxies. The slope of the abundance gradient does not significantly vary with galaxy type, though a shallower slope is found in barred spiral galaxies.

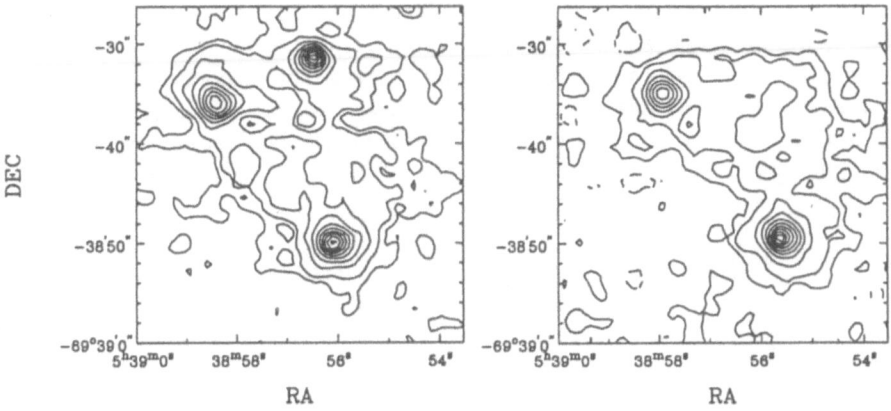

Fig. 3. [Ne II] image (left panel) and [S IV] image (right panel) of the H II region N160A in the Large magellanic Cloud

Most of these results, however, pertain only to oxygen abundances which are often based on the empirical method of Edmunds and Pagel (1984, see also Skillman 1989, McGaugh 1991) which uses the ratio [O II] 3727 + [O III] 4959/5007Å to Hβ to determine O/H when a measurement of the temperature sensitive line of [O III] 4363Å is unavailable. This method is not unambiguous unless the abundances are above solar. The problem with measuring the [O III] 4363Å line is that this line is generally weak. An additional problem may be that in the past the resolution of spectrophotometric observations has been poor (R ∼ a few 100), which makes it difficult to detect faint, narrow lines, in particular in the presence of strong continuum. Figure 4, showing spectra at resolutions of R ∼ 300 and R ∼ 1200 illustrates this point. Reliable O/H abundances are only available for 80 of the 400 H II regions observed in about 40 galaxies. Observations with high spectral resolution and high sensitivity, such as can be obtained with the VLT and FORS or even better FORS2, can improve this situation considerably.

Another limitation of such studies is the poor knowledge of the abundances and abundance gradients in other species such as N and S. Recent studies remain ambiguous as to whether N/H and S/H follow O/H (Garnett 1989, Torres-Peimbert et al. 1989). Though access to the wavelength range beyond 7000Å has provided measurements of the [S III] 9069, 9532 lines permitting more reliable abundances, there remains a need for measuring higher ionization stages to tighten the results. The 10.5 μm line of [S IV] is accessible from the ground. With the VLT equipped with a mid-IR spectrometer, [S IV] will be detectable in spiral galaxies out to distances of several Mpc and be very helpful in determining the ionization equilibrium. Though more difficult, one can also consider observing the 18 μm line of [S III]. The ratio S/O is determined by nucleosynthesis in stars during the lifetime of a galaxy and radial changes have important consequences for evolution models describing the star formation history. A change in S/O with metallicity could imply, for example, variations in the IMF for massive stars with metallicity.

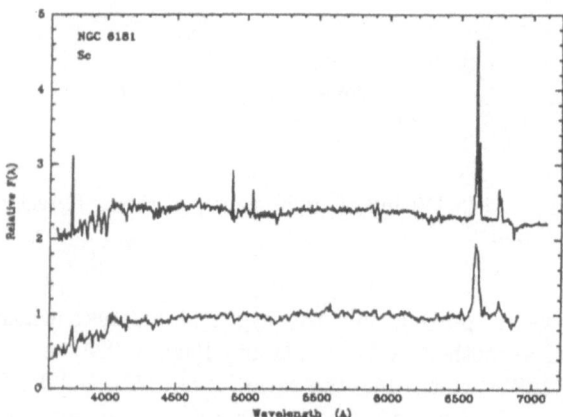

Fig. 4. Integrated spectra of the Sc galaxy NGC 6181 at resolutions of R $\sim$ 300 and R $\sim$ 1200, taken from Kennicutt (1992)

A third issue of interest which can be addressed with the VLT and FORS or UVES is the metallicity in H II regions in low surface brightness galaxies. These rather unevolved galaxies have low O abundances (McGaugh and Bothun 1993, van der Hulst et al. 1993, McGaugh 1994) but the present data are scarce and noisy and no measurements of the temperature sensitive [O III] 4363Å line exist. High S/N spectroscopy will not only provide more reliable abundance measurements but also establish the presence or absence of gradients in these objects. Chemical evolution models predict that abundance gradients develop rather rapidly. An interesting result in this respect is that the abundances in Virgo spirals are systematically higher than in field spirals (Shields et al. 1991, Skillman et al. 1993). This could be the result of suppression of inflow of metal poor gas in the Virgo spirals, implying that the abundances in the ISM in field galaxies is kept low through infall of fresh gas.

An additional question of interest which can be addressed is that of the primordial helium abundance. The primordial helium abundance can be determined from an extrapolation of the observed He/H versus O/H ratios from H II regions spanning a range of O/H (Figure 5, taken from Pagel et al. (1992)). Crucial for this determination is to have reliable helium abundances for low abundance objects which are truely unevolved. LSB galaxies are such objects. To date only measurements of Blue Compact Dwarf galaxies have been used to determine the primordial helium abundance, and it is very important to have an independent measure of the primordial helium abundance from another class of objects.

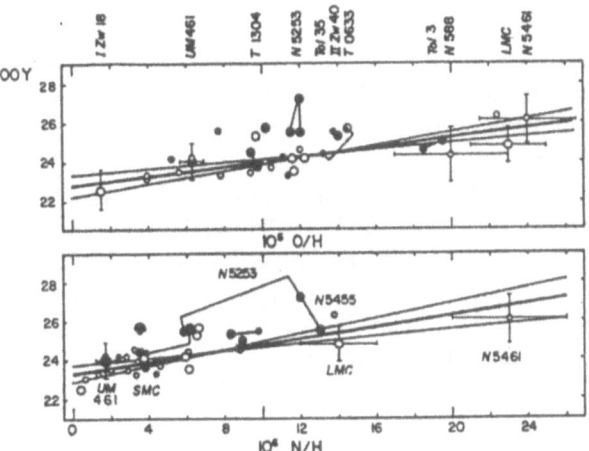

Fig. 5. Regression of helium against oxygen abundance for objects with WR features (filled circles) and without WR features (open circles) taken from Pagel et al. (1992). Object names are noted at the top.

4 Circumnuclear Regions

Quite a few spiral galaxies show enhanced star formation activity around the nucleus in a region of a few kpc in size. This enhanced star formation activity can be traced by enhanced radio continuum emission (van der Hulst 1991 and references therein), the enhanced near-IR emission (Telesco 1988, Telesco et al. 1993) and enhanced optical line emission (Keel et al. 1985). The main restriction for detailed studies of the circumnuclear star formation regions is the presence of large amounts of dust. A clear example is the hot-spot galaxy NGC 2903 where the ratio of Brα to Brγ emission from the central few 100 pc indicates an extinction of $A_v = 10 - 15$ (Simons et al. 1988).

It will be clear that the only possible way to penetrate into these regions is the near- and mid-IR windows. Operation of the VLT with tip-tilt correction will provide high angular resolution imaging and spectroscopy in the IR of the circumnuclear regions. Spectroscopy and imaging of H-recombination lines and IR fine structure lines such as [S IV], [Ne II] and [Ar III], will provide insight into the physical conditions of the circumnuclear gas. Existing work is extremely limited by sensitivity and the results so far are ambiguous. The ratios of the [Ne II], [Ar III] and [S IV] lines should be good diagnostic tools to separate starburst activity from AGN type excitation according to recent models by Spinoglio and

Malkan (1992). Starburst nuclei are expected to have lower excitation and therefore relatively stronger [Ne II] emission as compared to [Ar III] and [S IV]. The starburst galaxy NGC 5253, however, apparently has a much higher excitation than expected, as judged from its low [Ne II]/[S IV] ratio (see also Roche et al. 1991).

Observations of high excitation lines such as [Fe II] 1.64 μm can be used to trace shock and large scale winds. An example is the study of NGC 253 by Forbes et al. (1993). The large collecting area of the VLT and possibility of diffraction limited imaging in the IR will be exceptionally powerful for probing the physics in the nuclear regions of galaxies.

The kinematics of the circumnuclear gas is of interest for probing the central mass distribution and large scale flows and turbulence. The high resolution of the VLT is imperative here, provided it is coupled to a high resolution ($R \geq 20000$) spectrometer in the near- or mid-IR. Using the strongest emission lines it will be possible to make detailed studies of the kinematics of the circumnuclear gas. Such resolution is foreseen for the mid-IR spectrometer and could be provided in the near-IR by CRIRES.

References

Bloemen, J.B.G.M., 1991, *The Interstellar Disk-Halo Connection in Galaxies*, IAU symp. 144, Kluwer Acad. Publ. p.223

Courtès, G., Petit, H, Sivan, J.-P., Dodonov, S., Petit, M. 1988, Astron. Astrophys. **174**, 28

de Blok, W.J.G., van der Hulst, J.M., and Bothun, G.D. 1995 MNRAS, submitted

Dettmar, R.J. 1990 Astron. Astrophys. **232**, L15

Dettmar, R.A. 1992, Fund. Cosm. Phys. **15**, 143

Dinerstein, H. 1990, in *"The Interstellar Medium in Galaxies"*, ed. H. A. Thronson and J. M. Shull, Kluwer, p.257

Edmunds M.G. and Pagel B.E.J. 1984, MNRAS, **211**, 507

Ferguson H.C. and McGaugh S.S. 1994 Ap.J. in press.

Forbes, D. A., Ward, M. J., Rotaciuc, V., Blietz, M., Genzel, R., Drapatz, S., van der Werf, P. P. and Krabbe, A. 1993, Ap.J. **406**, L11

Field, G.B. Goldsmith D. and Habing, H.J. 1969 Ap. J. **15**, L49

Garnett, D. R. 1989, Ap.J. **345**, 282.

Heydari-Malayeri, M. and Testor, G. 1986 Astron. Astrophys. **162**, 180

Hodge, P.W. and Kennicutt R.C. 1983a A.J. **88**, 296

Hodge, P.W. and Kennicutt R.C. 1983b Ap.J. **267**, 563

Hunter D.A. and Gallagher J.S. 1990, Ap.J. **362**, 480

Hunter D.A. and Gallagher J.S. 1992, Ap.J. **391**, L9

Hunter D.A., Hawaly W.N. and Gallagher J.S. 1993, A.J. **106**, 1797

Keppel, J. W., Dettmar, R,Gallagher, J. S. I.,Roberts, M. S. 1991 Ap.J. **374**, 507

Kennicutt, R. C., Edgar, B. K. and Hodge, P. W. 1989 Ap.J. **337**, 761

Kennicutt, R. C. 1992, Ap.J. **388**, 310

Keel, W.C., Kennucutt, R.C., Hummel, E. and van der Hulst J.M. 1985, A.J. **90**, 708

Kunth, D. and Sargent, W.L.W. 1983, Ap. J. **273**, 81

Lequeux J., Rayo, J.F., Serrano, A. Peimbert, M and Torres-Peimbert, S. 1979, Astron. Astrophys. **80**, 155

Mathis, J.S. 1986, Ap.J. **301**, 423

McGaugh S.S. 1992, Ph.D. Thesis, Univ. of Michigan

McGaugh S.S. 1994, Nature **367**, 538

McGaugh, S.S. 1991, Ap.J., **380**, 140

McGaugh, S. S., Bothun, G. D., 1993 in in *The Evolution of Galaxies and Their Environment* NASA Conf. Proc. 3190, 88

Norman, C.A. and Ikeuchi S. 1989, Ap.J. **345**, 372

Pagel, B.E.J., Simonson, E.A., Terlevich, R.J., and Edmunds, M.G. 1992, MNRAS, **255**, 325

Rand, R.J., Kulkarni, S.R. and Hester, J.J. 1990 Ap.J. **352**, L1

Rand, R. J., Kulkarni, S. R. and Hester, J. J. 1992 Ap.J. **396**, 97

Rand, R.J. 1992, A.J. **103**, 815

Reynolds, R.J. 1988 Ap.J. **333**, 341

Reynolds, R.J. 1989 Ap.J. **339**, L29

Roche, P. F., Aitken, D. K., Smith, C. H. and Ward, M. J. 1991, MNRAS, **248** 606

Shields, G. A., Skillman, E. D., Kennicutt, R. C. 1991, Ap.J. **371**, 82

Shields, G. A 1990 Ann. Rev. Astron. Astrophys. **28**, 525

Sivan, J., Stansinska, G. and Lequeux, J. 1986 Astron. Astrophys. **158**, 279

Shull, J.M. 1987 in *Interstellar Processes*, ed. D.J. Hollenbach and H.A. Thronson, p. 225

Shull, J.M. and McKee C.F. 1979, Ap.J. **227**, 131

Simons, D. A., Depoy, D. L., Becklin, E. E., Capps, R. W. and Hodapp, K. 1988 Ap.J. **335**, 126

Skillman, E. D., Kennicutt, R. C., Shields, G. A. 1993 in *The Evolution of Galaxies and Their Environment*, NASA Conf. Proc. 3190, 244

Skillman, E.D. 1989, Ap.J. **347**, 883

Spinoglio L. and Malkan M.A. 1992, Ap. J. **399**, 504

Telesco, C. M., Dressel, L. L. and Wolstencroft, R. D. 1993 Ap.J. **414**, 120

Telesco, C.A. 1988, Ann. Rev. Astron. Astrophys. **26**, 343

Torres-Peimbert, S., Peimbert, M., and Fierro, J. 1989, Ap.J., **345**, 186.

van der Hulst, J.M. 1991 in *The Interpretation of Modern Synthesis Observations of Spiral Galaxies*, ed. N. Duric and P. C. Crane, ASP Conf. series No. 18, p. 215

van der Hulst, J. M., De Blok, W. J. G., McGaugh, S. S., Bothun, G. D 1993 in *The Evolution of Galaxies and Their Environment*, NASA Conf. Proc. 3190, 92

van der Werf, P. P. and Krabbe, A. 1993, Ap.J. **406**, L11

Vila-Costas, M.B. and Edmunds, M.G. 1992 MNRAS **259**, 121

Walterbos, R.A.M. and Braun, R. 1992 Astron. Astrophys. Suppl. **92**, 625

Walterbos, R.A.M. and Braun, R. 1995, Ap.J. in press.

Walterbos, R.A.M. 1991, in *The Interstellar Disk-Halo Connection in Galaxies*, IAU symp. 144, ed. J.B.G.M. Bloemen, Kluwer Acad. Publ. p.223

Walterbos, R.A.M. and Braun, R. 1992, Astron. Astrophys. Suppl. **92**, 625

Part 6

**Astronomical Targets
in Nearby Galaxies**

Supernovae and the VLT: More Light to Examine

Bruno Leibundgut and Jason Spyromilio

European Southern Observatory,
Karl-Schwarzschild-Strasse 2,
D–85748 Garching bei München,
Germany

Abstract. Supernovae constitute ideal laboratories for the investigation of the end phases of stellar evolution. The explosion mechanisms are fairly well understood in general, but the individuality of supernovae needs to be explored in more detail. The opportunity provided by these bright point sources to study the interstellar medium of external galaxies has not been fully exploited to date. The future 8m telescopes will provide the means for such investigations with profound implications for the chemical composition and evolution of galaxies. In addition to the expected extensions of existing observing programs to fainter and more distant supernovae, four specific projects for original observations of supernovae and the interstellar medium are described. Asymmetries in the explosions can be detected through spectropolarimetry and structure in emission lines. Interstellar dust properties can be inferred from polarimetry and light echoes, and very high-resolution spectroscopy probes gas in the halo, external galaxies, and intergalactic space. All these programs require the light gathering power of 8m class telescopes to investigate a fair sample of supernovae. Suitable instruments in the VLT program are identified.

1. Introduction

Characteristic signatures of supernovae (SNe) are their distinct light curves, spectral appearance and evolution, and high luminosities. Spectroscopy near maximum light forms the main classification mechanism (e.g. Harkness & Wheeler 1990). The presence (SNe II) or absence (SNe I) of hydrogen lines distinguishes the two main types. The absorption due to the Si II doublet (rest wavelength 6355 Å) near 6100 Å defines the subclass of SNe Ia and separates it from the other clear subtype of SNe Ib/c (Harkness & Wheeler 1990). A further separation into SNe Ib (He-rich) and SNe Ic (He-poor) is proposed (Wheeler & Harkness 1990, Wheeler et al. 1994), but an easy and clear identification scheme is still missing. Optical spectra are often too complicated to identify the He I series and the rather rapid spectral evolution makes classification after a few weeks past maximum a difficult proposition. The breakdown of the current classification system is illustrated by the association of SNe II and SNe Ib/c with core collapse explosions, while SNe Ia most probably result from thermonuclear explosions. The decoupling of the explosion mechanism from the light emission at the surface has been most readily demonstrated by the close resemblance of SN 1987K (Filippenko 1988) and SN 1993J (see Wheeler & Filippenko 1994 for a review of the early observations and models), which as a type II displayed many characteristics (light curves and spectra at late times) of SNe Ib/c. We are thus confronted with the problem of connecting the internal explosion physics with the external appearance (intimately related to the structure of the progenitor star). The creation of additional subclasses of supernovae does not contribute to the discussion of the underlying physics but may lead to atomization of the issues.

There exists fair agreement on the global picture, and variations among individual supernovae can be traced back to differences in the evolutionary stage of the progenitor star at explosion. The optical display from core collapse supernovae starts with the shock breaking out the surface of the progenitor star. Soft X-ray and extreme UV emission ionizes a large volume around the supernova (Fransson et al. 1994a). The object itself, i.e. the shock-heated stellar material, can brighten by $\sim$10 magnitudes in a few hours. The duration of this peak is determined by the size of the progenitor star before explosion (Falk & Arnett 1977, Klein & Chevalier 1978). After the passing of the shock, the envelope immediately starts to cool and the supernova bolometric luminosity drops rapidly. This phase has so far been observed only in SN 1987A (Arnett et al. 1989) and S-N 1993J (Schmidt et al. 1993a, Lewis et al. 1994, Richmond et al. 1994, Benson et al. 1994, Van Driel et al. 1993). The decline stops when the recombination wave, moving inward in mass coordinates, is balanced by the expansion of the envelope, which creates a spatially nearly stationary photosphere, and the light curve settles on a prolonged plateau (e.g. Eastman et al. 1994). Such a plateau can last for several months depending on the envelope size and mass, and the energy generated in the explosion (Chugai 1991, Popov 1993). The luminosity of the supernova until this phase is completely dominated by the interaction of the shock with the envelope. When the photosphere has receded deep enough into the envelope, radioactive heating begins to dominate the energetics. ^{56}Ni, produced in the shock moving through dense, hot material deep inside the progenitor star, decays to radioactive ^{56}Co, which in turn decays to stable ^{56}Fe. As a result the plateau phase might be extended for as long as the envelope remains optically thick to optical photons.

Once the ejecta have thinned out enough to allow thermalized photons to escape easily, we observe a brightness decline. For supernova progenitors with large envelopes almost all energy is converted to optical photons and the decline follows the radioactive decay time of ^{56}Co (111.3 days; the ^{56}Ni decay time [8.80 days] is so short that it all has been converted to ^{56}Co by this phase). Many SNe II have masses high enough to trap most of the energy, and display decline rates matching closely the cobalt decay times (Turatto et al. 1990, Schmidt et al. 1993b). In cases of low ejecta masses, γ-rays from the ^{56}Co decay can escape and only a fraction of the energy is deposited in the ejecta. Since the column density decreases due to the expansion, the escape fraction increases and the decline of the light curve is steeper than the decay times (Leibundgut & Pinto 1992). The contribution of e^+ depositing energy locally in the ejecta is a small fraction for several hundred days. A prime example is SN 1993J (Shigeyama et al. 1993, Woosley et al. 1994) and most SNe Ib/c (Vacca & Leibundgut 1994, Leibundgut 1994a).

More detailed descriptions have to include the amount of mixing which occurs when the shock crosses density steps in the envelope (Müller et al. 1991, Herant & Woosley 1994). Rayleigh-Taylor instabilities can bring radioactive material to the surface and some of the envelope is sunk into the core (Herant et al. 1992). The density variations also lead to clumping in the ejecta, which has been observed in a few supernovae (Stathakis et al. 1991, Spyromilio 1994).

Supernovae do not explode in a vacuum. The intense radiation field and shocks have profound environmental impacts. Many effects from such interactions have been observed, including narrow emission lines, intense X-ray and radio emission, and non-thermal continua. In some supernovae this interaction completely dominates the optical evolution (Leibundgut 1994b). Several years after explosion the conversion of shock energy into radiation can be observed (Fesen & Becker 1990, Leibundgut et al. 1991).

A dramatic change in the filter light curve occurred $\sim$500 days after explosion in SN 1987A (Suntzeff & Bouchet 1990), when dust formed in the ejecta. Most of the emission shifted to infrared wavelengths at this point.

The light curve tracks the energy deposition, since the time scale for energy release is much shorter than for the deposition. After about 900 days, however, the light curve of SN 1987A flattened out (Bouchet et al. 1991, Suntzeff et al. 1991). The recombination time for very thin material ionized at early phases becomes comparable to the age of the supernova, which causes the flattening of the light curve (Fransson & Kozma 1993).

The optical display of thermonuclear explosions (SNe Ia) is very similar to the one just described. Since the supernova here is the result of explosive burning and the progenitor star is a very compact object, the ejecta are not heated by a shock wave (Woosley & Weaver 1986). The light curve is powered solely by the decay of radioactive nuclei produced in the burning of material to nuclear statistical equilibrium. The photon diffusion in the expanding material creates a peak phase of about 40 days (Arnett 1982, Woosley et al. 1986). The small mass and rapid expansion of the ejecta, as well as the relatively complete burning of the progenitor star, result in small column densities for the γ-rays. Thus, only a small fraction of the available energy is thermalized (Leibundgut & Pinto 1992). The light curve at these phases is powered by exactly the same mechanism as in core collapse supernovae. A rapid cooling of the ejecta due to (infrared) fine structure lines of iron peak elements is expected at late phases (Fransson et al. 1994b). Also predicted is a freeze-out phase, similar to that in SN 1987A (Fransson et al. 1994b). Neither of these effects has been observed to date.

Current problems of supernova research are apparent from the above description. Prominently missing is the link between progenitor star evolution and the supernova explosion (Branch et al. 1991). Which stars blow up and what in their evolution determines the appearance of the supernova? Stellar evolution theory provides many ways to produce a star which *may*, when exploded, appear similar to what we observe (e.g. Woosley et al. 1993, 1994, Nomoto et al. 1994, Wheeler et al. 1994). Supernovae provide the unique opportunity to observe the results of stellar evolution. The stellar interior is exposed in the explosion and shock interaction with the remnant of the stellar wind unravels the mass loss history (Fransson & Chevalier 1989, Chevalier & Fransson 1994). The explosion models are not able to fully describe the propagation of the burning fronts for SNe Ia (Woosley 1990, Khokhlov 1994, Arnett & Livne 1994). The explosions produce instabilities in the ejecta, mixing various layers and favoring clumping (Müller et al. 1991). Detailed analysis of explosive nucleosynthesis is necessary to explore the chemical enrichment of galaxies. An important new field is the investigation of the effect supernovae have on their local environments. The radiation field and shock wave from the explosion interact with the circumstellar material in various observable ways (Fransson et al. 1994a, Cumming et al. 1994, Lundqvist 1994). Several years after outburst some supernovae are still emitting light through such processes (Fesen & Matonick 1994, Leibundgut 1994b).

For galaxies beyond the Local Group, supernovae are the only light beacons which allow us to study the local interstellar medium. This is one of the few ways we can learn about the chemical compositions of the ISM in external galaxies. The dust-to-gas ratio can be inferred from selective absorption (reddening) and interstellar lines in the spectra.

Supernovae have been in popular use as distance indicators (cf. Branch & Tammann 1991). One determination of the Hubble constant is based on the use of SNe Ia (see Hamuy et al. 1994 for the best controlled and most distant set of supernovae). With the availability of absolute calibrations (Saha et al. 1994) H_0 will be reliably determined in the next few years. An independent way towards H_0 is provided by the expanding photosphere method (Schmidt et al. 1994a).

2. Supernovae and ESO's Very Large Telescope

There are various obvious ways in which 8m-class telescopes will improve supernova studies. The spectral evolution at optical wavelengths can be followed to much later phases and for fainter supernovae. Infrared wavelengths will become as easily accessible as the optical region is today.

Important observations include the direct determination of the radioactive decay through the evolution of the [Co II] 10.52μm line (Aitken et al. 1988, Danziger et al. 1991), the Fe/Co ratio measured at late phases from the [Co II] ($\lambda 1.547\,\mu$m) and the [Fe II] ($\lambda 1.533\,\mu$m) lines (Varani et al. 1990) and the [Fe III] and [Co III] blends in the optical (Kuchner et al. 1994). The comparison of the the optical and IR multiplets will probe the physical conditions in the supernova ejecta. With some more spectral modelling, the Ni mass can be deduced (Ruiz-Lapuente & Lucy 1992). Detailed mapping of the 1–3 μm spectral range of SNe Ia (Spyromilio et al. 1994) is required for a further understanding of the opacities and temperatures in the ejecta. Direct observation of dust formation (Lucy et al. 1991), as well as the IR catastrophe and the freeze-out (Fransson & Kozma 1993, Fransson et al. 1994b), will further constrain the models for the ejecta and the close environment of the supernovae.

The list of supernovae observable several years past maximum (Leibundgut 1994b) will enlarge. One of the main questions here are the shapes of the emission lines which indicate the origin of the emission (forward shock in the CSM and reverse shock in the ejecta; Chevalier & Fransson 1994).

Spectroscopic studies of distant SNe will be commonplace. They will provide important clues on the cosmic distance scale well beyond most other distance indicators (e.g. Jacoby et al. 1992). The expanding photosphere method of SNe II and the empirical use of SNe Ia as distance indicators both require good spectral and photometric (not necessarily from the VLT) coverage to identify peculiarities and the decline parameter for SNe Ia (Phillips 1993, Hamuy et al. 1994). Determination of the Hubble constant beyond any currently known bulk flows could be achieved (Turner et al. 1992, Lauer & Postman 1994). The measurement of the deceleration of the Universe might be feasible with accurate data (cf. Cappellaro et al. in these proceedings).

To perform the studies mentioned above many of the foreseen and planned VLT instruments will be needed. We anticipate use of FORS, ISAAC, and MIIS. Observational requirements will be as large a wavelength coverage as possible, flexible and frequent observations, and accurate background subtraction.

In the following, we will concentrate on a few projects which are not simple extensions of existing programs but try to illustrate observations which become feasible with 8m telescopes.

2.1 Polarization

Current models for core collapse indicate large asymmetries emerging at the inner hot bubble above the proto-neutron star (Herant et al. 1992). How strongly these asymmetries imprint themselves on the surface is as yet unknown but certainly depends on how the shock is restarted by the neutrinos. Critical measurements of asphericities in the explosion are obtained with spectropolarimetry. Since any asymmetries are expected to dampen out very quickly, early phase observations are crucial.

To date spectropolarimetry has been obtained for a few objects only. The strong polarization found in SN 1987A (Bailey 1988, Cropper et al. 1988) is still mostly unexplained, but has been interpreted as due to an asphericity of the material (Jeffery 1991). The evolving polarization of SN 1993J requires an aspherical scattering surface below the region where Hα forms, i.e. inside the supernova (Trammel et al. 1993). Asymmetry has also tentatively been reported for a third core-collapse supernova, the type Ib SN 1983N (McCall 1985). All SNe Ia with spectropolarimetric observations appear to be spherical to the measurement limits (McCall et al. 1987, Spyromilio & Bailey 1994). For explosions of compact objects like SNe Ia, asymmetries are expected to disappear rapidly and only observations at very early phases can detect them. Such a measurement would be of paramount importance for the current discussion on the explosion mechanism for white dwarfs (e.g. He shell detonation models; Livne 1990, Woosley & Weaver 1994).

A second polarization component, which arises from interstellar dust grains, is often present in the observations. It provides a way to measure the size and orientation of interstellar dust grains in other galaxies (Trammel et al. 1993). Supernovae near maximum are for many galaxies the only available point sources to obtain such a measurement. The implication for our understanding of the chemical history of galaxies and the mechanisms orienting the grains is profound. All the supernovae mentioned above have determinations of the interstellar polarization along the line of sight. Most interestingly, the value for the total to selective absorption found from SN 1986G for the dust lane in NGC 5128 is lower than the canonical one for the solar neighborhood, indicating smaller characteristic grain sizes (Hough et al. 1987).

Spectropolarimetry will be possible for several supernovae per year with FORS. Polarimetry in the infrared would be advantageous since the interstellar contamination is low and a more direct measurement of the source is possible. Hence, the polarimetric functionality of ISAAC is an asset.

2.2 Probing the ISM

Interstellar (and intergalactic) narrow absorption lines are also seen in supernova spectra. The most prominent ones are Na I D and Ca II H and K. In addition, diffuse interstellar bands have been observed in at least two objects (SN 1987A: Vidal-Madjar et al. 1987, Pettini 1988; SN 1986G: di Serego Alighieri & Ponz 1987).

The spectral resolution of most supernova observations are not high enough to separate contributions of different absorbers along the line of sight. The few exceptions have yielded fairly surprising results. Towards SN 1987A up to 24 components have been identified in the Ca II line, some of intergalactic nature (Vidal-Madjar et al. 1987). These measurements probe the gas distribution in the outer Galaxy and the LMC. A similar study of SN 1993J detected intergalactic gas possibly stripped from a companion galaxy of M 81 (Vladilo et al. 1993). Currently, such observations can be obtained for SNe Ia at maximum in the Virgo cluster (SN 1991T; Meyer & Roth 1991). Surprisingly, the supernova sight line contains three additional, galactic absorption components compared with the one towards 3C 273, situated only 1.4° from the supernova (Meyer & Roth 1991). They might be the metal counterparts of high-velocity H I clouds (Robertson et al. 1991).

High resolution spectroscopy with high signal-to-noise ratio is required for such investigations. The 8m telescopes will substantially increase the limiting magnitudes. Not only more distant supernovae can be observed, also the time for which a supernova stays bright enough for such a study is extended. Supernovae in the Coma cluster can be observed with the same signal-to-noise (spectral resolution 20000, S/N∼10) with UVES in 1 hour (Dekker 1994), while it took 10 hours on a 4m telescope for a supernova in Virgo (Meyer & Roth 1991). For brighter supernovae, detailed studies at spectral resolutions of a few 100000 and S/N of better than 30 can be envisaged with a very high resolution spectrograph.

2.3 Clumping in the supernova ejecta

Evidence for fragmentation of the ejecta come from the early γ-ray detections and light curve shapes (e.g. Shigeyama et al. 1988). Models have identified Rayleigh-Taylor instabilities when the shock wave passes the O/He layer interface (Müller et al. 1991) as responsible for this mixing. The mixing scale length, however, remains largely speculative. Polarization can possibly trace large scale clumping, but an alternative method is detailed analysis of very high S/N spectral of emission lines (typically the [O I] ($\lambda\lambda$ 6300,6364 Å doublet) at late phases. This has been attempted for SN 1987A (Stathakis et al. 1991), SN 1985F (Filippenko & Sargent 1989), and SN 1993J (Spyromilio 1994). In all cases, several velocity components could be identified providing evidence for rather large (and dense) clumps. Connecting such observations with polarization measurements promise great advances in understanding the internal structure of the progenitor star.

Very high S/N is required for an unambiguous detection of the different velocity components. The resolution should be at least 3000, just at the limit of FORS 2. Higher resolution might be warranted for certain interesting cases, but no instrument for such a resolution ($\sim$10000) could be identified in the VLT program.

2.4 Light Echoes

Light we observe has not always reached us by the shortest possible route. Detection of very faint light echoes arising from dust, however, has rarely been observed. We can distinguish two kinds of echoes: radiation from dust heated by the intense radiation field; and scattered light. The two effects have distinctly different observational signatures. Heated dust primarily emits at infrared wavelengths and the light echo is observed at early times. Scattering tends to make the spectrum bluer at very late phases.

The first form of light echo has been observed in SN 1982E (Graham et al. 1983, 1986) and possibly SN 1980K (Dwek 1983). The infrared excess emission in these supernovae has been observed near maximum light and is associated with material close to the supernova.

In the case of SN 1987A scattered light has been observed as spatially resolved "expanding" rings. For certain objects the scattered light can add up to dominate the direct emission from the supernova after it has faded. SN 1991T seems to remain at B$\approx$21.3 and the observed spectrum after $\sim$900 days matches remarkably well that expected from scattering of the integrated light of the supernova through the peak phase (Schmidt et al. 1994b). Several years after explosion the extent of the echoes might be large enough to be resolved. The calculated angular size of the light echo around SN 1991T is of the order of 0.1 arcsecond three years after explosion. Although this is too small for ground-based telescopes, we might be able to resolve echoes from closer supernovae, like SN 1993J and SN 1994I, in a few years. The light echoes from SNe II will resemble their spectrum in the plateau phase.

Spectroscopic observations of (spatially unresolved) light echoes open a new window on studies of interstellar material in external galaxies. The scattering characteristics and hence the grain size can be determined from comparison of the observed spectrum with calculated integrated light spectra modified by a scattering model, thus yielding a completely independent method to infer dust properties in external galaxies.

The faintness of light echoes puts spectroscopic observations in the realm of 8m class telescopes. With a typical magnitude difference between peak light and the integrated light echo of $\sim$9 magnitudes (Schmidt et al. 1994b), we have to expect the light echoes below 20th magnitudes for almost all supernovae. The spectral evolution should be monitored to detect changes in the scattering properties of the dust as the light front expands. Low-resolution spectroscopy is the best we can hope for and FORS will certainly deliver the data required for detailed investigations.

3. Summary

Many of the proposed research projects reproduce observations which have been obtained for SN 1987A in the LMC and SN 1993J in M 81 for supernovae at significantly larger distances. Although this statement appears trivial, it is of vital importance for supernova studies. While with the current instrumentation we are able to perform a comprehensive analysis on objects which are among the half a dozen in the last century with apparent peak magnitude less than 10, the 8m telescopes will be able to observe supernovae with the same detail out to the Virgo cluster and beyond. In other words, we can expect to observe a supernova with a comparable wealth of information as was gathered for SN 1987A about once every year!

Rapid access to the telescopes through flexible scheduling will be a very important asset of the VLT. With a wide variety of instrumentation, a newly discovered supernova can be observed with several techniques simultaneously. The interesting early epochs have been quite elusive due to rapid evolution of supernovae and the difficulties encountered in supernova searches. As more searches come on line and communications improve, we can expect more supernovae to be discovered very early.

All core collapse supernovae with spectropolarimetric measurements exhibit intrinsic asymmetries. It is paramount for the explosion models to confirm these gross asymmetries. Similarly, the link between the evidence for clumping from spectral lines at late times and the polarimetric observations near maximum almost certainly will shed new light on the fragmentation mechanisms.

Supernovae will provide convenient light beacons for studies of the interstellar material in the outer halo of the Galaxy and in external galaxies. The determination of total to selective absorption is an important input for the chemical history of the galaxies. Independent determination of reddening and absorption also improve the knowledge of the supernova parameters (e.g. temperature, luminosity).

Supernova observations yield a heap of tangled information. Explosion ejecta, excited circumstellar matter, and scattered light from material in the line of sight are observed simultaneously. We can learn about the material synthesized in the explosion and by the progenitor star's internal processes, the mass loss history of the progenitor star and its evolutionary state at explosion, and interstellar as well as intergalactic material. To disentangle this information is the great trick. The VLT will provide a powerful platform to separate the effects and enhance the sample for which such studies can be performed.

References

Aitken, D. K., Smith, C. H., James, S. D., Roche, P. F., Hyland, A. R., & McGregor, P. J. 1988, MNRAS, 235, 19P

Arnett, W. D. 1982, ApJ, 253, 785

Arnett, W. D., Bahcall, J. N., Kirshner, R. P., & Woosley, S. E. 1989, ARA&A, 27, 629

Arnett, D., & Livne, E. 1994, ApJ, 427, 315

Bailey, J. 1988, PASAus, 7, 405

Benson, P. J., et al. 1994, AJ, 107, 1453

Bouchet, P., Danziger, I. J., & Lucy, L. B. 1991, AJ, 102, 1135

Branch, D., Nomoto, K., & Filippenko, A. V. 1991, Comm.Astrophys., 15, 221

Branch, D., & Tammann, G. A. 1991, ARA&A, 30, 359

Chevalier, R. A., & Fransson, C. 1994, ApJ, 420, 268

Chugai, N. N. 1991, SovAL, 17, 210

Cropper, M., Bailey, J., McCowage, J., Cannon, R. D., Couch, W. J., Walsh, J. R., Strade, J. O., & Freemann, F. 1988, MNRAS, 231, 695

Cumming, R., Meikle, P., Walton, N., & Lundqvist, P. 1994, Circumstellar Media in Late Stages of Stellar Evolution, eds. R. Clegg, I. Stevens, P. Meikle, (Cambridge University Press, Cambridge), 192

Danziger, I. J., Lucy, L. B., Bouchet, P., & Gouiffes, C. 1991, Supernovae, ed. S. E. Woosley (New York: Springer), 69

Dekker, H. 1994, Instruments for the ESO VLT, ed. A. Moorwood, (ESO: Garching), 20

di Seregho Alighieri, S., & Ponz, D. 1987, ESO Workshop on Supernova 1987A, ed. I. J. Danziger, (Garching: ESO), 545

Dwek, E. 1983, ApJ, 274, 175

Eastman, R. G., Woosley, S. E., Weaver, T. A., & Pinto, P. A. 1994, ApJ, 430, 300

Falk, S. W., & Arnett, W. D. 1977, ApJS, 33, 515

Fesen, R. A., & Becker, R. H. 1990, ApJ, 351, 437

Fesen, R. A., & Matonick, D. M. 1994, ApJ, 428, 157

Filippenko, A. V. 1988, AJ, 96, 1941

Filippenko, A. V., & Sargent, W. L. W. 1989, AJ, 345, L43

Fransson, C., & Chevalier, R. A. 1989, ApJ, 343, 323

Fransson, C., Houck, J., & Kozma, C. 1994b, IAU Colloquium 145: Supernovae and Supernova Remnants, ed. R. McCray, (Cambridge: Cambridge University Press), in press

Fransson, C., & Kozma, C. 1993, ApJ, 408, L25

Fransson, C., Lundqvist, P., & Chevalier, R. A. 1994a, ApJ, in press

Graham, J. R., et al. 1983, Nature, 304, 709

Graham, J. R., & Meikle, W. P. S. 1986, MNRAS, 221, 789

Harkness, R. P., & Wheeler, J. C. 1990, Supernovae, ed. A. G. Petschek, (New York: Springer), 1

Hamuy, M., Phillips, M. M., Maza, J., Suntzeff, N. B., Schommer, R. A., & Avilés, R. 1994, AJ, in press

Herant, M., Benz, W., & Colgate, S. 1992, ApJ, 395, 642

Herant, M., & Woosley, S. E. 1994, ApJ, 425, 814

Hough, J. H., Bailey, J. A., Rouse, M. F., & Whittet, D. C. B. 1987, MNRAS, 227, 1P

Jacoby, G., et al. 1992, PASP, 104, 599

Jeffery, D. J. 1991, ApJ, 375, 264

Khokhlov, A. 1994, ApJ, 424, L115

Klein, R. I., & Chevalier, R. A. 1978, ApJ, 223, L109

Kuchner, M. J., Kirshner, R. P., Pinto, P. A., & Leibundgut, B. 1994, ApJ, 426, L89

Lauer, T. R., & Postman, M. 1994, ApJ, 425, 418

Leibundgut, B. 1994a, The Lives of Neutron Stars, eds. A. Alpar, J. van Paradijs, (Dordrecht: Kluwer), in press

Leibundgut, B. 1994b, Circumstellar Media in Late Stages of Stellar Evolution, eds. R. Clegg, I. Stevens, & P. Meikle, (Cambridge: Cambridge University Press), 100

Leibundgut, B., Kirshner, R. P., Pinto, P. A., Rupen, M. P., Smith, R. C., Gunn, J. E., & Schneider, D. P. 1991, ApJ, 372, 531

Leibundgut, B., & Pinto, P. A. 1992, ApJ, 401, 49

Lewis, J. M., et al. 1994, MNRAS, 266, L27

Livne, E. 1990, 354, L53

Lucy, L. B., Danziger, I. J., Gouiffes, C., & Bouchet, P. 1991, Supernovae, ed. S. E. Woosley, (New York: Springer), 82

Lundqvist, P. 1994, Circumstellar Media in Late Stages of Stellar Evolution, eds. R. Clegg, I. Stevens, P. Meikle, (Cambridge University Press, Cambridge), 213

McCall, M. L. 1985, Supernovae as Distance Indicators, ed. N. Bartel (Berlin: Springer), 48

McCall, M. L., Reid, N., Bessell, M. S., & Wickramasinghe, D. 1984, MNRAS, 210, 839

Meyer, D. M., & Roth, K. C. 1991, ApJ, 383, L41

Müller, E., Fryxell, B., & Arnett, D. 1991, A&A, 251, 505

Nomoto, K., Yamaoka, H., Pols, O. R., van den Heuvel, E. P. J., Iawamoto, K., Kumagai, S., & Shigeyama, T. 1994, Nature, in press

Pettini, M. 1988, PASAus, 7, 527

Phillips, M. M. 1993, ApJ, 413, L105

Popov, D. V. 1993, ApJ, 414, 712

Richmond, M. W., Treffers, R. R., Filippenko, A. V., Paik, Y., Leibundgut, B., Schulman, E., Cox, C. V. 1994, AJ, 107, 1022

Robertson, J. G., Morton, D. C., Schwarz, U. J., van Woerden, H., & Murray, J. D. 1991, MNRAS 248, 508

Ruiz-Lapuente, P., & Lucy, L. B. 1992, ApJ, 400, 127

Saha, A., Labhardt, L., Schwengeler, H., Macchetto, F. D., Panagia, N., Sandage, A., & Tammann, G. A. 1994, ApJ, 425, 14

Schmidt, B. P., Kirshner, R. P., Eastman, R. G., Grashuis, R., Dell'Antonio, I., Caldwell, N., Foltz, C., Huchra, J. P., & Milone, A. 1993a, Nature, 364, 600

Schmidt, B. P., et al. 1993b, AJ, 105, 2236

Schmidt, B. P., Kirshner, R. P., Eastman, R. G., Phillips, M. M., Suntzeff, N. B., Hamuy, M., Maza, J., & Avilés, R. 1994a, ApJ, 432, 42

Schmidt, B. P., Kirshner, R. P., Leibundgut, B., Wells, L. A., Porter, A. C., Ruiz-Lapuente, P., Challis, P., & Filippenko, A. V. 1994b, ApJ, in press

Shigeyama, T., Nomoto, K., & Hashimoto, M. 1988, A&A, 196, 141

Shigeyama, T., Suzuki, T., Kumagai, S., Nomoto, K., Saio, H., & Yamaoka, H. 1994, ApJ, 420, 341

Spyromilio, J. 1994, MNRAS, 266, L61

Spyromilio, J., & Bailey, J. 1993, PASAus, 10, 263

Spyromilio, J., Pinto, P. A., & Eastman, R. G. 1994, MNRAS, 266, L17

Stathakis, R. A., Dopita, M. A., Cannon, R. D., & Sadler, E. M. 1991, Supernovae, ed. S. E. Woosley, (New York: Springer), 95

Suntzeff, N. B., & Bouchet, P. 1990, AJ, 99, 650

Suntzeff, N. B., Phillips, M. M., Depoy, D. L., Elias, J. H., & Walker, A. R. 1991, AJ, 102, 1118

Trammel, S. R., Hines, D. C., & Wheeler, J. C. 1993, ApJ, L21

Turatto, M., Cappellaro, E., Barbon, R., Della Valle, M., Ortolani, S., & Rosino, L. 1990, AJ, 100, 771

Turner, E. L., Cen, R., & Ostriker, J. P. 1992, AJ, 103, 1427

Vacca, W. D., & Leibundgut, B. 1994, in preparation

Van Driel, W., et al. 1993, PASJ, 45, L59

Varani, G.-F., Meikle, W. P. S., Spyromilio, J., & Allen, D. A. 1990, MNRAS, 245, 570

Vidal-Madjar, A., Andreani, P., Cristiani, S., Ferlet, R., Lanz, T., & Vladilo, G. 1987, A&A, 177, L17

Vladilo, G., Centurion, M., de Boer, K. S., King, D. L., Lipman, D., Stegert, J., Unger, S. W., & Walton, N. A. 1993, A&A, 280, L11

Wheeler, J. C., & Filippenko, A. V. 1994, IAU Colloquium 145: Supernovae and Supernova Remnants, ed. R. McCray, (Cambridge: Cambridge University Press), in press

Wheeler, J. C., & Harkness, R. P. 1990, Rep. Prog. Phys., 53, 1467

Wheeler, J. C., Harkness, R. P., Clocchiatti, A. Benetti, S., Brotherton, M., Depoy, D., & Elias, J. 1994, ApJ, in press

Woosley, S. E. 1990, Supernovae, ed. A. G. Petschek (New York: Springer), 30

Woosley, S. E., Eastman, R. G., Weaver, T. A., & Pinto, P. A. 1994, ApJ, 429, 300

Woosley, S. E., Langer, N., & Weaver, T. A. 1993, ApJ, 411, 823

Woosley, S. E., Taam, R. E., & Weaver, T. A. 1986, ApJ, 301, 601

Woosley, S. E., & Weaver, T. A. 1986, ARA&A, 24, 205

Woosley, S. E., & Weaver, T. A. 1994, ApJ, 423, 371

Supernovae with the VLT

Enrico Cappellaro[1], I. John Danziger[2], Paolo A. Mazzali[3], Massimo Turatto[1]

[1] Osservatorio Astronomico di Padova, vicolo dell'Osservatorio 5, I-35122 Padova, Italy
[2] European Southern Observatory, Karl-Schwarzschild 2 D-85748, Garching bei München, Germany
[3] Osservatorio Astronomico di Trieste, via Tiepolo 11, I-34131 Trieste, Italy

Abstract. A strategy for a SN search in clusters at redshift $z = 0.3 - 1$ and for the use of SNIa as distance indicators for cosmological tests is presented. It is stressed that, after peculiar SNIa are identified and excluded, the main uncertainty lies in the absorption correction. In particular we illustrate the problems of the widely applied method of $(B-V)$ color comparison for the estimate of the absorption. A fundamental step is the identification of the physical origin of the observed diversity of SNIa: to this aim the synthetic spectrum fitting technique is the most important tool available. The expected performance of FORS is adequate for our programme; the main requirement for VLT is flexibility of scheduling and operation

1 The ESO Key Programme on SNe

Since 1990 we have conducted at ESO a Key Programme for the optical monitoring of supernovae (SNe). With the outstanding exception of SN 1987A in the LMC, before the beginning of this programme, very few SNe had been observed at ESO. The main problem was that SNe are unpredictable, short time scale events and using the year-long proposal submission – scheduling process is not effective. Especially if one wants to study SNe at maximum, telescope time has to be allocated in advance of object discovery. Thanks to the fact that in the last decade the number of SN discoveries has grown to several tens per year, we could convince the OPC that at the time of observation there will actually be SNe to be observed. The KP has a three year life-time but telescope time is scheduled on a six month basis according to target priorities. Ideally for a SN monitoring program, an even higher flexibility would be required, with telescope time allocated, but not scheduled, and some freedom left to the investigators on how to distribute this time according to target priorities.

The telescopes and instruments available at La Silla are adequate for the study of the photometric and spectroscopic evolution in the early phases of SNe at z up to 0.05 or for the nebular phases of SNe up to the Virgo distance. However, based on the past experience, there are a number of topics that cannot be addressed with the available instrumentation but will be within the capabilities of VLT. Among these we are especially interested in the following:

1. searching for SNe in distant galaxy clusters and thereafter using SNIa as distance indicators for cosmological tests;

2. filling the gap between the SN event and the SN remnant phase. At present only a handful of objects, either type II linear or unclassified, have been observed up to several years after explosion. Most SNe, even in galaxies of the Virgo cluster, drop below the limiting magnitude of the largest existing telescope within a couple of years after the explosion.

2 SNIa as Distance Indicators

SNe are attractive distance indicators mainly because of their bright absolute magnitudes at maximum. In the literature two approaches have been followed: either to select a subclass of SNe showing homogeneous behaviour (typically SNe Ia) and assume that they reach the same maximum absolute magnitude, or to apply to SNII the so-called expanding photosphere method. The latter method has the advantage that it does not require local calibrators, but requires detailed photometric and spectroscopic observations of the photospheric phase of SNII, and it is therefore somewhat limited in distance (up to now the best case is SN1992am at $z = 0.05$, Schmidt et al. 1994) . Also, the basic assumption that SNII in the early phases radiate like black bodies needs to be corrected by introducing an *ad hoc* parameter, the distance correction factor, which is model dependent.

The homogeneous photometric behaviour of SNI, i.e. those SNe not showing hydrogen lines, was emphasized almost 30 years ago (see Branch & Tammann 1992 for a recent review). In the eighties, a subclass of SNI, which do not show a typical SiII spectral feature at 6350 Å, was identified. The progenitors of these SNe, called Ib/c, clearly originate from a different evolutionary path with respect to the more frequent SNIa. It is important to stress that SNIb/c are fainter and redder than SNIa: if photometry alone is available the difference may erroneously be attributed to extinction. After the discovery of this new class, the assumption was made that only Ia are standard candles. Then in 1991 two Ia SNe were discovered that have shaken this belief: the bright 1991T and the faint 1991bg. Since both showed distinctive spectral peculiarities it has been argued , once again, that they can be recognized and excluded.

But are *normal* SNIa really standard candles ?

Recent reports have undermined this conclusion: Phillips (1993) showed that the absolute magnitudes of SNIa are correlated with the luminosity decline rate, over a range of more than 2 mag, and van den Bergh & Pazder (1992) suggested that the absolute magnitudes of SNIa that exploded in ellipticals are different from those of SNIa in spirals. The latter conclusion was based on the assumption that SNIa have the same intrinsic $(B - V)_0$ colors and that the observed differences are only due to extinction. In the literature adopted values of the $(B - V)_0^{max}$ range from -0.30 to $+0.09$, which implies an indeterminacy of 1.6 mag in the absorption correction. In the past the most used value has been $(B - V)_0^{max} = -0.15$ (Branch & Tammann 1992), but recently a redder color has become more fashionable (e.g. Sandage & Tammann 1993).

To stress this crucial point, in Fig. 1 we report, for the SNIa exploding in the Virgo cluster, the apparent B magnitude and the B-V color at maximum, both corrected only for galactic absorption. Included are also SNe 1991T and 1991bg. From the figure, the following information can be gleaned: a) if *normal* SNIa have a unique colour at maximum this is at least as blue as that of SN1994D, $(B - V)_0 = -0.08$; b) even excluding peculiar SNIa, the dispersion of the extinction-corrected B magnitude is relatively large, $\sigma \simeq 0.3$ mag; c) alternatively there may exist an intrinsic relation between color and magnitude with cooler SNe being fainter. The least squares fit, including peculiar SNIa, mimics an absorption law with a value of the absorption to reddening ratio $R = A_B/E(B-V) = 3$, smaller than the standard value R=4.3. In addition we note that this may be the reason why anomalous R values are derived from SN photometry (Branch & Tammann 1992). The question can be solved only by using different techniques for the estimate of the interstellar extinction. Particularly promising, for calibrating the relations between interstellar line strengths and amount of extinction, is high resolution spectroscopy of SNe, a field were VLT+UVES will allow significant improvements.

Another way to tackle the problem is to try to understand the physical causes of the observed diversity. In the last few years and in the framework of the ESO-KP a spectrum synthesis code has been developed which is a powerful tool for the interpretation of the observed spectrum (e.g. Lucy 1987, Mazzali et al. 1992). The code computes the emergent spectrum of SNe in the photospheric phase, starting from different explosion models and using an extensive line list and the most sophisticated physics possible. Because spectra of good S/N are needed, at present the application of this technique has been confined to nearby SNe up to the Virgo distance. With some effort it can be extended, using the larger telescopes available today, to SNe at $z \sim 0.1$. This is an important step towards the goal of using VLT observations of SNIa up to $z = 1$ for cosmological tests.

3 Search of SNe in Distant Clusters

Despite the caveats mentioned above we stress that, once the most peculiar objects are excluded, SNIa remain among the best distance indicators at cosmological distances. An example of what could be accomplished beyond H_0 by observing SNIa at a redshift $z = 0.5$ is the determination of the deceleration parameter q_0. At a redshift $z = 0.5$ the difference in distance modulus between $q_0 = 0$ (empty universe) and $q_0 = 0.5$ (closed universe) is 0.3 mag. If the real dispersion in normal SNIa's luminosity at maximum light is 0.3, then one would need to accumulate good quality data on 20 SNIa in this redshift regime to distinguish between the above 2 models. When one allows for time dilation the light curves are such that in an interval of 1 month the SN would fade 1.5 magnitude below maximum. This provides one measure of the time scale on which work of this kind has to be accomplished.

The main observational limitation is that SNe must be caught within the short period of their maximum luminosity and therefore a SN search must be

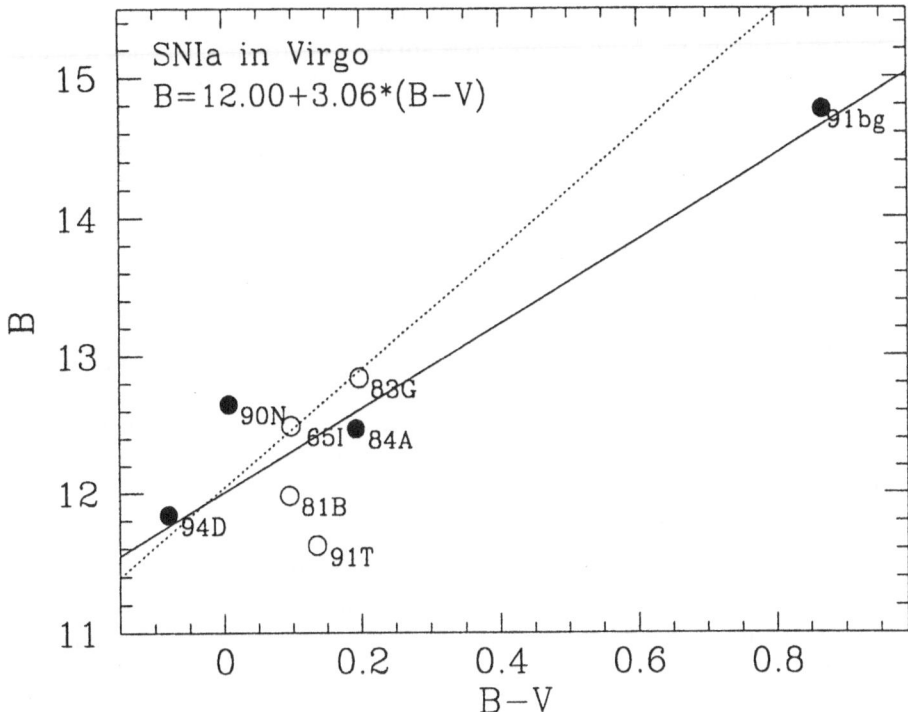

Fig. 1. Apparent B magnitude vs. $(B-V)_0$ color index for SNIa that have exploded in galaxies which are certain members (filled dots) or possible members (open circles) of the Virgo cluster. Magnitudes and colors have been corrected for the galactic extinction. The continuous line shows the least squares fit through all the data points, whereas the dotted line shows the standard absorption-reddening relation.

active. With enormous efforts, SN search programs in clusters at redshift $0.2 < z < 0.5$ have brought some success (Nørgaard-Nielsen et al. 1989, Perlmutter et al. 1994). However so far they have given only a marginal contribution to the distance scale problem mainly because, at present, it is impossible to obtain spectra of adequate S/N for SNe at these distances to confirm that the discovered SNe are *normal* Ia.

In the range $z = 0.3 - 1$, the expected SNIa maximum magnitude is $21 - 25$ and for an efficient search the telescope limiting magnitude must reach $25 - 28$. This is because: a) we must account for the chance that the SN is discovered not exactly at maximum; and b) to reduce the number of false alarms, the contrast with the background must be at least 2 mag. A possible strategy for a SN search in distant clusters would be to obtain one exposure per month of each of a dozen selected fields. Even if near maximum SNIa are intrinsically very blue, because of the large redshift, red band observations are most appropriate for such a search. Based on the recent estimate of SNIa rate ($0.05 - 0.25$ SNu (1 SNu = 1 SN per 100 yr per $10^{10}L_{B\odot}$), Cappellaro et al. 1993), and exploiting the wide field of view of FORS, or even better of the proposed Wide Field Direct Visual Camera (WFC), to include a large number of possible parent galaxies, we would expect from 3 to 15 SNIa per year.

The operational procedure might be the following:

1. one frame for each field is obtained;
2. within the next day the frame is compared with archive images and SN candidates are identified. The good sampling of the PSF of the WFC would allow a strong reduction of the false alarms;
3. within a couple of days a confirmation image is taken and thereafter, but not later than a week from discovery, a spectrum is obtained;
4. at least a couple of photometric points are obtained in the month following discovery to constrain the light curve.

We stress that the number of SNe discovered is in itself very important. There have been suggestions that, in order to account for the observed mass of iron in clusters, the past rate of SNIa had to be at least a factor of 10 higher than the present rate (Renzini et al. 1993). The proposed SN search program could be a direct test of this scenario.

4 Requirements

The proposed investigations can be tuned to the currently estimated VLT instrument capabilities. The only strong requirement is flexibility of scheduling and operation. It is necessary that images are available to investigators early on and that the required additional observations can be included in the telescope schedule at short notice (one day time-scale). On-line evaluation of the data is important and therefore the presence of one of the investigators during the observations is important and almost mandatory.

References

Branch, D., Tammann, G.A., 1992, ARA&A 30, 359
Cappellaro, E., Turatto, M., Benetti, S., Tsvetkov, D.Yu., Bartunov, O.S., Makarova, I.N., 1993, A&A 273, 383
Lucy, L.B., 1987, in ESO Workshop on SN 1987A, ed. I.J. Danziger, p.417
Mazzali, P.A., Lucy, L.B., Danziger, I.J., Gouiffes, C., Cappellaro, E., Turatto, M., 1993, A&A 269, 423
Nørgaard-Nielsen, H.U., Hansen, L., Jørgensen, H.E., Salamanca, A.A., Ellis, R.S., Cough, W.J., 1989, Nature 419, 52
Perlmutter, S. et al., 1994, IAU Circular no. 5956
Phillips, M.M, 1993, ApJ 413, L105

Renzini, A., Ciotti, L., D'Ercole, A., Pellegrini, S., 1993, 419, 52

Sandage, A, Tammann, G.A., 1993 ApJ 415, 1

Schmidt, B.P. et al., 1994, AJ 107, 1444

van den Bergh, S., Pazder, J., 1992, ApJ 390, 34

Late Stages of Supernovae with the VLT

Claes Fransson

Stockholm Observatory, S-133 36 Saltsjöbaden, Sweden

Abstract. Many of the most fundamental questions in relation to supernovae, like abundance determinations, isotope ratios, the IR-catastrophe, molecule and dust formation, mixing, possible pulsar input, and circumstellar interaction are best studied at epochs later than one year. The VLT offers new opportunities to study these, especially in the near and far-IR, to comparable levels as for SN 1987A. The most interesting VLT instruments for the study of the late phases are FORS for the optical spectrum, ISAAC for the near-IR and VISIR for far-IR photometry and low resolution spectroscopy.

1 Introduction

The VLT will make new types of observations of supernovae possible, especially at late phases, which is also the epoch when some of the most fundamental questions can be addressed. For reasons of space I will, in this review, limit myself to these phases, where the VLT can make the most obvious progress. There are, however, a number of problems at early phases where the VLT can offer new possibilities, e.g., high resolution observations and polarization (see Leibundgut & Spyromilio, these proceedings).

SN 1987A was in many ways unique. Its distance of ~ 50 kpc makes it a factor of ~ 70 closer than the second closest supernova during the last 50 years, the Type IIb SN 1993J in M 81 (distance ~ 3.5 Mpc) and the Type Ia SN 1972E at a distance of ~ 4.1 Mpc. Other recent nearby supernovae include SN 1994I in M 51 (distance ~ 7 Mpc). In a five year period we can therefore realistically expect 1 ± 1 supernova within 5 – 10 Mpc. The dilution factor is therefore at least 10^{-4}, compared to SN 1987A, or ~ 10 magnitudes. However, SN 1987A only represented one case and was also in some aspects peculiar. This was mainly related to the compact nature of the blue progenitor, resulting in e.g. a relatively faint peak magnitude, and a near-absence of a shock dominated diffusion phase during the first ~ 100 days. The circumstellar medium was also different from normal Type II supernovae, where the red supergiant progenitor is normally surrounded by a dense circumstellar medium. It is therefore clear that observations of a larger sample of supernovae, representing different classes, are necessary. Hopefully, these observations will match the quality of those of SN 1987A.

After ~ 100 days the light curve and spectral properties of SN 1987A converged to that of standard Type II supernovae. Since I will in this review mainly discuss the late phases, i.e. later than ~ 100 days, I will use the SN 1987A as a template for what can be expected for Type II supernovae. Also, the Type IIb and Type Ib/Ic supernovae fall into this class of core collapse object, originating from progenitors with masses larger than $\sim 10\ M_\odot$. The main difference between these classes and Type II's is the near or complete absence of a hydrogen envelope, and in the case of Type Ib supernovae, also the helium mantle. The mass of the oxygen core, where most of the nuclear processing has taken place, is similar in all cases, as well as the basic power supply, ^{56}Ni. The absence of a massive hydrogen envelope results in a higher expansion velocity for the core, and therefore a lower density and higher ionization in the ejecta.

Type Ia supernovae constitute a physically different class of objects, with the explosion energy coming from nuclear processing, rather than gravitational binding energy. The formation of the spectrum and light curve at late phases is, however, similar to Type II's, differing mainly in the relative fractions of iron-group elements to that of intermediate mass elements.

Drawing on our experience from SN 1987A, I will discuss a few specific problems, where the VLT, and similar size telescopes, can provide information of the same standard as obtained from SN 1987A. The final sections summarize the implications for the VLT instruments. Complementary reviews covering the physics in more detail can be found in Fransson (1994a, b), for general reviews of SN 1987A see Chevalier (1992) and McCray (1993).

2 Scientific Case

2.1 Type II Supernovae at Late Phases

The bolometric light curve reflects the integrated energy input and diffusion of energy from the supernova, and is therefore of special importance. After ~ 100 days the diffusion time-scale is short compared to the hydrodynamic expansion time-scale (i.e. the age of the supernova). Photons can then escape from the ejecta relatively freely, and the bolometric light curve directly reflects the energy production in the ejecta, and therefore the radioactive decay processes. The most important of these is the decay of ^{56}Ni into ^{56}Co on a time-scale of 8.8 days, and then ^{56}Co into ^{56}Fe on a time-scale of 111.3 days. In both decays most of the energy is emitted as γ-rays, which are thermalized in the ejecta into optical and IR photons. It is now well established from the bolometric light curve that $\sim 0.07\ M_\odot$ of ^{56}Ni was created in SN 1987A, and that this was responsible for most of the observed emission from the supernova during the first ~ 800 days. From ~ 100 days up to ~ 600 days a log $L_{bol} - t$ plot should give a straight line with a slope determined only by the decay time of ^{56}Co. In Fig. 1 L_{bol} is shown for SN 1987A, as given by Bouchet et al. (1994), and it is obvious that the ^{56}Co decay model is confirmed to a high degree of accuracy up to ~ 800 days.

Earlier than ~ 500 days most of the thermalized energy is emitted in the optical and near-IR, and L_{bol} is relatively easy to calculate. However, later than ~ 500 days the determination of the bolometric light curve is far from trivial. The main problem is that after this epoch most of the energy comes out in the much less accessible far-IR, with $\lambda \gtrsim 5\mu$m both for observational and theoretical reasons, and has basically two origins.

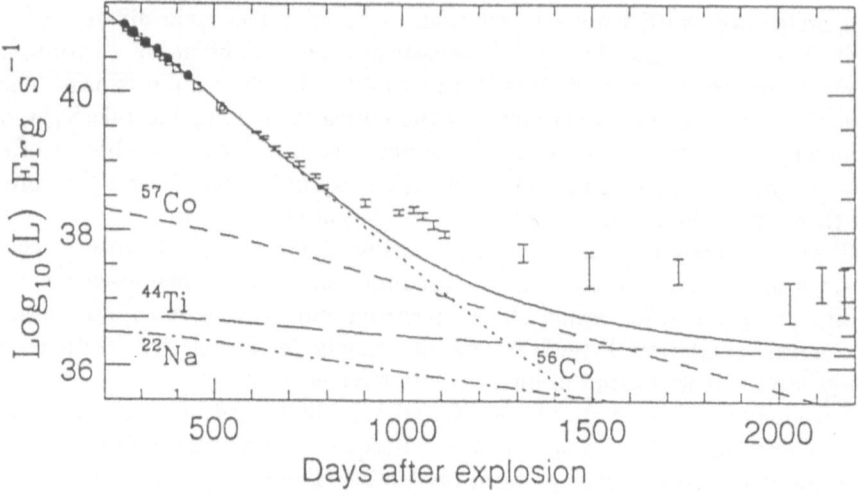

Fig. 1. Bolometric luminosity for SN 1987A from Bouchet et al. (1994). The solid line gives the predicted luminosity from Woosley et al.'s (1989) model, and the dashed lines the individual contributions from ^{56}Co, ^{57}Co, ^{44}Ti and ^{22}Na. Note the deficit after $\sim$ 800 days.

First, at $\sim$ 500 days the temperature in the inner oxygen rich regions drops rapidly from $\sim$ 2000 K to $\lesssim$ 500 K, as a result of a thermal instability (Fransson & Chevalier 1987, 1989; Fransson et al. Kozma 1994). This instability is commonly referred to as the IR-catastrophe (Axelrod 1980), and is a result of the gradual dominance of far-IR fine structure lines on the radiative cooling.

Although the IR-catastrophe has long been predicted, observational evidence has been scarce. However, clear evidence can be seen in observations of especially the [O I] $\lambda\lambda$ 6300, 6364 line at late times. In Fig. 2 we show the [O I] luminosity from Danziger et al. (1991) (dots), together with the bolometric luminosity from Suntzeff et al. (1991). The [O I] luminosity calculated by Fransson et al. (1994) is plotted together with the observations, as the solid line. If the [O I] flux is compared with the bolometric light curve, one sees that between $\sim$ 400 days and $\sim$ 700 days, the [O I] flux drops much faster than the bolometric flux. After this epoch the ratio of the two is approximately constant at a level of $\sim 2.9 \times 10^{-3}$. The rapid decrease is a result of the temperature drop, giving an exponential drop in the emissivity. The [O I] line would disappear completely if a new process did not enter. Fransson et al. (1994) show how modeling of the non-thermal part of the lightcurve can give a relatively model-independent determination of the oxygen mass. This is a direct measure of the progenitor ZAMS mass, and shows why late time observations are of prime interest for understanding supernovae. Additional evidence for the IR-catastrophe comes from observations and analysis of near and far-IR [Fe II] lines by Spyromilio & Graham (1992) and Li et al. (1993). Spyromilio & Pinto (1991) also show how the observation of the line ratio [O I] λ 6300 / [O I] λ 6364 provides an important possibility to determine the density in the ejecta. The observation of the IR-catastrophe and the non-thermal phase have only been possible for SN 1987A, but are within reach with the VLT for more distant supernovae. Direct observation of the nucleosynthesis is, of course, one of the major challanges for VLT.

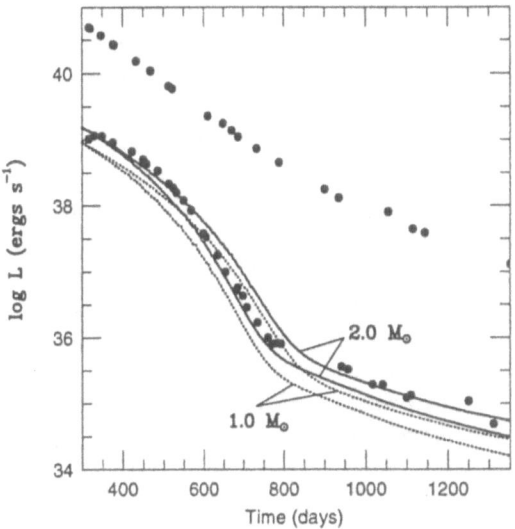

Fig. 2. Luminosity of [O I] $\lambda\lambda$ 6300 – 64 as a function of time, together with observations from Danziger et al. (1991). The lower curve for each mass corresponds to an expansion velocity of 2000 km s^{-1} and the upper to 1000 km s^{-1}. The bolometric luminosity from Suntzeff et al. (1991) is shown as the upper dots. Before 750 days the emission is dominated by thermal excitations, while after this epoch non-thermal excitations dominate. This results in a nearly constant ratio of the line luminosity to the bolometric, as is observed.

At the same epoch as the IR-catastrophe occurred, dust formation was observed in SN 1987A (Lucy et al. 1991). This was seen as a shift in the line peaks of the strong optical lines, like [O I]. Evidence for a dust continuum was first seen at $\sim$ 350 days (Meikle et al. 1993). The coincidence of the dust formation and IR-catastrophe is not surprising, because molecule and dust formation are natural consequences of rapid cooling (e.g. Kozasa et al. 1989, 1991). Both the IR-catastrophe and the dust has the effect of shifting the energy output into the far-IR. At temperatures above $\sim$ 2000 K excitations of forbidden, optical and near-IR lines dominate the cooling. Below $\sim$ 2000 K the strongest lines are fine structure transitions of [Ne II] 12.81μm, [Si II] 34.81μm, [Fe I] 24.05, 34.72μm, and [Fe II].

SN 1987A was extensively observed in the IR. Ground based observations mainly concentrated on medium resolution spectroscopy in the J, H and K bands, and narrow and medium wide band photometry in the L, M, N and Q bands (e.g. Bouchet et al. 1994; Suntzeff et al. 1991, 1992). In Fig. 3 we show the broad band photometry from days 616 to 2172 from Bouchet et al. Atmospheric absorption is obviously a large problem for a complete coverage. Especially the [Fe I] lines at 1.443, 1.498, 1.535μm and [Fe II] lines at 1.257μm and 1.644μm, together with the [Co II] 1.547μm line in the near-IR are interesting for the study of the radioactive decay of ^{56}Co and ^{57}Co (see below). In the K and L windows the most important features were the fundamental and first overtone vibration bands of CO at 2.2μm and 4.6μm, as well as SiO at 7.9μm. These appeared already at $\sim$ 100 days, and are likely to be formed in the carbon rich parts of the oxygen core. Also the [Ne II] 12.81μm line was strong, probably coming from processed material. In the far-IR the most prominent lines were [Co II] 10.52μm, observable from ground, and [Fe II] 17.94, 25.99μm and [Fe I] 24.05μm, only observable with the Kuiper Airborne Observatory (KAO) (e.g., Wooden et al. 1993). From fluxes and profiles of these lines, abundances and distribution of iron can be estimated. An important advantage with the IR lines is their insensitivity to density and temperature, and the relative freedom of line blending. Abundance determinations from these lines are therefore more reliable. Summarizing, it is clear that *the most important wavelength range for the energy budget after $\sim$ 500 days, as well as for understanding the ejecta structure, is from $\sim$ 1μm to $\sim$ 100μm.*

Returning to the bolometric light curve, we see from Fig. 1 that after $\sim$ 700 days the light curve flattens considerably. The most straightforward interpretation is that the next most abundant isotope, ^{57}Co, takes over as the dominant energy source. Although the mass is small, $\sim 10^{-3} M_\odot$, the longer decay timescale of 391 days causes it to dominate ^{56}Co after $\sim$ 800 days, if the ^{57}Ni/^{56}Ni ratio is 1.5 – 2 times solar (see below). Calculations of the nucleosynthesis indicate that $\sim 10^{-4} M_\odot$ of ^{44}Ti is also produced in the explosion, which should dominate after $\sim$ 1500 days. The decay time of ^{44}Ti is 78 years. Because of the efficient trapping of energy it should keep the ejecta observable for many years at roughly the same level. The ejecta of SN 1987A may therefore be observable even in the VLT era.

Although a reasonable fit to the light curve could be obtained, the mass of ^{57}Co determined from the light curve was considerably higher than in models. In addition, it also disagreed with the independent determination of the ^{57}Ni/^{56}Ni ratio from observations of line ratios in the IR. Varani et al. (1990) observed the ratio of the [Co II] 1.547μm lines and [Fe II] 1.533μm with FIGS at the AAT as a function of time, while Bouchet & Danziger (1993) used the low resolution CVF ($R \sim 60$) to observe the 10.52μm [Co II] line. Both these sets of observations implied an isotope ratio of ^{57}Ni/^{56}Ni $\approx$ 1.5 – 2. The ratios obtained by these groups were therefore in considerably better agreement with the nucleosynthesis

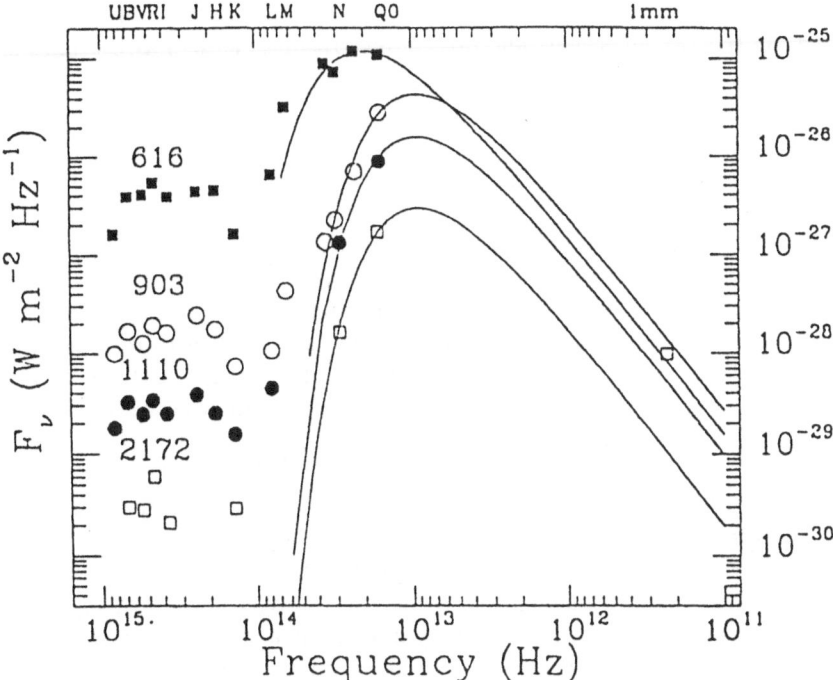

Fig. 3. Broad band photometry of SN 1987A from day 616 to day 2172 (from Bouchet et al. 1994).

models. Roughly the same ratio was indicated by γ-ray observations (Kurfess et al. 1992).

Observations of these line ratios for other supernovae would be extremely valuable, because of the sensitivity of the $^{57}Ni/^{56}Ni$ ratio to the temperature and density immediately outside the neutron star (e.g., Woosley & Hoffman 1991). In addition, the widths of the [Fe II] and [Co II] IR-lines give crucial information on the degree of mixing in the explosion. This is one of the few probes we have of the physics during the first few seconds of the explosion, in particular, the importance of neutrino-induced convection (e.g., Herant et al. 1994).

There are two observationally interesting complications to this picture in the form of additional energy sources and time-dependent effects. These may also explain the disagreement between the spectral observations and the light curve conclusions. The most obvious additional energy source, besides radioactive decay, is energy input from a neutron star. This can be either as a result of conversion of the rotational spin-down energy of a pulsar, or as a result of accretion of material which did not reach escape velocity in the explosion. The latter possibility has been discussed by Chevalier (1989), and could lead to the formation of a black hole. Accretion either onto a black hole or a neutron star results in X-rays, which are absorbed by the inner parts of the ejecta, and converted into optical and IR emission. In the case of a pulsar the electromagnetic radiation will be absorbed by the ejecta in the same way as in the Crab nebula. A signature of a central pulsar would be narrow ($V \lesssim 1000$ km s^{-1}) high-ionization lines, e.g. [O III] $\lambda\lambda$ 4959, 5007 and optical [Fe III] lines, as well as UV lines of e.g. C IV and O IV – VI (Chevalier & Fransson 1992).

As the density falls and the radioactivity decays, the recombination time-scale must become longer than the radioactive decay time-scale, τ, or even the expansion time-scale, t. This defines the freeze-out time (Fransson & Kozma 1993). The recombination time-scale varies quite strongly from region to region in the core, and is most important in the H and He rich regions, and especially in the envelope. For SN 1987A freeze-out occurred after 800 – 900 days. If the density in the ejecta is highly inhomogeneous, as is indicated from mixing calculations, freeze-out effects may dominate the evolution in the low density regions. The freeze-out leads to a nearly constant light curve and spectrum. It therefore has the same effect as an additional energy source. The discrepancy between the light curve calculations and the IR- and γ-observations for SN 1987A was explained by this effect by Fransson & Kozma (1993). Bouchet et al. (1994) find that this explains most of the deficiency of Fig. 1, but argue that a somewhat improved fit can be obtained by adding an additional source with luminosity $(2 - 5) \times 10^{37}$ erg s^{-1}. The significance of this is marginal, but illustrates the importance of late observations. Disentangling freeze-out, radioactivity and pulsar input for other supernovae is an observationally very interesting and demanding problem, which requires observations at epochs $\gtrsim 1000$ days.

2.2 Type Ia Supernovae

While considerable progress has been made on the late emission from Type II supernovae, the situation is worse for Type Ia supernovae both observationally and theoretically. With the exception of the dust echo dominated SN 1991T (Leibundgut & Spyromilio, these proceedings), good photometry and spectra beyond a year exists only for SN 1972E (Kirshner et al. 1973, 1975). Our ability to produce detailed synthetic spectra is also rather limited, in part because of the lack of high quality atomic data for iron ions.

Even neglecting the absence of hydrogen, the physical conditions in Type Ia supernovae are quite different from Type II's. For a Type Ia supernova produced by the explosion of a Chandrasekhar mass white dwarf, the total mass of $\lesssim 1.4$ $M_\odot$ is smaller than that of a massive stellar core by a factor ~ 3. The observed expansion velocity of this material ($\sim 10^4$ km s^{-1}) is considerably greater than the ~ 2000 km s^{-1} typical in Type II's, so that the total density is smaller by at least a factor ~ 400 from the Type Ia case. The level of ionization is also much higher than in a Type II at a similar epoch, e.g. in the form of [Fe III] and [Co III] lines.

Model DD4 from Woosley (1991), corresponding to the "delayed detonation" of a Chandrasekhar-mass white dwarf, produces 0.62 $M_\odot$ of ^{56}Ni, and gives the best fit to the observed optical spectrum of SN 1972E at 250 days. In this model, the onset of the thermal instability occurs at ~ 450 days; over the next 200 days the temperature rapidly decreases from ~ 3000 K to ~ 300 K (IR-catastrophe).

After the IR-catastrophe nearly all emission emerges as fine structure lines of [Fe II] 25.99μm, 35.35μm and [Fe I] 24.05μm.

In Fig. 4 we show the light curves from the UV to the IR for model DD4. We have here used a distance of 4 Mpc, appropriate for SN 1972E. For a more typical distance of 10 Mpc one should add 2 magnitudes. The most interesting feature is the marked decrease in the fluxes in the B, V and R bands between 450 and 600 days, coinciding with the IR-catastrophe. The slow decline in J occurs at the same time as the rapid decline in B and can be attributed to a shift in emission from the strong [Fe III] blend at ~ 5000 Å to the [Fe II] lines at 1.257μm and 1.644μm in the near-IR.

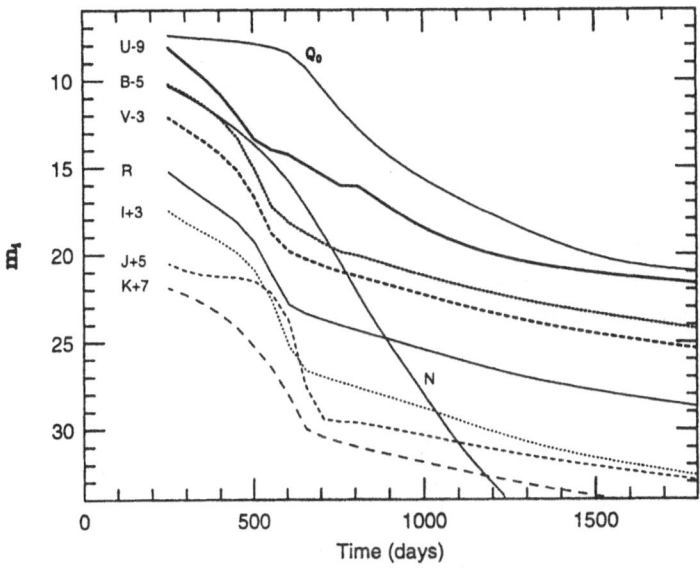

Fig. 4. Light curves for a Type Ia supernova, at a distance of 4 Mpc appropriate to SN 1972E. For 10 Mpc add two magnitudes. The decrease at ~ 600 days is due to the IR-catastrophe.

From this discussion it should be clear that both spectroscopy and photometry at phases later than a year offer important possibilities to test the ejecta structure. Because of the faint level however, photometry at $R \sim 26 - 27$ may be the best way to study these phases. The background from the host galaxy may be a problem. The best targets will therefore be those in the outer parts of the host galaxies.

2.3 Circumstellar Interaction

Massive stars are in most cases surrounded by dense circumstellar gas, originating from mass loss either in the form of a wind or due to e.g. binary mass loss. The interaction between the supernova and this gas gives rise to emission in all bands from radio to hard X-rays (see Chevalier 1990 or Fransson 1994b for reviews). The Hα line is of special importance because it can give information on both the X-ray emission and about the dynamics of the interaction from the line widths. The interaction gives rise to two shock waves, one propagating outwards into the circumstellar medium, and one inwards (in a Lagrangian sense, i.e. in velocity). The latter dominates the X-ray emission, and is responsible for the excitation of the optical and UV lines from the outer parts of the ejecta and from the shock region. Chevalier & Fransson (1994) find that there is a close relationship between the X-ray luminosity from the reverse shock propagating into the ejecta and that reemitted as Hα. Observations of the Hα-luminosity therefore give direct information about the evolution of the X-ray luminosity.

Monitoring the evolution of the line widths of especially Hα and [O III] shows how the maximum velocity in the unshocked ejecta evolves with time. This is determined by the position of the reverse shock, which in turn depends on the density distribution in the ejecta. The evolution of the line width is therefore an important probe of the ejecta structure. The further in time this monitoring is possible, the further into the ejecta we can study the density distribution, and therefore provide a constraint for explosion models. In addition, this maximum velocity can together with VLBI observations be used to determine the distance to the supernova, and therefore the Hubble constant (see Chevalier & Fransson 1985).

A unique opportunity for VLT will occur in connection of the collision between the SN 1987A ejecta and the circumstellar ring at ~ 0.6 light years distance. This is estimated to occur in the year 1999 ± 3 (Luo et al. 1994). Models show that the collision will result in both optical and IR emission from the ring and ejecta. The strongest lines are likely to be $H\alpha, H\beta$ and [O III]. These observations will throw new light on both the structure of the ring and the ejecta.

3 The Contribution of VLT

The VLT can improve appreciably on current observations of supernovae in several ways. The most obvious is by extending the available distance by a factor two, and the number of supernovae by a factor ~ 8. Perhaps more important, we can study supernovae to considerably fainter levels, and therefore to later times. The advance in instrumentation, especially in the IR, will probably be as important as the increase in surface area. As we have already seen, this is crucial, since much of the most interesting physics occurs later than ~ 400 days, where observations become difficult. For Type II's and Ib/Ic's this includes abundance determinations and isotope ratios, especially using the IR lines, the IR-catastrophe, molecule and dust formation, possible pulsar input, and circumstellar interaction. For Type Ia's roughly the same set of problems can be addressed. The main difference is that the expansion velocity is considerably higher, leading to a shorter evolutionary time-scale.

It is also important to realize that because the background from the host galaxy is often the limiting factor at late times, spatial resolution is important for all types of observations. However, to some extent the generally large widths of the lines help in discriminating against the background, e.g. H II regions.

A generally important aspect is that regular observations of supernovae will be greatly simplified by the running of VLT in service mode. This will hopefully also lead to a fast response for new, bright supernovae.

Both to study dust formation and the IR-catastrophe in Type II's the optical and near-IR lines have to be followed as late as possible, at least until ~ 800 days. With the exception of the dust echo of SN 1991T (see Leibundgut & Spyromilio, these proceedings) no spectra of Type Ia's have been obtained later than ~ 700 days, and the possibility of doing this provides an important test of the evolution of the ejecta. We estimate that using FORS with $R \sim 500$, it should be possible to follow the luminosity of e.g. the [O I] $\lambda\lambda$ 6300, 6364 lines to $\gtrsim 1000$ days for a 7 Mpc supernova. Observations of line profile asymmetries to follow the dust formation, require high S/N spectra with $R \sim 1000$ at epochs later than ~ 500 days.

For the study of clumping in the ejecta, as revealed by small scale fluctuations in e.g. the Hα or the [O I] line profiles, UVES will be of great interest. A resolution of 10 km s^{-1} is sufficient, which should be easy to at least 500 days. In addition, searches for narrow circumstellar emission lines, similar to those observed for SN 1987A (e.g., Wang 1991) and SN 1993J (Cumming et al. 1994) benefits greatly from high resolution. The contrast between the narrow ($V \lesssim 10$ km s^{-1}), weak circumstellar lines and the strong ejecta lines increases directly with the resolution.

ISAAC offers an important opportunity to follow the evolution of the [Fe I] 1.54μm, [Fe II] 1.257, 1.64μm and the [Co II] 1.547μm lines, and therefore also the decay of ^{56}Co, as well as determining the ^{57}Co/^{56}Co ratio. This is crucial for Type II supernovae (as well as Ib's and Ic's), and provides a very important constraint on Type Ia models. We estimate that with $R \sim 500$ this will be possible up to ~ 700 days, sufficient to determine the ^{57}Co/^{56}Co ratio.

The K and L bands are important for the study of the CO molecule vibrational bands, and therefore the first indications of dust formation, which can then be followed up in the optical (line profiles) and in the far-IR. The 2.2μm overtone band should be observable up to ~ 500 days with ISAAC.

For the far-IR lines, as well as for photometry, VISIR is of prime interest and offers unique possibilities, even compared to instruments on ISO. From SN 1987A we estimate that a 7 Mpc supernova should be detectable in the N and Q bands up to ~ 700 days. To follow the [Co II] 10.52μm line as late as possible, and therefore also the ^{56}Co to ^{57}Co transition, a variable narrow band filter ($R \sim 50$) is highly desirable. The planned $R = 300$ mode would permit spectroscopy of the 10.52μm line up to ~ 500 days, marginally sufficient for detecting the ^{57}Co transition.

For supernovae at cosmological distances, the possibility to extend the distance range of accessible supernovae by a factor of at least two is extremely valuable. The difference between cosmological models with $q_0 = 0$ and $q_0 = 0.5$ increases from 0.22 magnitudes at $z = 0.4$ to 0.44 magnitudes at $z = 0.8$. This is considerably larger than the dispersion among Type Ia supernovae. There are already efficient programs for discovering supernovae at these distances (e.g., Ellis 1991), and photometry of the supernovae should be fairly easy. Spectroscopy, to confirm the type of the supernova, is, however, more difficult, and will certainly require VLT class telescopes. Because Type Ia spectra fall off rapidly below ~ 4000 Å both photometry and spectroscopy must be done in the near-IR above ~ 8000 Å.

Acknowledgments: I am grateful to Robert Cumming for useful comments. This work is supported by the Göran Gustafsson Foundation for Research in Natural Sciences and Medicine.

References

Axelrod, T.S. (1980): Ph.D. thesis, Univ. of California, Santa Cruz

Bouchet, P., & Danziger, I.J. (1993): Astr. Ap., **273**, 451

Bouchet, P., & Danziger, I.J., Goiffes, C., Della Valle, M., & Moneti, A. (1994): in IAU. Coll. No. 145 *Supernovae and Supernova Remnants*, ed. R. McCray , Cambridge University Press, in press

Chevalier, R. A. (1989): Ap. J., **346**, 847

Chevalier, R. A. (1990): in *Supernovae*, ed. A. G. Petschek, Berlin, Springer, p. 91

Chevalier, R. A. (1992): Nature, **355**, 691

Chevalier, R. A., & Fransson, C. (1985): in *Supernovae as Distance Indicators*, ed. N. Bartel (Berlin Springer): p. 123

Chevalier, R. A., & Fransson, C. (1992): Ap. J., **395**, 540

Chevalier, R. A., & Fransson, C. (1994): Ap. J., **420**, 268

Cumming, R. J., Meikle, W. P. S., Walton, N. A., & Lundqvist, P. (1994): in *Circumstellar Media in the Late Stages of Stellar Evolution* eds. R. E. S. Clegg, W. P. S. Meikle, & I. R. Stevens, (CUP, Cambridge), p. 192

Danziger, I.J., Gouiffes, C., Bouchet, & P. Lucy L.B. (1991): ESO/EIPC Workshop *Supernova 1987A and Other Supernovae*, eds. I.J. Danziger & K. Kjär, p. 217

Ellis, R.S. (1991): in *Observational Tests of Cosmological Inflation*, eds. T. Shanks et al. (Kluwer, Dordrecht), p. 243

Fransson, C. (1994a): in Les Houches, Session LIV, 1990, *Supernovae* eds. J. Audouze, S. Bludman, R. Mochkovitch and J. Zinn-Justin, Elsevier Science Publishers B.V., p. 677

Fransson C. (1994b): in *Circumstellar Media in the Late Stages of Stellar Evolution*, eds. R.E.S. Clegg, I.R. Stevens & W.P.S. Meikle (Cambridge University Press, Cambridge) p. 120

Fransson, C., & Chevalier, R.A. (1987): Ap. J., **322**, L15

Fransson, C., & Chevalier, R. A. (1989): Ap. J., **343**, 323

Fransson, C., & Kozma, C. (1993): Ap. J., **408**, L25

Fransson, C. , Houck, J. and Kozma, C. (1994): in IAU. Coll. No. 145 *Supernovae and Supernova Remnants*, ed. R. McCray , Cambridge University Press, in press

Herant, M., Benz, W., Hix, W.R., Fryer, C.L., & Colgate, S.A. (1994): Ap. J., **435**, 339

Kirshner, R.P., Oke, J.B., Penston, M.V., & Searle, L. (1973): Ap. J., **185**, 303

Kirshner, R.P., & Oke (1975): Ap. J., **200**, 574

Kozasa, T., Hasegawa, H., and Nomoto, K. (1989): Ap. J., **344**, 325

Kozasa, T., Hasegawa, H., and Nomoto, K. (1991): Astr. Ap., **249**, 474

Kurfess, J.D. et al. (1992): Ap. J., **399**, L137

Li, H., McCray, R., & Sunyaev, R. A. (1993): Ap. J. **419**, 824

Luo, D., McCray, R. & Slavin, J. (1994): Ap. J., **430**, 264

Lucy, L.B., Danziger, I.J., Gouiffes, C., & Bouchet P. (1991): *Supernovae*, Proc. of the Tenth Santa Cruz Summer Workshop in Astronomy and Astrophysics, ed. S.E. Woosley, Springer Verlag, 82

McCray, R. (1993): Ann. Rev. Ast. Ap., **31**, 175

Meikle, W.P.S., Spyromilio, J., Allen, D.A., Varani, G.-F., & Cumming, R.J. (1993): M.N.R.A.S., **261**, 535

Spyromilio, J. & Pinto, P. A. (1991): in ESO/EIPC Workshop *Supernova 1987A and Other Supernovae*, eds. I.J. Danziger & K. Kjär, p. 423

Spyromilio, J., & Graham, J.R. (1992): M.N.R.A.S., **255**, 671

Suntzeff, N.B., Philips, M.M., Depoy, D.L., Elias, J.H., & Walker, A.R. (1991): A. J., **102**, 1118

Suntzeff, N.B., Philips, M.M., Elias, J.H., Depoy, D.L., & Walker, A.R. (1992): Ap. J., **384**, L33

Varani, G.-F., Meikle, W.P.S., Spyromilio, J., and Allen, D.A. (1990): M.N.R.A.S., **245**, 570

Wang, L. (1991): Astr. Ap., **246**, L69

Wooden, D.H., Rank, D.M., Bregman, J.D., Witteborn, F.C., Tielens, A.G.G.M., Cohen, M., Pinto, P.A. & Axelrod, T.S. (1993): Ap. J. Suppl. **88**, 477

Woosley, S.E. (1991): in *Gamma-Ray Line Astrophysics*, eds. P. Durouchoux & N. Prantzos, (New York: American Institute of Physics), 270

Woosley, S.E., & Hoffman R.D. (1991): Ap. J., **368**, L31

Woosley, S.E., Pinto, P.A., & Hartmann D. (1989): Ap. J., **346**, 395

Faint Planetary Nebulae as Mass Tracers for Early Type Galaxies

M. Arnaboldi[1], S. Beaulieu[1], M. Capaccioli[23], K.C. Freeman[1], P.J. Quinn[1]

[1] Mt. Stromlo and Siding Spring Observatories, Canberra ACT, Australia
[2] Dipartimento di Astronomia, Universita' di Padova, Padova, Italy
[3] Osservatorio Astronomico di Capodimonte, Napoli, Italy

Abstract. The recent spectroscopic observations of Planetary Nebulae (PNe) with the ESO New Technology Telescope (NTT) in the multi-object spectroscopy (MOS) mode have achieved very important results for understanding the dark matter content and the formation mechanism of the cD giant elliptical galaxy NGC 1399 in the Fornax cluster. The telescopes of the new generation will revolutionise the use of PNe as mass tracers for the dynamics of the outer regions of early type galaxies. With an 8-m telescope, a spectrograph similar to EMMI with the MOS mode, and a wavelength resolution of about 1.0 Å/pix, it will possible to detect and measure velocities of PNe as faint as $\log F_{5007}$ (ergs cm^{-2} s^{-1}) = -18.0. At the distance of the Virgo and Fornax clusters, this will allow us to measure about 600 PNe in the outer regions of the giant galaxies NGC 1399 and M87, which will give a good sampling of their potential wells. It will also be possible to measure at least 100 PNe in the outer regions of smaller ellipticals, down to 2.8 mag fainter than the cluster giants.

1 Tracing the Dynamics of the Outer Regions of Es and cDs Using Planetary Nebulae. Introduction

Our knowledge of the dynamical properties of elliptical galaxies is confined to their inner regions because of the lack of good dynamical mass tracers extending far out to the optical limits of the systems. Integrated light observations of the inner regions of giant Es indicate that most of them are slow rotators: their specific angular momentum is an order of magnitude lower than that of spirals with similar mass. This is in contrast with what is expected from cosmological simulations, which predict similar specific angular momentum for spirals and ellipticals. A possible solution to this problem is that the angular momentum in giant ellipticals lies mainly in the outer regions, which are not accessible via integrated light techniques. This angular momentum segregation is predicted by secondary infall models for the growth of dark matter halos. The gross dynamical properties of these dark halos are very similar to those of the luminous components of real E galaxies. This prediction about the angular momentum can be tested with the telescopes of the new generation, by using them to measure the radial velocities of planetary nebulae (PNe) in the outer regions of giant and normal ellipticals.

2 Planetary Nebulae as Mass Tracers

PNe are very powerful mass tracers for the kinematics of the outer regions of early type galaxies (Hui et al 1993, Arnaboldi et al. 1994). Using the Anglo Australian 4.m telescope and the fiber spectrograph, Hui et al. (1993) and Freeman et al. (1994) were able to measure the radial velocities of $\sim$ 500 PNe in NGC 5128 (Cen A) and in the Sombrero galaxy. The kinematics of giant Es could not be explored because the nearest systems are in the Fornax and Virgo clusters, and their PNe have fluxes $\simeq 10^{-17}$ erg cm^{-2} sec^{-1}, which are too faint to measure spectroscopically with the above instrument.

So far, the only giant cD galaxy for which it has been possible to measure the radial velocities of the PNe is NGC 1399 in the Fornax cluster (Arnaboldi et al. 1994), at a distance of 16.9 ± 1.1 Mpc (distance derived using the PN luminosity function, McMillan et al. 1993). The spectra of these faint PNe were obtained using the ESO New Technology Telescope (NTT) plus EMMI in the Multi-Object Spectroscopy (MOS) mode. Direct spectroscopy through the MOS masks, plus the excellent image quality at the NTT, allowed us to perform this difficult observation. With the NTT and the above instrumental setup, we can now begin to study the kinematics of the brightest PNe in the outer regions of the giant ellipticals in the Fornax and Virgo clusters. For these observations it is crucial to determine the PNe positions in the field very accurately, and to have a very stable spectrograph and telescope.

When making our observations of the PNe in NGC 1399 with the NTT and EMMI, we were able to obtain spectra for 37 out of the 61 known PNe in the $8' \times 8'$ field centered on NGC 1399. Our measurements were limited by charge trapping problems in the coated FA 2048 Loral CCD and by the wavelength resolution of 1.7 Å/pix.

3 Using the ESO VLT: Expected Developments

What gains could we expect for this kind of work by going to the VLT? First, there are significant gains to be made from the instrumentation. With a thinned CCD, without problems of charge trapping, we expect to gain a factor of 1.8 in the signal-to-noise ratio. Another factor of 1.7 can be achieved with the better wavelength resolution of 1.0Å/pix. of the VLT UV-visual focal reducer/spectrograph (FORS).

An 8-m telescope in an average seeing of 0.4" gives a further improvement in signal-to-noise ratio of about a factor 12, relative to our NTT observations. The overall improvement, with the better CCD, the higher dispersion, and the larger telescope, is about 35. Therefore we will be able to measure PNe that are 1.55 dex fainter than those that we were able to measure at the NTT. In NGC 1399, our NTT limit for the PNe is $\log F_{5007}$ (ergs cm^{-2} s^{-1}) = -16.4. With a 8-m class telescope and the above instrumentation we could detect PNe with $\log F_{5007}$ (ergs cm^{-2} s^{-1}) as low as -18.0. How many more PNe would we then expect to see in a galaxy like NGC 1399? There is no information about the luminosity function of PNe in elliptical galaxies down to such low fluxes (but see

Mendez et al. 1993 for more details), so we assume that their PNe luminosity function is proportional to that for the PNe of the Galaxy bulge (Acker et al. 1992, Strasbourg-ESO Catalogue of Galactic Planetary Nebulae); this seems reasonable because the bulge is an old population and its metallicity is probably similar to that in the outer regions of giant ellipticals. With this assumption, it will be possible with the VLT to measure 600 PNe in the outer regions of NGC 1399 and a similar number for the giant galaxy M87 at the core of the Virgo cluster, see Fig. 1. Such a large sample of PNe would allow us to make some quite detailed inferences about the dynamics of the outer regions of these very extended galaxies, including an estimate of the isotropy of the orbits.

It is already clear from earlier work with integrated light techniques that the dynamical properties of ellipticals change dramatically with decreasing luminosity. It seems likely that the formation processes for the largest ellipticals are different from those of the smaller galaxies. If we are to understand the dynamics and the formation of elliptical galaxies in general, it will be essential to study not only the largest of the giant ellipticals but also some of the less luminous ellipticals. In these smaller galaxies, we would want to measure at least 100 PNe in order to have acceptably small errors on the inferred kinematics in their outermost regions. With this constraint, of measuring at least 100 PNe in each of these fainter galaxies, we estimate that it will be possible with the VLT to study galaxies that are intrinsically 2.8 mag fainter than NGC 1399. (For this estimate it is necessary to take into account the different surface brightness profiles of cD and normal giant elliptical, and also the severe incompleteness of the PNe sample in the high surface brightness inner regions of the galaxies). The VLT will allow us to study a large sample of objects both in the Fornax and in the Virgo clusters for which we will be able to use PNe to trace the kinematics of the outer region for the first time.

The multi-object spectrographs forseen for the VLT and now under construction are FORS and MFAS, and they do present some limitations with respect to the ideal picture described above. FORS has a field of view of $7' \times 7'$ and only 19 slitlets can be allocated simultanously; a mask as already used for the EMMI spectrograph at the NTT will be a better solution. The fiber spectrograph MFAS has a limit on the minimum distance between fibers of $14''$ and this can put a constraint on the number of objects observable in a single exposure. The fiber spectrograph will work for this project with the resolving power $R = 10000$, provided that the throughput is good enough.

4 Conclusions

It is important to stress that at this moment the only telescope with which the faint emission of the planetary nebulae at a distance of 17 Mpc can are measured is the ESO NTT with the EMMI spectrograph in the MOS mode. Even if the velocity fields determined through the planetary nebulae with the ESO NTT are limited by statistics, it is of crucial importance to use the NTT to study the dynamics of the outer regions of early type galaxies with $D \geq 17$ Mpc: they are not accessible to any other telescopes. In the interval of time between now

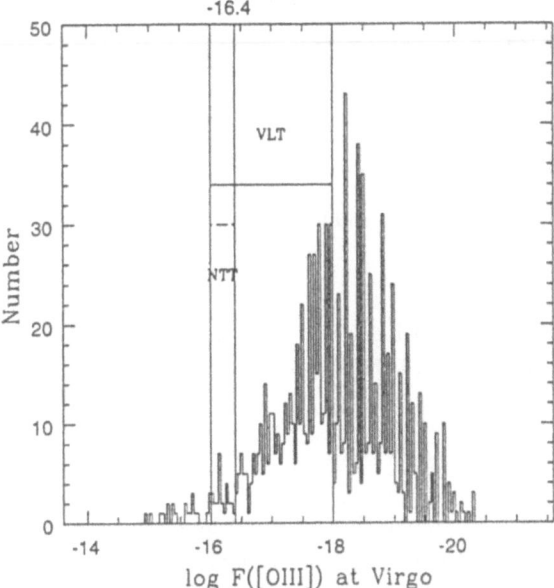

Fig. 1. The planetary nebulae luminosity function for the Galactic Bulge with the logarithmic flux of the [OIII] line scaled at the distance of Virgo (16.9 Mpc). On the diagram are indicated the portion of the luminosity function accessible with the ESO NTT, and the expected improvement once the observations will be performed with the ESO VLT.

and the VLT, the ESO NTT will be of primary importance for this challenging project.

We would like again to stress that a large study of the kinematical properties of the outer regions of elliptical galaxies, including cDs, giant and normal ellipticals, will shed great light on the unsolved problem of elliptical galaxy formation and the amount of dark matter in these galaxies.

References

Acker, A., Ochsenbein, F., Stenholm, B., Tylenda, R., Marcout, J., Schohn, C., (1992): Strasbourg-ESO Catalogue of Galactic Planetary Nebulae, ESO, Garching bei München, Germany

Arnaboldi, M., Freeman, K.C., Hui, X., Capaccioli, M., Ford, H. (1994): ESO Messenger **76** 40

Freeman, K.C. et al. (1994): preprint

Hui, X., Ford, H.C., Freeman, K.C., Dopita, M.A. (1993): preprint

McMillan, R., Ciardullo, R., Jacoby, G.H. (1993): ApJ **416** 62

Mendez, R.H., Kudritzki, R.P., Ciardullo, R., Jacoby, G.H. (1993): A&A **275** 534

Resolving Distant Galaxies Into Stars

D. Minniti and T.R. Bedding

European Southern Observatory, Karl-Schwarzschild-Str. 2, D-85748 Garching bei München, Germany

Abstract. We describe a project that uses VLT + CONICA to resolve distant galaxies (D $\geq$ 10 Mpc) into stars, obtaining IR color-magnitude diagrams.

1 Introduction

With telescopes such as the VLT, many projects of extragalactic astronomy will be transformed into stellar astronomy. The instrument specifications for CONICA at the VLT give a limiting magnitude of $m_H = 26.3$ (3σ in 1 hour integration). Taking a naive example, the brightest stars in the Galaxy reach $M_K = -11.2$ (Humphreys 1988), or $M_H = -11.5$. Therefore, in a perfect universe without any confusing sources, we would be able to detect one of these stars at a distance modulus of $m - M = 37.8$, or $D = 360$ Mpc. This is further away than Virgo — further even than Coma — and is approaching cosmologically interesting distances. In short, it is a **BIG** number. Of course, the universe is not perfect and sources are not isolated. In order to assess the capabilities of the VLT, we will discuss a more realistic example aimed at addressing a specific problem in galaxy formation.

Galaxy formation is an important area of study in modern astronomy. In particular, the formation of bulges of spirals and elliptical galaxies and the relationship between them remains an area of considerable uncertainty. Did they form shortly after the Big Bang in a short period of intense activity that consumed all the gas and prevented further star formation? Or did they have several episodes of star formation, somehow triggered by mergers or accretion processes? In this case, the observed light would come from both old and intermediate age stars.

The issue can be resolved by constructing luminosity functions (LFs) and colour-magnitude diagrams (CMDs) for the stars in ellipticals and bulges, and comparing these with predictions from stellar evolutionary theory. The theory is fairly well understood, having been tested and calibrated on open and globular clusters spanning a wide range of ages and metallicities. The project requires photometry to identify and measure luminosities for the stars at the top of the asymptotic giant branch (AGB). This method would complement the existing integrated spectroscopic data and would allow one to study the stellar distribution, to separate the different populations, and to estimate the ages, metallicities and distances of these populations.

2   Observing in the Near–Infrared

There are several advantages to obtaining CMDs in the near–IR:

• Giant stars have the peak of their spectral energy distribution here, increasing the contrast relative to the underlying fainter and bluer stars.

• We avoid the degeneracy in the optical colours of the red giant branch. This degeneracy makes it difficult with optical photometry to determine ages or metallicities, especially for the more metal–rich populations that dominate bulges and ellipticals.

• The transformation between the photometric observations and theory (Mbol, Teff) is easier in the IR than in the visible.

• Extinction and reddening due to dust are less $(A_K = 0.1A_V)$.

New IR arrays under development (both HgCdTe and InSb) should deliver higher quantum efficiency over an expanded wavelength range. The large formats will allow a reduction of the pixel scale on the sky and/or an increase in the area coverage (see Rieke 1994).

Adaptive optics in the IR will give two benefits. Firstly, it will allow a substantial gain in sensitivity for point-source photometry by reducing the diameter of stellar images, thus increasing the signal relative to the background sky. Secondly, the increased resolution reduces confusion in crowded regions. We should not forget that VLT + AO will have higher angular resolution than HST + Nicmos.

3   What Has Been Done Before?

The bulges of all 3 spirals of the Local Group (MW, M31 and M33) have been well studied in the near–IR, as has the dwarf elliptical M32. Figure 1 shows CMDs for the three Local-Group galaxies, which are all at about the same distance. In all three cases, the top of the AGB is clearly seen and has a sharp cutoff at $K \simeq 16$. Measuring this cutoff magnitude in more distant galaxies should allow one to determine the distance.

Here is a summary of results obtained from infrared CMDs:

• In the bulge of M31, Rich & Mould (1991) found that the brightest giants are brighter than stars in globular clusters, concluding that the M31 bulge must be younger than the globulars. These results were disputed by Renzini (1992), DePoy et al. (1993), and Davies et al. (1992). More recently, Rich et al. (1993) expanded their observations, supporting their original conclusions.

• In the dwarf elliptical M32, Freedman (1992), and Elston & Silva (1992) found evidence for a luminous intermediate-age population. This suggests an episode of star formation in the last 5 Gyr.

• The M33 bulge was detected and resolved by Minniti et al. (1993). They find a trace for a young stellar population in the inner regions of M33 on the basis of H-band LFs. Minniti et al. (1994) presented further IR CMDs, confirming the presence of a bluer plume of younger objects. With the help of the isochrones of Bessell et al. (1991) we can get an estimate of ages and metallicities of these populations.

The Local Group galaxies are the only ones where these kinds of studies have been done. Obtaining CMDs for more distant galaxies would clearly be a very significant result. In particular, this would give access to both normal and giant ellipticals, neither of which are are present in the Local Group (see Freedman 1994). In addition, M32 is anomalous in many respects (e.g., Rose 1994), so it may not be representative of the distant and more luminous elliptical galaxies.

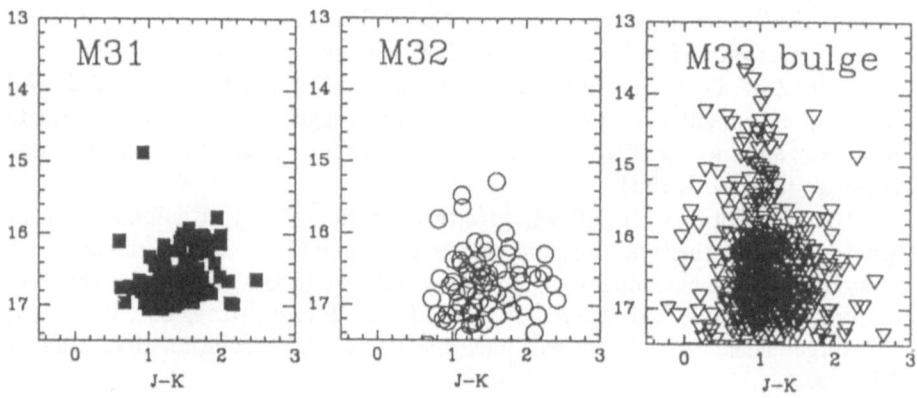

Fig. 1. Colour-magnitude diagrams for the brightest stars in three Local Group galaxies. Left: the bulge of M31 (Rich & Mould 1991). Middle: the dwarf elliptical galaxy M32 (Freedman 1992). Right: M33 (Minniti et al. 1994).

4   What Can Be Done Now?

Until now, it has not been possible to obtain photometry for individual stars in galaxies beyond $\sim$1 Mpc. Thanks to advances in adaptive optics, this should now be achievable with 4-m class telescopes. We will use COME-ON+ to obtain infrared photometry of the brightest stars in the a giant elliptical galaxy NGC 5128 (Cen A). This galaxy is of particular interest because of its special status as the closest radio galaxy and because it shows evidence of having undergone a recent interaction. We will obtain deep H and K' photometry in two fields centred on foreground SAO reference stars that are conveniently placed at 4.5 and 11 arcmin from the nucleus. With these observations, we hope to construct luminosity functions and colour-magnitude diagrams of the brightest stars in Cen A, which we will use to derive the distance to the galaxy and estimate ages for the youngest stellar populations.

5 What Can Be Done with VLT?

The VLT with CONICA and adaptive optics will extend the distance to which we can resolve stars in bulges and ellipticals by a large factor. The following estimates are based on specifications given in the VLT instrument booklet. The best filter for deep photometry is H, where the bolometric corrections are also straightforward (e.g. Bessell & Wood 1984). However, photometry in a second band is desirable to add colour information.

CONICA will reach a 3-sigma limiting magnitude of $H = 26.3$ in one hour, with a scale of 12.8 mas/pixel. Finding a bright star to use as wavefront sensor next to these galaxies should not be a problem in most cases, since the AO system at the VLT will use stars as faint as $V = 16$. ISAAC will be about three magnitudes less sensitive in H because of the absence of adaptive optics, although tip–tilt correction would give some improvement.

Based on these limiting magnitudes, CONICA will detect the brightest stars in bulges and ellipticals out to a distance modulus of $m - M = 34.8$ (D = 91 Mpc) and sample the top of the luminosity function out to $m - M = 33.8$ (D = 57 Mpc). These figures are based on an absolute magnitude of $M_H = -8$ for the brightest giants in spheroidal populations of Local Group galaxies (neglecting reddening). For comparison, for the Centaurus group has $m - M = 27.5 \pm 0.3$, Virgo has $m - M = 31.1 \pm 0.5$ (D = 17 Mpc, $Vr = 1000$ km/s) and Coma has $m - M \sim 35$ ($Vr = 4500$ km/s). Therefore, we should be able to perform a line-of-sight depth study of Virgo and Coma.

6 What is the Limit?

We cannot go as far as one wants, because of crowding and pixel size problems, and more important problems like contamination from the Milky Way and background galaxies. The limits will be dictated by the following effects, which will make these observations challenging:

• Field crowding $\longleftrightarrow$ Limited pixel size:

This is the main limiting factor. To overcome this, we can chose the appropriate surface brightness level so that crowding is not a problem. DePoy et al. (1993) have illustrated the effect for M31.

• Contamination:

The line of sight to distant objects goes through the halo of our Galaxy, which contains many low mass stars. These can be estimated down to very faint magnitudes using simple Galactic models, and turn out not to be important. The contamination by background galaxies is expected to dominate at very faint K magnitudes. In principle, this could be accounted for by observing control fields.

• We need bright stars conveniently located to use for wavefront sensing.

7 Final Remarks

The VLT will transform extragalactic astronomy into stellar astronomy. There will be no need to rely on models and observations of the integrated light to find the physical properties of distant galaxies. Our technical requirement is simply that CONICA achieves the performance described in the booklet. With this, we will be able to determine ages, metallicities and distances for the stellar populations of very distant galaxies. We might also be able to put important constraints on galaxy formation, and decide on the history of mergers for a particular galaxy, just by looking at its shells and shreds.

Finally, we note that pushing the distance limit to which we can resolve individual stars in the IR with the VLT also means pushing the limit where the surface brightness fluctuations method is useful. Additionally, considering the suggestion put forward by G. Jacoby at this meeting of using surface brightness fluctuations of globular clusters as a distance indicator, we might consider obtaining a 3–dimensional picture of the galaxy distribution within several hundred Mpc, independently of radial velocity measurements.

References

Bessell, M. S., Brett, J. M., Scholz, M., & Wood, P. R. 1991. AAS, 89, 335

Bessell, M. S., & Wood, P.R. 1984, PASP, 96, 247

Davies, R. L., Frogel, J. A., & Terndrup, D. A., 1991, AJ, 102, 1729

Elston, R., & Silva, D. R. 1992, AJ, 104, 1360

Freedman, W. L. 1992, AJ, 104, 1349

Freedman, W. L. 1994, in "Local Group Galaxies", eds. A. Layden & J. Storm (ESO: Garching) in press

Humphreys, R. M., 1988, in ASP Conf. Ser. IV, p. 103

Minniti, D., Olszewski, E., & Rieke, M. 1993, ApJ, 410, L79

Minniti, Olszewski & Rieke 1994 in "Local Group Galaxies", eds. A. Layden & J. Storm (ESO: Garching) in press

Renzini, A. 1992, in I.A.U. Symp. 153 "Galactic Bulges", eds. H. Dejonghe & H. Habing (Dordrecht:Kluwer), p. 151

Rich, R.M. & Mould, J.R. 1991, AJ, 101, 1286

Rich, R.M. & Mould, J.R., & Graham, J. R. 1993, AJ, 106, 2253

Rieke, M. 1994, in "Astrophysics with IR Arrays", ed. I. S. McLean, in press

Rose 1994, AJ 107, 206

Proper Motions for the Nearest Galaxies using ASTROCAM on the VLT

Christopher G. Tinney

European Southern Observatory, Karl-Schwarzschild-Str 2, D-85748, Garching, Germany.[1]

Abstract. In recent years CCD astrometry has demonstrated the ability to achieve precisions at the several-mas-level on modestly sized telescopes in less than optimal locations. In this contribution I show what sort of improvements can be expected from the VLT at its Paranal site, and discuss a particular scientific case which could be addressed by an astrometric camera – the proper motions of the nearest galaxies.

1 Introduction

Numerous papers have been presented at this meeting on the scientific prospects for using adaptive optics, speckle and interferometry, on the VLT to achieve high resolutions over small, iso-planatic-patch-sized fields-of-view (FOV). However, for a large number of scientific programs astrometry using these techniques cannot address the issues of most interest – either because the FOV required is too large, or because the targets of interest are too faint. In these cases the iso-planatic limit must be relaxed, and we are left in the 'old-fashioned' astrometric situation of simply trying to measure the positions of unresolved objects as accurately as possible. Traditionally, it has been assumed that the limit to the precision which can be obtained is several tenths of the seeing disk, or roughly 50 to 100 mas. However, this assumption is no longer true – largely because of the introduction of CCDs into astrometry.

2 Astrometry with CCDs

It turns out that CCDs make almost ideal detectors for performing relative astrometry over small angles (ie., several arcminutes) for faint objects (V $\gtrsim$ 15). Because CCD pixels have identical sizes to very high tolerances, they are able to measure positions which (in the absence of the atmosphere) are essentially photon-counting limited. And since they are *much* more sensitive than photographic plates, they collect many more photons. Lastly, they allow the astrometry to be easily carried out in a *differential* fashion - it is simple with a CCD to place the target object back in exactly the same location in the focal plane from epoch to epoch. This means that the zeroth and first order optical distortion

---
[1] Present Adress: Anglo-Australian Observatory, PO Box 296, Epping, 2121. Australia

terms cancel out, which for most telescopes means that astrometric distortion becomes unimportant to the final solution.

The atmosphere, therefore, now imposes the fundamental limit for ground-based astrometry – current parallax programs on 1.5m-class telescopes on sites of 1-2" seeing routinely achieve precisions of 10-20mas per exposure, or 5-10mas per epoch (see Tinney 1994, and references therein). It is also worth noting that there are applications for which CCDs aren't useful – in particular they are not useful for astrometry on bright objects, or for objects unresolved by the seeing. Ground-based CCD astrometry, HIPPARCOS astrometry and ground-based interferometry, therefore, all address well separated areas of observational phase space.

3 Pushing Back the Envelope

An obvious question to ask then, is what scientific goals can be achieved at higher astrometric precisions? What can we do with astrometry at the 1 mas/epoch (rather than 5 mas/epoch) level? A few of the more obvious prospects include; proper motions for globular clusters (with motions in the 5 mas/year range these could be measured in just a few years of observing); proper motions of the Magellanic Clouds and the Milky Way (MW) dwarf spheroidal satellite galaxies (motions of $\sim$ 1mas/yr); and the internal motions and dynamics of the Magellanic Clouds and MW globular clusters (motions of $\lesssim$ 1mas/yr). Rather than attempt to examine all these issues in the available space, I will focus for the remainder of this contribution on the second of these applications.

3.1 A Scientific Context : MW Satellite Proper Motions

Apart from being an exciting technical challenge, measurement of the proper motions of the nearest galaxies to the MW will tell us much about the structure and formation of our own galaxy. Most immediately, with full space motions (proper motions and radial velocities) the dSphs, at distances of 50kpc to 150kpc, can be used as test particles for exploring the potential well of the MW. Full space motions will also allow us to examine the history of these galaxies and their interaction with the MW. The recent discovery by Ibata et al. (1994) of a dSph in the constellation of Sagittarius in the very act of interacting with the MW, has highlighted the fact that these nearby galaxies and the MW *must* have an intricately linked history.

The projected orbits of the MW satellites, therefore, will tell us about their interaction history with the MW and with each other. Several authors (see e.g. Gilmore & Wyse 1993) have claimed that star formation in the MW may reflect past mergers with satellite systems. Indeed, it has been claimed that the MW halo itself may have been formed by the infall of satellite-galaxy sized objects, perhaps even after the main part of the MW had collapsed (e.g., see Majewski 1994 and references therein). There is also evidence to suggest that several of the dSph galaxies lie in a plane with the Magellanic Clouds, and are possibly systems formed from gas torn from them in recent interactions. (Indeed, Lin (1993) has claimed that the LMC is currently in the process of merging with the MW.) Lynden-Bell (1976) and Majewski (1994) (among others) have argued that there are *two* distinct streams resulting from the interaction of larger satellites with the MW. Alternatively, these streams (which are aligned with the Galactic poles) may exist because such orbits are more stable against tidal decay. Accurate proper motions would enable a discrimination between these alternatives (Majewski & Cudworth 1993), as well as providing much needed input on the other issues discussed above.

3.2 The Technical Challenge : How to Do Better at CCD Astrometry

Clearly there are sufficient reasons for pushing ground-based astrometry to new limits. But how can we go about dealing with the technical challenges? There are essentially four major problems which must be dealt with.

1. *Differential Colour Refraction (DCR)* : Target and reference objects will generally have different λ_{eff} through a given passband. Observations at different airmasses will therefore produce refraction-induced motions.
2. *Differential Seeing* : This is the major problem. Because the FOV exceeds the iso-planatic-patch size, the seeing motions induced for target and reference objects will be de-correlated.
3. *CCD sub-pixel non-uniformity* : CCD pixels are not uniformly sensitive. Measurements (Jordan et al. 1993) have shown that the sensitivity can vary by as much as 5-10% across a CCD pixel.
4. *An Inertial Reference Frame* : To measure proper motions of ~ 1 mas/year distant and unresolved reference objects are needed – ie. QSOs.

Excitingly – the VLT will allow us to address *all* these issues.

(1). The groundwork laid by current CCD programs has shown that DCR can be dealt with by observing close ($\lesssim 30^m$) to the meridian, using narrow, red passbands, and carrying out a program of calibration observations as targets rise and set. Because of the large apertures of the single VLT units, it will allow the necessary number of photons ($\gtrsim 10000$) to be collected through narrow ($\lesssim 100$Å) passbands down to V ≈ 22. Moreover, once passbands this narrow are allowed, regions in the far-red (~ 9000Å) can be selected which lie between the atmospheric airglow emission – this means that observations can be performed with a 'dark' sky even in the far-red.

(2) & (3). The effects of differential seeing can be reduced by; exposing longer (averaging more seeing motions), using a smaller FOV (so that seeing motions are more correlated), using a better seeing site, and using a larger aperture (so that more seeing cells are averaged). The VLT will address all of these issues. In particular, the VLT forces a small pixel scale on us in any case (so sub-pixel non-uniformity will be well oversampled and the FOV small); at both Nasmyth and Cassegrain focii the image scale will be $\lesssim 0.02$"/pixel giving a FOV of $\lesssim 30$" for a typical 2048 Tektronix CCD.

(4). Lastly, and most importantly, the fact that the VLT can collect the required number of photons down to faint magnitudes $V \approx 22$, means that the QSO number density is sufficient to make detecting QSOs behind a given large galaxy or globular cluster is very likely. (These QSOs can be identified using either CCD UV-excess search techniques, or A number of collaborators and I are currently carrying out programs to do this at La Silla.) It should be noted that there will be only one QSO per VLT FOV. The technique for performing astrometry in this situation is to 'invert' the normal system, and measure the *reflex* proper motion of the QSO relative to a reference frame of stars in the target galaxy. Each QSO therefore gives one mean proper motion estimate. Clearly multiple QSOs per target will be required as a 'sanity' check on any motion measured.

3.2   The VLT Context : The ASTROCAM Instrument

Given that the measurements are feasible, which instrument should be used? Unfortunately the answer is "None of the planned ones". FORS is not suitable for this program as it contains movable re-imaging optics – experience has shown that precision astrometry must be as differential as possible. The only thing you want to change from epoch to epoch is the position of the target object. A system such as FORS which is subject to flexure and optical realignment is simply not stable enough.

In fact, it turns out that the ideal system would be almost identical to the Superb Seeing Imager (SUSI) currently installed on the ESO's New Technology Telescope (NTT) – ie. a direct imaging system mounted at the Nasmyth focus of one of the units, which for the moment I denote ASTROCAM. At this focus a 1024×1024 CCD with $24\mu m$ pixels would have 0.014" pixels and a 14" FOV on the sky. A 2048×2048 CCD would have a 28" FOV. This may seem small, however it should be remembered that the target objects will be very rich at 22nd-23rd magnitude and easily provide sufficient stars in the target galaxy – in any case a small FOV provides higher precision astrometry. Like SUSI such an instrument could be permanently mounted on the side of one of the Nasmyth adapters, with light directed into it by a flat M4. ASTROCAM would certainly be the simplest and cheapest instrument ever mounted on the VLT.

4   Experience with VLT Unit 5

All of which may sound fine - but it is a huge leap to make from a 1.5m telescope to an 8m telescope. Is there any evidence to indicate that the results I've described are obtainable?

Test observations for this program are currently being carried out on the VLT test-bed – the 3.5m NTT. A field in the globular NGC6752 (centred on QSO1908-6002) is being observed with SUSI (0.13" pixels, 2.2' FOV). To date observations have been obtained in May and June 1994. Preliminary reductions of this data have shown that for 15^m I-band exposures in 0.5" seeing, the residuals about a linear transformation of a *single frame* in one epoch onto a *single frame* from the other epoch achieve a precision of $\sigma_x, \sigma_y \approx 2.5$mas. Moreover, these reductions do not yet include corrections for DCR. It seems clear, therefore, that precisions of 2 mas per epoch (recall that an epoch usually includes several individual frames) can be achieved with the NTT. This makes it almost certain that the VLT will be able to achieve the goal, described above, of 1 mas/epoch

5 Conclusions

It is clear then that a suitably designed ASTROCAM instrument for the VLT will give us access to 1 mas/epoch relative astrometric precisions for objects as faint as 22nd magnitude. Given such a system the proper motions for the nearest Galactic satellites should be obtainable in around 5 years – in the process superceding almost 100 years worth of photographic astrometry on these objects. Moreover, it should be remembered that in 10 years that CCD astrometry will be twice as good, while the existing data will be around $\frac{1}{20}$th better. No other 8-m telescope is planning on exploring this observational technique. Now is the time for us to begin ...

References

Gilmore, G. & Wyse, R.F.G, 1994, in Galactic and Solar System Astrometry, (Cambridge University Press, Cambridge).

Jordan, P., Deltorn, J.-M., Oates, P., 1993, Gemini, 41, p. 1.

Ibata, R. A., Gilmore, G., Irwin, M. J., 1994, Nature, 370, 194.

Lin, C.C., BAAS, 25, 783.

Lynden-Bell, D., 1976, MNRAS, 174, 695.

Majewski, S.R., 1994, ApJ, 431, L17.

Majewski, S.R. & Cudworth, K., 1993, PASP, 105, 987.

Tinney, C.G., 1994, in The Bottom of the Main Sequence and Beyond, ed. C.G.Tinney (Springer-Verlag) *in press*.

Quantitative Spectroscopy of Luminous Blue Stars in Distant Galaxies

R.P. Kudritzki, D.J. Lennon, J. Puls

Universitäts-Sternwarte München, Scheinerstrasse 1, D-81679 München, Germany

Abstract. Hydrodynamic non-LTE model atmospheres, applied to the interpretation of the spectral signatures of strong stellar winds from blue stars, provide the key for the determination of their intrinsic luminosities. Applying these methods to the interpretation of Hα emission profiles and infrared excesses of OBA stars in the Milk Way, the Magellanic Clouds, M31 amd M33, it is shown that the wind-momentum versus luminosity relationship is well defined. This relationship may thus be as an important new distance determination method, and with FORS II and the VLT it will be possible to use the brightest blue supergiants as distance indicators to galaxies as remote as Virgo. Furthermore these stars may be used to determine metallicities, in itself of inherent interest but also important given the metallicity dependence of the wind-momentum versus luminosity relationship discussed below.

1 Introduction

Because of their considerable brightness, luminous blue stars can provide the ideal means of investigating the young stellar populations in distant galaxies. With absolute visual magnitudes of Ia or Ia-O supergiants and Luminous Blue Variables ranging between -7 and -10 it is already possible, with 4m-class telescopes, to take spectra of sufficient resolution and signal-to-noise of such objects in galaxies of the Local Group. This is illustrated by Fig.1 where blue spectrograms of early B-supergiants in the Milky way, LMC, SMC, M31, M33 are compared. It is obvious that the VLT will allow us to go far beyond the Local Group.

On the other hand, luminous blue stars are also notorious among extragalactic astronomers as being entirely useless for extragalactic work, in particular for the determination of distances. However, this is caused by the fact that, so far, these objects have been used only in the most blunt of possible astronomical ways; looking for strict correlations between simplified spectral types, luminosity classes and absolute magnitudes. These attempts have indeed more or less failed, the physical reason being that the atmospheres of these objects are dominated by strong stellar wind outflows that can lead to significant modifications of the observable spectra.

However in truth the outlook is much more positive. We will show that the reason for this complication - the strong stellar winds - provide a unique tool to determine directly the stellar luminosities and therefore the distances, provided the appropriate techniques of quantitative spectroscopy using **HYDRODYNAMIC NLTE MODEL ATMOSPHERES** are applied.

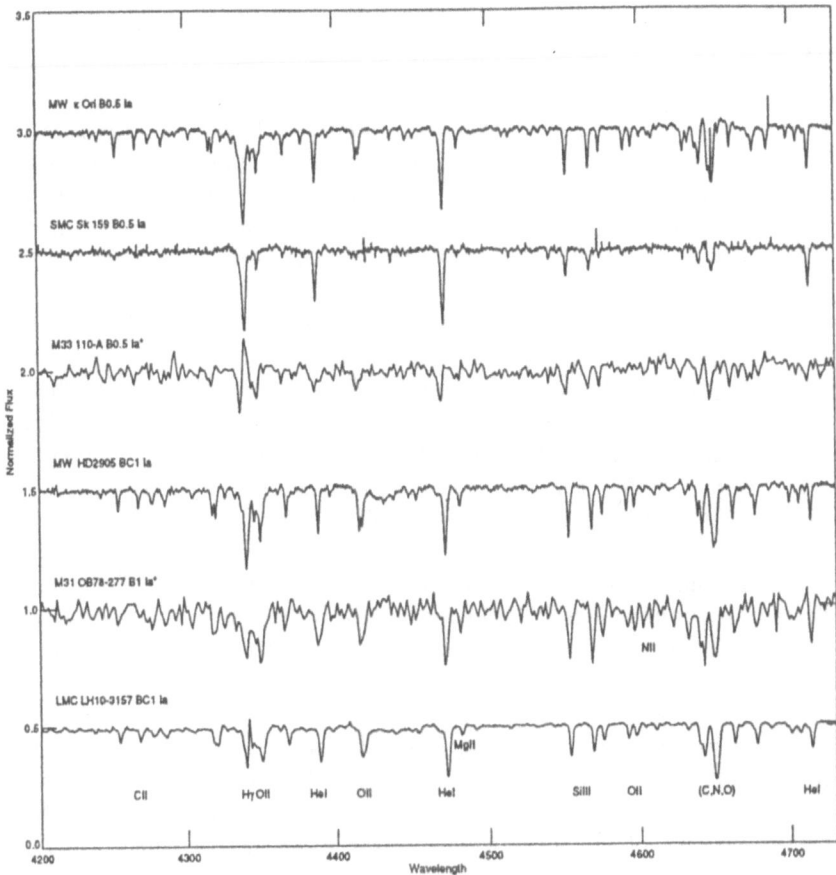

Fig. 1. Blue spectra of some early B-type supergiants in some Local Group galaxies. The data are taken from Lennon et al. (1991, 1992) and Herrero et al. (1994), where the individual objects are discussed in more detail. Note the evident range in metallicity from the metal poor SMC to the metal rich M31.

Our paper is organized as follows: In section 2 we briefly review the status of the theory of hydrodynamic NLTE model atmospheres for hot luminous stars. The diagnostic techniques for the determination of stellar parameters (temperature, gravity, mass-loss rate, terminal velocity, abundances, luminosity, mass) are described in section 3. Section 4 introduces a very recent result obtained by these techniques, the **WIND MOMENTUM-LUMINOSITY RELATION**. We discuss this relation and its potential for extragalactic distance determinations. Section 5, finally, suggests an observing program using FORS I and II attached to the VLT units.

2 Hydrodynamic NLTE Model Atmospheres for Hot Luminous Stars

All hot stars with initial masses larger than about 20 solar masses are subject to significant stellar winds throughout their lifetime. The signatures of these winds are found in all spectral windows as radio- and IR- excess of the continuous energy distribution, as IR and optical emission lines, as UV P-Cygni profiles or as shock heated gas in the X-ray emission. As has been proven convincingly by Lucy and Solomon (1970), Castor et al. (1975), Abbott (1979, 1985) and Pauldrach et al. (1986) these winds are basically radiation driven by absorption of photon momentum due to UV metal lines.

Consequently, hydrodynamic model atmospheres are needed to describe the formation of spectral lines properly. In addition, because of the low particle densities and the comparatively high temperatures, it is necessary to allow for departures from Local Thermodynamic Equilibrium in the entire atmosphere (for a review see Kudritzki and Hummer, 1990, and Kudritzki et al. 1991).

Using current methods and with present computing resources, it is now possible to compute full NLTE models accounting for some 10^5 metal lines, including the effects of hydrodynamics of radiation driven winds and spherical extension. The approach is to iterate between atmospheric structure and stellar wind calculations to achieve a consistent solution - a 'unified' model atmosphere and a sophisticated representation of the stellar wind. This work has been reported on extensively in the literature, see for example, Pauldrach (1987), Puls (1987), Pauldrach et al. (1990, 1992, 1994), Gabler et al. (1989, 1991, 1992) and Schaerer and Schmutz (1994). These references contain further details of the methods plus many examples of the applications. To illustrate the complexity such calculations have attained, the Munich Observatory stellar wind code includes in the (NLTE) rate equations a total of 26 elements and 142 ionization stages with very detailed ionic models, these data are continually being updated with special emphasis being put on iron group elements (see Pauldrach et al. 1994).

3 Diagnostic Methods

The first step in obtaining information about the stellar parameters of luminous blue stars is to use the photospheric absorption lines in the visual that are only weakly contaminated by stellar wind emission. For O-stars, for example, Hγ, HeI 4471, 4923, 4713 , HeII 4200,4524Å yield effective temperature, gravity and helium abundance quite precisely (see Kudritzki and Hummer, 1990; Herrero et al. 1992, also Sellmaier et al. 1993 and Schaerer & Schmutz, 1994). For B- and A-supergiants the ionization equilibria of silicon and magnesium replace HeI/II to give information about the effective temperature (Groth et al. 1990; Husfeld, 1994).

In the next step information about the mass-loss rate ($\dot{M}$) is extracted from a fit of those lines in the visual that are strongly affected by stellar wind emission, Hα for instance. A typical example is given in Fig.2. Such a fit requires knowledge about v_∞, the terminal velocity of the stellar wind. For O-stars and early B-supergiants v_∞ can be measured from the shape of the UV resonance lines observable with IUE or HST (see Fig.2). For A-supergiants v_∞ is obtained either from the velocity shifts of the strongest UV Fe II lines or from the fit of the Hα P-Cygni profile, which then yields both $\dot{M}$ and v_∞ (see Stahl et al. 1991). Finally, abundances can be obtained from both UV and optical lines.

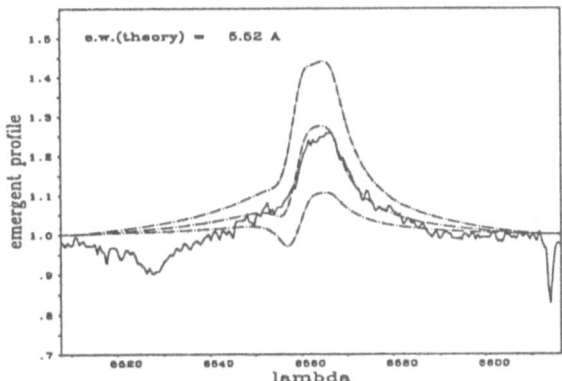

Fig. 2. Hα fit of the O5 I f star HD14947 for different rates of mass-loss (5.25, 7.50, 9.75 10^{-6} M$_\odot$/yr) (from Puls et al. 1994).

For O-stars and early B-supergiants UV-spectrum synthesis can yield the abundances of iron group elements. Optical spectra as displayed in Fig.1 provide complementary information about light elements such as C, N, O, Si, Mg (see Lennon et al. 1991). For late B-supergiants and A-supergiants the optical spectra become dominated by singly ionized ions of the iron group (see Fig.3) and make the determination of metallicity possible from optical spectra alone.

As shown recently by Kudritzki et al. (1992a), the measurement of T_{eff}, $\dot{M}$, v_∞ together with the metallicity (z) can then be used to determine directly the stellar masses and radii without any knowledge about the distance. This is done by applying the theory of radiation driven winds which predicts both v_∞ and $\dot{M}$ to be functions of mass (M) and radius (R) as described by Kudritzki et al.,1989;

$$v_\infty = f_1(T_{\text{eff}}, M, R, z)$$

$$\dot{M} = f_2(T_{\text{eff}}, M, R, z)$$

This allows a determination of M and R once the other quantities are known. Having T_{eff} and R it is then easy to deduce the distance modulus from the dereddened apparent V-magnitude. Examples for this new method of distance determination for stars in the Galaxy and the LMC are given in Kudritzki et al. (1992a, b).

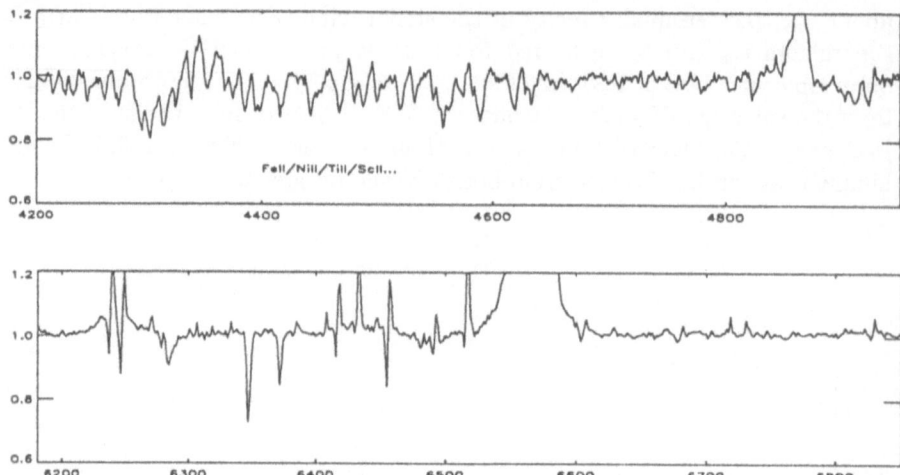

Fig. 3. The remarkable spectrum of the A5 Iae-O supergiant BS 324 in M33. The object has an absolut visual magnitude of -10 mag. Note the presence of iron group elements in the spectrum which will allow a determination of metallicity. The wings of the hydrogen lines are attributed to electron scattering.

IR spectroscopy can also provide important information about the properties of luminous blue stars, particularly useful in cases of strong extinction. A beautiful example is the very recent quantitative analysis of the enigmatic He I emission line stars in the Galactic Center by Najarro et al. (1994). Using the line profiles of all the important hydrogen and helium lines in the K-band (see Fig.4) it was possible to prove that these objects are pre-Wolf-Rayet stars and to determine the age of the starburst in the Galactic Center to be 5 million years.

4 The Wind Momentum-Luminosity Relation

The theory of radiation driven winds (Castor et al. 1975; Kudritzki et al. 1989) predicts a relation between the momentum contained in the stellar wind and the stellar luminosity

$$\dot{M}v_\infty \sim \frac{1}{R^{0.5}}L^x \left\{ M\left(1 - \frac{L}{L_E}\right)\right\}^{1.5-x} \tag{1}$$

For galactic metallicity the exponent x is predicted to be close to 3/2 so that we expect to observe a relation

$$\dot{M}v_\infty R^{0.5} \sim L^x \tag{2}$$

Indeed, if one combines our recent results on galactic O-stars (Puls et al. 1994) with published data on B- and A-supergiants (Barlow and Cohen, 1977) one finds a striking confirmation of this relation holding over several orders of magnitude

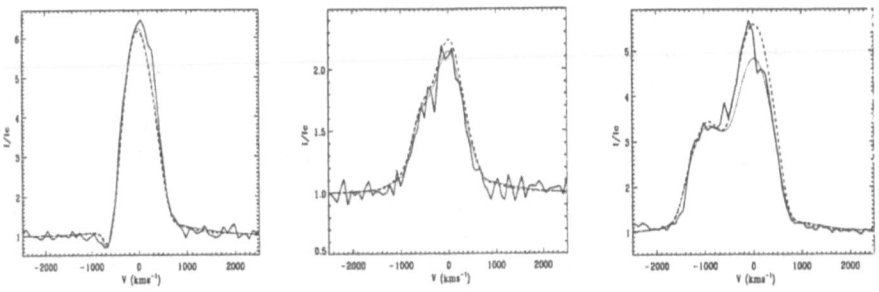

Fig. 4. Fit of the IR lines of the brightest Galactic Center He I emission line star: He I 2.058 μm, Bγ/He I, Pα/He I (from Najarro et al. 1994).

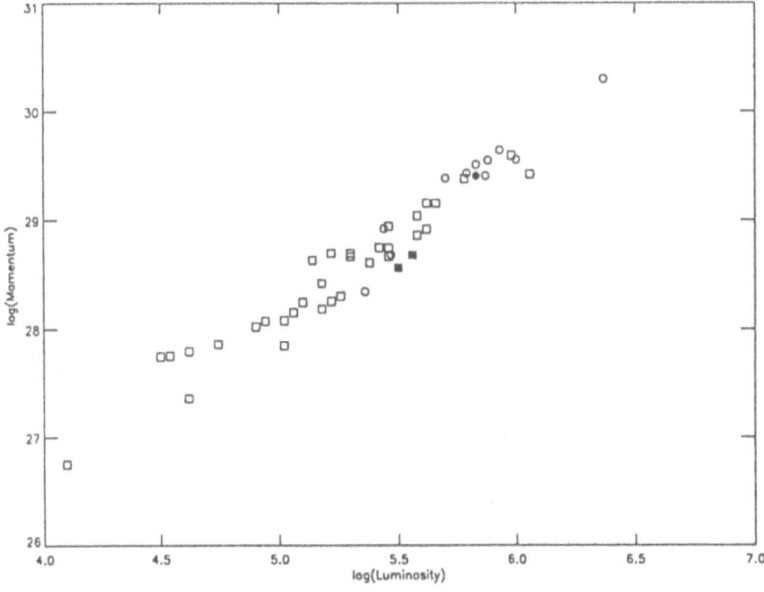

Fig. 5. The $\log(\dot{M}v_\infty R^{0.5}) - \log(L/L_\odot)$ relation for galactic A,B-supergiants (open squares) and O-giants and supergiants (open circles). The objects in M31 (filled squares) and M33 (filled circles) discussed in the text are also shown.

for objects of completely different spectral type and evolutionary status (Fig.5). We believe that much of the scatter in the relation for the A,B-supergiants is caused by the fact that the information about $\dot{M}$ in the pioneering paper by Barlow and Cohen is based solely on the analysis of the continuum IR and radio free-free emission of the stellar winds. We are presently determining mass-loss rates for a large sample of galactic B,A-supergiants using Hα profiles and expect a clear improvement of the relation.

It is obvious that the observational relation shown in Fig.5 has a great potential for distance determinations. The recipe is in principle to determine the wind momentum, which together with T_{eff} would give, using the above relation, the stellar radius and thus the distance. It is very important to realize that the most promising objects for extragalactic applications using visual wavelengths are the most luminous supergiants of spectral type A, since their bolometric correction is small and their absolute visual magnitude is brightest.

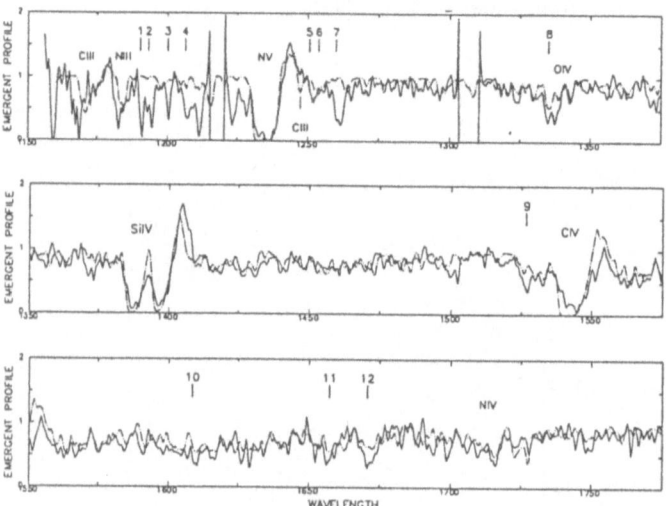

Fig. 8. HST UV spectrum of the M31 O8.5 Ia(f) supergiant OB78-231 and a spectrum synthesis fit confirming solar abundance for iron group elements. (see Haser et al. 1994)

It is, of course, extremely important to investigate whether the supergiants in M31 and M33 confirm the relation. At the moment, we can only use three (still somewhat uncertain) objects, the analysis of which has been carried out using optical spectra by Herrero et al. (1994) and HST-FOS UV spectra (Bianchi et al. 1994; Haser et al. 1994): OB78-277 (B1 Ia, $T_{\text{eff}} \sim 20000$K, $R \sim 50$R$_\odot$, $v_\infty = 700$km/s, $\dot{M} \sim 1.3 10^{-6}$M$_\odot$/yr) and OB78-231 (O8.5 Ia(f), $T_{\text{eff}} \sim 32000$K, $R \sim 20$R$_\odot$, $v_\infty = 1700$km/s, $\dot{M} \sim 10^{-6}$M$_\odot$/yr, see Fig.6) in M31 and BS324 (A5 Iaep-O, $T_{\text{eff}} \sim 8000$K, $R \sim 430$R$_\odot$, $\dot{M} \sim 10^{-5}$, $v_\infty \sim 200$km/s, see Fig.3 - note that the Hα profile for this star exhibits strong electron scattering wings) in M33. However, the result is quite encouraging and we will add a larger number of more reliable data points to Fig.5 in the near future.

One of the immediate questions is whether the relation (2) does depend on metallicity. Indeed, the theory of radiation driven winds predicts a steeper slope for lower metallicity (Kudritzki et al. 1987). Using the Magellanic Clouds as a test

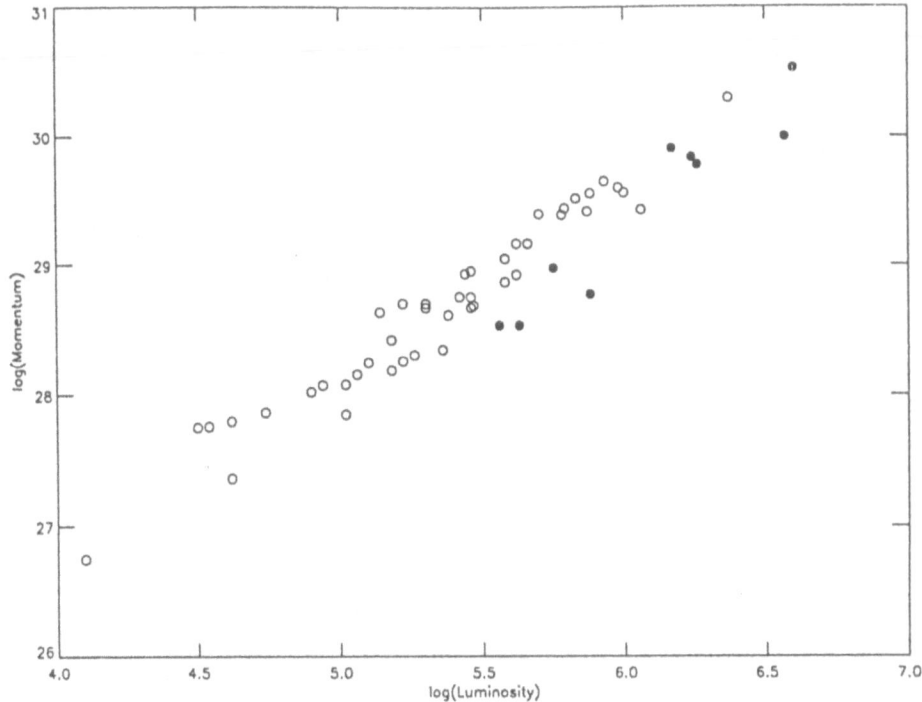

Fig. 7. $\log(\dot{M}v_\infty R^{0.5})$ versus $\log(L/L_\odot)$ for galactic (open circles) and LMC blue supergiants (filled circles).

laboratory it is possible to investigate empirically the metallicity dependance. We have used optical data obtained with the ESO 3.6m, and UV spectra taken with HST, to determine luminosity, wind momentum and metallicity for OB-supergiants in the LMC and SMC. The results are shown in Figures 7 and 8 (note that we have also added one A-supergiant for LMC and SMC from the paper by Stahl et al.,1991). The result is striking. For the LMC, where our HST analysis of O-stars yields metallicities around one half solar, only a small change in the slope is evident. For the SMC, however, where our O-star HST spectra indicate one tenth solar metallicity, there is a clear effect.

5 The VLT and Quantitative Spectroscopy beyond the Local Group

As outlined in the introduction quantitative spectroscopy within the Local Group can be done with 4m-telescopes. The VLT, however, will for the first time enable us to go one step beyond. With a spectral resolution of 3000 (sufficient for the work descibed in the preceeding sections) FORS II will allow quantitative spectroscopy down to mv = 22 mag. With absolute magnitudes of the brightest blue supergiants between -7 to -10 mag, this means that we can reach galaxies with distance moduli between 29 to 32 mag. Thus, before we dare the step to the Virgo cluster, we intend to investigate galaxies such as NGC 5253, an

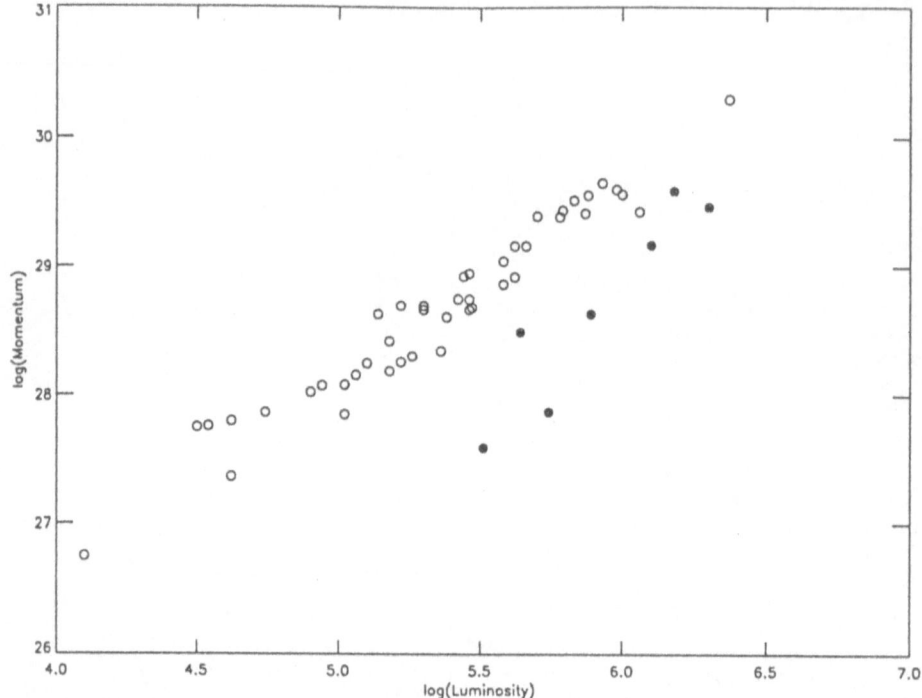

Fig. 8. $\log(\dot{M} v_\infty R^{0.5})$ versus $\log(L/L_\odot)$ for galactic (open circles) and SMC blue supergiants (filled circles).

S0/pec galaxy about 5Mpc distant containing 2 type Ia SN and for which a Cepheid distance should become available from HST. NGC 5253 is a primary calibrator for the SN distance scale. For this and other key galaxies we will obtain metallicities of the young population and, if the wind momentum-luminosity relation proves to work, we will be able to carry out a new independent check of the Cepheid distances.

A crucial problem for the distance determination (not for the determination of abundances) is of course crowding and possible stellar multiplicity. Our argument against this problem compromising the usefulness of the method is as follows. The brightest stellar targets that we will detect will be the most extreme A-supergiants. For such objects, multiplicity with similar spectral type is extremely unlikely because of evolutionary time scales (as for the Cepheids). Multiplicity with a large number of objects of different spectral type will be detectable from the spectra and will lead to a rejection of such targets (see for instance Herrero et al. 1994). In addition, the VLT, hopefully, will provide also high spatial resolution by Adaptive Optics with artificial guide stars, Speckle Interferometry and Interferometry. This will allow us to check for multiplicity at least in doubtful cases in not too distant galaxies.

References

Barlow, M.J., Cohen, M., 1977, ApJ, 213, 737

Bianchi, L., Hutchings, J.B., Massey, P., Kudritzki, R.-P., Herrero, A., Lennon, D.J., 1994, A&A, in press.

Castor, J., Abbott, D.C., Klein, R., 1975, ApJ, 195, 157

Gabler, R., Gabler, A., Kudritzki, R.-P., Puls, J., Pauldrach, A.W.A., 1989, A&A, 226, 162

Gabler, R., Kudritzki, R.-P., Mendez, R.H., 1991, A&A, 245, 587

Gabler, R., Gabler, A., Kudritzki, R.-P., Mendez, R.H., 1992, A&A, 265, 656

Groth, H.G., Kudritzki, R.P., Butler, K., Becker, S., Humphreys, R.M., 1990, in Properties of Hot Luminous Stars, A.S.P. Conf. Ser. 7, 151

Haser, S.M., Lennon, D.J., Kudritzki, R.-P., Puls, J., Pauldrach, A.W.A., Bianchi, L., Hutchings, J.B., 1994, A&A, in press

Herrero, A., Kudritzki, R.-P., Vilchez, J.M., Kunze, D., Butler, K., Haser, S., 1992, A&A, 261, 209

Herrero, A., Lennon, D.J., Vilchez, J.M., Kudritzki, R.-P., Humphreys, R.H., 1994, A&A, 287, 885

Husfeld, D., 1994, Habilitation thesis, University of Munich

Kudritzki, R.P., Pauldrach, A.W.A., Puls, J., Abbott, D.C., 1989, A&A, 219, 205

Kudritzki, R.-P., Hummer, D.G., 1990, Ann. Rev. Astron. Astrophys., 28, 303

Kudritzki, R.-P., Gabler, R., Kunze, D., Pauldrach, A.W.A., Puls, J., 1991, in Massive Stars in Starbursts, STScI Sym. Ser. 5, 59

Kudritzki, R.-P., Hummer, D.G., Pauldrach, A.W.A., Puls, J., Najarro, F., Imhoff, J., 1992a, A&A, 257, 655

Kudritzki, R.-P., et al, 1992b, in Science with the Hubble Space Telescope, ESO Conf. and Workshop Proc. 44, 279

Lennon, D.J., Kudritzki, R.-P., Becker, S.R., Butler, K., Eber, F., Groth, H.G., Kunze, D., 1991, A&A, 252, 498

Lennon, D.J., Dufton, P.L., Fitzsimmons, A., 1992, A&AS, 94, 569

Lucy, L.B., Solomon, P., 1970, ApJ, 159, 879

Najarro, F., Hillier, D.J., Kudritzki, R.-P., Krabbe, A., Genzel, R., Lutz, D., Drapatz, S., Geballe, T.R., 1994, A&A, 285, 573

Pauldrach, A.W.A., Puls, J., Kudritzki, R.-P., 1986, A&A, 164, 86

Pauldrach, A.W.A., 1987, A&A, 183, 295

Pauldrach, A.W.A., Kudritzki, R.-P., Puls, J., Butler, K., 1990, A&A, 228, 125

Pauldrach, A.W.A., Kudritzki, R.-P., Puls, J., Butler, K., Hunsinger, J., 1994, A&A, 283, 525

Puls, J., 1987, A&A, 184, 227

Puls, J., Kudritzki, R.-P., Herrero, A., Haser, S., Lennon, D.J., Gabler, R., Voels, S.A., Wachter, S., 1994, in prep.

Schaerer, D., Schmutz, W., 1994, A&A, in press

Sellmaier, F., Puls, J., Kudritzki, R.-P., Gabler, R., Gabler, A., Voels, S., 1993, A&A, 273, 533

Stahl, O., Aab, O., Smolinski, J., Wolf, B., 1991, A&A, 252, 693

Distance Indicators
and the Expansion Field of the Universe

Lukas Labhardt and G.A. Tammann

Astronomical Institute, University of Basel,
Venusstrasse 7, CH-4102 Binningen, Switzerland

Abstract. Several distance indicators are reflected against the potential of VLT. Distances to many galaxies, reliable peculiar motions, the accuracy of H_0, and even the determination of q_0 will enormously benefit from higher signal-to-noise and angular resolution, using mainly – in order of increasing range – RR Lyraes, Cepheids, brightest stars, globular clusters, novae, the rotation-luminosity relation of disk galaxies, the D_n-σ relation, SNe II, SNe Ia, and gravitational lenses.

1 Introduction

Distances to galaxies are the first step to their understanding. A number of sufficiently large distances will, moreover, provide the large-scale value of H_0, which not only provides the distances to all galaxies beyond $\lesssim 10\ 000$ km s^{-1}, but which is also required by cosmologists to better than 10%. Optimum distances with reliable error determinations to large numbers of galaxies at intermediate distances are needed for a detailed mapping of the expansion field. This offers the only possibility to test for dark matter over different scales. A considerable fraction of VLT observing time is required for these programs

2 Variables as Distance Indicators

2.1 Cepheids

Cepheids are widely accepted as the fundamental distance indicators because they follow an extremely well defined period-luminosity relation (Sandage & Tammann 1968; Feast & Walker 1987; Madore & Freedman 1991). In order to determine the absolute distance to a late-type galaxy, a suitable number of Cepheids must be observed over an adequate span of time. Most Cepheids are found to have periods in the interval 3 to 30 days and, correspondingly, average M_V of -2.5 to -5.5. A series of repeated deep images of a nearby, highly resolved galaxy down to $V \approx 27$ will, in principle, reveal most of the Cepheids in the field of view. Successful discovery, photometry and period determination of Cepheids will thus be feasible for distances up to 8 Mpc. To reduce the intrinsic dispersion in the PL relations, a sample of about 20 Cepheids per galaxy is required if the final accuracy of the distance is to be $\pm$ 5%.

Besides adhering to a well designed timing scheme that eliminates aliasing problems in the period determination, the detectability of variable stars crucially depends on the spatial resolution of the stellar images obtained at every single epoch. In highly crowded regions, there is the particular problem of faint neighbors which can easily be included into the bright component and thus mimic an increased stellar brightness.

For galaxies outside the Local Group, distances from Cepheids cannot be accomplished from the ground in an efficient and convincing way, even when using ESO's New Technology Telescope under the most promising circumstances (Tammann et al. 1991). Considerable progress can be expected from improving the image quality of ground-based telescopes by adaptive optics technology and/or performing image restoration of well-sampled images. As a matter of fact Pierce et al. (1994) reported recently the discovery of variable stars in NGC 4571 based on observations obtained with an image-stabilization, high-resolution camera on the Canada-France-Hawaii Telescope (McClure et al. 1989).

As their light amplitudes decrease with increasing wavelength, Cepheids are best searched for in the optical. According to the expected imaging performance of an 8m VLT unit telescope equipped with adaptive optics (ESO 1987), single stars will be imaged with a diameter of ≥ 0.3 arcsec in the optical. Deep imaging with FORS (Appenzeller et al. 1992) or the Wide Field Direct Visual Camera proposed by Wampler (1994) will provide fine enough sampling to perform accurate photometry by using PSF fitting techniques.

Compared to ground-based investigations, space-based searches for variable stars are more expensive but they do have a number of advantages like the absence of seeing variations, the potential optimization of the observing schedule irrespective of weather, time of day, moon light, etc., and the reacquisition of identical telescope orientation. Observations taken with the Wide Field Camera of the Hubble Space Telescope before and after the Servicing Mission in December 1993, with an imaging scale of 0.1 arcsec/pixel, provided almost ideal data for the discovery of Cepheids and subsequent distance determination in the nearby galaxies IC 4182 (Sandage et al. 1992, Saha et al. 1994), NGC 5253 (Sandage et al. 1994a, Saha et al. 1995) and M81 (Freedman et al. 1994) as well as in the more distant galaxies M100 (Freedman et al. 1994), NGC 4496 and NGC 4536 (both Sandage et al. 1994b). These galaxies do not yet settle the question of the distance to the Virgo cluster *core* because of the considerable depth effect of the cluster *spirals*. Since this year it is appropriate to use HST's improved detection power for probing distances to Virgo cluster galaxies. To close the distance gap between the Local Group and Virgo, VLT and its foreseen and planned instrumentation will become the major tool to rapidly increase the number of good Cepheid distances to a few dozen S and Im galaxies.

2.2 RR Lyrae Stars

RR Lyrae stars are fainter than typical Cepheids by roughly 4–5 mag. Their applicability and their distance range is correspondingly more restricted. In spite of that, they are important to check the distances from Cepheids. The P-L relation of the latter is slightly dependent on metallicity, but quite sensitive to possible changes of the helium abundance (Sandage 1994). The zero point and the metallicity dependence of the RR Lyrae calibration is now well determined (Sandage 1993a), giving for four Local Group galaxies excellent agreement with the Cepheid distances (Table 1 below). The Cepheid versus RR Lyrae check should be extended with VLT to nearby galaxies outside the Local Group.

Moreover, RR Lyrae stars are unrivalled distance indicators for dE galaxies. One might even reach with VLT these variables in S0 galaxies like NGC 404, NGC 5102, and NGC 5128, the latter two being members of the Cen A group. The RR Lyrae stars are here expected around 28.5 mag. Their discovery would be a major breakthrough, because they would give the first precision distances to normal early-type galaxies.

2.3 Novae

Novae have been found out to the distance of Virgo cluster ellipticals (Pritchet & van den Bergh 1987). They can be fitted to the rather well known absolute magnitude-decay rate relation of the novae in M31 (Capaccioli et al. 1989) or even to the Galactic calibration (Cohen 1985), which eventually may still be improved. Thus it will be possible to establish a completely independent distance scale out to at least the Virgo cluster using only novae.

The price of novae is the large amount of observing time required to discover a statistically meaningful number and to determine their decay rate. The Virgo ellipticals require a minimum of 20 consecutive dark and grey nights with an 8m telescope. The rewards are good distances to Virgo ellipticals which define the core of the cluster much better than late-type galaxies. Also for the field, beyond the reach of RR Lyraes, novae are the only primary distance indicators if one does not want to wait for a SN Ia event.

2.4 Supernovae of Type II

They have widely different luminosities, and these must be derived from their spectra by the expanding-photosphere method. A difficulty here is to account, as first pointed out by Wagoner (1980), for the dilution factor by which the emergent flux from rapidly expanding atmospheres differs from a blackbody. Considerable progress has recently been reported (Branch 1995). If these results are applied e.g. to the SNe II of Schmidt et al. (1994) one obtains a small Hubble constant, i.e. $H_0 \approx 53$. It may be expected that this purely physical method will soon provide competitive distances.

SNe II are intrinsically by far the most frequent supernovae (Tammann 1994), although all classified *distant* objects are of type Ia due to their very high luminosity. Searches with sufficiently large telescopes will, however, also reveal the distant SNe II. The follow-up work can then easily be done with VLT out to $z \approx 0.2$.

2.5 Supernovae of Type Ia

Normal SNe Ia are known to have uniform luminosities at maximum. This is shown, independent of redshift, by galaxies, groups and clusters with double or multiple events (n=20; Sandage & Tammann 1994a) as well as by the small scatter of their Hubble diagram extending to 30 000 km s^{-1} (Tammann & Sandage 1994). From this it follows that the intrinsic scatter of M(max) is, allowing for some photometric errors and peculiar motions, < 0.3 and possibly < 0.2 mag.

It is therefore possible to determine an excellent *large-scale* value of H_0 with only a few luminosity-calibrated SNe Ia. An ongoing HST program, using Cepheids as primary distance indicators, is mentioned in §2.1. As of this writing, three calibrators are available (Sandage et al. 1992, 1994a; Saha et al. 1994, 1995). The resulting absolute magnitudes M(max) are in excellent a-greement ($\lesssim 0.1$ mag) with purely physical determinations from the expanding-photosphere method and the ^{56}Ni mass by light curve fitting (Branch 1992, 1995; Müller & Höflich 1994; Nugent et al. 1995; Branch & Khokhlov 1995). The calibration corresponds to $H_0 = 55 \pm 7$ (external error).

The usefulness of SNe Ia has occasionally been questioned on the basis of abnormal SNe Ia, which are very red and subluminous by $\lesssim 1$ mag, and therefore easy to recognize. Moreover, at large distances they are automatically missed due to their much reduced discovery chance. In case they would contaminate the nearby sample of calibrators, the true value of H_0 could only be *smaller*. Phillips (1993) and Hamuy et al. (1994) have suggested that M(max) depends on the light curve decay rate and that – statistically quite unlikely – the local calibrators are overluminous. A rediscussion of the proposal shows the effect, however, to be marginal (Tammann & Sandage 1994).

By the time the VLT comes into operation the well observed, nearby SNe Ia with accessible Cepheids will foreseeably be covered by HST, except, of course, newly discovered nearby SNe Ia. Yet VLT, if flexibly scheduled, can do very important follow-up work (to guard against any remaining peculiarities and intrinsic absorption) on faint SNe Ia and thus provide distances to within 10–15% for very distant individual galaxies.

In fact, VLT is perfectly suited to tackle the next fundamental problem of cosmology, i.e. the determination of q_0. The program using SNe Ia as standard candles has originally been proposed for HST (Tammann 1979). At a cosmological useful distance of $z = 0.5$ SNe Ia are still as bright as $M_V(\text{max}) \approx 24$ mag. Only ten SNe Ia at this distance would already give a significant result on q_0.

Finally, well observed light curves of distant SNe Ia would provide a unique test for the time dilution factor (Tammann 1979) and hence *prove* the nature of redshifts. The only other known proof, the so-called Tolman test, comes from the relation of *metric* diameters with redshift (Sandage & Perelmuter 1991).

3   Brightest and Largest Objects as Distance Indicators

Few kinds of the largest objects offer themselves as standard rods and hence as distance indicators. The size (Sérsic 1960; Sandage 1962; Sandage & Tammann 1974a) and the Hα flux (Kennicutt 1981 and references therein) of H II regions have been used for late-type galaxies, but their properties depend on galaxy luminosity, i.e. sample size. Use of the diameters of the largest Sb and Sc galaxies (Sandage 1993b,c) poses primarily the problem of the definition of *distance*-limited samples, and hence does not call for particularly large telescopes.

3.1   Brightest Stars

The brightest stars were Hubble's primary distance indicators (cf. also Sandage 1962). Only the bluest $[(B - V) < 0.4$, including Hubble-Sandage variables] and reddest $[(B - V) \gtrsim 2.0]$ ones can be distinguished from foreground stars on the basis of photometry alone. The luminosity of the brightest blue stars depends on the galaxy size, whereas $M_V = -8.0$ for the brightest very red stars seemed to be quite stable (Sandage & Tammann 1974b), but a weaker dependence on galaxy size was later found also here (Sandage & Carlson 1985; Saha et al. 1994b). These brightest stars can therefore be used only with caution.

Crowding around luminous stars and correspondingly *resolution* becomes a severe problem beyond ≈10 Mpc. An illustration is the brightest, seemingly stellar object in NGC 1569, which spectroscopy shows to consist of a compact association of many O and B stars (Arp & Sandage 1985). Essentially all "single" stars, as seen with the Palomar 5m telescope in Virgo cluster galaxies, turn out to be multiple with HST. Spectroscopy of the more distant brightest stars, including those of intermediate color, is therefore mandatory. The great potential of this method is described by Kudritzki (these proceedings).

A powerful method to determine distances emerges also from the I-magnitudes of the tip of the red-giant branch (TRGB) of old populations (Mould, Kristian, & Da Costa 1983; Mould & Kristian 1986; Da Costa & Armandroff 1990; Lee, Freedman & Madore 1993). From six globulars in the Galaxy we derive

$$M_I(\text{TRGB}) \quad = \quad -4.02 + 0.1 \cdot ([\text{Fe/H}] + 1) \quad , \tag{1}$$

i.e. only a weak dependence on metallicity. This still preliminary calibration yields distances for 10 galaxies, which agree with the best Cepheid and/or RR Lyrae distances to 0.1 mag on average (cf. Table 1). For the VLT here is a very rich field of application.

Table 1. Comparison of distances from Cepheids, RR Lyraes, and the tip of the red giant branch

	[Fe/H]	$(m-M)_{Cepheid}$	$(m-M)_{RRLyr}$	$(m-M)_{TRGB}$
LMC	−1.2	18.50	18.60	18.43
NGC 6822	−1.8 :	23.62	–	23.62
NGC 185	−1.2	–	24.05	23.96
NGC 147	−0.9	–	24.06	24.09
IC 1613	−1.3	24.42	24.32	24.26
M31	−0.8	24.44	24.34	24.40
M33	−2.0	24.63	24.70	24.87
WLM	−1.6 :	24.92	–	24.89
NGC 205	−0.8	–	24.74	24.38
NGC 3109	−1.6	25.72	–	25.55

3.2 Globular Clusters

After first experiments with the very brightest globular clusters (GC), the turn-over of their luminosity function (LF) for Virgo cluster ellipticals was first obtained and applied as distance indicator by Harris (1991). The B and V calibration of the turnover with new RR Lyrae distances of Galactic GC and with the Cepheid-calibrated GC in M31 is fully consistent, and provides a Virgo cluster modulus of (m-M) = 31.71±0.07 (Sandage & Tammann 1994b). This value which agrees well with most other determinations supports the underlying assumption, also supported by circumstantial arguments (Harris 1991), that the turn-over luminosity is the same for spirals and ellipticals. Definite proof will come from VLT, once the first fundamental distances to early-type galaxies become available through RR Lyrae stars and novae, and/or through Cepheid distances of groups containing late- *and* early-type galaxies.

The turn-over of the GCLF in Virgo lies at $B = 24.7$ and $V = 24.0$. It should therefore be relatively easy for VLT to go to twice the Virgo distance and to provide most useful distances to many E/S0 galaxies.

3.3 Planetary Nebulae

Following a suggestion by Ford (1978) to use the $\lambda\,5007$ Å luminosity of the shells of PNe as a distance indicator, it has been postulated that the corresponding LF has a sharp cutoff (cf. Jacoby & Ciardullo 1993 for a summary) although these authors found 9 PNe which are brighter than the supposed cutoff by up to one magnitude. Consequently it was pointed out that the available data are better fitted by a (exponential) bright tail of the LF (Bottinelli et al. 1991; Tammann 1993). The large scatter was also confirmed by Wagner & Tammann (1993) and, semi-empirically, by Mendéz et al. (1994). The ensuing dependence of the maximum shell luminosity on sample size is accompanied by a severe compression of the distance scale which – even if allowed for – impairs the utility of PNe as distance indicators.

PNe are, even in the 5007 Å line, by no means easy objects. Different teams disagree even on the existence of the very brightest PN in two galaxies (NGC 3377, 5128). If VLT is used for PNe, it should test the applicability of the method in many E/S0 galaxies in the Virgo cluster core and then provide the distance of a closer calibrator of the same Hubble type and of comparable size.

3.4 Surface Brightness Fluctuations

The count-surface brightness method by Baum & Schwarzschild (1955) has been pushed beyond resolution by Tonry & Schneider (1988). They infer the number of brightest stars (presumably at the TRGB) in E/S0 galaxies from pixel-to-pixel intensity variations in the I passband. Tonry, Ajhar & Luppino (1990) have published distances for 13 E/S0 galaxies in the Virgo cluster. The distances of these highly clustered galaxies diverge by a factor of 2; moreover they are a strong function of metallicity (Tammann 1992). A revision by Tonry (1991), by empirically introducing $(V - I)$ colors as an additional parameter, does not significantly improve the situation (Lorenz et al. 1993). The latter authors emphasize that the method is sensitive – in addition to foreground stars, background galaxies, and globular clusters – to any internal structure the galaxy may have.

The resulting mean Virgo cluster distance from this method is $\approx 30\%$ too small, moreover other cluster distances *relative* to Virgo are smaller than D_n-σ distances by 10-30% (Sandage & Tammann 1994a). The method at present does not warrant the application of VLT.

4 Galaxy Properties as Distance Indicators

Several distance-independent, global galaxy parameters are known which correlate with the galaxian luminosity and thus can serve as distance indicators. These parameters have also been combined to find Brosche's (1973) "fundamental plane", i.e. the optimum correlation with luminosity. For brevity we discuss here only the two most widely used correlations.

4.1 Rotation velocity vs. luminosity

Öpik (1922) was the first to use the rotation velocity of a disk galaxy (M31) to determine its distance. The method was employed in a grand style by Tully & Fisher (1977) who derived the rotation velocity simply from the 21 cm-line width. An extensive literature on the method in different optical and IR wavebands has since been presented, discussing zeropoints and slopes of the relation, inclination and absorption corrections, as well as scatter and ensuing, very important selection bias. The method yields distances to $\approx 15\%$ for *fast* rotators, while it essentially fails for slow rotators (Federspiel, Sandage, & Tammann 1994).

More recently Mathewson, Ford, & Buchhorn (1992) and Mathewson (1995) have demonstrated on very large samples that it is advantageous and easy to derive the rotation velocity from *optical* spectra. The latter offer the possibility, with sufficient spatial resolution, to eventually determine also a dynamical radius, which as a second parameter is expected to reduce the scatter. If confirmed, VLT can provide good distances for thousands of (sufficiently inclined) disk galaxies to $\approx 10\ 000$ km s^{-1}, which would be essential for the mapping of the not so local expansion field.

4.2 The D_n–σ method

A combination of the central velocity dispersion σ – mass (luminosity) relation (Minkowski 1962) with the surface brightness-luminosity relation (Oemler 1974, 1976; Kormendy 1977) of E galaxies has led to the diameter (D_n – σ relation of E/S0 galaxies (Dressler et al. 1987; Djorgovski & Davis 1987) and of the bulges of spirals (Dressler 1987). The quite extensive data on the method are very promising. In the case of E/S0 galaxies the method lacks a local calibrator. This can be provided, as discussed before, by VLT. Moreover, this telescope can provide D_n diameters and high signal-to-noise velocity dispersions σ to at least $10\ 000$ km s^{-1}. This offers the best possibility to map the velocity field of E/S0 galaxies and of spirals with low inclination, for which no rotation velocities can be derived.

5 Purely Physical Distance Determinations

The weight of purely physical distance determinations is constantly increasing. Theoretical luminosities of SNe II and SN Ia have been discussed in §§ 2.4 and 2.5. The Sunyaev-Zeldovich effect provides distances for a steadily increasing number of X-ray clusters, giving consistently values of H_0 <50 (Birkinshaw & Hughes 1994; Jones 1995). The firm upper limit of H_0 <70 (strictly speaking of a combination of H_0 and q_0) from a gravitational double quasar (Dahle, Maddox, & Lilje 1995) can be much improved by a more detailed mass model of the deflector. VLT will have to provide optimum images of the deflectors of this and other gravitationally lensed quasars as well as, in the case of very close double quasars, the required high-accuracy photometry of their variability.

6   Concluding Note

We cannot refrain from noting that, quite independent of all distance determination programs, the recent marginal detection of (large amounts of) ^{2}D in a quasar absorption spectrum (Carswell et al. 1994; Songaila et al. 1994) opens up a new, fascinating possibility. With its enormous collecting power VLT will almost necessarily see the ^{2}D line in many high-z clouds and thus provide a unique test for the *chemical* homogeneity of the early Universe.

The authors thank the Swiss National Science Foundation for support.

References

Appenzeller, I., et al. 1992, in Progress in Telescope and Instrument Technologies, ed. M.H. Ulrich (Garching: ESO), p. 577

Arp, H. & Sandage, A. 1985, AJ 90, 1163

Baum, W.A., & Schwarzschild, M. 1955, AJ 60, 247

Birkinshaw, M., & Hughes, J.P. 1994, ApJ 420, 33

Bottinelli, L., Gouguenheim, L., Paturel, G.H., & Teerikorpi, P. 1991, A&A 252, 550

Branch, D. 1992, ApJ 392, 35

Branch, D. 1995, in Proc. Seventh Marcel Grossmann Conf., eds. M. Kaiser & R.Jantzen (Singapore: World Scientific), in press

Branch, D., & Khokhlov, A.M. 1994, Physics Reports, in press

Brosche, P. 1973, A&A 23, 259

Capaccioli, M., Della Valle, M., D'Onofrio, M., & Rosino, L. 1989, AJ 97, 1622

Carswell, R.F., Rauch, M., Weyman, R.J., Cooke, A.J., & Webb, J.K. 1994, MNRAS 268, L1

Cohen, J.G. 1985, ApJ 292, 90

Da Costa, G.S., & Armandroff, T.E. 1990, AJ 100, 162

Dahle, H., Maddox, S.J., & Lilje, P.B. 1995, ApJ Letters, in press

Djorgovski, S.B., & Davis, M. 1987, ApJ 313, 59

Dressler, A. 1987, ApJ 317, 1

Dressler, A., Lynden-Bell, D., Burstein, D., Davies, R.L., Faber, S.M., Terlevich, R.J., & Wegner, G. 1987, ApJ 313, 42

European Southern Observatory 1987, Proposal for the Construction of the 16m VLT, VLT Report No. 57 (Garching: ESO)

Feast, M., & Walker, A.R. 1987, ARA&A 25, 345

Federspiel, M., Sandage, A., & Tammann, G.A. 1994, ApJ 430, 29

Freedman, W.L., et al. 1994, ApJ 427, 628

Freedman, W.L., et al. 1994, Nature 371, 757

Harris, W.E. 1991, ARA&A 29, 543

Jacoby, G., & Ciardullo, R. 1993, IAU Symp. 155, 503

Jones, M. 1995, Ap. Letters & Comm., in press

Kennicutt, R.C. 1981, ApJ 247, 9

Kormendy J. 1977, ApJ 218, 333

Lee, M.G., Freedman, W.L., & Madore, B.F. 1993, ApJ 417, 553

Lorenz, H. Böhm, P., Capaccioli, M., Richter, G.M., & Longo, G. 1993, A&A 277, L15

Madore, B.F., & Freedman, W.L. 1991, PASP 103, 933

Mathewson, D.S. 1995, IAU Symp. N0. 168, in press

Mathewson, D.S., Ford, V.L., & Buchhorn, M. 1992, ApJS 81, 413

McClure, R.D., Grundmann, W.A., Rambold, W.N., Fletcher, J.M., Richardson, E.H., Stillburn, J.R., Racine, R., Christian, C.A., & Waddell, P. 1989, PASP 101, 1156

Mendéz, R.H. Kudritzki, R.P., Ciardullo, R., & Jacoby, G.H. 1994, A&A 275, 534

Minkowski, R. 1962, in Problems of Extragalactic Research, IAU Symp. No. 15, ed. G. C. McVittie (New York: Macmillan), p. 112

Mould, J., & Kristian, J. 1986, ApJ 305, 591

Mould, J., Kristian, J., & Da Costa, G.S. 1983, ApJ 270, 471

Müller, E., & Höflich, P. 1994, A&A 281, 51

Nugent, P., Baron, E., Hauschildt, P., & Branch, D. 1995, ApJ, in press

Oemler, A. 1974, ApJ 194, 1

Oemler, A. 1976, ApJ 209, 693

Öpik, E. 1922, ApJ 55, 406

Pierce, M.J., Welch, D.L., McClure, R.D., van den Bergh, S., Racine, R., & Stetson, P.B. 1994, Nature 371, 385

Pritchet, C.J., & van den Bergh, S. 1987, ApJ 318, 507

Saha, A., Labhardt, L., Schwengeler, H., Macchetto, F.D., Panagia, N., Sandage, A., & Tammann, G.A. 1994a, ApJ 425, 14

Saha, A., Sandage, A., Labhardt, L., Schwengeler, H., Tammann, G.A., Panagia, N., & Macchetto, F.D. 1994b, ApJ, in press

Sandage, A. 1962, in Problems of Extragalactic Research, IAU Symp. No. 15, ed. G.G. McVittie (New York: Macmillan), p. 359

Sandage, A. 1993a, AJ 106, 703

Sandage, A. 1993b, ApJ 404, 419

Sandage, A. 1993c, ApJ 402, 2

Sandage, A. 1994, unpublished

Sandage, A., & Carlson, G. 1985, AJ 90, 1464

Sandage, A., & Perelmuter, J.-M. 1991, ApJ 370, 455

Sandage, A., Saha, A., Tammann, G.A., Panagia, N., & Macchetto, F.D. 1992, ApJ 401, L7

Sandage, A., Saha, A., Tammann, G.A., Labhardt, L., Schwengeler, H., Panagia, N., & Macchetto, F.D. 1994a, ApJ 423, L13

Sandage, A., et al. 1994b, in preparation

Sandage, A., & Tammann, G.A. 1968, ApJ 125, 435

Sandage, A., & Tammann, G.A. 1974a, ApJ 190, 525

Sandage, A., & Tammann, G.A. 1974b, ApJ 191, 603

Sandage, A., & Tammann, G.A. 1994a, in Current Topics in Astrofundamental Physics, ed. N. Sanchez, in press (Basel Preprint No. 76)

Sandage, A., & Tammann, G.A. 1994b, ApJ, in press (Basel Preprint No. 73)

Schmidt, B.P., Kirshner, R.P., Eastman, R.G., Phillips, M.M., Suntzeff, N.B., Hamuy, M., Maza, J., & Avilés, R. 1994, ApJ 432, 42

Sérsic, J.L. 1960, Zeitschr. f. Ap. 50, 168

Songaila, A. Cowie, L.L., Hogan, C.J., &Rugers, M. 1994, Nature 368, 599

Tammann, G.A. 1979, in Astronomical Uses of the Space Telescope, eds F. Macchetto, F. Pacini, & M. Tarenghi (Geneva: ESA/ESO), p. 329

Tammann, G.A. 1993, IAU Symp. 155, 515

Tammann, G.A. 1994, in Supernovae, Les Houches LIV, eds. S.A. Bludman, R. Mochkovitch, & J. Zinn-Justin (Amsterdam: North Holland), p.1

Tammann, G.A., et al. 1991, The Messenger 61, 8

Tammann, G.A., & Sandage, A. 1994, ApJ, in press (Basel Preprint No. 70)

Tonry, J.L. 1991, ApJ 373, L1

Tonry, J.L., Ajhar, A.E., & Luppino, G.A. 1990, AJ 100, 1416

Tonry, J.L., & Schneider, D.P. 1988, AJ 96, 807

Tully, R.B., & Fisher, J.R. 1977, A&A 54, 661

Wagner, S.J., & Tammann, G.A. 1993, Astron. Ges. Abstr. Ser. 8, 163

Wagoner, R.V. 1980, in Physical Cosmology, Les Houches XXXII, eds. R. Balian, J. Audouze, & D.N. Schramm (Amsterdam: North Holland), p. 180

Wampler, E.J. 1994, in Instruments for the ESO VLT, ed. A.F.M. Moorwood (Garching: ESO), p. 54

Distance Indicators: Beyond Distances - A Return to Astrophysics

George H. Jacoby

Kitt Peak National Observatory, P.O. Box 26732, Tucson, Arizona, 85726, USA

Abstract. Continuing progress in the development, application, and physical understanding of distance indicators is likely to yield a generally accepted value for the Hubble constant during the VLT construction years. Thus, the VLT's contribution to that highly visible subfield is unlikely to be pivotal. Nevertheless, the accuracy of distance indicators has increased to the point where discrepancies between methods are astrophysically meaningful; that is, residuals in the methods tell us new information about the physics of the galaxies and their components. The large aperture, extensive instrumentation, and four-telescope availability of the VLT promises to enhance our ability to apply distance indicators as astrophysical research tools. A sample of recent and potential applications are described here.

1   Introduction: Why Are Distance Indicators Important

Rarely is distance the parameter that astronomers seek. Distances are, however, vital intermediate answers in devising astrophysical models to compare with observations. For example, distances are needed to measure the luminosities of stars, galaxies, and quasars, that are used, in turn, to judge the viability of various energy sources postulated to power these phenomena. Similarly, distances provide the means to measure the sizes of objects, such as the diameters of galaxies and clusters, and consequently the time scales for dynamical events.

The most visible and newsworthy role for distance indicators historically has been to derive the Hubble constant, H_0. The tremendous interest in H_0 is rooted in defining the basic parameters of the Big Bang, where the expansion rate should be the easiest parameter to measure. Without a firm value for H_0, it becomes more difficult to derive the other parameters, q_0, Ω_0, and the cosmological constant, Λ. Most importantly, though, we reasonably expect that the age of the universe, T_0, to be older than the ages of the oldest stars. Unfortunately, under the preferred inflationary Big Bang cosmology, this expectation demands that $H_0 < 50\ H_0$ (Fukugita et al. 1993) and has driven some to argue for $H_0 \sim 30$ (Bartlett et al. 1994) on theoretical grounds alone.

2 Recent Progress in Distance Indicator Development

Despite the 65 year effort to measure H_0 accurately, the astronomical community continues to be subjected to debate about this elusive parameter. Over the past 20 years, the struggle has intensified, with reported values being distributed nearly bimodally near 50 and 100 H_0. Progress during the last 10 years has led to upper values ∼80. This is a reduction in the discrepancy, but the factor of 1.6 in distance (1.0 mag in distance modulus) remains unsatisfactory.

2.1 Technological Advances

It is advances in technology that will end the distance scale controversy; the impact of technology is apparent already. The most dramatic improvements during the last 5 years stem from telescope improvements. Foremost among these is the availability of the Hubble Space Telescope. Contributions from ground-based telescopes such as the Canada-France-Hawaii-Telescope also are impressive (Pierce et al. 1994). It is possible now to derive distances using Cepheid variables to spiral galaxies as far as the Virgo Cluster (Freedman et al. 1994).

Additionally, thanks to advances in detector technology (e.g. CCDs, IR arrays with near-perfect DQE), computers and software (e.g. numerical methods for error analysis, Monte Carlo simulations), and the consequent invention of new distance indicators such as surface brightness fluctuations (Tonry & Schneider 1988), planetary nebula luminosity functions (Jacoby et al. 1988), and the expanding photosphere method for SN II (Eastman & Kirshner 1989), most modern methods for measuring extragalactic distances are in excellent agreement when compared on a galaxy-by-galaxy basis (Jacoby et al. 1992). Since there is no perfect "reference galaxy" whose distance in known a priori, we are forced to rely on intercomparisons between the various methods both to build confidence and to identify inconsistent techniques.

2.2 An Implication of Solving the Distance Scale Controversy

Once we can measure distances without controversy, at least relative ones, then we can explore the use of distance indicators as astrophysical tools. Any distance discrepancy among the methods serves as a flag to a physical effect; that is, if the methods work, then barring observational error, a disagreement in distance indicates that a physical process is taking place that is not accounted for in the assumptions of one of the methods. In the following sections, I will present some examples to demonstrate this line of research and to identify several projects based on using distance indicators as astrophysical probes.

3   The Case For Distance Indicators

During the 20th century, many distance indicators have been proposed. Today's most important ones are listed Table 1. Columns 2 and 3 are estimates of the range over which these methods can be used with the VLT and the exposure times needed to achieve the maximum distances. Column 4 indicates the primary instruments required.

Table 1. Commonly Used Distance Indicators

Indicator	VLT Range, Mpc	Exposure, Hr	Instrument
Radar	...	...	...
Parallax	0.001	0.01–4	VLTI, CCD
RR Lyr	2	2/epoch	FORS
Cepheid	15	1/epoch	FORS/AO
PNLF	30^1	2	FORS+NB filters
SBF	45^1	3	FORS/ISAAC
GCLF	70^1	4	FORS
D_n-σ	300	3	FORS/slitlets
Tully-Fisher	300	3	FORS/ISAAC/slitlets
SN II	400	2/epoch	FORS
SN Ia	400	1/epoch	FORS

[1]Increase by 2.5X if wide field camera yielding 0.1″ resolution is available.

Of these methods, only radar (included for completeness as it defines the parsec) might be considered perfect; its accuracy is $10^7 \times$ better than any other method, but is limited to the solar system. Each of the others is weak in some way when one considers that a good distance indicator...

1. Has a good zero point calibration (using multiple calibrators)
2. Has a good prescription for metallicity correction
3. Has a good prescription to correct for effects of stellar ages
4. Can be corrected for effects of foreground and internal extinction
5. Has a good physical rationale
6. Can be compared with other methods

The additional methods shown in Table 2 are either more difficult to apply than the primary methods and have not been used much, or they suffer from multiple weaknesses. Distance measurements via gravitational lenses, for example, have been applied to only one system, and that distance has yet to be compared with another method.

The ultimate test for a distance indicator is whether it gives the right answer (criterion 6). Methods yielding distances that cannot be compared are as useful as theories that cannot be tested.

Table 2. Other Notable Indicators

Novae	Red Giant Branch Tip	Red Supergiants
HII Region L_β-σ	HII Region Diameters	Blue Supergiants
Luminosity vs Hubble Type	Gravitational Lens	Miras
Long Period Variables	Galaxy Diameters	SZ Effect

Jacoby et al. (1992) presented intercomparisons using the SBF method as the reference technique. Additional data and new methods are included in this paper. Ciardullo et al. (1993a) pointed out that good agreement between methods constrains the severity of existing systematic errors, such that:

1. Dust is not affecting the distances because each method behaves differently in the presence of dust (especially SBF with respect to the others).
2. An error in the application of one method (e.g. adopting an incorrect fitting function for PNLF or GCLF) is unlikely, since it must be mirrored by unrelated effects in other methods.
3. Metallicity effects must be small since each method responds differently to abundance variations (e.g. Cepheids and GCLF assume no effects, SBF explicitly corrects for colors, PNLF explicitly corrects for low metallicity).
4. Population age effects must be small, since many galaxies sample a wide range of age.

3.1 Planetary Nebulae Luminosity Functions

The planetary nebula luminosity function (PNLF) method (Ciardullo et al. 1989) relies on the observed uniformity of the bright end of the $\lambda5007$ luminosity functions of PN in external galaxies. Relative distances to galaxies are derived by comparing the maximum brightnesses of the PN ensembles (see summary in Jacoby & Ciardullo 1993), and absolute distances are obtained by scaling to a calibrator galaxy, usually M31 whose Cepheid distance is known (Freedman & Madore 1990). Theoretical support for the method comes from Jacoby (1989), Dopita et al. (1993), and Mendez et al. (1993).

It is essential, though, that the survey for PNe extend faint enough to sample the shape of the PNLF, which over the very brightest 0.5 mag can be approximated by a power law (Bottinelli et al. 1991). By extending down 0.8 mag or deeper, any confusion with a power law is avoided (Jacoby et al. 1990; Mendez et al. 1993). It is desirable, though, to extend more than 1 mag down the PNLF, and *this has yet to be done for Virgo or Fornax galaxies, but requires only 1-2 hours with the VLT!*

Fig. 1 and Table 3 illustrate that the PNLF method yields distances that are indistinguishable from Cepheid/RR Lyr distances. The Cepheid distance to NGC 5253 (Sandage et al. 1994) has been adjusted to E(B-V)=0.06 (Burstein & Heiles 1984) and the PNLF distance has been corrected for low metallicity as prescribed by Ciardullo & Jacoby (1992). The LMC distance is from SN 1987A (Panagia et al. 1991).

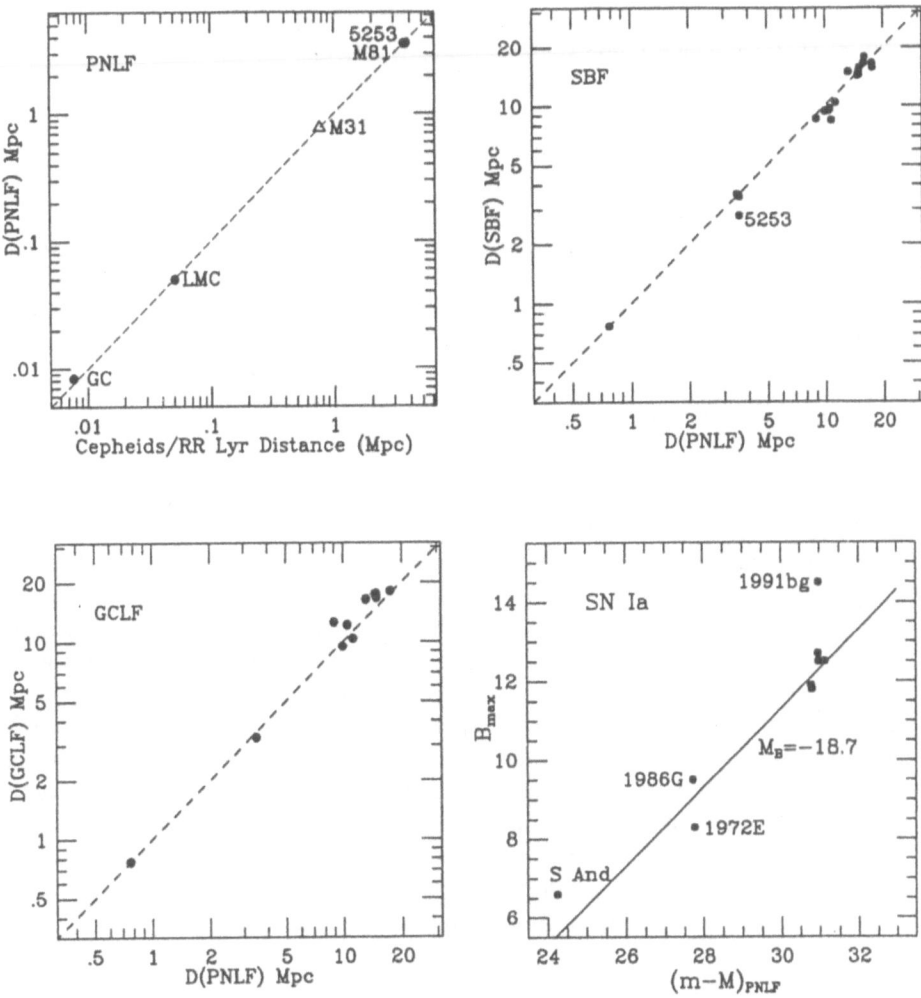

Fig. 1. A comparison between the PNLF and Cepheid/RR Lyr distance scales (upper left), SBF and PNLF scales (upper right), GCLF and PNLF scale, (lower left), and SN Ia and PNLF scales (lower right). The 1:1 diagonal line is shown in each case; the SN Ia line suggests a peak absolute luminosity of M(B)=-18.7. Outlying objects of special interest are noted in the SBF and SN Ia panel.

In this and subsequent comparisons, I will concentrate on the most discrepant outliers from which to draw astrophysical conclusions. Here, there are no statistically significant outliers, although the most discrepant case is the distance to the Galactic Center. The PNLF distance is 8.4 kpc, about 8% larger than the

RR Lyr distance of 7.7 kpc from Reid (1989). Others (e.g. Walker 1992) have noted that Cepheids tends to yield slightly longer distances than the RR Lyr scale, and there is evidence here to support that conclusion.

Future astrophysical projects for planetary nebulae: (1) Derive the 3D distribution of nearby (e.g. Virgo, Fornax) clusters by obtaining accurate distances to hundreds of galaxies. (2) Derive galaxy mass distributions using PN found in (1) as kinematic test particles (e.g. 433 PN velocities in Cen A - Hui 1992; 29 in NGC 3379 - Ciardullo et al. 1993b; 37 in NGC 1399 - Arnaboldi et al. 1994; $\sim$ 800 in M31 - Hui 1995). This requires $\sim$ 5 VLT hours per galaxy with FORS/slitlets. (3) Derive metallicity (Ciardullo 1995) either by counting PN ($\sim$ 2 hours per galaxy), or by averaging spectra from dozens of PN (Kaler 1995).

3.2 Surface Brightness Fluctuations

The SBF technique relies on the statistical fluctuations in high signal-to-noise data arising from the slightly different number of stars projected on the pixels in an image of a galaxy (Tonry & Schneider 1988). There is good theoretical support for the population independence of the technique in the I-band (Worthey 1993), although population effects appear in the K-band (Luppino & Tonry 1993).

Ciardullo et al. (1993a) discussed the excellent agreement between the S-BF and PNLF techniques for 16 galaxies, where the largest discrepancies were accounted for by errors in the adopted extinction (usually Burstein & Heiles 1984). The unusual behavior of the SBF method, where greater extinction yields larger derived distances, is due to a reddening of the galaxy V-I color. This, in turn, requires a brighter calibration for the fluctuations, and forces the galaxy to be at a larger distance for the apparent fluctuation brightness. Since the other techniques follow the usual behavior (greater extinction implies a closer galaxy), an extinction value usually can be found to bring an SBF distance into better agreement with the PNLF distance through the relation:

$$(m - M)_{\mathrm{PNLF}} - (m - M)_{\mathrm{SBF}} = 6.8\Delta E(\mathrm{B} - \mathrm{V})$$

Fig. 1 illustrates the SBF-PNLF comparison, but now with 19 galaxies. The dominant outlier is NGC 5253, being 22% closer than either the PNLF or Cepheid distances. Data quality is not an issue for this nearby galaxy. The large discrepancy indicates that something is unusual. Adopting the precept that disagreements tell us something physical, one might conclude that the reddening adopted to NGC 5253 is too small and should be E(B–V)=0.15. Alternatively, the stellar population may be extremely unusual in some way, but there is no evidence for population effects influencing SBF to this degree. Increased reddening seems the best path to resolution, although NGC 5253 is a very odd galaxy and could harbor a strange population mix.

Thus, a very sensitive technique to measure reddening to galaxies is to compare the distances from SBF and PNLF and find the reddening which forces agreement. This requires $\sim$ 1 VLT hour per galaxy.

Other future projects for SBF: (1) Search for intermediate age populations by comparing the I-band and K-band distances (Luppino & Tonry 1993; Pahre

& Mould 1994). This requires ≤ 1 hour per galaxy. (2) Improve the accuracy of large scale flow studies by deriving 5% distances out to 3500 km/s.

Table 3. Indicator Comparison and H_0 Summary

Indicator	Nr Galaxies	Offset	RMS	H_0	Notes
PNLF	5	+ 2%	5%	79	1
SBF	18	- 3%	9%	83	2,3
GCLF	11	+10%	15%	72	2
SN Ia	9	...	19%	69	2,4,5
D_n-σ	11	...	20%	...	2,6
D_n-σ_{10}	11	...	11%	...	2,6,7
SN II	12	+13%	12%	73	8,9
T-F	2	...	...	85	10

[1]relative to Cehpeids/RR Lyr
[2]relative to PNLF
[3]excludes NGC 5253
[4]H_0 from SN 1972E, 1937C (Hamuy et al. 1995; Pierce & Jacoby 1995)
[5]excludes SN 1991bg
[6]best slope is 76 km/s/Mpc; no zero-point available for H_0
[7]σ_{10} is velocity dispersion at $10''$ from nucleus (available for 6 galaxies)
[8]relative to T-F; H_0 from Schmidt et al. (1994)
[9]excludes SN 1973R; adopting Baron et al. (1994) distance for SN 1993J
[10]only common galaxies with PNLF are calibrators - no offset or RMS

3.3 Globular Cluster Luminosity Functions

This technique assumes that all galaxies have identical peaks in their GCLF. The theoretical arguments are weak (Jacoby et al. 1992) and the evidence is mixed. The widths of the GCLF are not constant (Secker 1992; Secker & Harris 1993), but the M_V peak does seem to be close to universal (Jacoby et al. 1992). The evidence in Fig. 1 is highly suggestive that the method works.

If the GCLF peak is truly constant, then the lack of disagreement provides a severe astrophysical constraint for formation theories of GC systems. For example, Harris (Jacoby et al. 1992) argues that the uniformity in the GCLF peak must be due to the original mass spectrum of the GC system rather than a result of dynamical interactions.

Future projects for GCs: the clusters provide good kinematical test particles to study galaxy mass distributions (e.g. Mould et al. 1990) in the same spirit as planetaries, but they extend much further out in the galaxy halos.

3.4 Type Ia Supernovae

The similarities in peak B luminosity for SN Ia suggests that they are good standard candles (Leibundgut & Tammann 1990). Phillips (1993) and Hamuy et al. (1995), however, have shown that there is a range of several mag in M_B in which the luminosity correlates with other observables such as decline rate (fast decliners are faint), color (red SN Ia are faint), and a variety of spectral properties. The dispersion in Fig. 1 is, in part, intrinsic to the SNe. SN 1991bg, SN 1986G, and S And are all red, fast decliners and fall to the faint side of the fit line. SN 1972E is a blue, slow decliner (Hamuy et al. 1995) and falls to the bright side of the fit line. The remaining 7 SNe (in Virgo and Fornax galaxies) are very close to the fit line.

Phillips (1993) used this kind of intercomparison to identify the astrophysical result that not all SN Ia are the same and that they can be placed into a one-parameter family of luminosities. One explanation for the existence of a family is that SN Ia derive from a range of progenitor masses rather than the canonical Chandrasekhar mass white dwarf. One might expect then, that SN Ia were brighter in the past due to the greater presence of young massive main sequence stars.

Future projects for SN Ia: If a sample of SN Ia observed at $z \sim 0.4$ is brighter than nearby SNe, then SN Ia derive from sub-Chandrasekhar mass stars (Schommer, priv. comm.). With 2 nights on the VLT separated by 2 weeks, an imaging survey will find SN Ia (to $R \sim 24$) out to $z \sim 0.5$. Low resolution (R=1000) spectra are required to confirm the SN Ia identifications. SN Ia that derive from sub-Chandrasekhar progenitors will be ~ 0.5 mag more luminous at z=0.5. Once this effect is calibrated, it may be possible to attempt to measure q_0 with SN Ia.

SN Ia are among the most far-reaching distance indicators, and given a high detection rate, they can be used to measure large scale flows over a vast volume. It may be possible to constrain Ω_0 in the same manner as D_n-σ has been used.

3.5 D_n-σ

One projection of the fundamental plane of elliptical galaxies (Faber et al. 1987) is D_n-σ (i.e. diameter at the enclosed average surface brightness of B=20.75, and nuclear velocity dispersion) relationship. Other parameters (e.g. luminosity, surface brightness, metallicity) also correlate well (de Carvalho & Djorgovski 1992). Gregg (1992) and de Carvalho & Djorgovski (1992) recommend using D_n-σ with caution due to systematic differences seen in a variety of environments.

The upper panels in Fig. 2, however, show that the method can be improved by exercising care in its application. By comparing the D_n-σ velocities to the PNLF distances, 3 highly discrepant galaxies emerge: M87, having a central black hole, NGC 1399, having a central velocity spike (a black hole candidate), and NGC 4494, having a kinematically distinct rotating core (Jedrzejewski & Schechter 1989). One should not expect the distinct objects at the centers of these galaxies to reflect galactic global properties. Consequently, adopting a central velocity dispersion is risky; instead, I've adopted a velocity dispersion measured $10''$ off the nucleus, σ_{10}. Data were available for 6 of the 11 galaxies, and the comparison in Fig. 2 and Table 3 indicates that this modification is an improvement.

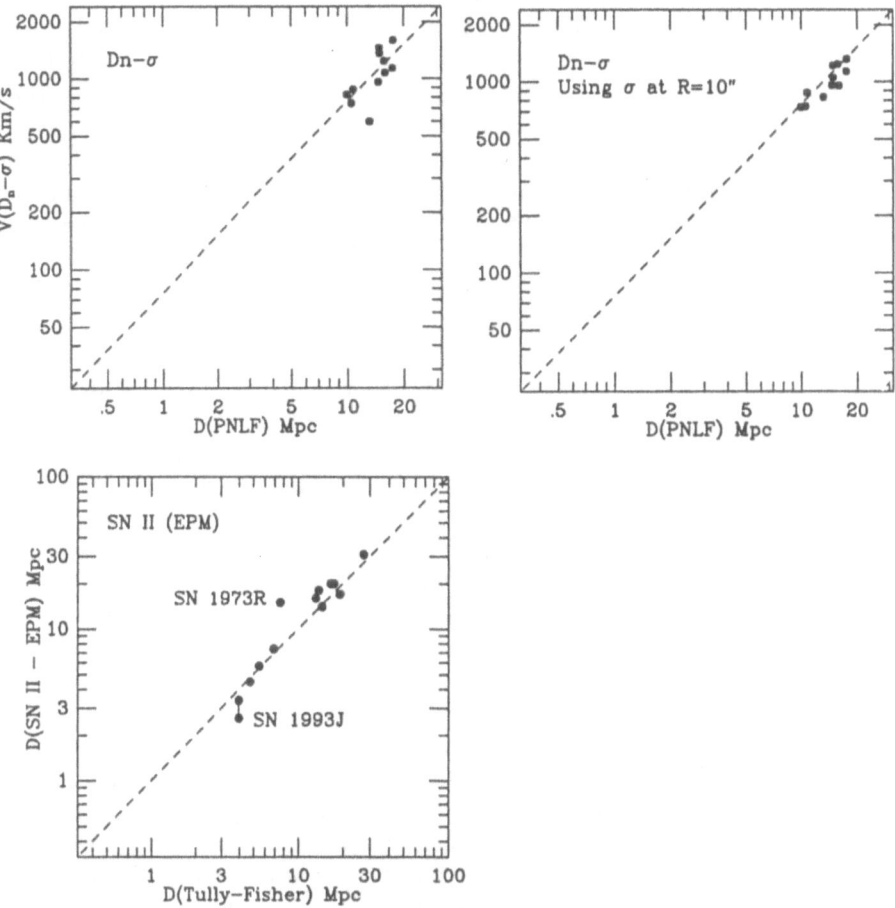

Fig. 2. A comparison between the PNLF and D_n-σ "distances" before (upper left) and after (upper right) correction for nuclear velocity anomalies. The corrected values are based on σ_{10}, the velocity dispersion measured $10''$ from the nucleus (Franx et al. 1989; Jedrzejewski & Schechter 1989). The lower left panel compares the SN II distances against Tully-Fisher distance (from Schmidt et al. 1994). Outlier SN 1993J in M81 is shown for EPM distances from Schmidt et al. (1993) and Baron et al. (1994).

Future projects for D_n-σ: improve large scale flow solutions by application of the σ_{10} modification to a large sample. This requires slit spectra (R $\sim$ 10,000) and VLT exposures of 15 min/galaxy. Alternatively, apply the PNLF method to a sample of D_n-σ galaxies to identify galaxies having bizarre central kinematics.

3.6   SN II Expanding Photosphere Method

This method is similar to the Baade-Wesselink method (Kirshner & Kwan 1974) where each event must be modelled uniquely. EPM has been applied to ~ 20 SN (Schmidt et al. 1994) up to 180 Mpc to derive H_0=73 H_0. The comparison in Fig. 2 is made with the Tully-Fisher method since SN II and PNLF are applied in different galaxies (i.e. SN II occur in late type spirals, while PNLF suffers badly from confusion with HII regions and so those galaxies are avoided). Pierce (1994) and Schmidt et al. (1994) have made similar comparisons, but without the SN 1993J results (Schmidt et al. 1993; Baron et al. 1994).

The agreement between these methods is good except for SN 1973R and SN 1993J. The former is a rare case where the observational data limit the quality of the comparison and the discrepancy reveals little; the error bars (not shown) are nearly 50% of the EPM distance. SN 1993J is more interesting; the distance discrepancy suggests a genuine physical cause. In fact, 2 EPM distances are plotted; the lower one from Schmidt et al. (1993) is based on an overly general model, whereas, the upper one from Baron et al. (1994) accounts for the physical peculiarities of the specific supernova (e.g. depleted hydrogen envelope of progenitor, asymmetry). Here is an excellent example where comparisons can be used to flag odd situations. Had SN 1993J occurred in a distant galaxy, its peculiarities would have been far less obvious, but would be indicated in a comparison with a Tully-Fisher distance.

Future projects for SN II: like SN Ia, SN II can be applied over very large distances to derive information about large scale flows and possibly Ω_0 and q_0.

3.7   Tully-Fisher Method

The Tully-Fisher (also known as the HI line width – Luminosity) method is an empirically based one relying on the correlation between luminosity of a spiral galaxy and its peak-to-peak rotational velocity (Tully & Fisher 1977). Usually, the velocity is measured in HI, but optically derived rotation at $H\alpha$ works nearly as well (Schommer et al. 1993; Bernstein et al. 1994). The original prescription for T-F was based on B band mags and was improved using H band mags (Aaronson et al. 1979). The I band also provides a tight relation (Pierce & Tully 1988) and is easy to apply with CCDs. The improvement with redder passbands is primarily due to the reduced sensitivity to errors in the internal extinction corrections and inclination measurements.

Burstein & Raychaudhury (1989) explored the sources of error in T-F (e.g. cluster depth, magnitude errors, poor sample selection, inclination errors, and velocity width errors). Any of these would cause a visible distance discrepancy when comparing to another method such as SN II EPM.

Future projects: The large wavelength range available to VLT instruments (0.3-10 μ) provides an opportunity to explore the properties of dust in a variety of galaxies. By measuring distances to galaxies using U-band through K-band luminosities, one can explore the extinction and dust scattering laws in individual galaxies. With the VLT and slitlets, one can apply the Hα version of T-F efficiently to galaxies in clusters at large distances (z ∼ 0.2) to study large scale flows. At slightly larger distances, interacting and merging galaxies are thought to become more common, and departures from the T-F relation will occur. These departures can be used to identify starburst galaxies.

4   Summary

Distance indicators have been surveyed briefly for use as astrophysical tools. By comparing the distances derived using different indicators but to the same galaxies, we find that:

1. The primary distance indicators agree to 15% or better.
2. Disagreements are significant; they imply physical causes rather than failures of the indicators. These disagreements can be exploited to learn new information about galaxies and their components.

Many of the projects presented in this paper use distance indicators to investigate the physics of the objects under study, rather than simply deriving their distances. The VLT offers unique instrumentation and large apertures to apply these astrophysical probes efficiently.

References

Aaronson, M., Huchra, K., & Mould, J. 1979, ApJ 229, 1

Arnaboldi, M., Freeman, K.C., Hui, X., Capaccioli, M., & Ford, H. 1994, Messenger 76, 40

Baron, E., Hauschildt, P.H., & Branch, D. 1994, ApJ 426, 334

Bartlett, J.G., Blanchard, A., Silk, J., & Turner, M.S. 1994, Nature, submitted.

Bernstein, G.M., et al. 1994, AJ 107, 1962

Bottinelli, L., Gougenheim, L., Paturel, G., & Teerikorpi, P. 1991, A&A 252, 550

de Carvalho, R.R., & Djorgovski, S. 1992, ApJ 389, L49

Burstein, D., & Heiles, C. 1984, ApJS 54, 33

Ciardullo, R., 1995, IAU Highlights of Astronomy, 1994, in press

Ciardullo, R., & Jacoby, G.H. 1992, ApJ 388, 268

Ciardullo, R., Jacoby, G.H., & Dejonghe 1993a, ApJ 414, 454

Ciardullo, R., Jacoby, G.H., & Tonry, J.L. 1993b, ApJ 419, 479

Ciardullo, R., Jacoby, G.H., Ford, H.C., & Neill, J.D. 1989, ApJ 339, 53

Dopita, M.A., Jacoby, G.H., & Vassiliadis, E. 1993, ApJ 389, 27

Eastman, R.G., & Kirshner, R.P. 1989, ApJ 347, 771

Faber, S., Dressler, A., Davies, R., Burstein, D., Lynden-Bell, D., Terlevich, R., & Wegner, G. 1987, in Nearly Normal Galaxies, ed. S. Faber (New York: Springer), p. 175

Franx, M., Illingworth, G., & Heckman, T. 1989, ApJ 344, 613

Freedman, W.L., & Madore, B.F. 1990, ApJ 365, 186

Freedman, W.L., & The Distance Scale Key Project Team 1994, Nature, 371, 757

Fukugita, M., Hogan, C.J., & Peebles, P.J.E. 1993, Nature, 366, 309

Gregg, M. 1992, ApJ 384, 43

Hamuy, M., et al. 1995, AJ, in press

Hui, X. 1992, PhD Thesis, Boston University

Hui, X. 1995, IAU Highlights of Astronomy, 1994, in press

Jacoby, G.H. 1989, ApJ 339, 39

Jacoby, G.H., et al. 1992, PASP 104, 599

Jacoby, G.H., & Ciardullo, R. 1993, in IAU Symposium 155, Planetary Nebulae, ed. R. Weinberger & A. Acker, (Dordrecht, Kluwer), p. 503

Jacoby, G.H., Ciardullo, R., & Ford, H.C. 1988, in The Extragalactic Distance Scale, PASPC, Vol 4, ed. S. Van den Bergh & C.J. Pritchet (Provo, UT, ASP), p. 42

Jacoby, G.H., Ciardullo, R., & Ford, H.C. 1990, ApJ 356, 332

Jedrzejewski, R., & Schechter, P.L. 1989, AJ 98, 147

Kaler, J., 1995, IAU Highlights of Astronomy, 1994, in press

Kirshner, R.P. & Kwan, J. 1974, ApJ 193, 27

Leibundgut, B., & Tammann, G.A. 1990, A&A 230, 81

Luppino, G.A. & Tonry, J.L. 1993, ApJ 410, 81

Mendez, R.H., Kudritzki, R.P., Ciardullo, R., & Jacoby, G.H. 1993, A&A 275, 534

Mould, J.R., Oke, J.B., de Zeeuw, P.T., & Nemec, J.M. 1990, AJ 99, 1823

Pahre, M.A., & Mould, J.R. 1994, ApJ, in press

Panagia, N., Gilmozzi, R., Macchetto, F., Adorf, H.-M., & Kirshner, R.P. 1991, ApJ 380, L23

Phillips, M.M. 1993, ApJ 413, L105

Pierce, M.J. 1994, ApJ 430, 53

Pierce, M.J., & Jacoby, G.H. 1995, AJ, submitted

Pierce, M.J. & Tully, R.B. 1988, ApJ 330, 579

Pierce, M.J., Welch, D.L., McClure, R.D., van den Bergh, S., & Racine, R. 1994, Nature, 371, 385.

Reid, M.J., 1989, in IAU Symp. 136, The Center of the Galaxy, ed. M. Morris (Dordrecht, Kluwer), p. 37

Sandage, A., Saha, A., Tammann, G.A., Labhardt, L., Schwengeler, H., Panagia, N., & Macchetto, F.D. 1994, ApJ 423, L13

Schmidt, B.P., et al. 1994, ApJ 432, 42

Schmidt, B.P., et al. 1993, Nature 360, 600

Schommer, R.A., Bothun, G.D., Williams, T.B., & Mould, J.R. 1993, AJ 105, 97

Secker, J. 1992, AJ 104, 1472

Secker, J., & Harris, W.E. 1993, AJ 105, 1358

Tonry, J.L., & Schneider, D.P. 1988, AJ 96, 807

Tully, R.B., & Fisher, J.R. 1977, A&A 54, 661

Walker, A.R. 1992, ApJ 390, L81

Worthey, G. 1993, ApJ 409, 530

Part 7

Active Galactic Nuclei

High Resolution Infrared Imaging and Spectroscopy with the VLT: Galactic Nuclei

R.Genzel, A.Eckart, S.Drapatz and A.Krabbe

Max–Planck Institut für extraterrestrische Physik Garching, FRG

Abstract. High spatial resolution infrared imaging and spectroscopy with the **VLT**, using adaptive optics and interferometry, will result in very significant progress in our understanding of galactic nuclei. Extrapolating from recent progress in the field and the technological advances in infrared instrumentation expected in the near future an outlook is attempted of what the **VLT** may achieve. Obscured nuclear star clusters and bright individual stars can be studied in detail. The distribution and dynamics of interstellar gas disks and flows on scales of a few parsecs or less can be investigated. The environment of massive central black holes can be studied on scales of less than 1000 Schwarzschild radii. Starbursts in high redshift galaxies can be spatially resolved.

I. General Considerations

The following general comments explain why high resolution infrared imaging and spectroscopy will likely be of particular importance for future studies of galactic nuclei.

Extinction

Observations of far–infrared dust emission and radio molecular line emission have now demonstrated that the nuclei of spiral galaxies are often embedded in large columns of interstellar material. Evidence is also increasing that nuclear interstellar dust and gas clouds strongly influence the visible, UV and soft X–ray characteristics of active galactic nuclei (AGNs). For a Galactic interstellar extinction curve, the hydrogen column density equivalent to unity dust optical depth is 10 times larger at 2μm than at V=5500Å/ (N(H)/τ (V)=2.3x10^{21} cm^{-2}) and about 60 times larger at 7 to 20μm. Unlike visible, UV and soft X–ray measurements, near– and mid–infrared observations thus can penetrate column densities of $10^{22.5}$ to $10^{23.5}$cm^{-2} , quite typical for the nuclear molecular cloud layers in gas rich spirals and starburst galaxies, or for the dense dust tori postulated to be present in many AGNs.

Angular Resolution

Classical narrow line regions, or nuclear starburst zones/molecular cloud layers (100 pc) have angular sizes of $\leq$200/D(100 Mpc) milli–arcsecs. Nuclear bulges ($\leq$1kpc) at redshift z$\geq$1 have diameters of $\approx$100 milli–arcsecs. Central nuclear stellar clusters and dense dust tori in AGNs have sizes of 1 to 100 pc, corresponding to 2 to 200/D(100 Mpc) milli–arcsecs. An accretion disk around the possible central black hole in the Galactic Center may also have milli–arcsec size. It is thus fairly clear that some of the crucial new insights in the physics of galactic nuclei must come from high resolution imaging and spectroscopy with 1 to 100 milli–arcsec resolution. Most of that parameter space cannot even be probed with the Hubble Space Telescope (**HST**) and thus requires larger ground–based telescopes and interferometry. High resolution imaging from the ground becomes substantially easier as one goes from the visible into the infrared and adaptive optics and interferometry will most easily work there. The seeing-limited angular resolution decreases with wavelength λ approximately as $\lambda^{-1/5}$ or faster, the coherence time of the atmosphere increases with $\lambda^{6/5}$ and the isoplanatic angle, over which the phase distribution is constant also increases with $\lambda^{6/5}$. The individual 8m **VLT** telescopes will have diffraction limited angular resolutions of θ (FWHM)=25 $\lambda(\mu m)$ milli–arcsecs, about 30/ λ (μm) better than the seeing limit. **VLT** interferometry (**VLTI**) will result in resolutions of a few milli–arcsec, another order of magnitude greater yet than single 8m telescopes.

Sensitivity

Order of magnitude improvements in sensitivity of broad band photometry and spectroscopy of compact sources (such as nuclei of external galaxies) can be expected with the **VLT** in the infrared, above and beyond that provided by the large collecting area. This is demonstrated in Figs. 1a and b that show the 10σ, 10min, point source flux density (1 Jy=10^{-26} W m^{-2} Hz^{-1}) sensitivity for broad band measurements (R=$\lambda/\Delta\lambda$=5) and the 10σ, 10min, point source line flux sensitivity (erg s^{-1} cm^{-2}) for R=2000 spectroscopy, respectively. As ground–based, broad band measurements throughout the infrared are background limited by the combination of OH airglow (at $\lambda \leq 2\mu m$) and thermal emission of telescope and atmosphere (at $\lambda \geq 2\mu m$), a seeing limited 8m telescope gains about a factor of 2 over a 4m class telescope. For spectroscopy that gain will be between 2 (at the background limited, longer wavelengths) and 4 (at short near–infrared wavelengths and at very high spectral resolution). With realistic assumptions about achievable Strehl ratios, **adaptive optics** probably will result in an additional **order of magnitude gain** in sensitivity for compact sources for both broad band photometry and spectroscopy (Figs.1a and b, see Beckers 1993 and Beuzit et al. 1994 for overviews of the present state of adaptive optics). The basic reason for this substantial improvement is the more than hundred–fold concentration of the light of a point source when compared to the seeing limited case. Finally, in the 1 to 2μm region **suppression of the OH airglow background** is possible by removing the OH emission lines in a suitable spectrometer (e.g. Maihara et al. 1993), leading to another factor of 2 or so improvement. Taken together with the still increasing quality and formats of infrared detector arrays, Figs.1a and b indicate that infrared spectroscopy and imaging on the VLT with state of the art instruments, such as **ISAAC, CONICA, MIIS** and **CRIRES**, plus instruments that include OH suppression, imaging spectroscopy (such as the MPE spectrometer **3D**, Weitzel et al. 1994) and multi–object spectroscopy (**NIRMOS**) will result in **up to two orders of magnitude** improvement in sensitivity as compared with what is now available on 4m telescopes.

SENSITIVITY OF IR TELESCOPES FOR BROAD BAND MEASUREMENTS

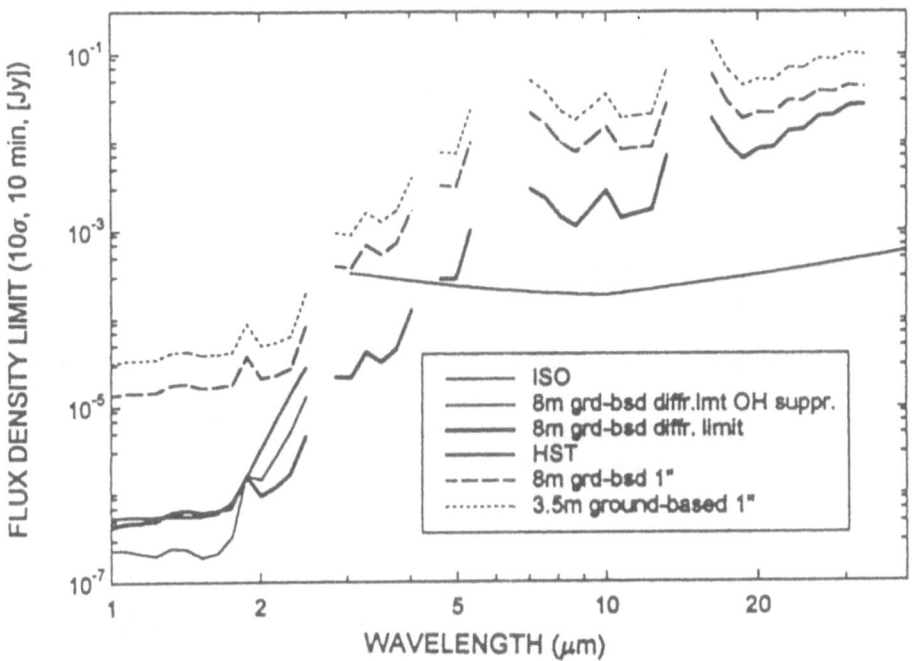

Fig. 1a. Broad–band flux density sensitivity ($R=\lambda/\Delta\lambda\approx5$) for point sources as a function of wavelength λ for different telescopes.

II. Key Scientific Issues

As examples of the impact of the many possible high resolution infrared observations we have selected in the following three specific areas: the investigation of the nuclear stellar component; the study of the interstellar medium; and the investigation of the evolution of galactic nuclei and the relationship between starburst galaxies and AGNs.

The Stellar Component

The **HST** has already begun resolving star clusters in external galaxies into individual stars. Fig. 2 shows that with **adaptive optics** it is now possible for the first time possible to obtain infrared images of similar quality and resolution from the ground with 4m class telscopes. The figure shows a direct comparison of images of the core of the R136 star cluster in the 30 Doradus nebula of the LMC taken with the repaired **HST** in the visible and with the ESO adpative optics system **COME–ON–PLUS** on the ESO 3.6m (Brandl et al. 1994). Basically all stars apparent in the HST image can also be seen in the infrared. A comparison of the two data sets allows the best analysis of the properties of the brightest stars (K≤18) in this dense young star cluster to date. Another example of the power of high resolution near–infrared imaging is given in Fig. 3 which shows 0.15" J– and K–band images of the central ≈7" (0.3pc) of our own Galaxy, taken with the MPE speckle camera **SHARP** on the ESO **NTT** in very good (≤0.4") seeing . In the Galactic Center, as in many external galaxies, the nuclear star cluster is obscured in the visible so that only infrared observations can penetrate the veil of foreground interstellar dust. The Galactic Center observations now resolve the emission into many hundreds of stars of different characteristics, indicate from number density counts that the core density is likely very high ($\geq 10^{7.5}$ Mo pc^{-3}) and for the first time, show a source(s) at the position of the compact radio source SgrA* (see Eckart et al. 1992, 1993). The centroid of the stellar distribution in fact is on or very near SgrA* and the most recent images shown in Fig. 3 suggest that there may be a concentration of fainter (K≈14-15) stars near SgrA*, both findings strengthening the common interpretation that the radio source is a massive object at the very center.

While the present observations of R136 and the Galactic Center only sample the most massive stars in such clusters, high resolution near–infrared imaging with the **VLT**, reaching to much lower flux limits (K≈21), because of the larger collecting area and lower source confusion, will for the first time sample **solar mass stars**. It will then be possible to directly derive, from number counts of different types of stars, the evolution of the star cluster, and adress questions about the formation of stars at very high stellar densities and the role of stellar collisions and mergers (see Genzel, Hollenbach and Townes 1994). One can also test evolution models of massive stars and determine whether the initial mass function in starbursts favors massive stars, clearly fundamental issues in the investigation of such star clusters. In somewhat more distant systems containing an AGN, infrared imaging with adaptive optics will be capable of unambiguously detecting and measuring the flux, spectral characteristics and spatial extent of such clusters in the presence of a very bright AGN. This is demonstrated in Fig. 4 where a simulation of a 5 Mpc distant galactic nucleus containing a Galactic Center type nuclear cluster, as well as a bright compact nuclear source is shown. If this same nucleus is then observed with the milli–arcsec resolution of the **VLTI** even the brightest individual stars can be identified and studied. To successfully carry out such challenging observations, that will present a quantum leap in the capability of astronomical observations, it will be necessary that

SENSITIVITY OF IR TELESCOPES FOR R=2000 SPECTROSCOPY

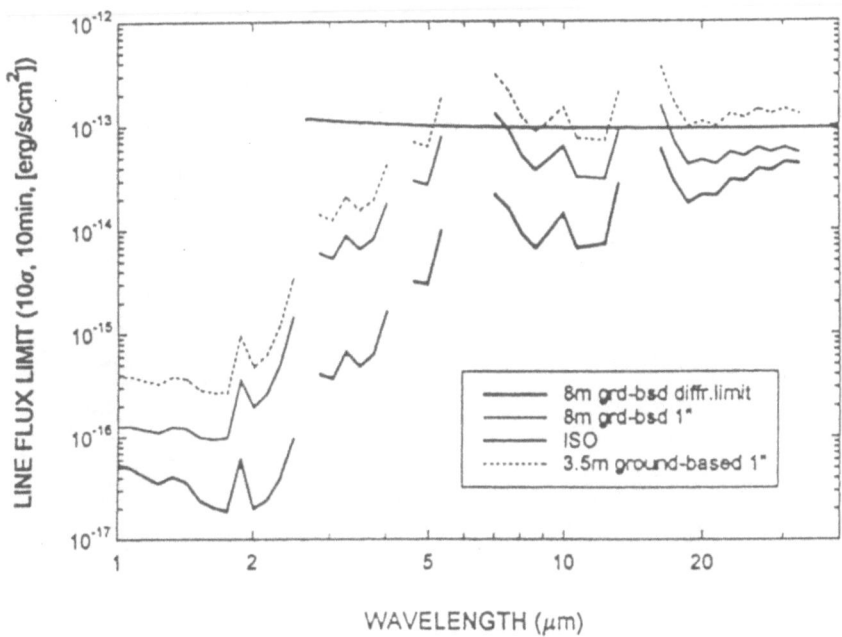

Fig. 1b. R=2000 spectroscopic sensitivity for point sources.

30 DORADUS

WFPC2
HUBBLE SPACE TELESCOPE

COME-ON+/SHARP2
3.6m ESO TELESCOPE LA SILLA

3" (0.8 pc)

Fig. 2. Comparison of HST V–band image (after repair, left) and COME–ON–PLUS/SHARP 2 K–band image (right, Brandl et al. 1994) of the core of the R136 cluster in 30 Doradus. The K–band image has a resolution of 0.15".

0.15" IMAGES OF THE GALACTIC CENTER WITH SHARP AT ESO-NTT

K-BAND (2.2μm)

J-BAND (1.2μm)

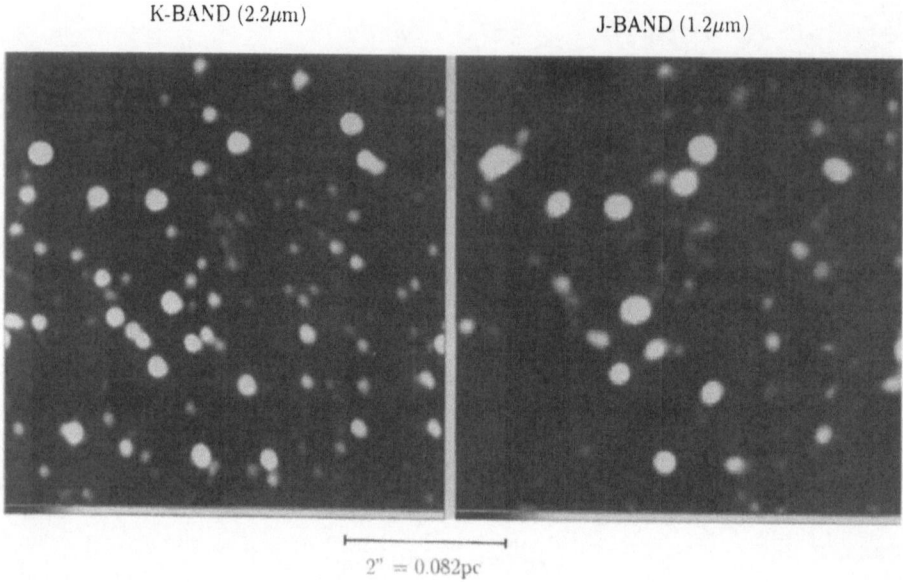

2" = 0.082pc

Fig. 3. 0.15" K–band (left) and J–band (right) images of the central ≈0.1pc of the Galaxy, taken with the MPE **SHARP** camera on the ESO **NTT** in April 1994 under exceptional seeing (≈0.4") conditions. The images were constructed from coadded shift–and–add , short exposure frames that were then "CLEANED" to remove the seeing beckground (cf. Eckart et al. 1993).

the interferometer has a **non–zero field of view** so that co–phasing of the interferometer telescopes can be done on the bright AGN. Otherwise it will be very difficult to interferometrically detect the faint (K>20) stars. In addition, **laser beacons** will be required in most galaxies to obtain good quality adaptive optics correction.

The issue of the presence of massive black holes is another fundamental issue on which high resolution infrared imaging and spectroscopy with the VLT will have a major impact. In our own Galactic Center the gas and stellar dynamics on 0.2pc scale suggests the presence of a 1 to 3×10^6 Mo central mass concentration, possibly a black hole (Genzel, Hollenbach and Townes 1994). Using the **SHARP** camera on the **NTT** a program is now well underway taking K–band images like the one shown in Fig. 3 at least once every year and trying to measure the proper motions of the ≈30 stars with K≤13 within 2" (0.08pc) of SgrA*. Analysis of the

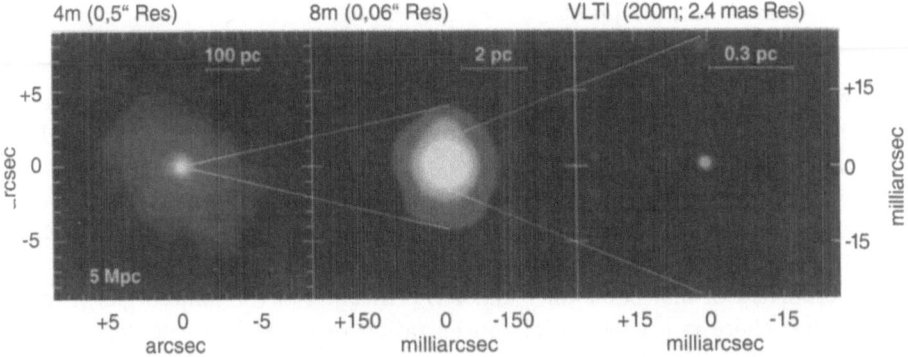

Fig. 4. Simulating the VLT: 1" image of the K–band emission of a galaxy (left, adapted from a 0.4" SHARP image Δof NGC 1068); 0.06" adaptive optics image (middle); and 0.003" VLTI image (right). The 5 Mpc galaxy was assumed to have a central point source (=AGN) as well as a star cluster like the one observed in our own Galactic Center.

first 5 periods covering 2.5 years or so are already fairly promising and indicate that proper motions should be detectable within about half a dozen years if there is indeed a $\geq 10^6$ Mo black hole present. Considering the better sensitivity and 4 times greater spatial resolution (at 1μm) available with adaptive optics on the **VLT**, proper motions of a statistically significant sample of stars within a yet smaller radius (0.5" or 0.02pc) of the dynamic center will be detectable with **CONICA** in only one or two years. If it could be demonstrated that there is indeed a dark mass concentration of $\approx 2\times 10^6$ Mo within the central arcsecond or less, the corresponding mass density of $\geq 10^{10}$ Mo pc^{-3} would prove beyond doubt the existence of a massive black hole there. With interferometry and a resolution of a few milli-arcsecs, it will be possible to study the radio synchrotron emission region of SgrA* and perhaps elucidate the properties of the central accretion disk on scales of a few thousand Schwarzschild radii (see Tacconi–Garman et al. elsewhere in this volume). In the nearest starburst galaxies and AGNs, imaging spectroscopy of the 2.3μm CO overtone bands (with **ISAAC and CONICA**) will allow determination of stellar radial velocity dispersions on spatial scales of just under 1pc, thus extending the massive black hole searches currently undertaken with ground–based telescopes and the **HST** to obscured nuclei.

Interstellar Gas and Dust

The **VLT** will allow for the first time the study at high spatial resolution of the distribution, dynamics and chemical/physical properties of the interstellar medium in galaxies, especially the characteristics of the **neutral (molecular) gas and dust** that cannot be probed in the visible and UV. As an example of some of the recent results emerging in this area, Fig. 5 shows maps of 2.1μm H_2 and 10μm dust emission toward the nucleus of the Seyfert 2 galaxy NGC 1068 (Tacconi et al. 1994, Cameron et al. 1993). These data, along with mm interferometry of molecular lines and **HST** imaging of the ionized gas (Fig. 5) show that there is a substantial concentration of neutral interstellar gas and dust associated with the narrow line region on ≤ 100 pc scale. They confirm the notion (Antonucci and Miller 1985) that dense molecular gas and dust may be the obscuring agent of the broad line region and central ionizing source in NGC 1068. The narrow line clouds observed with the **HST** may be the surfaces of dense molecular/dust clouds exposed to the nuclear ionizing radiation and wind. Imaging spectroscopy with adaptive optics and interferometry with the **VLT** will improve on the resolution of the measurements shown in Fig. 5 by up to **3 orders of magnitude!** High resolution near– and mid–infrared observations with the **VLT** will tell unambiguosly where the obscuration and perhaps also the anisotropy of the ionizing radiation of AGNs is created, and how interstellar matter is fed into nuclei of galaxies. It will then be possible (with **CONICA** and **MIIS**) to determine by direct imaging, whether there is a parsec scale, thick molecular torus/accretion disk in basically all types of AGN as proposed by current 'unified' scenarios (e.g. Antonucci 1993). Fig. 6 shows a simulation how such a torus/accretion disk might look like in a nearby AGN with **VLTI** resolution. Line mapping (with **CONICA** and a possible future '3D'–spectrometer) will elucidate the dynamics of the dense, neutral circum–nuclear gas and how it relates to the ionized gas in the narrow line and coronal line region, both of which can also probed by high resolution near–infrared data. The **infrared coronal line emission** from species as Si VI, Si VII, Ca VIII etc. recently discovered to be strong in a number of AGNs (Fig. 7, Oliva et al. 1994, Moorwood et al. 1994) is particularly interesting as it probes, at high spatial resolution, material that is associated with the X–ray emission zone. These lines may be unique fingerprints of obscured central AGNs. Such measurements will be undertaken with **ISAAC**.

Evolution of Galaxies

A third set of key questions about galactic nuclei has to do with their evolution, the possible relationship between starburst galaxies and AGNs, and the nature of galaxies in the early Universe. For all these issues infrared observations play a key role. Sanders et al. (1988) and others have proposed that starburst galaxies and luminous AGNs (QSOs and Seyferts) are just two phases of the evolution of the same galaxy that has undergone a strong disturbance by colliding/merging with another galaxy. To test this plausible and attractive scenario it is necessary

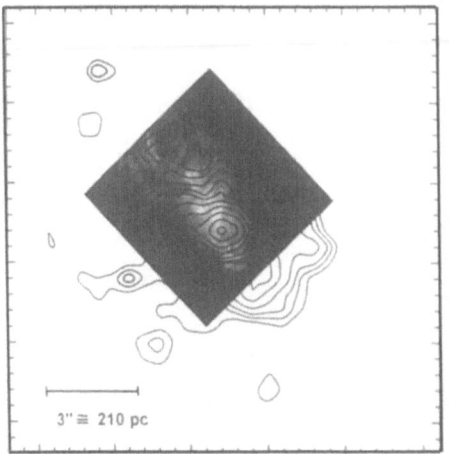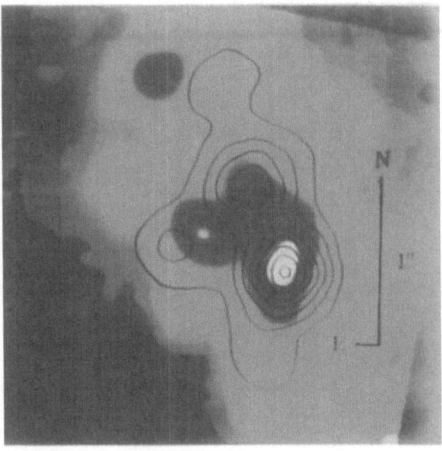

Fig. 5. Observing the central "torus"/molecular cloud distribution in NGC 1068:Contour map of 2μm H₂ S(1) emission (Tacconi et al. 1994) superposed on a false color **HST** image of 5007Å[O III] (left); false color image of 10μm dust continuum superposed on contours of 5007Å[O III] HST map smoothed to the same resolution (Cameron et al. 1993).

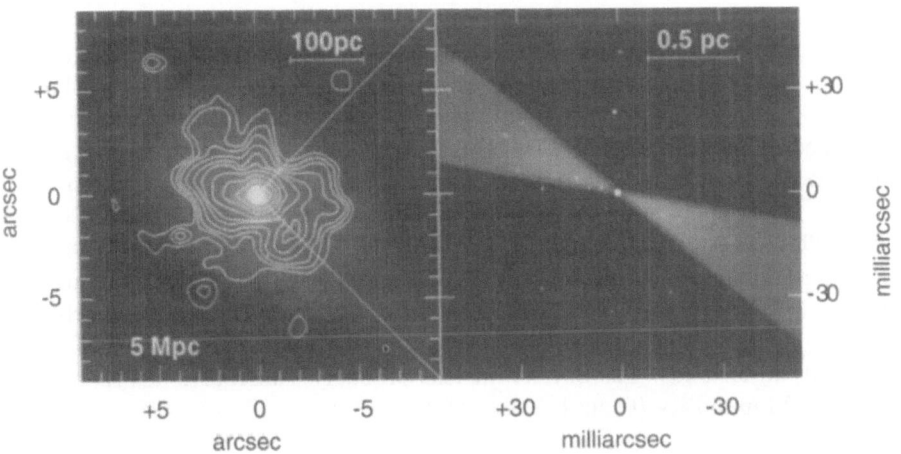

Fig. 6. Simulating the VLT: 1" image of the 2.2 μm H₂ S(1) emission (left, adapted from the NGC 1068 data of Tacconi et al. 1994); and 0.003" observations of the central accretion disk/torus taken with the VLTI.

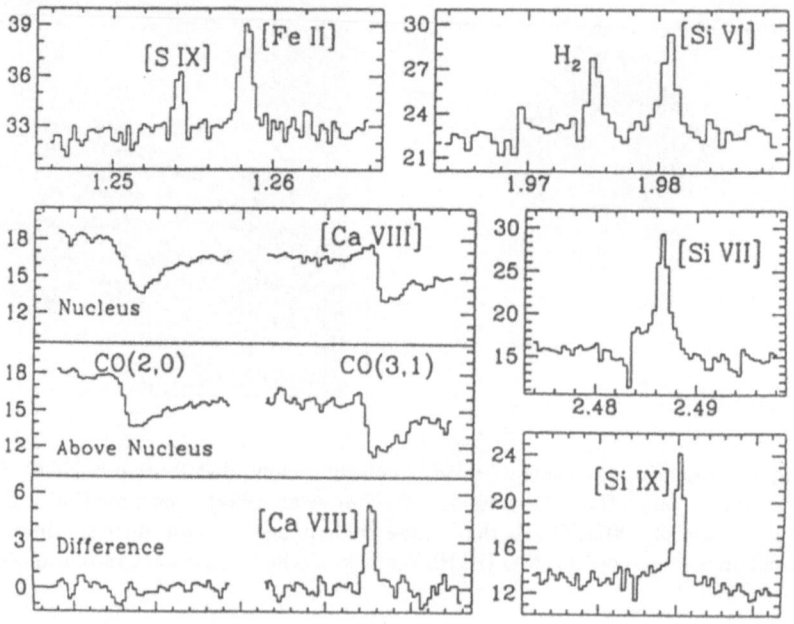

Fig. 7. Infrared coronal lines in the hidden Seyfert Circinus: a new tool for exploring the conditions in dust embedded nuclei (from Oliva et al. 1994).

to quantitatively determine how much of the luminosity of a luminous galaxy actually emerges from a starburst vs. an AGN. (This replaces the earlier but probably ill posed question of whether a galaxy is either entirely a starburst or entirely an AGN.) In relatively nearby systems, this question can already be tackled at present with high resolution infrared imaging and imaging spectroscopy. As an example Fig. 8 shows a 0.4" K–band image of the core of the Seyfert 1 galaxy NGC 7469 taken with **SHARP** on the ESO **NTT** (from Genzel et al. 1994). These data, combined with **3D** and **FAST** imaging spectroscopy, mid–infared imaging and mm–line interferometry clearly show that most of the bolometric luminosity of NGC 7469 comes from a 500 pc (1.5") radius starburst ring that is embedded in and heavily extincted by a massive concentration of molecular/dust clouds. The Seyfert nucleus probably contributes less than 30% of the total luminosity (Genzel et al. 1994).

NGC 7469 presents two lessons. One is, of course, that the classification of a galaxy as an AGN (Seyfert 1 in this case), solely on the basis of classical optical line ratios or X-ray emission, needs to be treated with some caution. Second, intrinsic dust extinction may play an important role in many galactic

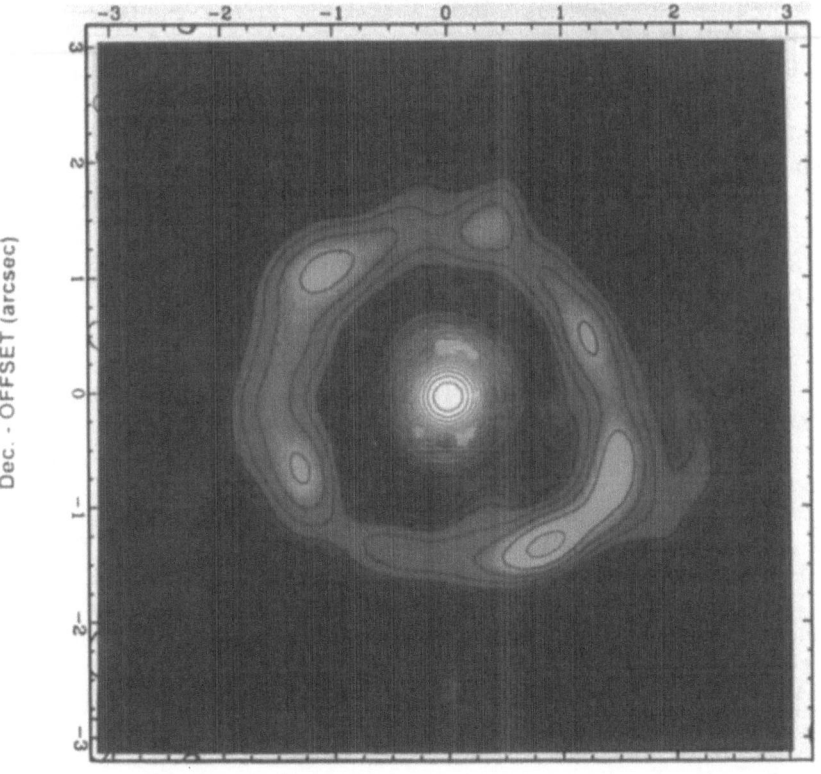

Fig. 8. 0.4" SHARP image of the nucleus of the Seyfert 1 galaxy NGC 7469 (from Genzel et al. 1994).

nuclei so that reliable estimates of the relative contributions of AGN vs. starburst components require measurements at (infrared!) wavelengths where that dust is reasonably transparent. The NGC 7469 data also show that the starburst and AGN components cannot be easily separated purely on the basis of large aperture, multi–wavelength photometry and spectroscopy; high resolution imaging is crucial. With adaptive optics, **CONICA** on the **VLT** will be able to resolve a starburst ring of the type seen in NGC 7469 even at $z \approx 1!$. The nature of high redshift, hyperluminous galaxies and QSOs, such as IRAS F10214+4724, can then be studied in detail. As a technical point it should be noted that the typical sizes of the emitting regions of high redshift objects and their immediate (cluster?) environment is ≥ 10". It is very important to cover such fields simultaneously and include nearby faint stars (that are often present in the isoplanatic angle of 10 to 20" radius in the infrared) as direct calibrators of the point spread function delivered by the adaptive optics. Therefore **it is highly desirable**

for this (and other applications) to have 1024^2 arrays incorporated in
CONICA (and **ISAAC**).

In addition to detailed studies of individual sources, statistical investigations of larger samples of faint galaxy spectra are obviously of great importance for the problem of galaxy evolution (see other contributions elsewhere in these proceedings). Recent work (e.g. at the **Keck**) already shows the importance of the near–infrared for such programs and the ESO user community should actively pursue the development of multi–object spectrometers such as **NIRMOS** or the '**3D**'–concept. OH airglow suppression technology appears very attractive for optimizing the sensitivity of near–infrared deep imaging of high redshift galaxies.

References

Antonucci, R.R.J. 1993, Ann.Rev.Astr.Ap. 31, 473

Antonucci, R.R.J. and Miller, J.S. 1985, Ap.J. 297, 621

Beckers, J. 1993, Ann.Rev.Astr.Ap. 31, 13

Beuzit, J.L. et al. 1994, The Messenger 75, 33

Brandl, B., Sams, B., Eckart, A., Hofmann, R. and Genzel, R. 1994, in prep.

Cameron, M., Storey, J.W.V., Rotaciuc, V., Genzel, R., Verstraete, L., Drapatz, S., Siebenmorgen, R. and Lee, T.J. 1993, Ap.J. 419, 136

Eckart, A., Genzel, R., Krabbe, A., Hofmann, R., van der Werf, P.P. and Drapatz, S. 1992, NATURE 335, 526

Eckart, A., Genzel, R., Hofmann, R., Sams, B. and Tacconi–Garman, L.E. 1993, Ap.J. 407, L77

Genzel, R., Hollenbach, D. and Townes, C.H. 1994, Rep.Progr.Phys. 57, 417

Genzel, R., Weitzel, L., Tacconi–Garman, L.E., Blietz, M., Cameron, M., Krabbe, A., Lutz, D. and Sternberg, A. 1994, Ap.J. in press

Maihara, T., Iwamuro, F., Yamashita, T., Hall, D.N.B., Cowie, L.L., Tokunaga, A.T. and Pickles,A. 1993 PASP 105, 940

Moorwood, A.F.M. et al. 1994, in prep

Oliva, E., Salvati, M., Moorwood, A.F.M. and Marconi, A. 1994, Astr.Ap. in press

Sanders, D., Soifer, B.T., Elias, J., Madore, B., Mattews, K., Neugebauer, G. and Scoville, N.Z. 1988, Ap.J. 325, 74

Tacconi, L., Genzel, R., Blietz, M., Harris, A.I. and Madden, S.C. 1994, Ap.J. 426, L77

Weitzel, L., Cameron, M., Drapatz, S., Genzel, R. and Krabbe, A. 1994, in "Infrared Astronomy with Arrays (3)", ed. I.S.McLean, Kluwer (Dordrecht), 531

Resolving Extragalactic Nuclei with the VLT

L.E. Tacconi-Garman, A. Eckart, S. Drapatz, R. Genzel, R. Hofmann, M. Löwe, and A. Quirrenbach

Max-Planck-Institut für extraterrestrische Physik, Postfach 1603,D-85740 Garching bei München, Germany

Abstract. Recent high spatial resolution studies of the Galactic Center with the ESO NTT have demonstrated the presence of different stellar populations in the central cluster of the Galaxy. These studies have also allowed us to derive the core radius of the central stellar cluster and have revealed for the first time a near-infrared source at the radio position of Sgr A*, a prime candidate for a supermassive black hole. In the case of the Seyfert 1 galaxy NGC 7469 (D = 66 Mpc), similar high resolution near-infrared imaging has revealed a powerful circumnuclear starburst ring which is responsible for 2/3 of the bolometric luminosity of the entire galaxy.

With the high spatial resolution of the ESO 8m dishes (60 mas at $2.2\,\mu$m) and the VLTI (2.4 mas at $2.2\,\mu$m on a 200 m baseline) these kinds of studies can be extended to extragalactic nuclei. In nearby galaxies (D $\leq$ 10 Mpc) we will be able to resolve the nuclei into individual stars or star clusters, perhaps revealing central very massive objects. In addition, probing the near-nuclear regions of active galaxies at cosmologically interesting distances will be made possible via the VLT(I) resolution.

In this contribution we present simulated images of these nuclei based on our current knowledge of the structure of the Galactic Center and nearby galaxies.

1 Introduction

High spatial resolution imaging in the near-infrared is essential in order to study extragalactic nuclei and the center of our own Galaxy. In many galaxies the radiation originating in the central stellar bulges is dominated by a massive starburst, a nonstellar active nucleus or a combination of the two. The investigation of the innermost circumnuclear regions of galaxies is important for the understanding of galaxy evolution and star formation processes.

High spatial resolution imaging is crucial for the following reasons. In the case of star-dominated galactic nuclei, the core radii of the central stellar clusters may range from less than a light year to a few parsecs. In addition, the nonstellar component of active galactic nuclei (AGN) are unresolved but there interaction with the interstellar medium will take place in regions with linear scales of one or a few parsecs.

Although active galactic nuclei are bright with respect to their host galaxies, their distances make them appear faint and compact. In most of the cases they are embedded in a large amount of molecular gas and dust and are therefore highly extincted. From the ground high spatial resolution imaging and spectroscopy in the infrared allows us to study these objects in detail, owing to the fact that the near-infrared extinction is about a factor of ten lower than in the optical. Moreover, the atmospheric and technical conditions are more favorable for high angular resolution studies in the near infrared than in the optical.

2 Findings in the Galactic Center, and Beyond

Over the past few years, we have obtained diffraction-limited images (0.15" resolution) in the near-infrared (K-band [2.2 μm]) using the MPE speckle camera SHARP on the NTT (Eckart et al. 1993). The results of these observations are manyfold.

These high resolution images allow us to discern that most of the flux in earlier seeing-limited images comes from about 340 unresolved stellar sources with K magnitudes $\leq$14. Indeed, the images show the IRS 16 and IRS 13 complexes to be composed of about 25 and 6 sources, respectively, a number of which are probably luminous hot stars. Most of the fainter stars in the central parsec are likely M- rather than K-giants. We confirm the presence of a blue near infrared object (K$\approx$13) at the position of the compact radio source Sgr A*.

The spatial centroid of the number distribution of compact sources is consistent with the position of Sgr A* but not with a position in the IRS 16 complex. The stellar surface density is very well fit by an isothermal cluster model with a core radius of 0.15±0.05 pc, with the central stellar density being a few times $10^7 M_\odot$ pc^{-3}. This high stellar density results in the very probable buildup of massive stars by collisional merging of lower mass stars and collisional disruption of giant atmospheres in the central 0.2 pc.

The spatial resolution achievable with the VLTI, 2.4 mas at 2.2μm on a 200 m baseline *using full adaptive optics correction* (Section 4), corresponds to 0.1 milliparsec (20 A.U.). With such a resolution we will be able to probe the very inner edges of an accretion disk around a central black hole in the Galactic Center should one be present.

In Figure 1 (left) we present a recent diffraction-limited K-band mosaic of the Galactic Center made with SHARP on the NTT. The axes represent R.A. and Dec. offsets from the position of Sgr A*. The light is seen to break up into hundreds of individual sources. The panel in the upper right of Figure 1 shows a simulation of a snapshot observation using the four 8 m telescopes of the VLT of one of these sources. The spatial information in this image is at the resolution of the small individual features. Figure 1 (lower right) shows a schematic picture of the central regions (1 milliparsec on a side) of an accretion disk surrounding a black hole. VLTI observations of the region surrounding the putative black hole Sgr A* will allow us to sample thoroughly size scales such as those depicted here, with ten or more resolution elements in each of the spatial dimensions. That is, we will be able to resolve the region whose radio emission originates at $\sim$10^3 Schwarzschild radii. Such observations are imperative for not only determining the physics of accretion disks around black holes, but also for understanding the evolution of the Galactic Center as a whole.

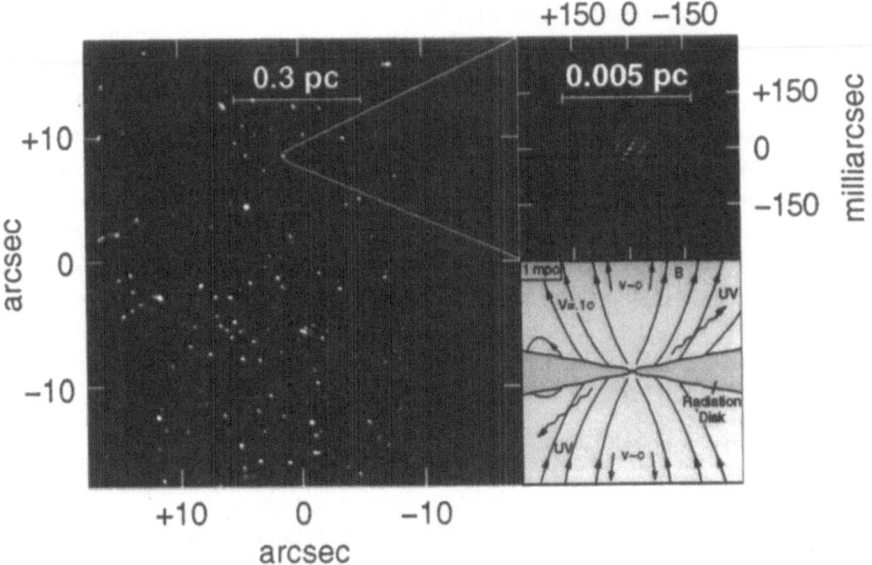

Fig. 1. Left: Diffraction-limited K-band mosaic of the central parsec of the Galaxy. Offsets on the axes are with respect to the position of Sgr A*. Upper Right: Simulated VLTI snapshot. Lower Right: Schematic diagram of regions resolvable with the VLTI.

3   Extrapolation to Distant Galaxies

A   Nearby Systems (D ~ 5–10 Mpc)

The present state-of-the-art techniques in near-infrared continuum and line imaging allow us to achieve spatial resolutions of $\sim$0.5–1″. Such resolution allows us to probe, *but not resolve*, the central regions of the nearest starburst and active galaxies. With the VLT, we will be able to resolve partially the near-infrared continuum light in nuclei of these systems into the respective stellar and non-stellar components. Further, with the VLTI we will be able to extend studies like the present-day investigation of the Galactic Center to these more active systems. VLTI observations of the near-infrared line emission of such species as H_2 (which has a prominent line at 2.12 μm) will probe and resolve the central gaseous disks surrounding the central engines in the nearer AGN. See Genzel et al. (these proceedings) for illustrations of VLT(I) observations as described here.

B Further Systems (D ∼ 10–25 Mpc)

With the SHARP camera on the NTT, we have been able to achieve routinely spatial resolutions of ∼0.5″ on a wide variety of extragalactic nuclei, both active and starburst. Indeed, in the case of NGC 1808, a starburst system at a distance of 11 Mpc, we have been able to resolve the K-band light in the inner half kiloparsec into a large number of bright starforming knots (Genzel et al. 1994, Tacconi-Garman et al. 1994). These knots represent extra large 30 Doradus-like star formation sites. A vexing issue which still remains, however, is what the initial mass function (IMF) of the star formation is in these sites. Further resolving these knots with the VLTI, and perhaps in the process isolating an individual supernova event, would allow us to probe the IMF at scales which are presently only possible for 30 Doradus itself.

C Distant Systems ($z \sim 0.2$)

It is becoming clearer with time that the central regions of active galaxies are powered by some combination of non-stellar and stellar emission, with the former being the dominant one in most cases. However, the nature of the stellar component in Seyfert galaxies, has only recently been addressed through high resolution infrared observations. In the specific case of the Sy 1 NGC 7469 (D = 66 Mpc), Genzel et al. (1994) find that the K-band light of the nuclear region is due equally to the non-stellar nucleus and a clearly separated, knotty, circumnuclear ring of star formation. This observation sheds light on the interaction of the active nucleus with its environs. To test whether there is any evolution in this interaction we must extend high resolution near infrared studies out to galaxies at much larger distances. With the present achievable resolution this is simply not possible. Figure 2 (left) shows how NGC 7469 would look in the K-band if it were observed with SHARP on the NTT and it was at $z = 0.2$. Clearly the circumnuclear ring is unresolved from the nucleus. The 2.4 mas resolution obtainable with the VLTI will make it possible to resolve circumnuclear regions in order to investigate the effect of the nucleus on its surroundings, and vice versa, in systems as far away as $z = 0.2$. Figure 2 (right) shows the effect of observing the same "$z = 0.2$ NGC 7469" with the VTLI plus full adaptive optics correction.

4 Technical Aspects

Adaptive optics on the 8 m telescopes will be essential for high quality, high spatial resolution observations of AGN. A number of bright sources will be self-referencing. That is, their intensity in the visible is large enough to measure the atmospheric wavefront. For fainter objects, laser guidestars probably provide the only solution to obtain this information. Adaptive optics will also provide the maximum amount of intensity (single speckle mode) for interferometry. Thus, the sensitivity of the VLTI will be unprecedented in the near-infrared.

For z = 0.2 with 0.4″ (NTT) Res.   For z = 0.2 with 0.06″ (8 m + AO) Res.

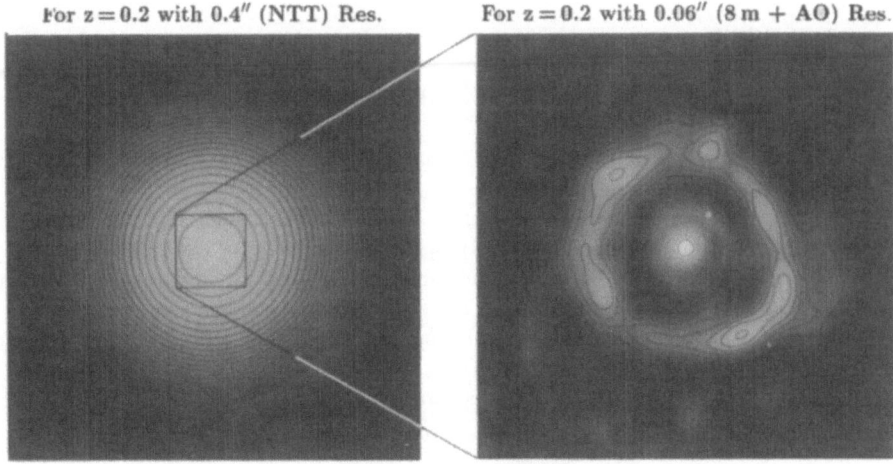

Fig. 2. A simulation of the Seyfert 1 galaxy NGC 7469 placed at $z = 0.2$ and viewed with the NTT (left). The same system now resolved via VLTI "observations" (right).

The planned imaging beam combination will allow one to obtain high resolution interferograms over an 8″ field of view. This FOV is particularly well suited for the Galactic Center, since the Sgr A* counterpart candidate as well as the bright reference source IRS 7 can be observed simultaneously. Since the reference is contained in the image they can also also be taken in snapshot mode and cleaned individually. The 8″ FOV is also quite well suited for studying the inner regions of AGN, since the angular extent of the regions of interest is, in many cases, less than 5″. In order to simulate the imaging beam combiner at the coherent focus of the ESO VLTI in multi-speckle mode or with partial or full adaptive optics correction we have built a COmbiner SImulator (COSI; Katterloher et al. 1994). Recent results from COSI can be found in Eckart et al. (1994).

As a final note, interferometric imaging over a larger field of view will require large detector arrays or (most likely) mosaics of these arrays. The simplest case would be a two element array mosaic, in which one array can be positioned on the program source and one on the reference (if contained in the field) or on a different source in the same FOV.

References

Eckart, A., Böker, T., Hofmann, R., Katterloher, R., Quirrenbach, A., Löwe, M., and Cruzalèbes, P. 1994, *SPIE*, **2200**, 458.

Eckart, A., Genzel, R., Hofmann, R., Sams, B.J., and Tacconi-Garman, L.E. 1993, *Ap.J.Lett.*, **407**, L77.

Genzel, R., Weitzel, L., Tacconi-Garman, L.E., Blietz, M., Cameron, M., Krabbe, A., Lutz, D., and Sternberg, A. 1994, *Ap.J.*, submitted.

Genzel, R., Eckart, A., Hofmann, R., Quirrenbach, A., Sams, B., and Tacconi-Garman, L., *ESO Messenger*, Issue No. 75, 17 (March 1994).

Katterloher, R., Böker, T., Eckart, A., Hofmann, R., Jakob, G., Quirrenbach, A., and Löwe, M. 1994, *SPIE*, **2200**, 469.

Tacconi-Garman, L.E. et al. 1994, in preparation.

Galaxy Dynamics, Black Holes and FUEGOS: Prospects for Studying Stellar Dynamics

Massimo Stiavelli[1], Philippe Crane[2], Paul Felenbok[3]

[1] Scuola Normale Superiore, I56126 Pisa, Italy
[2] European Southern Observatory, D85748 Garching bei München, FRG
[3] Observatoire de Paris Meudon, F92195 Meudon Cedex, France

Abstract. The FUEGOS instrument currently being developed for the VLT will provide greatly expanded opportunities to study the dynamics of stellar systems in the central regions of galaxies with the aim of probing the dynamical nature of decoupled cores and investigating the presence of disks and the influence of putative black holes on the stellar dynamics. The prospects for determining the stellar rotation and velocity dispersion fields as well as line profiles are analyzed for simulated observations of a typical elliptical galaxy core at the distance of Virgo. The relation to theoretical models is discussed to show that only by measuring the absorption line profiles can the dynamical state of the system be disentangled.

1 The Complexity of Galaxy Cores

Up until the early eighties, it was usually argued that cores of galaxies were relatively simple systems, characterized by a smooth light profile, which was altered and made steeper by the occasional presence of massive black holes. This simple picture was shown to be incorrect by the first observations probing with high resolution the properties of cores. Spectroscopic observations showed that cores could contain components kinematically decoupled from the main galactic body (Bender 1988, Jedrzejewski and Schechter 1988, Franx and Illingworth 1988). It was later found that decoupled cores were characterized by a higher metallicity stellar population than the underlying galaxy (Bender and Surma 1992, Davies, Sadler and Peletier, 1993, Carollo and Danziger 1994).

High resolution imaging showed that the cores of many massive galaxies, usually characterized by boxy isophotes at large radii, had disky isophotes in their cores (Nieto et al. 1991a,b). In addition it was found that most resolved galaxy cores have substructure (Møller, Stiavelli, Zeilinger 1993, Crane et al. 1993).

The origin of this variety of phenomena is buried deeply into the processes of galaxy formation and evolution. Indeed important clues to these processes can be learned from studying galaxy cores. In addition, firmly identifying the frequency and mass function of black holes in galactic cores would constrain AGN models (see, e.g., Zamorani, this conference).

However, galaxy cores are very complex systems. This is well illustrated by a few cases for which unusually good observations both from space and from the ground are available. Two such cases are M31 (Crane et al. 1993, Bacon et *al.* 1994) and NGC 4594 (Crane *et al.* 1993, Bacon *et al.* 1994), which show the presence of several components, some of which even lack central symmetry. M31 has a double nucleus and the peak in the velocity dispersion is not coincident with either the rotation velocity center or with any of the two peaks in the light profile. The core NGC 4594 is a multicomponent system with a spheroidal component, an outer disk, an inner disk, a stellar cusp, and a central blue point source.

As has become customary in the global dynamical modelling of elliptical galaxies (Binney, Davies, and Illingworth 1990), the study of galactic cores with this level of complexity will be possible only through the measurement of two–dimensional kinematical fields.

2 Simulated Observations of a Test Case

In order to investigate the kind of observations that might disentangle the dynamical state of a complex core, we have constructed a galaxy model which has the same light profile as observed in NGC 4472 (Stiavelli and Crane, unpublished). This light profile has been decomposed in three different ways:

i) an "isothermal" core plus a massive black hole cusp ($M_{BH} \simeq 1.5 \times 10^9$)

ii) an "isothermal" core plus a cold, rotating, weak stellar disk

iii) an "isothermal" core plus a cold, rotating, weak stellar disk with a central hole filled in by a massive black hole cusp ($M_{BH} \simeq 7 \times 10^8$)

These models were characterized by identical light profiles, very similar isophotal shapes but utterly different physical properties. However, simulated velocity fields appear very similar and probably indistinguishable once seeing and resolution effects are taken into account. In fact model *i)* is non rotating, while models *ii)* and *iii)* rotate only very mildly. Without a knowledge of the line profiles (hereafter LP), the rotation due to the disk could be easily mistaken for a mild rotation of the core itself.

The velocity dispersion fields of the three models appear very different when no seeing effects are included. However, if one realizes that for the black hole models the most extreme velocity dispersion values are confined to very small radii, it is readily seen that, even under good seeing conditions, the velocity dispersion is easily smeared out to a value that could be obtained by radial anisotropy of stellar orbits (Binney and Mamon 1983). As an illustration of the observational difficulties, we have considered in more detail the black hole model *i)*, by considering a wide range of observational setups; namely, we assume that we take spectra through apertures with diameters in the range 0.2 to 0.8 arcsec, under a seeing in the range 0.3 to 1.0 arcsec. The model has been constructed by assuming that the underlying galaxy is characterized by gaussian LPs, therefore the non-gaussian LP shape is due to the presence of the black hole. In Fig. 1 we show the LPs obtained for this model. Note how the shape becomes rapidly gaussian as we move from the best seeing case with the smallest aperture to the worst seeing case with the largest aperture.

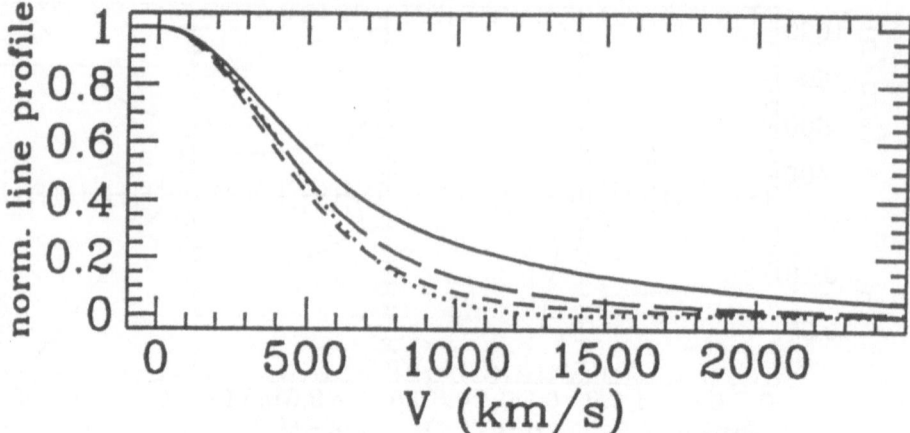

Fig. 1. We plot the line profile for a model with a central massive black hole as obtained through a 0.2 arcsec aperture under 0.3 arcsec seeing FWHM (solid line), a 0.4 arcsec aperture under 0.65 arcsec seeing FWHM (short dashed), and a 0.8 arcsec aperture under 1 arcsec seeing FWHM. The dotted line represents a gaussian and is plotted for reference.

In order, to better illustrate this we have derived for a number of models the h_4 term of the Gauss-Hermite expansion of the LP which measures the departures of the LP from a Gaussian (see, *e.g.*, van der Marel and Franx 1993). In Fig. 2 we show how the derived values of h_4 and of the (fitted) velocity dispersion σ are affected by aperture size and location, and by seeing. Given that typical error bars in h_4 measurements are of about ± 0.02, it is readily seen that miscentering the aperture by 0.4 arcsec reduces the detection from a 7 sigma to a 2 sigma or less. Similarly, increasing the aperture diameter from 0.2 to 0.4 arcsec reduces the detection to 4 sigma unless the seeing is exceptional. Seeing and instrumental resolution also rapidly reduce the detection confidence level, especially for the larger apertures.

These tests lead us to believe that only the two–dimensional measurement of line profiles with high wavelength and spatial resolution can allow us to study galaxy cores. The effect of miscentering of the slit is eliminated with such an instrument and very small aperture sizes can optimize the detection significance as long as the seeing is subarcsecond.

We note in passing, that two–dimensional spectra also allow the measurement of line strength fields and thus of a more precise identification of the metallicity to be attributed to the various components.

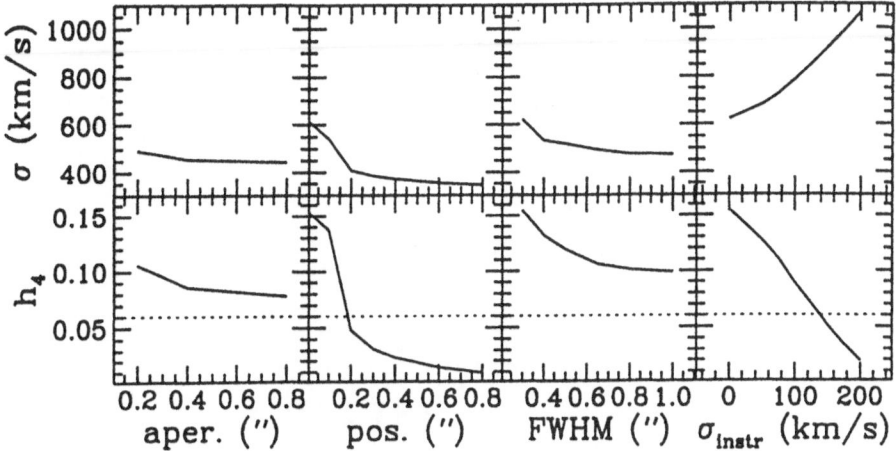

Fig. 2. From left to right, the panels show how the velocity dispersion and the h_4 parameter measuring departures from a gaussian line profile are affected by the aperture diameter (for a seeing FWHM = 0.65 arcsec), the dependence on aperture location (for a seeing FWHM = 0.2 arcsec), the effect of seeing (for an aperture of 0.2 arcsec), and the effect of instrumental resolution. The thin dotted line gives the three sigma limit for detection.

3 FUEGOS

FUEGOS is a fiber fed spectrograph to be installed on the VLT UT3 in the second half of year 2000. The spectrograph has two operation modes: MEDUSA and ARGUS. While MEDUSA allows several spectra of objects to be collected simultaneously throughout the whole field available at the Nasmith focus of the VLT, ARGUS makes use of a densely packed array of fibers to produce two–dimensional spectra with a spatial sampling of either 0.2 or 0.7 arcsec. We expect that ARGUS will be the prime instrument for spectroscopic studies of galaxy cores, with significant secondary imaging capabilities.

The typical photon flux from a typical Virgo galaxy with a central brightness of about $\mu_V \simeq 14$ arcsec^{-2} is expected to be of about 1 photon$\mathring{A}^{-1}s^{-1}$ at the CCD. In order to obtain two–dimensional spectra in the 5 arcsec field with a resolution in the range 5000 to 10000 with a signal–to–noise of 100, as required to measure detailed line profiles, one needs about 10 hours of integration per galaxy.

For the study of cores, it is important to stress the importance of high spatial resolution. Since ARGUS will be able to exploit the optical performance of the VLT down 0.2 arcsec (although undersampled below 0.4 arcsec), it would be

desirable to be regularly able to reach such a resolution. In addition to relying on the expected very good conditions at Paranal, it is therefore to be hoped that FUEGOS will be able to exploit the benefits of a tip–tilt correction system.

Acknowledgements

MS acknowledges financial support from the Italian Space Agency under contract ASI 299/6 FAE.

References

Bacon et al. (1994): Astron. Astrophys. **281**, 691.

Bender, R. (1988): Astron. Astrophys. **202** L5.

Bender, R., Surma, P. (1992): Astron. Astrophys. **258** 250.

Binney, J. J., Davies, R. L., Illingworth, G. D. (1990): Astrophys. J. **361** 78.

Binney, J. J., Mamon, G. A. (1982): Mon. Not. Roy. astron. Soc. **200** 361.

Carollo, C. M., Danziger, I. J. (1994): Mon. Not. Roy. astron. Soc. in press.

Crane, P., Stiavelli, M., King, I.R. et al. (1993): Astron. J. **106**, 1371.

Davies, R. L., Sadler, E. M., Peletier, R. (1993): Mon. Not. Roy. astron. Soc. **262** 650.

Franx, M., Illingworth, G.D. (1988): Astrophys. J. Lett. **327** L55.

Jedrzejewski, R. J., Schechter, P. L. (1988): Astrophys. J. Lett. **330** L87.

Møller, P., Stiavelli, M., Zeilinger, W.W. (1993): in "Structure, Dynamics, and Chemical Evolution of Elliptical Galaxies", ed. I. J. Danziger et al. (ESO, Garching), p. 131.

Nieto J. L., Bender, R., Arnaud, J., Surma, P. (1991a): Astron. Astrophys. **244** L25.

Nieto J. L., Bender, R., Surma, P. (1991b): Astron. Astrophys. **244** L37.

van der Marel, R. P., Franx, M. (1993): Astrophys. J. **407**, 525.

The VLT Contribution to Development of AGN Unified Schemes

Martin J. Ward[1]

[1] Astrophysics, Nuclear Physics Building, Keble Road, Oxford OX1 3RH, England

Abstract. The era of 8 meter class telescopes will naturally have a major impact on the study of AGN. In this review I focus on the promise of diffraction limited observations in the near and mid-infrared regions. An important pillar supporting the currently popular unified model for AGN, is the existence of a dusty molecular torus. The so-called AGN-starburst connection, is another area of intensive study. High spatial resolution observations with the VLT will observationally quantify the properties of these components.

1 Introduction

Futurology is a dangerous business, as the benefit of hindsight continually shows us. Take for example the scientific cases made for the construction of 4 meter class telescopes in the late 1960's and early 70s. In sections discussing active galactic nuclei nowhere will one find comments concerning *Unified Models, ionization cones* or *dusty molecular tori*. The reason for this is obvious, these subjects resulted from discoveries actually made using the then new 4 meter telescopes. The situation regarding predictions of the impact of 8 meter telescopes is somewhat easier, partly because a 10 meter telescope is already in operation, and also because state-of-the-art detectors are available on 4 meter telescopes and it is therefore less difficult to extrapolate the sort of scientific results one might anticipate.

In order to place the proposed VLT observations in some context, I will give a highly compressed sketch of the current standard model of AGN. Accretion of matter onto a black hole is now the almost universally accepted model for the primary energy source. This is required for reasons of efficiency of energy generation needed to explain the most luminous examples, and also explains the compactness at X-ray frequencies (inferred from short term variability). Although the compact continuum source and the region emitting the high velocity broadened emission lines (BLR) cannot be spatially resolved, we do know something of its size and geometry by indirect means eg. flux variability, and time lags in changes of the continuum and broad line fluxes (Alloin *et al.*, these proceedings). Detailed modelling of their infrared/optical/ultraviolet continuum distributions has been used to infer the existence of a hot thermal component dominant in the UV, associated with an accretion disc which *feeds* material onto the black hole. Further out, on scales of 100's of parsecs to kiloparsecs, we can spatially resolve the geometry directly. These regions emit the narrow lines, and in some cases may also contain a region of circum-nuclear star formation.

This is the model for Seyfert 1s and quasars, however many AGN have high excitation emission lines, but lack a BLR or strong continuum component (Seyfert 2s). The simplest version of the Unified Scheme, holds that whether or not we observe the innermost regions of the nucleus depends on the geometry of the nuclear regions. If this model is correct then it follows that there must be something blocking our view at certain angles. This component, which absorbs the optical/UV continuum and the BLR, is believed to be a dusty molecular torus.

2 High Resolution Near-Infrared Imaging

2.1 Technical Issues

There are several methods to achieve high resolution images: speckle interferometry; aperture masking; multi-aperture interferometry. A recent ESO conference (Ulrich 1992), and indeed contributions within these proceedings, describe the techniques in some detail. I will simply take as a bench mark performance aim, that adaptive optics on the VLT will provide full wavefront compensation at near-infrared wavelengths $> 2.2\mu$m. This will result in a resolution of 60 mas at this wavelength. Optical imaging and spectroscopy of AGN with the VLT will of course be essential. Also optical spectro- and imaging polarimetry, which is often *photon starved* on 4 meter telescopes, will benefit greatly provided that suitable instrumentation is available. Regrettably one cannot cover all these opportunities in a single article, hence I will concentrate specifically on the near to mid-IR regions.

2.2 Components of Interest

Consider an AGN, $z = 0.01$, 1 arcsec = 300 parsecs

($H_0 = 50$)

Principal Nuclear Components

Accretion disk	$\sim 3\mu$ arcsec
BLR (outer)	~ 0.1 milli-arcsec
NLR (inner)	~ 0.03 arcsec
Obscuring Material	$\sim 0.01 - 0.2$ arcsec
Circumnuclear Starburst	$\sim 0.2 - 3$ arcsec

The nearest example of an AGN is a matter of some debate. Cen A is certainly a strong candidate, at a distance of 4 Mpc. Other possibilities are M81 and the Circinus galaxy (1409-45), at about the same distance. In these cases 1 arcsec corresponds to 20 parsecs, and the diffraction limit at 2.2μm, to just over 1 parsec. These are obvious targets for high resolution imaging with the VLT. However, I do not wish to restrict possible projects to only the nearest and brightest examples, so in the following insert I list the sizes of the various components within an AGN at a redshift of 0.01. This is not a purely arbitrary choice. There are just over 100 AGN listed in the Veron and Veron catalogue (6th edition) within this distance limit, excluding those classified spectroscopically as HII regions. The majority of these nearby AGN are Seyfert 2s, and most have visual magnitudes from 13-14. Therefore on general grounds of brightness they represent feasible targets, and the projected sizes of some components of interest are within the regime accessible with the VLT.

2.3 The Torus, and Scattering Regions

The theoretical properties of the torus have been discussed in some detail (Pier and Krolik 1993). Infrared observations are particularly appropriate for several reasons. A common feature of AGN is a rise in their continuum distribution from $1 - 2μm$, peaking at about $5μm$, which it is suggested is associated with thermal emission from hot dust at about 1300 K. Hence observations at $> 2μm$ are matched to the increasing flux from this component. Also, this hot region may lie behind considerable amounts of dust, but at longer wavelengths the attenuation is less severe. Finally, the diffraction limit of an 8 meter telescope should be attainable at $> 2μm$. The size of the torus is uncertain, since it will have an inner radius determined by the intrinsic luminosity of the AGN which is responsible for destruction of dust grains within the zone of their sublimation temperature limit. This region would not be resolved, except in the very nearest AGN. However its outer radius could be much larger. Recent work has suggested that star-formation may occur within the torus. If so, this extended heating source may result in it being resolved in AGN up to a redshift of 0.01. A competing idea is that the near IR component originates predominantly from the inner walls of the torus, and that any extended emission would in fact be seen in the polar direction by reflection, in the same way as the scattered BLR components are seen in polarized light in many Seyfert 2s. Thus these two models predict an extended hot IR component in orthogonal directions. A crucial test is simply to resolve the non-stellar near IR emission in the K and L bands on sub-arcsec scales.

A serious problem with imaging the central regions of AGN at near infrared wavelengths, is that through the broad bands of J,H, and K, the bulge stellar light will be a major contributor. What is required is a way to increase the contrast between any torus-related component and the stellar component. One way to do this is to use colour ratio diagrams. For thermal components such as stars and re-radiating hot dust, the colour indexes eg. (H-K) and (K-L) are a function of temperature. By use of such colour/colour maps, it may be possible to detect tori which are otherwise lost within the starlight of the nucleus.

Figure 1, shows such a colour ratio map for the Seyfert 2 galaxy NGC 5728. The grey scale image does not show the features to best effect (darker shades

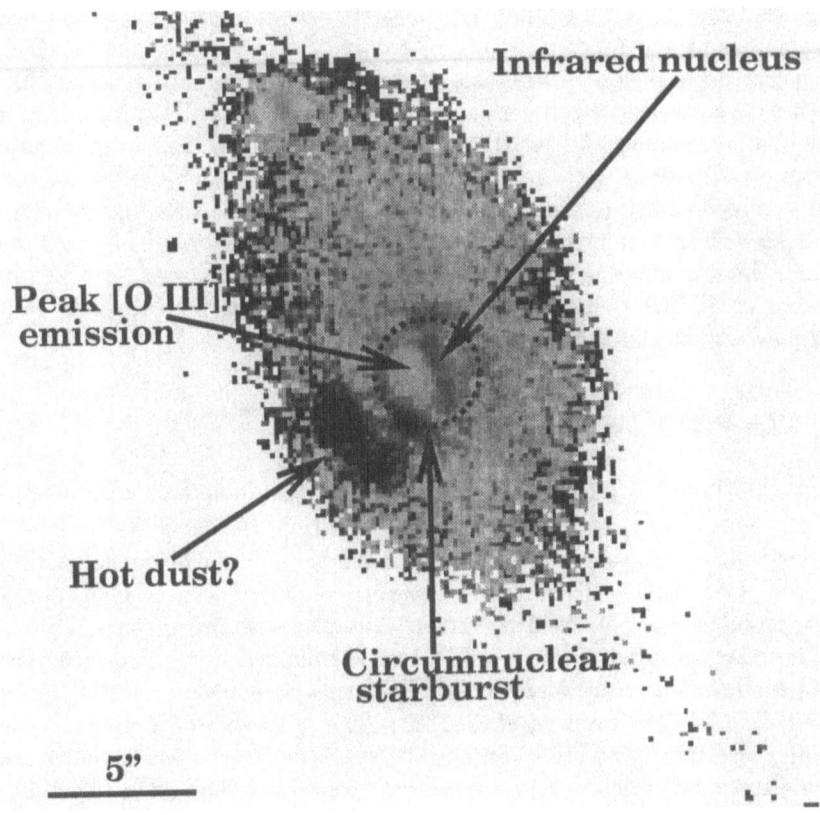

Fig. 1. NGC 5728, showing a colour ratio map made by dividing an H image by a K image. 1 arcsec corresponds to 270 parsecs. N is at the top, E to the left. (From Tsvetanov *et al.* 1995.)

are red, lighter shades are blue) but nevertheless we see a linear red distribution across the nucleus. The size of this feature is much larger than the classic torus dimension, but, as mentioned previously, the outer regions of the obscuring material are not well defined, and it could be related (IR spectroscopy to obtain dynamical information might settle this question). To the east of the nucleus there is a much bluer region coincident with the peak of the [O III] line emission. This could be a region of scattering, hence the relatively bluer colour. We also see faintly a partial circle around the nucleus, which might be associated with circum-nuclear star formation. Finally there is a very red (dark) region labelled *Hot Dust?* But without spectroscopy and additional continuum measurements this is speculation.

This example illustrates the power of colour ratio mapping in enhancing the contrast of various components within AGN. This image was obtained in roughly arcsecond seeing on a 4 meter telescope, so an equivalent map made using an 8 meter, at more than 10 times this spatial resolution would reveal far more about the inner structure of the red linear feature, and the nature of the scattering region. A foretaste of the detail that may be revealed, albeit at optical wavelengths, can be gained by inspecting the HST images of the green/red continuum, and [O III] and Hα emission lines, in the same galaxy (Wilson *et al.* 1993).

A complementary approach to enhance the contrast of any putative torus would be to use a tuneable Fabry-Perot instrument to image in the 2.12μm emission line of molecular hydrogen. Recent work in this area on NGC 1068 (Tacconi *et al.* 1994), has yielded some very interesting if perplexing results from the viewpoint of the torus model.

3.0 The AGN-Starburst Connection

The possibility of an evolutionary link between starbursts and AGN has long been the subject of debate. New impetus came from the discovery by IRAS of galaxies which emit quasar-like luminosities in the mid-far infrared, yet otherwise exhibit none of the other classic signatures of quasar activity. One approach to this problem is to observe less extreme examples of so-called *composite* activity. These are galaxies in which an AGN is surrounded by a powerful burst of star formation. There are many examples of this phenomenon, one of the best studied is NGC 7469 (Wilson *et al.* 1991). The properties of a circum-nuclear starburst are difficult to study because it is so close to the AGN. Information on sub-arcsecond scales is very sparse, yet this is just the region that will be most influenced by the presence of the AGN, or indeed *vice versa.* There follow two examples of how the properties of compact star-forming regions within the nucleus, can be studied.

3.1 Mid-IR Spectroscopy of Dust Related Features

A region accessible to ground-based observation is between 8 − 13μm, which fortuitously covers both diagnostically useful absorption and emission features. AGN in general, particularly those with broad lines in their optical spectra, usually have rather smooth featureless spectra from 8 − 13μm (Roche *et al.* 1991). It is supposed that the hard UV continuum emitted from the nucleus destroys the dust grains responsible for production of the emission features. On the other hand, most HII/starburst nuclei and a few Seyferts do show broad silicate absorption, and some have dust related emission features. The problem for studies of circum-nuclear star formation is that these regions plus the AGN, are both included within the spectrograph aperture. One nearby exception is NGC 1365 (fig. 2), where the two regions are just far enough apart to be separated. Note the broad absorption and the 11.3μ narrow emission band.

Clearly, if this object were more distant the two components would blend together, and it would then be impossible to say much about the silicate absorption or dust emission properties. Long-slit spectroscopy using an 8 meter, could trace the starburst properties from the kiloparsec scale down to the inner NLR, where the AGN could have a major influence via jet induced star formation, and shocks caused by collisions between molecular clouds in rapid motion around the nucleus. At shorter wavelengths stellar absorption features in the H and K bands can be used to estimate the stellar population as a function of distance from the centre. As with the torus, infrared observations are well suited because of the presence of dust and hence absorption.

3.2 Supernovae in and around the Nucleus

Any model proposed in which *all* Seyfert activity is powered by supernovae and their remnants, runs into difficulties when confronted with X-ray variability and collimated radio emission. However, there is no doubt that supernova occur in and around some nuclei. The [Fe II] lines at $1.26\mu m$ and $1.644\mu m$ are useful in that they distinguish between regions of OB stars, and regions containing SNRs. A high ratio of [Fe II] to Paschen-β or Brackett-γ is indicative of SNRs. Figure 3 shows an image in the [Fe II] $1.644\mu m$ line of NGC 1808. Numerous *blobs* are seen, which are not visible in H-band continuum images. Furthermore, in many but not all cases there is remarkable positional agreement between the [Fe II] regions and discrete peaks in the 6cm radio emission, strengthening the SNR hypothesis. A factor of 10-20 better resolution will show if these [Fe II] blobs split up into multiple components, and whether SNRs are found within the currently unresolved nucleus. High resolution spectroscopy of the [Fe II] line will give profile information, and a measure of the expansion velocity.

Finally on this subject, looking further into the future, if it is possible to extend VLT interferometry to the brightest and nearest AGN where a resolution of 5 mas will be available. Then individual SNRs will be resolved. The radio equivalent is the work on M82 (Muxlow *et al.* 1994), in which parsec sized remnants were resolved. If the VLTI resolution is attainable, then we will truly have information on star formation within the nucleus!

4 The Challenge from Space

This is a meeting about use of the VLT, but to use it effectively we must identify the parameter space where it is uniquely placed to make the most significant gains over other facilities. In this regard it is worth noting two satellite facilities which will be available in the relatively near future. For comparison I will only consider those wavelengths of overlap for both the VLT and these two space missions.

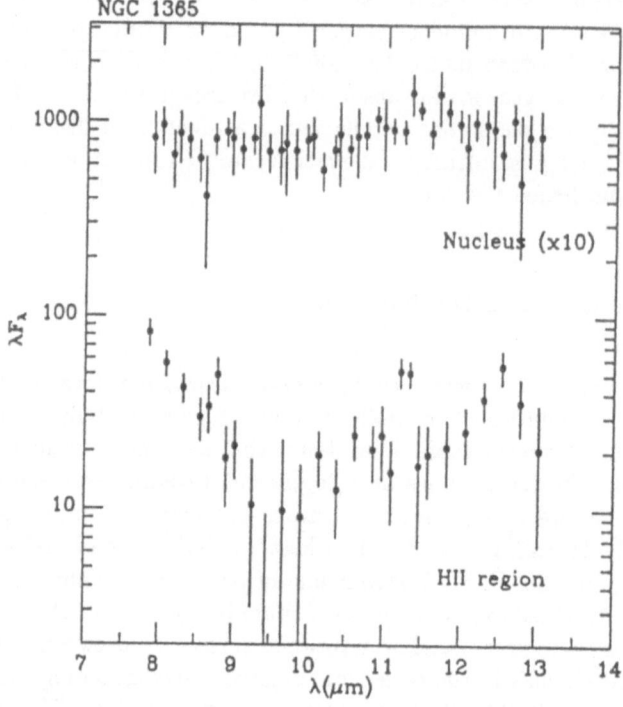

Fig. 2. Mid-IR spectra of the composite nucleus in NGC 1365, showing the Seyfert nucleus (top) and off-nucleus star formation (bottom). The spectral resolution is 50.

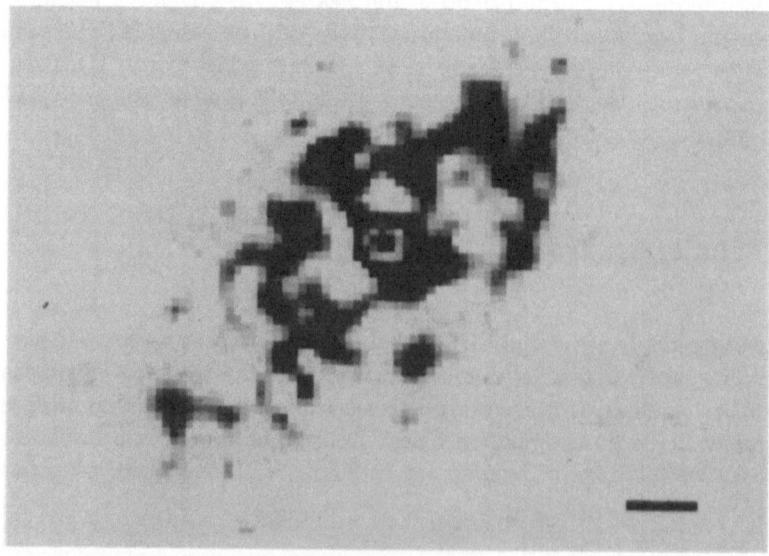

Fig. 3. An emission line image of NGC 1808, in [Fe II] 1.644μm. The scale bar is 2 arcsec, corresponding to 180 parsecs. N top, E left.

4.1   The Infrared Space Observatory - ISO

This cryogenically cooled 60cm aperture telescope is currently scheduled to be launched in September 1995. It will have a nominal lifetime of about 1.5 years. The ratio of a single VLT mirror to the ISO aperture is about 180, but because of the low thermal background emission in space, ISO will be about a factor 10 more sensitive than an 8 meter for imaging point sources at 10μm, and several orders of magnitude more sensitive for imaging extended sources. However, ISO will provide no spatial information on sources smaller than 3-4 arcseconds because of its diffraction limit.

ISO launch 1995 + 2   60cm Cooled Telescope

Overlap Spectroscopy with VLT:

PHT - S	$2.5 - 12.0\mu$	$R \sim 90$
CAM-CVF	$2.5 - 16.5\mu$	$R \sim 50$
SWS-Grating	$2.5 - 45.0\mu$	$R \sim 900\text{-}2{,}700$
SWS-FP	$15 - 35\mu$	$R \sim 20{,}000$

Imaging: CAM 32 x 32 pixels of 1.5, 3.0 or 6.0 arcsec.

Short wavelength channel $2.5 - 5.2\mu$

Long wavelength channel $4.0 - 17\mu$

Diffraction limit (Airy disc) is $\lambda(\mu)/3$ in arcsecs.

For thermal IR $(3.5-13\mu$m) spectroscopy, even at low resolution, the 8 meter telescopes will have significantly greater sensitivity for point sources, but not for extended sources. The moral for AGN observations is that ISO wins for obtaining spatially integrated data on the extended circum-nuclear star formation, but for imaging and spectroscopy of the torus itself, better use an 8 meter. Of course, this statement only applies to the wavelength region of overlap.

4.2   The HST plus NICMOS

The schedule for the next instrument replacement mission to the HST is uncertain. Nevertheless the current plan is to install an optical/uv imaging and spectroscopy instrument (STIS), and a near-IR instrument predominantly for imaging, but with some spectroscopic capability (NICMOS), by the end of the century (Thomson 1992). In the $1 - 1.5\mu$m J and H bands, NICMOS will have somewhat better sensitivity than the VLT (about 1 magnitude) for compact objects. Whereas at 2.2μm (K-band), about the longest wavelength accessible with NICMOS, its sensitivity will be significantly worse than an 8 meter. NIC-MOS spectroscopy will not suffer from the OH emission lines and the variable sky conditions that limit ground-based observations, but development of instruments that suppress the OH airglow, can in theory improve the sensitivity of ground-based spectroscopy in the J and H windows by 1-1.5 magnitudes (Maihara *et al.* 1993).

HST 2.4m + NICMOS/STIS 1997 - 1999 ?

NICMOS 0.8 - 2.5μ 256 x 256

Pixel Scales 0.04, 0.08, 0.2 arcsec.

Spatial Resolution 0.1 - 0.2 arcsec.

Slitless Grisms: spectra of all objects in field of view

Long Slits: $\lambda/\Delta\lambda = 100 - 10,000$, Echelle $\lambda/\Delta\lambda = 23,000$

The diffraction limit of the VLT is 3 times better than HST, but it should be noted that the stable PSFs across the whole 20 arcsec field of view of NICMOS (with 0.08 pixels) will be an advantage for imaging extended regions of complex morphology on these scales.

5 Do We Have the Instruments to Fully Exploit the VLT?

Having 8 meter telescopes capable of delivering excellent images is only one requirement along the road to achieving our scientific objectives. Fortunately the VLT project has carefully considered the suite of instruments needed. Below I list some suitable AGN projects for the VLT, together with the instruments that would be used. For acronym decoding see elsewhere in these proceedings.

a. Near-IR diffraction limited imaging of the nuclear regions. Colour ratio continuum mapping, and IR emission line mapping - **CONICA**.

b. $1 - 5\mu$m spectroscopy of reddened line emitting regions - **ISAAC**, and for high resolution velocity field and line profile work - **CRIRES**. Same instruments for inner starburst studies, stellar population gradients, and mapping of the 3.28μm dust related band.

c. $10 - 20\mu$m imaging and spectroscopy of dust emission from the nucleus - **MIIS**. A low resolution option (R=100) would be useful.

d. Interferometry of the inner regions of the torus, and resolving the shells of S-NRs in circum-nuclear regions of the nearest AGN. This is not an instrument but a technique - **VLTI**.

At the beginning of this article I said that the closeness in time of first light on the VLT, makes predicting what it may achieve an easier task, like forecasting the weather tomorrow. But the accompanying danger is that one may be too incremental in outlook. The projects I have mentioned are indeed extensions of current work, which is inevitably how things must start out. However, given an order of magnitude increase in resolution, or 2 orders if and when VLTI can be applied to AGN, it is likely that serendipitous discoveries will equal or exceed the promise of any projects that we can currently imagine.

References

Tacconi, L.J., Genzel, R., Blietz, M., Cameron, M., Harris, A.I., Madden, S., *Ap.J. Lett.* **426**, L77 (1994)

Maihara, T. *et al.*,*P.A.S.P* **105**, 940 (1993)

Muxlow, T.W.B., Pedlar, A., Wilkinson, P.N., Axon, D.J., Sanders, E.M., *M.N.R.A.S.* **266**, 455 (1994)

Pier, E.A., Krolik, J.H., *Ap.J.* **418**, 673 (1993)

Roche, P.F., Aitken, D.K., Smith, C.H., Ward, M.J., *M.N.R.A.S.* **248**, 606 (1991)

Thomson, R., *Space Sci Rev.* **61**, 69 (1992)

Ulrich, M-H., *Ed. ESO conference on Progress in Telescope and Instrumentation Technologies, ESO Workshop Proceedings No. 42* (1992)

Wilson, A.S., Braatz, J.A., Heckman, T.M., Krolik, J.H., Miley, G.K., *Ap.J. Lett.* **419**, L61 (1993)

Wilson, A.S., Helfer, T.T., Haniff, C.A., Ward, M.J., *Ap.J.* **381**, L79 (1991)

Subarcsecond Observations of Galactic Nuclei

Guy Monnet[1], Roland Bacon[1], Pierre Ferruit[1],
Emmanuel Pécontal[1], Eric Emsellem[2]

[1] Observatoire de Lyon, 69561 Saint-Genis-Laval, France
[2] Leiden Observatory, 2300 RA Leiden, The Netherlands

Abstract. Subarcsecond two-dimensional imagery and spectrometry of the stellar component of galactic nuclei show complex systems, with multiple components and the possible presence of supermassive black-holes. Strategies for observation with the VLT, both with and without adaptive optics, are discussed, in terms of presently planned or future instrumentation.

1 Why Bother?

While small in terms of global mass relative to that of their host, nuclei of galaxies do play a crucial role in unravelling many aspects of the formation and evolution of galaxies. Acting as "magical dumping places" (Nieto 1992), they still keep the memory of violent relaxation, merging processes, recurrent nuclear activity and late bursts of star formation. In addition, active galaxies certainly harbour some kind of central engine, generally hypothesized as a supermassive black-hole. Currently non-active galaxies may well contain a similar monster in its present, non-feeding phase. The latter, being observationally much more numerous, offer a handful of nearby candidates, where studies of the stellar dynamics can better pinpoint the putative black-hole (Kormendy 1988).

2 What Is Required?

Nuclei are typically about 10 pc in size, which corresponds to only 0.15 arcsec at the distance of the Virgo Cluster. Furthermore, the nearest nuclei, e.g. that of M31, show considerable kinematical complexities (sec e.g. Dressler and Richstone 1988, Kormendy 1988, Bacon et al. 1994) at the pc level or less. Getting the best spatial resolution, and in any case to better than 1 arcsec, is thus a prerequisite.

Both photometic and spectrometric information are needed to unravel these complex structures. A bi-dimensional sampling on the sky is required: a) because, as stressed above, rotationnal symmetry is not generally present; b) deconvolution from the seeing and from projection effects cannot be achieved without such a coverage.

Recent observations with a so-called three-dimensional (3D) spectrograph have been made at CFHT, and have attained spatial resolutions from 0.6 to 0.9arcsec. They have indeed revealed the extent of the complexity of three of the closest nuclei: M104 with multiple components, including a central disk; M32 with tantalizing clues for a triaxial shape; M31 with an offsett central mass concentration (Emsellem et al. 1992). These data were obtained with a relatively small spectral resolution of about 2,000, classically considered as adequate in view of the large radial velocity dispersion (200 to 400 kms^{-1}) in these objects. Recently however, Van der Marel et al. 1994 have shown that much new information on the dynamical state of nuclei (e.g. the presence of a central black-hole) can be obtained from measuring and modelling the deviations of the line of sight velocity profiles from pure Gaussians. Attaining that level of sophistication requires good signal to noise ratios (40-60) at substantially higher spectral resolutions, up to 10,000.

3 How To Get It?

Galactic nuclei are relatively bright objects: Peak V luminosity in the central arcsec can reach a values as high as 12-14 mag. arcsec^{-2}, for the nearest objects. Taking into account the required spatial (down to 0.1 arcsec or even less) and spectral resolutions (up to 10,000) , the need for an 8 m-class telescope becomes nevertheless imperative. This is of course especially true for the spectrometric observations, which in addition will suffer less from competition with the HST than pure imagery.

A word of caution may be needed here, as the HST is certainly not absent from the spectrometric scene. Recently HST observed two small spots at 0.25 arcsec from the center of the brightest elliptical in the Virgo Cluster, M87. They show emission, displaced by ±500 kms^{-1} and interpreted as the signature of a central 2×10^9 M$_\odot$ black-hole (harms et al. 1994). 3D spectrographic data, obtained with the VLT, would give the two-dimensional velocity field of this gas (as well as that of the stellar component), and hence a direct proof (or disproof) of the existence for this central mass concentration.

Presently planned or proposed VLT instruments, which offer potentially useful devices, are :

- CONICA and VHARC will give imaging capabilities, down to the diffraction limit, but of course will suffer from heavy competition from the HST;
- FUEGOS, with ample spectral resolution, a 3D spectrographic capability in its ARGUS mode, and the convenient spectral range of 0.37 to 0.9 μm, will be a powerful tool. Especially useful would be a fast tip-tilt capability, which could boost its spatial resolution down to about 0.4 arcsec In that respect, it may be worthwhile to point out that, with the use of state-of-the-art detectors (e.g. avalanche photodiodes), the VLT may be able to achieve fast guiding, using reference stars as faint as V 19-20 mag.;
- a dedicated instrument would use such a so-called 3D spectrographic instrument (see Monnet 1993 for a short review) coupled to an adaptive optics bonnette. This would fully open the sub-0.5 arcsec range, but would ultimately require a laser-guide system to reach 0.1 arcsec or less.

Such an instrument would not be restricted to this fascinating, but narrow, astrophysical field, but would be invaluable for detailed studies of any type of structured object. This potentially ranges from the surface of telluric planets to quasar environments, encompassing e.g stellar jets, active galaxies, etc. It could be considered as a way to provide a spectrographic follow-up to the VLT (and HST) imaging capabilities.

References

Bacon R., Emsellem E., Monnet G., Nieto J.L., 1994, Astron. Astrophys. 281, 691.

Dressler A., Richstone D.O., 1988, Ap.J. 324, 701.

Emsellem E., Bacon R., Monnet G., Nieto J.L., 1992, in ESO/EIPC Workshop "Structure, Dynamics and Chemical Evolution of Elliptical Galaxies", p. 147.

Kormendy J., 1988, Ap.J. 325, 128.

Harms, R. J., Ford, H. C., Tsvetanov, Z. I., Hartig, G. F., Dressel, L. L., Kriss, G. A., Bohlin, R., Davidsen, A. F., Margon, B., Kochhar, A. K., 1994, Ap. J., 435, L35.

Monnet G., 1993, in IAU Colloquum n 149 "Tridimensional Optical Spectroscopic methods in Astrophysics", in press.

Nieto J.L. ,1992, in "Morphological and Physical Classification of Galaxies", G. Longo et al. eds, p. 69.

Van den Marel R.P., Evans N.W., Rix H.W., White S.D.M., de Zeeuw P.T., 1994, MNRAS 268,521.

The AGN Environment

Sperello di Serego Alighieri

Osservatorio Astrofisico di Arcetri, Largo E. Fermi 5, I–50125 Firenze, Italy

Abstract. After a general discussion about the gains of large telescopes, I review the capabilities that are required for the VLT to study the AGN environment, with particular regard for the polarimetric ones. I compare the capabilities with what is foreseen in the present instrument plan and suggest some improvements.

1 Introduction

Scientific discovery is by itself a process which requires much imagination. I am not sure that I can imagine now which important discoveries our imagination will lead us to with the VLT. However I will try here to make some small extrapolations from what I know, with the consciousness that whatever I say now will only set a lower limit to what the VLT will do and that reality will certainly not cease to surprise my imagination.

I have selected the topic of my talk not only because it is my main area of research, but also because it is an important field of study for the insight it can provide on the formation and the early evolution of galaxies, both in their stellar and ISM content (McCarthy 1993), for the indirect information it can give on nuclear activity through the study of ionization cones and scattered light (Antonucci 1993) and, finally, for the application to problems of fundamental physics, like a test of the Einstein equivalence principle (Cimatti et al. 1994).

This area of research is one for which 4m class telescopes have already been pushed to their limits and it is not by chance that one of the first scientific observations obtained with the 10m Keck telescope was devoted to it (Graham et al. 1994).

2 The Gains of Larger Telescopes

Before launching into the detailed topic of my talk, I would like to remind you of the gains to be expected with the VLT, in particular with its increase in collecting aperture diameter D, on the limiting magnitude, i.e. on the faintest flux which is observable with a given signal–to–noise ratio (S/N). For CCD observations of point sources (or of unresolved parts of extended objects) the S/N is given by:

$$S/N = \frac{(\pi/4)D^2 F_o E\Delta\lambda t}{\sqrt{(\pi/4)D^2 F_o E\Delta\lambda t + (\pi/4)D^2 S_s(\pi/4)\theta^2 E\Delta\lambda t + (r_n^2 + Tt)n}}, \quad (1)$$

where F_o is the photon flux density from the point source, S_s is the photon surface brightness (density) of the sky background, E is the total throughput (in counts per photon) including atmosphere, telescope, instrument and detector, $\Delta\lambda$ is the bandwidth, t is the exposure time, θ is the diameter (FWHM) of the seeing disk, r_n is the CCD readout noise, T is the rate of production of thermal electrons per pixel and n is the number of pixels illuminated by the PSF.

If the readout noise and the thermal noise of the CCD are negligible, as is the case for imaging with modern CCDs, particularly on a large telescope, then:

$$S/N = \frac{DF_o\sqrt{(\pi/4)E\Delta\lambda t}}{\sqrt{F_o + S_s(\pi/4)\theta^2}},$$

which shows that S/N increases linearly with telescope diameter and with the square root of exposure time. More interesting is to examine the dependence of the limiting flux on the observational parameters. This can be done easily in two cases:

(1) if the flux from the source is much smaller than the flux from the sky within the PSF ($F_o \ll S_s(\pi/4)\theta^2$, the sky–limited condition), then:

$$F_{lim} \propto \frac{(S/N)\theta\sqrt{S_s}}{D\sqrt{E\Delta\lambda t}};$$

(2) if the flux from the source is much larger than the flux from the sky within the PSF (photon–limited condition, which is the case for example when a very high S/N is needed), then:

$$F_{lim} \propto \frac{(S/N)^2}{D^2 E\Delta\lambda t}.$$

First we note that the source flux separating the two cases does not depend on telescope diameter. In the sky–limited condition the limiting flux decreases inversely with telescope diameter only (a single VLT unit and the whole VLT should reach 0.87 and 1.62 mag. fainter than the 3.6m telescope respectively), while it decreases inversely with the square of the diameter in the photon–limited condition (the gains in magnitudes are then doubled and become 1.73 and 3.24). Therefore the gains of a larger telescope are higher for those applications that require a high S/N like high resolution spectroscopy and polarimetry.

It is also clear that in the sky–limited condition good seeing is as important as a large telescope (e.g. di Serego Alighieri 1986): for example a 4m telescope with 0.5 arcsec seeing is as good as an 8m telescope with 1.0 arcsec seeing. Many talk about the advantages of large telescopes for the IR and, most appropriately, all large telescope projects foresee the use in the IR. However we should not forget that in the IR one is mostly working in the sky–limited condition, where the gains of larger telescopes are smallest.

¿From equation (1) the impression could be gained that a sufficiently long exposure time could compensate for a smaller telescope and that, given the rapid increase of total project cost with telescope size, this might even be a cost–effective solution. However we all know that this is not true, mainly because there is a practical limit to the exposure time which is set by the density of cosmic ray signatures that one is prepared to accept and by the duration of the periods of best observing conditions (i.e. seeing, sky transparency and brightness, zenith distance). Many of us have experienced the fact that it is often worthless to combine exposures taken with different conditions: one is better off by taking just the data obtained with the best conditions and throwing away all the rest.

3   Observing the AGN Environment

AGN are the most powerful compact energy producers in the Universe, after the Big Bang itself. However the AGN proper is always unresolved and often invisible directly: therefore most of the information available on the phenomenon is obtained by studying secondary processes in the environment. This is made of stars, warm gas emitting broad and narrow lines, dust emitting in the IR and absorbing and scattering the strong nuclear light, cold absorbing gas, hot electrons emitting X-rays and radio emitting plasma. The plethora of physical processes and observable quantities that are involved around an AGN goes beyond the scope and size of this talk, and I will have to concentrate on a few examples, which I choose to select on the basis of the observing technique.

The fundamental observational goal is to get the spectral energy distribution and the polar diagram of the radiation emitted by the different components mentioned above over the broadest possible wavelength range, and to obtain their spatial distribution around the AGN. The observational tools available for this are high resolution imaging, long–slit spectroscopy and imaging- and spectro–polarimetry. As many of you in the audience would have predicted, I shall devote most of the rest of my talk to try to convince you of the beauty of polarimetry and of its usefulness to study the AGN environment.

3   The Beauty of Polarimetry

The importance of polarimetry can be best emphasized by remarking that if you are not measuring polarization, you are throwing away three quarters of the information $(I(\lambda), P(\lambda), \theta(\lambda), V(\lambda))$ that photons carry to us from celestial objects. And, a part from a few disturbing cosmic rays and several elusive neutrinos, photons are all you can ever get from our beloved sources, so we better use them as completely as we can!

Now for those of you who dislike general statements, I will illustrate in some detail what polarimetry can do for the understanding of the AGN environment and what are the anticipated advantages of the VLT for this technique.

3.1 Geometrical Information

Polarimetry can give information on the geometry of the AGN even if it is unresolved. If polarization is due to scattering of anisotropic nuclear radiation, as is the case in high redshift radio galaxies (e.g. Cimatti et al. 1993), then the position angle of the E–vector is perpendicular to the mean scattering plane and therefore to the preferred direction of nuclear emission. Spatially resolved polarization can also give more information on the polar diagram for the nuclear radiation, such as the opening angle of the radiation cones. Since the cross–sections of forward and backward scattering are different, the study of the relative polarization and intensity of oppositely directed cones can provide information on the inclination of the mean axis of the cone with respect to the plane of the sky. Combining this with radial velocity measurements can discriminate between inflow and outflow. If the polarization measurements can be made independently in the continuum and in emission lines, both broad and narrow, as in di Serego Alighieri et al. (1994), then we can get information on the geometry of the continuum source and of the emitting gas, in particular on the relative sizes of the featureless continuum nuclear source, of the BLR, of the NLR and of the obscuring material. Therefore, polarization can resolve the inner structure of very distant AGN on scales which not even space interferometry will ever be able to resolve.

The most distant radio galaxy for which linear polarization has been measured, by pushing 4m class telescopes close to their limit, is at a redshift of 2.6 (Cimatti et al. 1994). Taking into account the Hubble diagram of radio galaxies, a single VLT unit should be able to make a comparable measurement for a radio galaxy at z=4.1: close to that of the most distant radio galaxies and quasars. Combining several units together will allow a substantial decrease in exposure time and/or a higher precision and would give the possibilty to do simultaneously the measurements at the 3 or 4 different position angles which are necessary to get full polarization data and which are normally taken sequentially. The advantage here would be to have a more uniform set of data taken in the same observing conditions. This is, however, not essential, since a complete measurement of one Stokes component is obtained from a single frame when using a beam–splitting analyser like the Wollaston prism (di Serego Alighieri 1989), while the other components can be measured independently.

3.2 Spectral Decomposition

In order to get reliable information both on the stellar and on the active component in distant radio galaxies it is imperative to separate the continuum spectral energy distribution of the stellar and of the scattered light independent of the direct line emission. We have already shown that this can be done effectively only by combining spectral and polarization data (di Serego Alighieri et al. 1994). The search for broad polarized emission lines is necessary to unveil the nature of the hidden nuclear spectrum, discriminating between a type 1 quasar and a featureless blazar spectrum. The changes of polarization with wavelength provide information on the spectrum of the diluting stellar radiation. For example, this technique has given the most convincing evidence of the presence of a strong 4000Å break in the diluting radiation. The clean SED of the stellar light obtained in this way gives a reliable measurement of the age of the oldest stellar population and therefore a lower limit to the formation redshift. As I said in the previous paragraph, the VLT will push these studies to the highest redshift available and therefore give very stringent constraints on the galaxy formation redshift and on the cosmological parameters H_0 and q_0.

The difference in the degree of polarization between a broad emission line and the adjacent continuum can set an upper limit to the amount of diluting radiation. This technique, if applied to a UV line like MgII2800, can constrain the presence of hot young stars, which has been claimed to explain the UV alignment effect, but never directly demonstrated.

With the VLT, it will be very important to extend the polarimetry to the near IR. Our simple 3-component model (old stars + scattered nuclear light + narrow line emission) predicts that the alignment and the polarization in the IR should be negligible. However, some alignment has been observed, and there are some preliminary claims of IR polarization — although at a lower level and at a different angle to that in the optical (Jannuzi, priv. comm.) — which might be due to other mechanisms than scattering, like transmission through aligned dust grains. A detailed study of IR polarization in the continuum will help to solve this issue. Also, the $H\alpha$ line falls in the near IR for redshifts between 1 and 3 and we would like to check that it has a broad, polarized component in radio galaxies.

3.3   Information on the ISM

When polarization is due to scattering, its dependence on wavelength can give information on the nature of the scatterers. In fact the scattering cross section is wavelength independent for electrons, while it decreases with wavelength for dust grains. However, it is important to realize that a change in the cross–section with wavelength does not necessarily imply a corresponding change in the degree of polarization if there is no dilution. If electrons are responsible for the scattering then the broadening of scattered emission lines gives information on the temperature of the electrons. In this way we have been able to exclude the possibility that scattering in two radio galaxies is due to a hot gaseous halo (di Serego Alighieri et al. 1994).

The spectral deconvolution obtained from the polarization can be used to derive the column density of the scattering material and, with some assumption about its radial density distribution, also its total mass. This, in turn, can help discriminate between dust and electron scattering since the latter has a much lower cross–section per unit mass.

Surprisingly, the dust scattering properties in the UV are still poorly known: for example it is not clear whether the 2200Å feature is present also in pure scattering or only in absorption. Distant radio galaxies are easily observed in the UV from the ground and their clean scattering geometry allows us to address these issues. Moreover, studying the polarization and spectral properties of the scattered light gives information on the size distribution of dust particles (e.g. Fosbury et al. 1990), providing unique clues to the properties of the interstellar medium in other galaxies and at other epochs. Indeed observing the polarization properties as a function of redshift is a tool to study the evolution of the ISM (Cimatti & di Serego Alighieri 1994). Again the VLT, by reaching the most distant galaxies, will give us the possibility to follow the history of the ISM back to the time of galaxy formation.

3.4 Polarimetry with the VLT

As we have seen in the previous sections, large gains are expected with the VLT for polarimetry over what is possible at the moment, especially for objects brighter than the sky. Indeed, most of the planned VLT instruments will have polarimetric capabilities in order to exploit these gains. I shall review here these capabilities and discuss improvements and suggestions for new instruments.

FORS, an imager/spectrograph designed to work at the Cassegrain focus between 330 and 1100 nm with spectroscopic resolution up to 2000, will have both imaging– and spectro–polarimetry modes. These are provided by rotatable retarders and a Wollaston prism to be used in combination with filters or grisms and focal plane masks or slitlets. It is anticipated that the degree of linear polarization will be measured with an accuracy of 1% in one hour down to U, B, V and R magnitudes of 22–23 in imaging and down to V=17.3 in spectroscopy with 2.5Å resolution.

ISAAC, the IR imager/spectrograph for the Nasmyth focus, will work between 1 and 5μm with spectroscopic resolutions in the range 300–10,000. It will do imaging polarimetry using a fixed analyser in one of the filter wheels, to be used in combination with filters and with rotation of the whole instrument. Although the original design of ISAAC foresees the use of wire grid analysers, the possibility of replacing them with a Wollaston prism is being considered.

Similarly, imaging polarimetry will be possible with CONICA, the high spatial resolution, near–IR (1–5μm) camera. It is designed to work at the Nasmith focus in combination with adaptive optics (AO) but the effects of the AO optical train on polarization measurements have yet to be carefully assessed. CONICA will have both wire–grid analysers and Wollaston prisms to be used in combination with focal plane masks.

UVES is an echelle spectrograph for the Nasmyth with a spectroscopic resolution of 40,000. The possibility of doing spectropolarimetry with a polarization analyser in the pre-slit optical train has been investigated, but is not in the present plan because of possible difficulties with the polarization induced by M3, the image slicer, the image derotator and the spectrograph. Nevertheless these problems are not insoluble and polarimetry with UVES would be useful, for example, to study the line polarization structure in AGN.

The possibility of doing both imaging– and spectro–polarimetry with MIIS, the 8–24μm Cassegrain imager/spectrograph, has been studied and is feasible with a rotatable retarder and a wire–grid analyser. However, the polarization components are not in the present design, although room is kept for them. For distant radio galaxies the radiation emitted by stars, dust and non-thermal nuclear processes in the near IR (K–band) is shifted into the MIIS range. Therefore I would strongly recommend that the polarimetric capabilities of MIIS are actually implemented to study interstellar polarization by transmission through aligned dust and scattering of nuclear radiation by electrons, whose cross–section does not depend on wavelength.

Table 1: Polarimetric capabilities of VLT instruments.

Instr.	λ range (μm)	$\Delta\lambda/\lambda$	Mode	Polarizers	Limiting mag. ($\sigma_P=1\%$)
FORS	0.33–1.1		imag.	Woll., retard.	22–23 (U,B,V,R)
"	"	$\leq$2000	spectr.	" "	16–17 (1.3Å/pix)
ISAAC	1–5		imag.	Fixed analyser	K~18.5
CONICA	1–5		imag.	Woll., wire-grid	K=21
UVES*	0.33–1.1	$\geq$40000	spectr.	?	
MIIS*	8–24		im+sp	Wire-grid, retar.	

* Polarimetry is possible but not foreseen in the present design.

3.5 How to Improve Polarimetry with the VLT?

As a general rule, to be followed as much as possible by instrument designers, I would like to emphasise that, especially for faint distant galaxies, the polarization analyser should be of the beam–splitting type, like Wollaston prisms. These devices provide both orthogonally polarized beams for measurement and therefore the polarization parameters are derived from ratios of intensities measured on the same frame, independently of variations in the observing conditions. On the other hand, polarimetry by single beam analysers, like polarizing sheets and wire grids, is affected by such variations, particularly in the infrared where the sky is bright and variable.

Beam–splitting analysers are good both for imaging and for spectroscopy. However, for spectropolarimetry of extended objects, one would also need a rotatable half–wave plate in order to avoid having to rotate the whole instrument. This is because rotating the instrument necessarily changes the fraction of the object which falls in the slit and therefore does not give a consistent polarization measurement at all position angles.

The main lack of polarimetric capability in the current complement of VLT instruments is in the area of spectropolarimetry in the near IR. At a redshift of 4, the spectral region around 3000Å, where we have seen the peak of the scattered nuclear radiation for radio galaxies at z~1, is shifted into the near IR. We have shown that spectropolarimetry in the region between 2000 and 5000Å is essential to disentangle the stellar from the active component and derive the age of the oldest stars (di Serego Alighieri et al. 1994). By doing this on the most distant galaxies, IR spectropolarimetry will provide the most stringent constraints on the epoch of galaxy formation. Furthermore, for redshifts between 0.6 (where the radio/optical alignment effect sets in) and 2.8, Hα is shifted to the near IR. This line is the brightest line from the BLR, particularly for objects where some dust extinction weakens Lyα, and therefore is of the utmost importance to verify the presence of broad lines in the scattered radiation. IR spectropolarimetry will then be essential to improve the unified model of AGN. One way to allow this technique with the VLT would be to have a Wollaston prism and a rotatable half–wave plate in ISAAC: I would strongly recommend that this possibilty is studied carefully and that, if it is not viable, some alternative is found for near IR spectropolarimetry. Clearly the ideal instrument to complement FORS for polarimetric studies of the AGN environment would be an IR imaging– and spectro–polarimeter (IRISP?) to work at Cassegrain focus between 1 and 5 μm with spectral resolving power R~500–2000.

As already mentioned and justified above, I would also recommend to implement polarimetric capabilities in UVES and MIIS.

4 The Other Techniques

Of course polarimetry is not the only way, and probably not even the best one, to tackle the open questions on the AGN environment. It is a complicated technique with which I have some experience and this is why I have dedicated most of this review to it. However I am not forgetting the more classical techniques and wish to devote a few words to them as well.

4.1 Long–Slit Spectroscopy

Of the many ways in which spectroscopy of the AGN environment is useful, I wish to recall two of them. First, much better information than is now available from the study of the SED would become available on the stellar population, and on the interstellar medium, by the observation of absorption lines in distant radio galaxies. Possible detections of stellar (Chambers & McCarthy 1990) and interstellar (di Serego Alighieri et al. 1994) absorption have been obtained with very long integrations on 4m class telescopes. The situation however is far from safe and clear, and only waits for a larger telescope. Second, I would like to emphasize the importance of studying the emission line profiles, to detect broad lines or broad wings underlying narrow components, to determine the velocity structure of the narrow lines looking for possible differences between the high and low ionization lines, and to obtain good line ratios for the different components. Given the redshift range of the interesting AGN, these studies should be made with the VLT both in the optical and in the infrared, and it seems to me that FORS and ISAAC are well equipped for them.

4.2 High Resolution Imaging

Last but not least a few words on the importance of high resolution imaging of the AGN environment. Actually much better than my words is the HST+WFPC2 image of 3C 324 obtained by Mark Dickinson, which Bob Williams has shown at this conference. It resolves a fuzzy ground based picture into many bright clumps aligned along an S shaped structure extending over many tens of kiloparsecs and suggesting the possibility that scattering is occuring in optically thick clouds as large as galaxies. Adaptive optics with the VLT in the near IR will clearly be a very useful complement to the superb angular resolution of the HST in the optical. Even when the HST has an IR camera, the diffraction limit of a single VLT will be more than 3 times better than that of the HST.

5 Concluding Remarks

I hope to have convinced you that, although our eyes are nearly blind to it, polarization contains a wealth of information and, in particular, has the power to disentangle unresolved geometries and mixed spectral components. The AGN environment, being so rich in interstellar medium, anisotropic radiation and non-thermal emission, is full of polarizing phenomena: if we lived there, we would have probably developed some better capability to see polarization. Living on Earth, we are lucky that soon we shall have the VLT to bring us great advantages in polarimetry.

Acknowledgements

I wish to thank George Djorgowski and Hugo Schwarz for allowing me to view their data in advance of publication, Bob Williams for inspiring me by showing Mark Dickinson's HST image of 3C 324 before publication, Bob Fosbury for reading the manuscript and various ESO staff members for providing information on the VLT instruments.

References

Antonucci, R. (1993): ARAA, 31, 473

Chambers, K.C., McCarthy, P.J. (1990): ApJ, 354, L9

Cimatti, A., di Serego Alighieri (1994): MNRAS, submitted

Cimatti, A., di Serego Alighieri, S., Field, G.B., & Fosbury, R.A.E. (1994): ApJ, 422, 562

Cimatti, A., di Serego Alighieri, S., Fosbury, R.A.E., Salvati, M., & Taylor, D. (1993): MNRAS, 264, 421

di Serego Alighieri, S. (1986): in "ESO's Very Large Telescope", ed. by S. D'Odorico & J.-P. Swings (ESO, Garching bei München), p. 173

di Serego Alighieri, S. (1989): in "1st ESO/ST-ECF Data Analysis Workshop", ed. by P. Grosbøl et al. (ESO, Garching bei München), p. 157

di Serego Alighieri, S., Cimatti, A., & Fosbury, R.A.E. (1994): ApJ, 431, 123

Fosbury, R.A.E., di Serego Alighieri, S., Courvoisier, T.J.-L., Snijders, M.A.J., Tadhunter, C.N., Walsh, J., & Wilson, W. (1990) in "Evolution in Astrophysics", ESA SP-310, p. 513

Graham, J.R., Matthews, K., Soifer, B.T., Nelson, J.E., Harrison, W, Jernigan, J.G., Lin, S, Neugebauer, G., Smith, G, & Ziomkowski, C. (1994): ApJ, 420, L5

McCarthy, P.J. (1993): ARAA, 31, 639

Part 8

Quasars

Quasar Surveys and the VLT

Paul C Hewett

Institute of Astronomy, Madingley Road, Cambridge CB3 0HA, UK

Abstract. In the last decade a veritable explosion in the numbers of known quasars has occurred, and much has been learnt about the properties and evolution of the quasar population at radio, optical and X-ray wavelengths. Notwithstanding the increase in our knowledge, there remain significant portions of the luminosity-redshift plane, intrinsically faint quasars at intermediate redshifts $z \sim 2 - 3$ for example, where information remains limited or non-existent. Once suitable quasar samples have been defined the VLT, will be well-suited to the investigation of a number of these areas, while for others, such as luminous quasars at the highest redshifts, this is less true. Prospects for employing the VLT to study several outstanding questions relating to the evolution of the quasar population are discussed. Using quasars as probes of material lying along the line-of-sight will be a key area in the era of 8-metre telescopes. Likely progress towards defining quasar samples, suitable for investigations by the VLT of intervening hydrogen and metal absorption systems, is reviewed briefly.

1 Introduction

Quasar surveys will have a dual role in the era of 8-metre class telescopes: firstly, for experiments driven by a desire to increase our knowledge of the quasar population itself, and secondly, as a means of providing particular objects, or classes of objects, suitable as targets for investigation of other astrophysical phenomena. A number of contributions to these proceedings discuss quasar-related projects for the VLT, prerequisites for which are the availability of samples of candidate, or confirmed, quasars. In the vast majority of such cases, the task of the VLT will be to obtain detailed information, usually spectroscopy of some form, for the targets. The competition for time on the VLT will be such that rarely, if ever, will one have the luxury of employing the VLT itself to identify the quasars, and in any case 8-metre class telescopes are not well suited to identification spectroscopy or imaging of substantial numbers of relatively bright objects spread over large areas of sky. It follows that new samples of quasars, or quasar candidates, will be required in order to undertake certain projects with the VLT and the definition of such samples may require a significant investment of telescope and human resources. Where a proposed VLT project is believed to be sufficiently important, some thought should be devoted to ensuring the prerequisites for the project are completed in good time. Given the projected date of first light for the VLT, such efforts will be required very soon and substantial allocations of time on 2-4-metre class telescopes are necessary.

2 Quasars as Probes

Most notable of the class of studies employing quasars as tracers or probes of other phenomena, is the study of intervening absorption systems. This research area is itself now wide-ranging, encompassing studies of the physical properties and evolution of the Lyman-α forest, higher column density Lyman-limit systems, Mg II $\lambda2798$ and C IV $\lambda1549$ metal systems, and the very highest column density damped Lyman-α systems. As with the evolution of the quasar population itself, the qualitative behaviour and statistical properties of the various classes of absorbing systems are largely established. The emphasis has moved towards understanding the physical conditions and composition of the systems themselves and connections between the different systems and how they relate to the familiar structures, particularly galaxies and their associated halos, visible in the nearby universe. An important recent example of such work is the study of the gas-phase abundances of Zn and Cr in a sample of damped Lyman-α systems by Pettini et al. (1994).

There is no published equivalent of the northern Palomar-Green bright quasar survey (Green, Schmidt & Leibert 1986) in the south and there is a consequent lack of very bright $m \lesssim 17$ quasars suitable for high signal-to-noise ratio, high–resolution studies of intervening hydrogen and metal systems. Several groups are tackling the formidable task of surveying the thousands of square degrees required and preliminary reports of the Edinburgh-Cape UBV survey (Stobie 1994) indicate that a significant number of very bright quasars will be available on the timescale required by the VLT. Such bright, colour-selected samples, employing blue magnitudes, are not particularly effective at locating quasars at redshifts $z > 2$ suitable for studies of the Lyman-α forest and the higher column density systems. However, specific projects to identify such samples are also underway (e.g. Sealey et al. 1994). Thus, it appears likely that a substantial number of targets, for what is one of the key areas for exploitation of the new generation of 8-metre telescopes in extragalactic astronomy, will be available for the VLT.

Quasar samples at fainter magnitudes also offer great potential and Petitjean in these proceedings describes an experiment employing the high surface density of faint quasars to provide three-dimensional information on the clustering of intervening absorption systems, notably those exhibiting either Mg II $\lambda2798$ or C IV $\lambda1549$ absorption. Suitable quasar samples do not yet exist for this type of investigation, which can also be applied to the Lyman-α forest in order to investigate the clustering behaviour of the hydrogen clouds. The intergalactic ionizing flux can also be ascertained from the amplitude and scale of the proximity-effect close to luminous quasars (see Crotts (1989) for a description of a first attempt at such a project). The imminent commissioning of the Anglo Australian Telescope's two-degree diameter, multi-fibre spectrograph (2dF) should provide the means to identify quasar samples for a number of such VLT projects. Several groups are planing to generate very large (10 000 object) quasar samples using United Kingdom Schmidt Telescope plates (or increasingly Kodak Tech Pan films) to perform an ultraviolet excess selection for objects to $m_B \lesssim 21.5$ over an area of some 200 square degrees. Such a project is now quite feasible given the large field, > 3 square degrees, and impressive number of fibres, 400, feeding the 2dF spectrographs; surface densities of ~ 50 objects per square degree for quasars with redshifts $0.3 \lesssim z \lesssim 2.2$ should be obtained. The primary goal of the 2dF projects is an improved knowledge of the spatial clustering of quasars themselves; the resulting increased numbers of quasar-quasar pairs with separations $\lesssim 30h^{-1}$ Mpc should produce radical improvements in our knowledge of the strength of quasar-quasar clustering and its evolution with redshift. The typical quasar to quasar separation of ~ 10 arcmin provides probes separated by $5 - 10$ Mpc in comoving coordinates for reasonable cosmologies[1] and over a wide range in redshift. The VLT will have a unique opportunity to exploit the resulting catalogues to perform projects of the type described by Petitjean.

Significant advances towards providing samples of quasars relevant to a number of important VLT projects can be expected prior to first light.

3 The Quasar Population

Advances in our understanding of the behaviour of the quasar population as a function of lookback time have been considerable over the last five to ten years. This is due in large part to a series of projects to define samples of quasars and active galactic nuclei (AGN) suitable for statistical analysis, and prominent examples include the Palomar-Green survey (Green et al. 1986), the Durham ultraviolet excess survey (Boyle et al. 1990), the Einstein Medium Sensitivity Survey (Gioia et al. 1990) and the Parkes Selected Region investigations (Dunlop & Peacock 1990)). Boyle (1993) provides a succinct recent review of progress.

Locating known quasars in a plot of absolute magnitude versus redshift, or equivalently luminosity versus lookback time, illustrates the strengths and limitations of existing surveys in the optical regime. Figure 1 shows such a plot and includes quasar samples published prior to July 1992; the situation has not changed significantly since that time. There are more than 2000 quasars plotted and details of the specific surveys included can be found in Table 1 of Boyle (1993).

The direct correlation between redshift and absolute magnitude combined with the steep intrinsic luminosity function of quasars means that flux limited surveys produce samples of objects that populate a distinctive swathe-like region of the figure - the > 1050 quasars making up the Large Bright Quasar Survey (Hewett, Foltz & Chaffee 1993) for example, form the densely populated band running from $M_B = -22$, $z = 0.2$ to $M_B = -28$, $z = 3$. Such samples contain quasars with a very restricted range of luminosity at fixed redshift and, for any significant spread in redshift, the luminosity ranges become disjoint.

In some regimes the lack of large well-defined samples is not amenable to solution, for example the dearth of very luminous quasars with small redshifts

---

[1] A cosmology with $q_0 = 0.5$, $H_0 = 50 \, \mathrm{km s^{-1} Mpc^{-1}}$ and $\Lambda = 0$ is employed throughout, unless otherwise stated

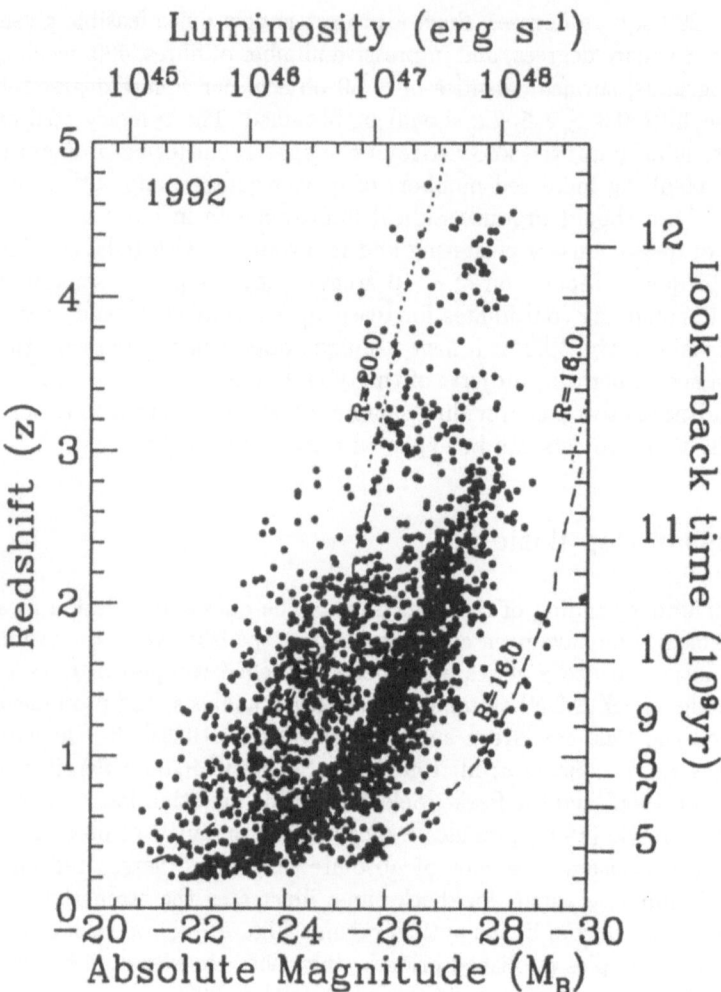

Fig. 1. Absolute magnitude versus redshift plot showing optically selected quasars i-
dentified in the compilation of surveys by Boyle (1993). Lookback time and approximate
bolometric luminosities are also indicated. Loci corresponding to apparent magnitudes
$m_B = 16.0$ and 20.0 (for $z < 2.2$) and $m_R = 16.0$ and 20.0 (for $z > 2.2$) are shown.

arises primarily because of the small volume in which such objects could be
found. Quasars from the most extensive survey at bright apparent magnitudes,
the Palomar-Green survey, can be seen hugging the dashed line that corresponds
to an apparent magnitude limit of $m_B = 16$. While further objects will be added
to this portion of the plot from the work of Stobie (1994) and others, the small
number of such objects over the whole sky sets a hard constraint.

Setting aside such fundamental limitations the observational goal is to populate the plane as extensively as possible and this process in practice involves locating fainter quasars and also quasars with higher redshifts - fainter and further. The successful prosecution of such observational projects allows the quasar luminosity function to be determined over a substantial dynamic range in luminosity at fixed redshift - a horizontal cut through Figure 1, and the intercomparison of quasars of similar luminosity over an extended range in lookback time - a vertical cut through the figure. This may be considered as stating the obvious, but the difficulties in building a more complete picture of the evolution of the quasar population, based on samples of objects from flux-limited surveys, cannot be stressed too highly.

4 The Highest Redshifts: $z > 5$

Identification of quasars with $z > 5$ will allow studies of intervening absorption systems to be extended to earlier epochs, constrain theories for galaxy formation by requiring that they explain the presence of massive bound objects at early epochs, and provide information on the source and form of the background ionizing radiation field. Prospects for the direct involvement of the VLT in the discovery of samples of $z > 5$ quasars are not good, but interest in pushing the redshift limit ever further appears undiminished and it is worthwhile outlining some of the requirements for successful surveys. The earlier search for quasars at redshifts $z > 4$ undertaken in the late 1980s can be regarded as something of a success in that specific searches employing a variety of techniques (e.g. Warren et al. 1991, Schneider et al. 1994, Hook 1994) have generated well-defined samples of quasars, the analysis of which has provided quantitative estimates of the behaviour of the quasar population in a regime where little was known hitherto (e.g. Warren, Hewett & Osmer 1994).

This success has wider implications in that the $z > 4$ surveys are the first in the quasar field to tackle what is not only a very difficult observational problem successfully, in the sense of locating examples of a (very rare) population, but more importantly they involved a considerable degree of what in physics research would be termed experimental design. An essential feature of good experimental design is the ability to translate the raw observational results, in this case the detection of quasars over redshifts to $z \lesssim 4$, into quantitative information on the physical properties under investigation, here the comoving space density as a function of lookback time and the luminosity function of the quasars within the volumes surveyed. This work has been criticised by some on the basis that the procedures for performing the translation, a major portion of which involves the precise determination of a selection function, may be complicated and time consuming. However, such an assessment misses the key point, that the ability to perform the translation is a prerequisite for constraining the physical properties under investigation, and also assumes that, because the procedure is complex, it is inherently unsound. The use of modern digital detectors and automated selection techniques means that, while the relation between the raw results and

the physical quantities of interest may involve numerous steps, the entire procedure can be quantified precisely and realistic error estimates on results obtained.
In contrast, the literature abounds with the results of surveys for quasars that
have employed very large quantities of time on the largest telescopes but where,
notwithstanding claims made in the papers, the ability to perform a quantitative transformation of the raw observations into physically interesting results is
lacking.

The success in the $z > 4$ regime bodes well for attacks on the $z > 5$ problem.
However, at first sight, the direct involvement of the VLT in discovery of such
objects does not appear likely. In common with all the planned 8-metre class
telescopes the VLT is not a wide-field instrument, at least not in the context
of searches for distant quasars. The comoving space density of quasars at high
redshift is low and the volume per unit redshift at $z > 3$ is small and decreases
further still as redshift increases. Optimising the chances of success thus requires
surveying very large areas, to increase the volume, or probing to the faintest possible apparent magnitudes, to maximise the space density of the objects sought
by including as much of the luminosity function as possible.

Table 1a. Survey Area ($^{\circ 2}$) Required to Locate One Quasar $5 < z < 6$, $q_0 = 0.1$

	A	B	C
18	—	1800	6500
19	3600	80	140
20	570	13	5
21	160	3	0.5

Table 1b. Survey Area ($^{\circ 2}$) Required to Locate One Quasar $5 < z < 6$, $q_0 = 0.5$

	A	B	C
18	—	1100	1400
19	3300	70	40
20	650	14	2
21	200	4	0.4

The scale of the task in acquiring a sample of $z > 5$ quasars is evident from the
predictions of their surface density on the sky for several plausible models of the
behaviour of the luminosity function. Tables 1a and 1b tabulate the area of sky,
in square degrees, containing one quasar with redshift $5 \leq z \leq 6$ to a specified
I-band magnitude limit, for three different models of the evolution of the quasar
luminosity function. Model "A" is that derived by Warren et al. (1994) based
on their multicolour survey for quasars in the redshift range $2 \leq z \leq 4.5$ and

corresponds to a substantial decline in the comoving space density of quasars at redshifts $z > 3.5$. Model "B" is also derived from Warren et al. by setting the decline in space density at the 2σ upper limit to their best-fit model, i.e. the decline is approximately the minimum allowed that gives consistency with the data. Model "C" corresponds to a fixed luminosity function with constant space density for all redshifts $z > 2.1$ and thus gives the most optimistic outcome at the fainter magnitudes of the three models.

Unless one of the key assumptions is grossly in error, then the prospects of acquiring a sample of any size from observations over small fields of view is poor. Extrapolating the models fainter is of dubious worth given the lack of constraints, but, unless the faint portion of the quasar luminosity function becomes much steeper at high redshift, increasing the depth of any survey will produce an increase in surface density of only a few per magnitude. At this conference, Djorgovski reported preliminary results of deep imaging from the Keck telescope around known bright $z > 4$ quasars, looking for objects of similar colours to the more luminous confirmed quasars. His results support the assumptions behind these predictions, i.e. no evidence for a rapid rise in the number of quasars with increased depth was found.

While stressing the difficulties posed by the low surface density of the target population there are a variety of techniques that can be employed to identify $z > 5$ quasars. One already employed successfully by Hook and McMahon (Hook 1994) at redshifts $z > 4$, employs an optical "filter" applied to a radio catalogue to target the small subset with the optical properties expected of high-redshift quasars. Shaver, Wall and collaborators are pursuing a similar strategy to identify $z > 5$ quasars. In outline the technique involves taking a catalogue of flat spectrum radio-sources, likely to be dominated by quasars, and examining the optical colours of the sources - with high precision radio positions and the availability of all-sky digital optical catalogues, this is now a relatively standard procedure. The effect of the dramatic decrease in flux caused by the cumulative effect of the Lyman-α forest and Lyman-limit systems means that the B and V magnitudes of even the most luminous $z > 5$ quasars will be extremely faint. Eliminating all sources visible on the blue Schmidt sky surveys immediately reduces the number of candidates substantially. CCD observations to provide a colour sensitive to the spectral discontinuity across the Lyman-α line (e.g. $B - I$) allows a large fraction of the remaining candidates to be eliminated. The final candidate list consists of a group of stellar objects with very red $B - I$ (or similar) colours. The vast majority of these are optically faint and spectroscopy in the far-red $\lambda\lambda 7000 - 9500$ is required for an identification. The strength and variability of the night-sky emission makes optical spectroscopy in the far red notoriously difficult. The VLT's combination of increased aperture and excellent image quality, allowing the use of a small slit, should enable follow-up spectroscopy with FORS to be undertaken some two magnitudes fainter than possible with existing 4-metre telescope/instrument combinations.

5 The Faintest Quasars

There is an almost complete lack of known quasars at redshifts $z > 3$ and fainter than $M_B = -26$ - Figure 1. More generally, with the exception of the Durham ultraviolet excess survey of Boyle et al. (1990), which extends to $m_B \sim 21$, there are few samples that provide significant numbers of objects at apparent magnitudes $m > 20$. Consequently, at redshifts $z > 3$ we are still confined to studies of the brightest quasars, and even for redshifts $z \gtrsim 1.5$ there is little information available on objects which populate the luminosity function close to the (conventionally adopted) transition luminosity between quasars, Seyfert galaxies and other AGN. Studies of the far more numerous population of lower luminosity quasars and brighter AGN at significant redshifts will have direct relevance to understanding the origin of the X-ray background, unified schemes for AGN and possibly the relation between galaxy interactions, star-formation and nuclear activity. Scientific aspects of such a study, with particular relevance to the X-ray background, are discussed in more detail in the contribution by Zamorani.

The quasar luminosity function has a distinctive two-power law form (see Boyle 1993) with a very steep increase in the number of objects with decreasing luminosity at bright absolute magnitudes, and a transition to a much shallower slope at fainter absolute magnitudes. Flux limited surveys, that probe the steep bright portion of the function at the redshifts of interest, thus benefit dramatically (in terms of the number of quasars detected), by pushing the detection limit fainter. However, for the absolute magnitudes and redshifts of interest in studying intrinsically fainter quasars, the surveys probe the shallower portion of the luminosity function, and the increase in numbers as the flux limit is decreased is modest - factor ~ 2 per magnitude - and large survey areas are required.

The VLT MFAS facility is capable of acquiring identification spectra of a sample of quasars to $m_R \sim 23$ and FORS will be available to reobserve objects with weak features that can not be identified in the necessarily rather short MFAS exposures. MFAS spectroscopy of ~ 1 square degree per night should be achievable and thus a survey of an area of $10 - 20$ square degrees is not an unreasonable goal - achieving the order of magnitude improvement in numbers over existing (and ongoing) surveys to this magnitude limit in a regime of great scientific interest.

A variety of quite efficient selection techniques are available to identify quasar candidates from normal Galactic stars, and at brighter magnitudes $m_R \lesssim 20$ these work well. At $m_R = 23$ the problem is more difficult because the source counts are dominated by galaxies which possess a wide range of spectral energy distributions and the presence of significant numbers of low metallicity halo subdwarfs broadens the range of properties exhibited by Galactic stars. Consider the requirements for a broadband multicolour selection in $BVRI$, and possibly U. To achieve sensitivity to quasars with a wide range of spectral energy distributions and redshifts, while avoiding the candidate list becoming swamped by Galactic stars and galaxies, high precision magnitudes are required and good seeing is necessary to eliminate a large fraction of the compact galaxies (e.g.

elliptical galaxies at redshifts $z > 0.3$). To generate an object catalogue over $10 - 20$ square degrees to $m_R = 23$ in (at least) four colours, with magnitude errors ≤ 0.05, and seeing conditions of $\lesssim 1$ arcsec, to allow morphological classification adequate to eliminate distant compact galaxies, is a major undertaking. ESO has the facilities to obtain such a data set, which need not occupy a contiguous area of sky and which would have substantial benefits for many other scientific programs, but the time allocation on 2 or 4-metre telescopes, equipped with a 2048×2048 CCD providing adequate spatial sampling in conditions of *good* seeing, are substantial.

6 Conclusions

This contribution has concentrated deliberately on a few areas in which there is considerable current interest, with the aim of highlighting some of the survey requirements necessary to undertake various VLT projects. The outlook for the provision of samples for undertaking projects of intervening absorbers is relatively good. On the other hand, to enable real advances to be made in understanding the evolution of intrinsically faint quasars, the prospects are not so good. Surveys of small areas of sky, enabling VLT projects of limited scope to be undertaken, will doubtless be completed in the near future. However, to provide the source material for projects that gain the order of magnitude improvement over existing work, that one hopes the VLT will facilitate, requires substantial investment of 2 to 4-metre telescope time. As first light approaches, the more integrated use of existing ESO telescopes, to allow major projects to be undertaken, is worth serious consideration.

Acknowledgements

I am most grateful to Stephen Warren, for providing the information contained in Table 1, and to Brian Boyle for making Figure 1 available.

References

Doyle, D.J. (1993): in *The Evolution of Galaxies and their Environment*, 3rd Teton Summer School, ed. H. Thronson and J.M. Shull (Dordrecht, Kluwer), p. 433
Boyle, B.J., Fong, R., Shanks, T., Peterson, B.A. (1990): MNRAS **243** 1
Crotts, A.P.S. 1989: ApJ **336** 550
Dunlop, J.S., Peacock, J.A. (1990): MNRAS **247** 19
Gioia, I.M., Maccacaro, T., Schild, R.E., Wolter, A., Stocke, J.T., Morris, S.L. Henry, J.P. (1990): ApJS **72** 567
Green, R.F., Schmidt, M., Leibert, J. (1986): ApJS **61** 305
Hewett, P.C., Foltz, C.B., Chaffee, F.H. (1993): ApJ **406** L43
Hook, I.M. (1994): Ph.D. Thesis, University of Cambridge
Pettini, M., Smith, L.J., Hunstead, R.W., King, D.L. (1994): ApJ **426** 79

Schneider, D.P., Schmidt, M., Gunn, J.E. (1994): AJ **107** 1245

Sealey, K, Drinkwater, M.J., Webb, J. (1994): in The Future Utilisation of Schmidt Telescopes, IAU Colloquium 148, (Dordrecht, Kluwer), in press

Stobie, R.S. (1994): in The Future Utilisation of Schmidt Telescopes, IAU Colloquium 148, (Dordrecht, Kluwer), in press

Warren, S.J., Hewett, P.C., Osmer, P.S., Irwin, M.J. (1991): ApJS **76** 1

Warren, S.J., Hewett, P.C., Osmer, P.S. (1994): ApJ **421** 412

QSO Absorption Line Systems

Patrick Petitjean

Institut d'Astrophysique de Paris, 98bis Boulevard Arago, F75014 Paris, France

Abstract. 10 m class telescopes and their instrumentation will greatly benefit the study of QSO absorption line systems. Two main subjects are discussed: the detection of the associated galaxies; and the determination of absorption line system physical properties (kinematics, physical state, abundances) and their evolution. The possibility of studying large scale structures in the universe using absorption line systems is also discussed. This latter project will only find a complete development when the VLT comes into operation.

1 Introduction

Although steady progress has been made towards detecting fainter objects in emission, the high redshift objects detected in this way up to now are drawn from a particular population of powerful emitters. On the contrary, absorption may reveal any standard object such as a normal galaxy or intergalactic gaseous clouds close enough to the line of sight to the QSO.

Absorption line systems observed in QSO spectra are generally divided into three categories:

(1) the heavy–element systems in which a large number of elements, in different ionization stages, are observed, from CI to CIV or OI to OVI. They are most certainly closely related, at any redshift, to galaxies. The latter have been successfully detected by direct imaging and follow-up spectroscopy at low or intermediate redshift (Bergeron & Boissé 1991; Bergeron et al. 1992; Steidel 1993). Studying the evolution in redshift of their number density and physical properties (kinematics, ionization state, abundances) is then a unique tool to trace galaxy formation;

(2) the Lyα lines with no detectable metal lines at the same redshift, are most certainly of intergalactic origin at high redshift (e.g. Sargent et al. 1980) but could somehow be associated with galaxies at low redshift (Lanzetta et al. 1994). Information on these systems (column density, Doppler width, kinematics, clustering) is all derived from determination of the profile of the line. Fitting of the latter requires high resolution, high quality data which are difficult to obtain on 4m class telescopes and some controversy has aroused about the results (Pettini et al. 1990, Rauch et al. 1993). Although much attention has been dedicated to observing and modelling of these lines, no clear picture of what their structure might be has emerged;

(3) the broad absorption line (BAL) systems are characterized by impressive absorption troughs from different ions of low and high excitation, extending from 0 up to 60000 km s^{-1} outflow velocity relative to the emission redshift of

the QSO. It is widely accepted that the gas is very close to the centre and may be part of the broad emission line region. They are thus intimately related to the AGN phenomenon.

To emphasize the invaluable benefit of using 10 m class telescopes and its instrumentation in this field, I have chosen to explore two directions which are of first interest for the near future: the detection and study of the associated galaxies, the detailed study (kinematics, physical state) of absorption line systems and their evolution. I will also discuss the possibility of studying large scale structures in the universe using absorption line systems.

2. Detection of the Objects Responsible for the Absorption

Search for galaxies associated with low z (~ 0.6) MgII systems has started a long time ago (Weymann et al. 1978; Carswell et al. 1984), but it is only with the advent of sensitive detectors that positive detections have been reported (Bergeron 1986). About fifty galaxy/MgII system pairs are known by now (Bergeron & Boissé 1991; Steidel 1993). The corresponding sample is unique in the sense that information on both the stellar content of the galaxy and the extent of the gaseous halo is known. From the observations, it has been shown that (i) the cross-section of the halos does not evolve in time and is of the order of $35h_{100}^{-1}(L/L_*)^{0.2}$ kpc; (ii) the galaxies are fairly bright ($L > 0.25L_*$) and the luminosity distribution at $z \sim 0.6$ is consistent with a mild evolution (by one magnitude) for the present day luminosity distribution.

Very recently UV observations with HST have revealed the presence of low z metal poor Lyα systems (Morris et al. 1991, Bahcall et al. 1991). Surprizingly their number density is about five times what is expected from extrapolation of what is known about high z Lyα systems. Moreover, Lanzetta et al. (1994) have claimed that they may be associated with galaxies. From imaging of the field around quasars previously observed with HST, and spectroscopic follow-up of the field galaxies, the latter authors have found nine galaxies at the same redshift as Lyα-only systems and with a separation between the line-of-sight to the quasar and the galaxy less than 150 h_{100}^{-1} kpc.

The nature of these absorption lines and their relation to galaxies is not that clear however: it is difficult to imagine how such huge halos can be maintained; it is possible also that the absorption arises through gas in much fainter objects, undetectable with present day instrumentation (Yanny et al. 1990, Philipps et al. 1992). To have a good understanding of what are the structure and properties of the absorbers, one needs to perform deeper imaging of the field around QSOs and, more importantly, to do multi-object spectroscopy up to magnitudes comparable to those of the blue galaxy population ($B \sim 24$–26). This is typically what can be achieved with FORS. These studies will also answer the question of what is the probability that a line of sight goes through a galactic halo without any detectable absorption.

At higher redshift ($z > 1$), two absorbing galaxies have been identified in the optical (Bergeron et al. 1992). It is clear however that these searches should be done in the infrared (Aragón–Salamanca et al. 1994). The future availability of good infrared capabilities and in particular ISAAC will revolutionize this domain. It must be noted also that most of these QSO fields will be observable with the adaptive optics system and CONICA, giving the exciting opportunity to study the morphology of the absorbing galaxies (see Petitjean et al. this volume).

3. Properties of the Absorbing Gas

Increasing the spectral resolution and S/N ratio of data using VLT will result in better determination of the width, the profile and the depth of the lines. This will give information about the temperature, kinematics and column densities, hence abundances, of the gas. I give below three examples showing the limits of what can be done with 4 m class telescopes.

3.1 The CIV Forest

One obvious task for VLT is to decrease the equivalent width limit for detection of weak lines. A crucial question for the discussion of the origin of the Lyα forest is that of the metal content; whether the gas is primordial or not and in this case what is the fraction of systems with metals. To address this question, one has to look for the CIV counterpart of the Lyα forest. In Fig. 1 are shown simulated spectra of S/N ratio infinite, 250 and 150 from top to bottom respectively and a spectral resolution $R \sim 10000$. The CIV lines correspond to Lyα systems at redshift $z \sim 3.5$ drawn at random from a population of clouds with a number density per unit redshift taken as $dN/dz = 4.5(1+z)^{2.5}$, a neutral hydrogen column density distribution $\propto N(\text{HI})^{-1.5}$, abundances 10^{-3} the solar value (e.g. Lu & Savage 1993 and references therein), HI/H $= 10^{-5}$ and CIV/C $= 1$. The latter assumption should be valid at high redshift where a break at 54.4 eV is expected in the ionizing spectrum, due to absorption by intervening HeII. The S/N ratios given above are thus a lower limit for detection of what could be called the CIV forest.

3.2 Physical Conditions in BAL

Broad absorption lines (BAL) are observed in 12% of the QSOs (Foltz et al. 1990). The absorption generally starts near the emission redshift of the QSO and extends up to 60000 km s^{-1} outflow velocities. Although the origin of the gas and the ejection mechanism are unclear, there is some evidence that this gas lies very close to the nucleus.

In the best data available at present (Wampler et al. 1994; see Fig. 2), structures are seen in the CIV, AlIII, AlII, SiII, SiIII, FeII and FeIII broad absorptions. Several distinct regions of very similar ionization state but different velocities

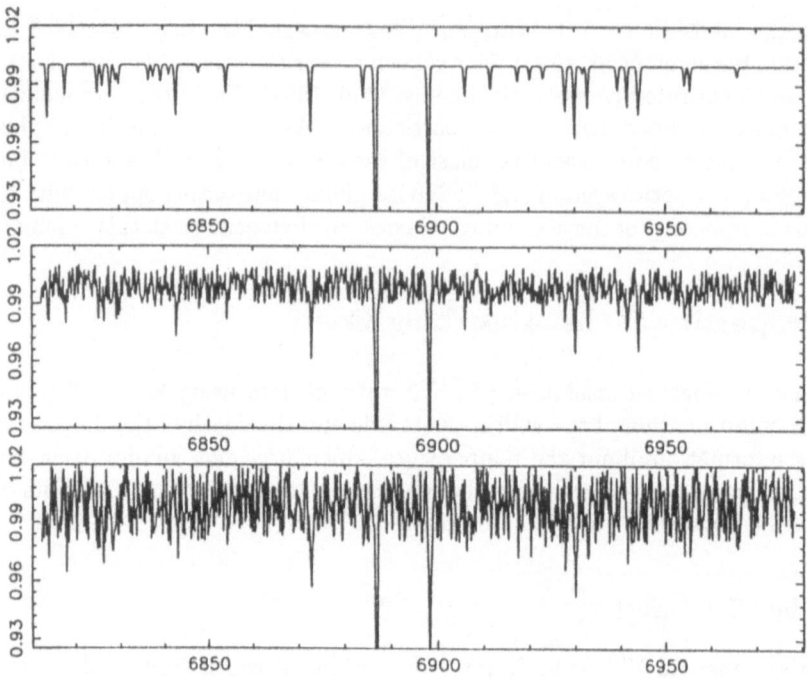

Fig. 1. Simulated data with S/N ratio infinite (top), 250 (middle) and 150 (bottom) and resolution $FWHM = 0.5$ Å showing what could be the detection of the CIV forest.

are present. Detailed photoionization modelling suggests that the ionization parameter (ratio of the density of ionizing photons to the gaseous density) is very high; that the carbon abundance is about solar and that iron may be enhanced by possibly a factor of ten.

Moreover, narrow ($b \sim 20$ km s^{-1}) FeII absorption lines are present, produced by gas at low temperature ($T < 10^4$ K) and with a small amount of turbulent motion, embedded in the BAL flow. The fact that the FeII lines do not go to the zero continuum level implies that the corresponding clouds do not cover the continuum source and should be very small. This is consistent with the high density ($n > 10^6$ cm^{-3}) needed to explain collisional excitation of atomic levels up to 4.5 eV above the ground state from which absorption lines are detected.

It is clear from this that there may exist some link between the BAL gas and the broad emission line region. In particular ionization parameters and abundances seem to be very similar (Hamann & Ferland 1993). The detailed study of BAL objects is thus a unique tool to investigate the very center of QSOs. Moreover BAL QSO's are only found in radio–quiet objects, whereas associated systems of high abundances and high ionization (Petitjean et al. 1994) are seen both in radio–quiet and radio–loud objects. Both types of systems must be studied to understand the fundamental reason why 10% of the QSOs are radio-loud.

Such data as those shown in Fig. 2 are very difficult to obtain with 4m class telescopes since the QSOs are faint ($m_V > 17$). A somewhat higher resolution than that foreseen for UVES is needed to definitively resolve the structures ($R \sim 3\ 10^5$).

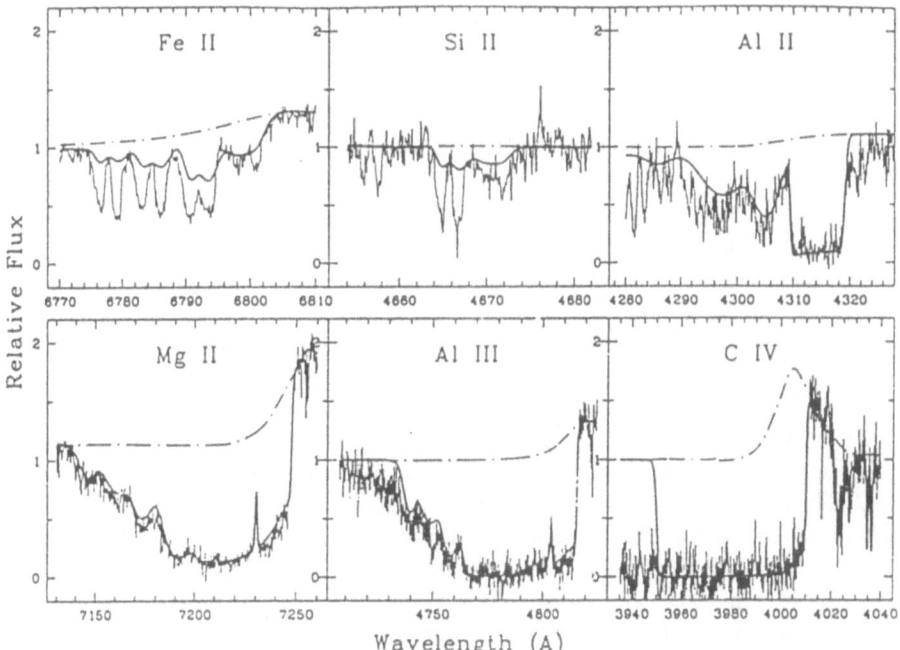

Fig. 2. Broad absorption lines in Q0059–2735. The best fit is overplotted together with the assumed continuum. Note that the wavelength scale is not the same for all spectra. Additional narrow FeII lines can be seen.

3.3 Abundance Determination

The determination of metallicities in absorption line systems is a difficult task since one needs good determination of the column densities of HI and several ionization stages of different elements to perform the ionization correction. It is therefore imperative to carefully select the systems in which the lines are neither strongly blended nor badly saturated.

In case the HI column density is very large as in the damped systems, hydrogen can be considered neutral and the ionization correction is negligible. Moreover, most of the elements are neutral or in their singly ionized state. These dominent ions cannot be used however since their lines are heavily saturated and one has to search for weak lines of ZnII and CrII (Meyer and York 1987, Pettini et al. 1990, Bergeron & Petitjean 1991). The abundances then derived spread a large range from 10^{-3} $Z_\odot$ to 1 $Z_\odot$ (Pettini et al. 1994). In the case when the HI column density is not that large, a photoionization code can be used. Abundances of the order of $0.01 Z_\odot$ have been found by Steidel (1990) in individual LLS systems at $z > 3$; Bergeron & Stasińska (1986) and Petitjean et al. (1992) derived statistically abundances larger by about a factor of 3 at somewhat smaller redshift ($z \sim 2.5$).

In a recent work, Petitjean et al. (1994) have used high quality high resolution ($R = 15000$) data to derive abundances in nine systems observed in two QSOs Q0424-131 and Q0450-131. The HI, CIV, NV and SiIV absorption spectra of the $z_{abs} = 2.231$ system in Q0450-131 are shown in Fig. 3. The best fit is overplotted as a solid line. The blue wing of the HI absorption is so steep that it strongly constrains the HI column density. The CIV and NV lines need two components whereas the SiIV line has only one. The zero level has been adjusted since it is clear, even from the 2D spectrum, that the lines although saturated do not go to zero. The Doppler parameters are taken as $b = b_{turb} + b_{th}$ where the first term represents the turbulent motions in the gas and is the same for every line; the second term represents the thermal broadening consistently tied to the temperature given by photoionization models constructed for that particular system. This method must be used since, for abundances larger than solar, the temperature is very sensitive to the abundances. In this system, carbon is found to have solar abundance and nitrogen and silicon to be enhanced by a factor ten with respect to carbon. The results for the nine systems are shown in Fig. 4 where the logarithm of the carbon abundance relative to solar is plotted versus the velocity difference of the system and the QSO. It can be seen that the systems within 10000 km s^{-1} from the QSO have abundance in excess of solar and thus are probably associated with the QSO. The other systems have typical abundances of Lyman limit intervening systems of the order of 10^{-2} solar.

As for the work presented in the previous section, this kind of data typically need 12 hours integration time with a 4m class telescope. Although the data are good to do some work it is clear that higher resolution is needed to resolve the lines and to do a more systematic study. UVES with resolution $R \sim 2\ 10^5$ should boost the activity in this field.

4 Large Scale Structures

The space distribution of the absorbers can be investigated on any scale and at any redshift and two approaches can be considered. On the one hand, spectra of QSOs with various projected separations can be searched for occurences of absorption lines at the same redshift. For small separations (< 100 kpc), a striking similarity between Lyα absorptions along different lines of sight is observed (Smette et al. 1991). It has been shown that the transversal dimensions of the absorbers giving rise to the Lyα forest are larger than 50 kpc. Few detections have been claimed for separations of the order of 1 Mpc (Robertson et al. 1986;

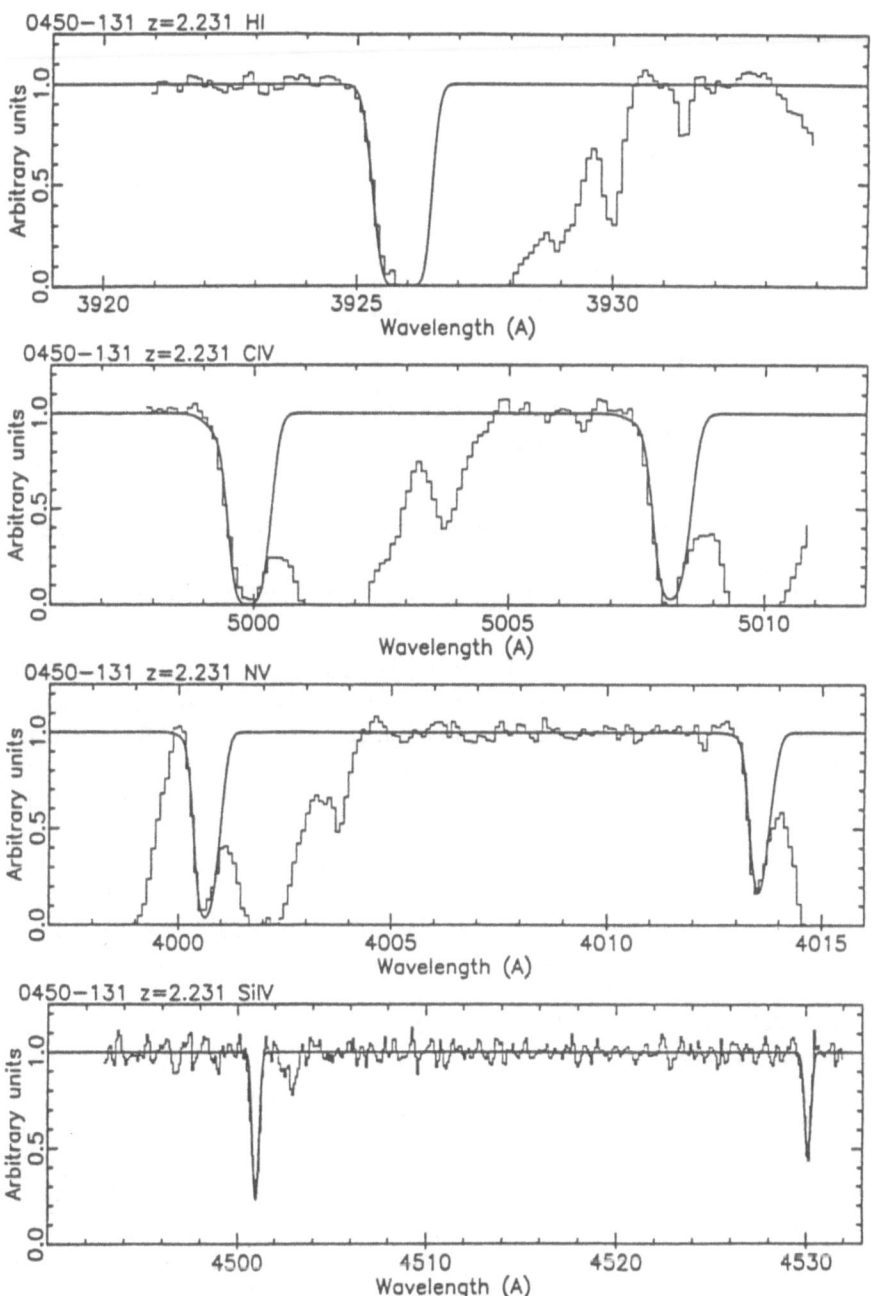

Fig. 3. Model fits of HI, CIV, NV and SiIV transitions of the $z = 2.2302$ systems in Q0450–131

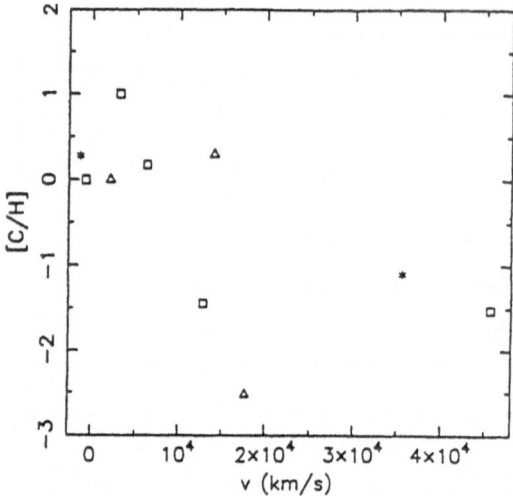

Fig. 4. Logarithm of the carbon abundance relative to solar versus ejection velocity from the QSO. Squares and triangles are for systems in PKS0424–131 and Q0424–131 (Petitjean et al. 1994), stars for systems in Q000–2619 (Savaglio et al. 1994)

Jakobsen et al. 1986). On the other hand, large samples of absorption line systems can be searched for 3D clustering. York et al. (1991) tentatively concluded that there is little evidence for clustering on scales of 300-600 Mpc but that a much larger sample was needed. Using a sample of 268 MgII systems distributed over 60 % of the sky in the redshift range 0.1-2.0, Tytler et al. (1993) found no significant periodicity on any scale from 10 to 210 h^{-1} Mpc.

The studies done up to now in this field are tentative since the amount of data needed is much larger than that yet available. MFAS or FORS2 attached to VLT will routinely allow simultaneous observations of QSOs up to magnitude 21. At this level, about ten high redshift QSOs are expected per square degree (Hartwick & Shade 1990). At $z \sim 2$, the resulting mean separation is about 10 Mpc. Of course deeper observations will be performed in peculiar fields to probe smaller scales.

To investigate how the gas responsible for the absorptions might trace the potential field and its structures and how the gas may be related to galaxies, we have modified a N–body simulation code so that it is possible to follow the thermal history of the gas, including cooling and photoionization by the UV background (Petitjean, Mücket & Kates 1994). If we assume that structures in the Universe form under the action of gravitation dominated by dark matter, we expect part of the baryonic matter to have collapsed in the center of the deepest potential wells where, as a consequence, stars form. However the presence of the ionizing background prevents the gas from cooling too fast when away from the highest density regions (Efstathiou 1992). We expect this gas to be orbiting in the potential wells of filamentary structures in the dark matter distribution. A slice $25 \times 25 \times 2$ Mpc^3 of the simulation is shown in Fig. 5. The redshift is $z = 0.5$. The dark matter, represented by dots, shows the usual structures, dense haloes

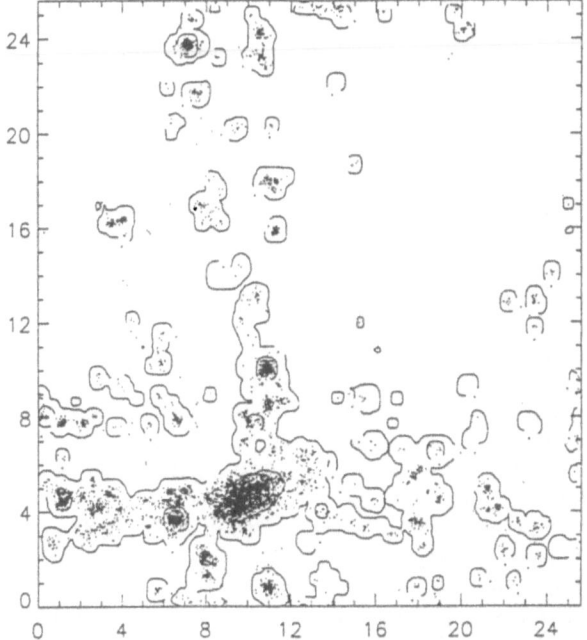

Fig. 5. $25 \times 25 \times 2$ Mpc^3 slide at $z = 0.5$ of a N-body simulation where dark matter (represented by dots) shows the usual structures of dense haloes connected by filaments. Overplotted are the contours for HI column density larger than 10^{17} and 10^{14} cm^{-2}, suggesting the existence of a pervasive inhomogeneous medium with irregular boundaries.

connected by filaments. Overplotted are the contours for HI column densities larger than 10^{17} and 10^{14} cm^{-2}. It is clear that: (i) the medium is smoothly distributed (although our resolution is somewhat small) and scales of 50 kpc to 5 Mpc should be observationally probed in order to have a good representation of its structure; (ii) Lyα gas can be found from time to time in the vicinity of a galaxy but not as a general rule.

References

Aragón–Salamanca, A. (1994): ApJ 421, 27

Bahcall, J.N., Januzzi, B.T., Schneider, D.P., Hartig, G.F., Bohlin, R., Junkkarinen, V. (1991): ApJL 377, 5

Bergeron, J. (1986): A&AL 155, L8

Bergeron, J., Boissé, P. (1991): A&A 243, 344

Bergeron, J., Cristiani, S., Shaver, P. (1992): A&A 257, 417

Bergeron, J., Petitjean, P. (1991): A&A 241, 365

Bergeron, J., Stasińska, G. (1986): A&A 169, 1

Carswell R.F., Morton, D.C., Smith, M.G., Stockton, A.N., Turnshek, D.A., Weymann, R.J. (1984): ApJ 278, 486

Efstathiou (1992): MNRAS 256, 43P

Foltz, C.B., et al. (1990): BAAS 2, 806

Hamann, F., Ferland, G. (1992): ApJ 391, L53

Hartwick, F.D.A., Shade, D. (1990): Ann. Rev. Astron. Astroph. 28, 437

Jakobsen et al. (1986): ApJ 303, L27

Lanzetta, K.M., Bowen, D.V., Tytler, D., Webb, J.K. (1994): preprint

Lu, L., Savage, B.D. (1993): ApJ 403, 127

Meyer, D.M., York, D. (1987): ApJL 319, L45

Morris, S.L., Weymann, R.J., Savage, B.D., Gilliland, R.L. (1991): ApJL 377, 21

Petitjean, P., Bergeron, J., Puget, J.L. (1992): A&A 265, 375

Petitjean, P., Mücket, J., Kates, R. (1994): in preparation

Petitjean, P., Rauch, M., Carswell, R.F. (1994): A&A in press

Petitjean, P., Théodore, B., Hubin, N. (1994): this volume

Pettini, M., et al. (1994): preprint

Pettini, M., Boksenberg, A., Hunstead, R.W. (1990): ApJ 348, 48

Pettini, M., Hunstead, R.W., Smith, L.J., Mar, D.P. (1990): MNRAS 246, 545

Philipps, S., Disney, M.J., Davies, J.I. (1993): MNRAS 260, 453

Rauch, M., Carswell, R.F., Webb, J.K., Weymann, R.J. (1993): MNRAS 260, 589

Robertson et al. (1986): MNRAS 219, 403

Sargent, W.L.W., Young, P.J., Boksenberg, A., Tytler, D. (1980): ApJS 42, 41

Savaglio, S., D'Odorico, S., Møller, P. (1994): A&A 281, 331

Smette, A., et al. (1991): ApJ 389, 39

Steidel, C.C. (1993): "The Environment and Evolution of Galaxies", in Proc. of the
    Third Tetons Summer School, ed. by J.M. Shull and H.A. Thronson Jr. (Kluwer,
    Dordrecht), pp. 263–293

Steidel, C.C. (1990): ApJS 74, 37

Tytler et al. (1993): ApJ 405, 57

Wampler, E.J., Chugai, N.N., Petitjean, P. (1994): ApJ in press

Weymann, R.J., Boroson, T.A., Peterson, B.M., Butcher, H.R. (1978): ApJ 226, 603

Yanny, B., York, G.G., Williams, T.B. (1990): ApJ 351, 377

York et al. (1991): MNRAS 250, 24

Part 9

The Distant Universe

Observational Cosmology With Faint Galaxies and a (9 ± 1)-meter Telescope

S. G. Djorgovski

Palomar Observatory, Caltech, Pasadena, CA 91125, USA

Abstract. The new generation of 8 to 10-meter class telescopes opens exciting new possibilities for observational cosmology. A brief overview is given of several types of studies of galaxy formation and evolution. Deep galaxy counts and redshift surveys will probe the evolution of normal field galaxies out to redshifts well in excess of 1, and L_* galaxies should be detectable in the infrared out to $z \sim 4 - 5$. It is possible that new types of galaxies or AGN may be found in the deep infrared surveys. Detailed studies of galaxian properties and their correlations (e.g., the fundamental plane correlations for ellipticals) may be possible out to $z \sim 1$. Infrared Hubble diagrams for brightest cluster ellipticals and moderate-power radio galaxies may again become a viable cosmological test. Finally, a population of primeval galaxies undergoing their first major bursts of star formation should be detectable, possibly through their nebular oxygen and hydrogen Balmer line emission, now redshifted to the near infrared.

1   Introduction

We may be on the threshold of a golden era in observational cosmology. A new generation of bigger and better telescope is now under construction. The first Keck 10-m telescope is already producing interesting new results, with the VLTs, Keck-2, Gemini, Subaru, and other 9-m class telescopes soon to follow. Armed with the modern large-format CCDs and infrared (IR) array detectors and a variety of image improvement techniques, and supported by the various space observatories, these new machines will lead us into the unexplored territories.

While virtually every field of astronomy can benefit from these technological advances, perhaps the one field which stands to gain the most is observational cosmology. The cosmological frontier has always been at the faintest reachable flux levels. Our empirical knowledge of galaxy evolution is still very sketchy, and our knowledge of galaxy formation almost nil. There is more to this than a simple increase in the collecting area. Large telescopes have a particular advantage in the IR, since the image diameters improve faster at longer wavelengths, and the Adaptive Optics (AO) is easier to implement. IR studies will be especially valuable for cosmology, in part simply because we know relatively little about the deep IR sky. There are also astrophysical reasons: redshifting of the energy emitted by stellar photospheres to the longer wavelengths; a lower sensitivity to dust extinction (this may be critical for the emission lines used as tracers of star formation, e.g., Lyα vs. Hα); and a lower sensitivity to a flicker of star formation which may dominate the light on the shorter wavelengths, while involving only a miniscule fraction of the galaxian stellar mass.

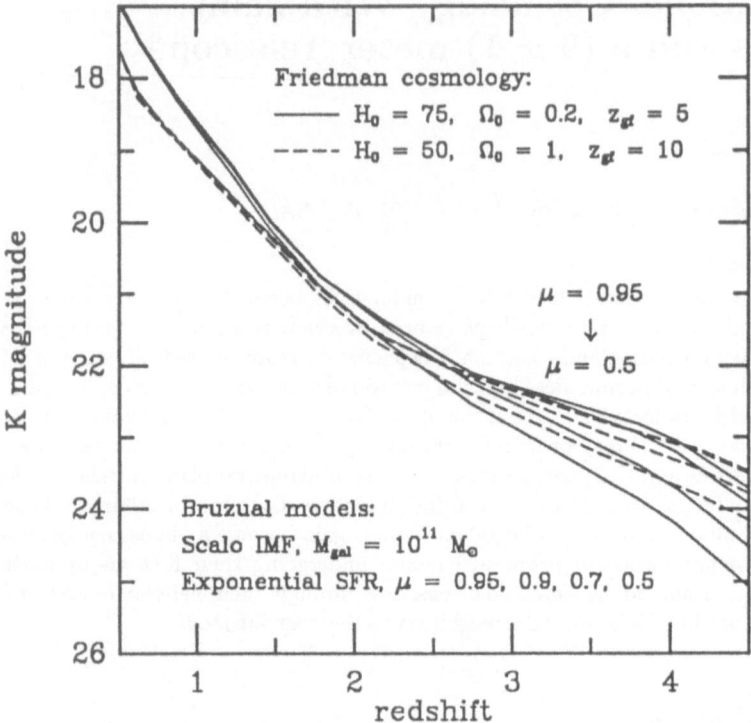

Fig. 1. IR Hubble diagrams in the K band for two cosmologies and redshifts of galaxy formation, as indicated. An updated version of Bruzual (1983) was used, for a galaxy with a total baryonic mass of $10^{11} M_\odot$, a Scalo IMF, and an exponential star formation rate with the e-folding times of 0.33, 0.43, 0.83, and 1.44 Gyr (Bruzual $\mu = 0.95$, 0.9, 0.7, and 0.5, respectively). These correspond to normal ellipticals and S0's, or even Sa's, in their present-day spectral energy distributions.

The modern IR observing technology is up to the task. We can, in principle, already detect normal, $\sim L_*$ galaxies out to $z \sim 3 - 5$, assuming a reasonable range of cosmologies and stellar population synthesis models (Figure 1). Detections of objects as faint as $K \approx 24^m$ are now being done with the Keck telescope, and it is expected that the next generation IR camera working with a full AO image compensation could reach $K \approx 27^m$. Note that for an average faint field galaxy $(B - K) \sim 5^m$, so these are very faint galaxies indeed.

Several lines of inquiry which may be pursued with the new generation of large telescopes are briefly described here. This is by no means an exhaustive list, and there are many other possibilities, some of them addressed by other contributors to these proceedings. The trick is to identify unique projects which simply would not be practical or even possible with a smaller telescope.

2   Deep Galaxy Counts and Redshift Surveys in the IR

Deep imaging surveys and redshift surveys of complete samples of faint galaxies can be used directly to constrain evolution models of normal, field galaxy populations, and perhaps the cosmological models as well. The subject has been reviewed extensively, e.g., by Koo & Kron (1992), and references therein.

The optical surveys have reached levels of $B \sim 28^m$ (Metcalfe *et al.* 1993), and produced the evidence for the notorious excess of faint blue galaxies. We now know from the deep redshift surveys that these are mostly starbursting dwarfs at the redshifts of a few tenths. Deep redshift surveys of blue-selected samples have so far failed to turn up any evidence for a high-z tail, and indicate at most a very mild evolution of field galaxies out to $z \sim 1$ (Koo & Kron 1992). Similar conclusions have been reached in the surveys of galaxies selected as absorbers near lines of sight to random quasars (Steidel 1993). While the mysterious blue dwarfs are very interesting on their own, we would really like to probe the evolution of normal galaxies at large look-back times.

Deep IR surveys may provide this information (c.f. Cowie *et al.* 1994, and references therein). The observed K band samples the light from more evolved stellar populations which should contain most of the mass, and should be much less sensitive to minor bursts of star formation. The initial deep K band imaging surveys with the Keck telescope have now reached to $K \approx 24^m$ (Soifer *et al.* 1994; Djorgovski *et al.* 1995). The counts continue to rise, with no sign of a turnover, and can be fitted with models with little or no evolution, and a low Ω_0.

Extrapolating the counts to fainter levels, we see that cumulative surface densities in excess of 10^6 galaxies/degree^2 may be reached by $K \geq 26^m$ (Figure 2). We estimate that $K \sim 27^m$ may be reachable with fully AO-compensated images. At that flux level, uncompensated images would be confusion limited; image improvement inherent to the AO is thus necessary. At such faint flux levels, we should be sampling a fair portion of the normal galaxy luminosity function out to $z \sim 5$, or even higher. This would then open the field for a complete, systematic mapping of field galaxy evolution, starting with their formation.

However, deep imaging alone is unlikely to contain a sufficient information to disentangle various evolutionary and cosmological effects. Redshift surveys of complete, IR-selected samples will be necessary (c.f. Songaila *et al.* 1994). A project along these lines in now under way by a Caltech group.

Another interesting result from deep K band surveys was the discovery of extremely red galaxies, with $(R - K) \geq 5^m$, or even redder (Soifer *et al.* 1994; Hu & Ridgway 1994; Cowie *et al.* 1994; Djorgovski *et al.* 1995). The nature of these objects remains unknown. A reasonable possibility is that they are normal ellipticals at $z \sim 1 - 2$. Figure 3 shows the predicted $(R - K)$ colors for dustless galaxies from Bruzual stellar population synthesis models, which are consistent with this suggestion. It is also possible that these objects represent some heretofore unknown population, and some of them may even be moderately dusty young galaxies at large redshifts (both dust and strong emission lines could modify the colors). The extreme redness of these objects will make their optical

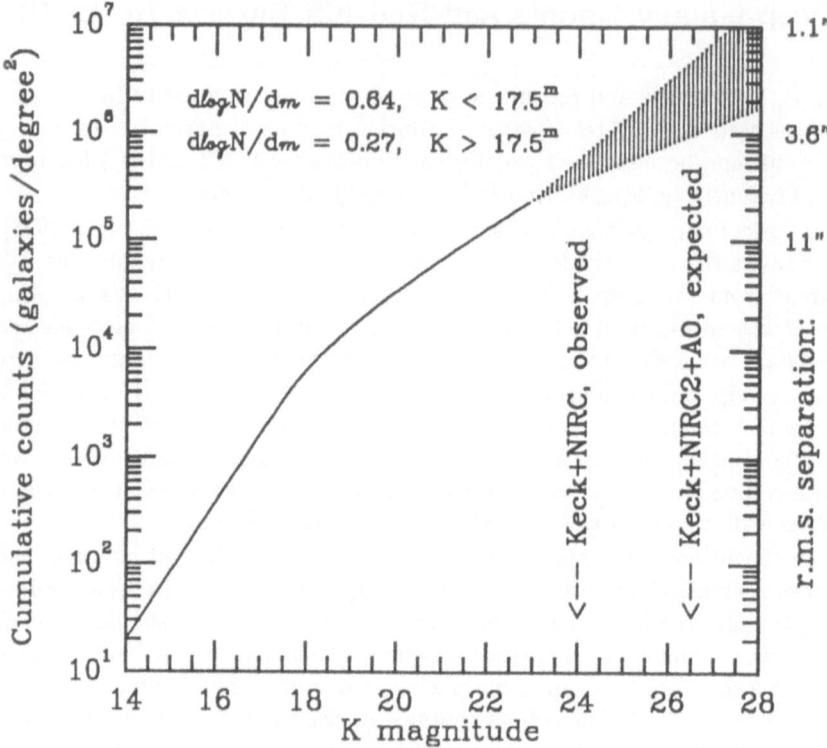

Fig. 2. The cumulative galaxy counts in the K band, from a broken power-law fit to the differential counts, as indicated in the upper left. The actual observations now extend down to $K \approx 24^m$. The shaded region indicates the estimated uncertainties in the extrapolation to the fainter magnitudes, and the completeness corrections at $K \sim 23^m - 24^m$. NIRC2, the next generation IR camera for the Keck telescopes, designed to be used with a full AO image compensation, is expected to reach to $K \sim 27^m$.

spectroscopy extremely difficult. A new generation of IR spectrographs, now forthcoming at large telescopes, will solve this problem.

A completely new territory opens up in the L (3.6μm) band, and perhaps the other mid-IR bands ($\lambda \sim 5 - 10 \ \mu$m) as well. These are the wavelengths in which the relative gain in sensitivity of the larger telescopes will be the highest, since in addition to the increase of the collecting area the image diameters will be smaller and essentially all of the noise is the foreground thermal emission, so the fewer pixels are covered by an image, the higher S/N will be.

The L band is particularly interesting, as it corresponds to the cosmic IR background window at $\lambda \sim 2 - 6 \ \mu$m, between the scattered zodiacal light, and the thermal emission from the zodiacal dust (the other, far-IR window is at $\lambda \sim 200 - 800 \ \mu$m, between the the thermal emission from the zodiacal and interstellar dust, and the CMBR). It is possible that redshifted stellar continuum and line emission (e.g., Hα at $z \sim 5$) from protogalaxies may be found here.

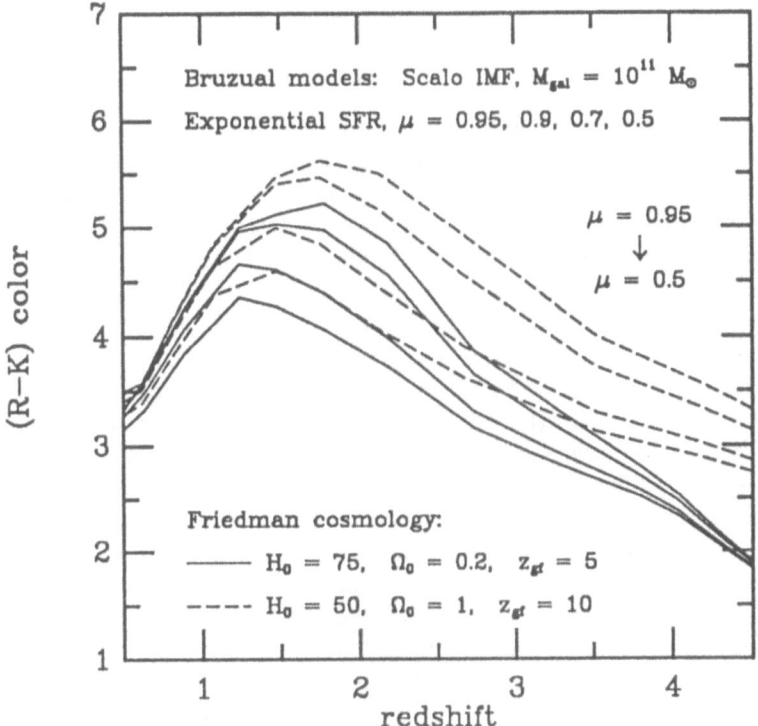

Fig. 3. Predicted galaxy colors as a function of redshift, for the same set of Bruzual population synthesis models and the two cosmologies as in Fig. 1. The models with longer SFR e-folding times (smaller μ), and with the shorter cosmological time scales ($H_0 = 75$, $\Omega_0 = 0.2$, $z_{gf} = 5$) have bluer colors at any given redshift. While $(R-K) \approx 5^m - 6^m$ can be easily reproduced by oldish ellipticals at $z \sim 1.5 - 2$, colors redder than $(R-K) \geq 6^m$ would suggest an additional reddening due to dust, and/or the presence of strong emission lines in the K band.

NIRC instrument at the Keck telescope is expected to reach $\sim 5\sigma$ detections in several hours of integration down to $L \sim 20^m - 21^m$ (fluxes $\sim 10^{-6}$ Jy). NIRC2 instrument, working with a full AO compensation, should go at least another 1^m or 2^m deeper. For comparison, ISOCAM instrument on ISO is expected to reach a 5σ confusion limit at $L \sim 17^m - 18^m$, in a comparable integration time (however, ISO is expected to excel at the longer wavelengths). Source counts models by Franceschini et al. (1991) predict a cumulative number density of sources of $\sim 10^2$ arcmin^{-2} at the flux levels we expect to reach with the Keck – unless some new, additional population of objects shows up.

In addition to the cosmological studies, these deep IR surveys will also place severe constraints on the faint stellar and brown dwarf content of the Galactic halo (cf. Hu *et al.* 1994).

3 Galaxy Dynamics and Fundamental Correlations at High Redshifts

While simple photometry and low S/N spectroscopy can be used to probe the evolution of bulk properties of galaxies, such as their luminosities, much more detailed information can be obtained from studies of *correlations* of galaxian properties over a range of look-back times. These correlations, such as the Tully-Fisher relation for disks, or the fundamental plane correlations for ellipticals, effectively define in a quantitative way the normal galaxy families. These correlations are products of the (astro)physics of galaxy formation, and they contain important clues about the processes which determined the global properties of galaxies (cf. Djorgovski 1992b, 1992c, Djorgovski *et al.* 1988, Kormendy & Djorgovski 1989, and references therein). Thus, they provide us with a powerful tool to probe the *evolution of systematic properties of galaxy families.*

Most of the interesting correlations require detailed spectroscopic information, e.g., velocity dispersions or line widths, or strengths of various absorption features. Thus, they require a higher S/N than just the redshift measurements, and cannot be pushed as faint. A heroic attempt in a cluster at $z = 0.18$ was made by Franx (1993). Therein lies the importance of the large telescopes: this is a clearly a photon starvation limited experiment.

For example, studies of the fundamental plane correlations involving velocity dispersions should be practical out to $z \sim 0.5$ or so, i.e., the look-back times of several Gyr. This includes the correlations with line strength indices, which reflect both the metallicities and the average stellar ages, and thus can be used to constrain directly the aging of stellar populations, and perhaps even the epoch of galaxy formation. Surface photometry information needed for these studies could be best obtained in the K band, where the seeing is best (and the AO might help as well), and the effects of dust and line blanketing are minimized. Alternatively (or in addition), CCD surface photometry using the HST can be used; a combination of the resolving power of the HST, and the light gathering power of the 9-meter class telescopes could be very effective.

As a byproduct of this work, we should be able to perform the Tolman test for the expansion of the universe, in which K-corrected surface brightness is expected to scale as $(1+z)^{-4}$. The fundamental plane correlations can be cast in the form in which surface brightness is one of the coordinates, and the intercept on that axis would define "a standard fuzz", a surface brightness which would be much more constant than, say, an average surface brightness of ellipticals (cf. Sandage & Perelmuter 1990). An early attempt, with a single radio galaxy, was made by Rigler & Lilly (1994). Once the universal expansion is (hopefully) confirmed, any residuals from the $(1 + z)^{-4}$ trend could be used as a direct measure of the evolution of the luminosity density of stellar populations of ellipticals.

4 The Return of the Hubble Diagrams?

The Hubble diagram is probably *the* classical test of cosmology (cf. Sandage 1988, for a review). The central problem with this test is the definition of a sample of objects, which: (*a*) should have a small intrinsic range of luminosities at any given redshift; and (*b*) whose intrinsic changes in luminosity over a range of look-back times can be understood and modeled reliably (standardizing the candle). Traditionally, the brightest cluster members (BCM), invariably giant ellipticals or cD's, were used. Empirically, they seem to satisfy the condition (*a*), and there is a reasonable hope that the present stellar population synthesis models are good enough to take care of the condition (*b*). The problem then reduces to discovering galaxies at high redshifts, where the cosmological discriminatory power of the test is higher.

For the past couple of decades, radio galaxies have been the most distant non-quasar objects known, and are relatively easy to find, thanks to the absence of any correlation of observed radio flux and redshift. At *low* redshifts, most of these objects appear to be giant ellipticals, even if sometimes disturbed looking. Since there is no evidence for any sudden change in the types of objects with an increasing redshift, the hope for many years was to use high-z radio galaxies in a Hubble diagram test, perhaps along with BCMs. While the visible light Hubble diagrams for high-z powerful (mostly 3C) radio galaxies indicated strong evolutionary effects and had a large scatter, the IR (K band) Hubble diagrams behaved wonderfully: there was very little scatter even out to the highest redshifts sampled ($z > 3$); no obvious segregation due to differences in the radio power; and an apparently seamless stitching with the Hubble diagram for BCMs at low z's (Spinrad & Djorgovski 1987, Lilly 1989).

Unfortunately, for the last several years we have learned enough about the powerful radio galaxies at high z's, to make them strongly suspect in their role as standard candles. The prominent radio-optical alignment effects, the presence of polarized light in some of them, extended high-ionization line emission, etc., all suggest that there is an important, and still poorly known component in the light we see from these objects, which may be due to their central engines; we are not dealing with simple progenitors of present-day BCMs. For an excellent review and references, see McCarthy (1993). Given these effects, it is remarkable that the scatter in the K band Hubble diagrams for these objects is as low as observed, and this is an unresolved mystery by itself.

However, there may be some hope, in the form of moderate-power radio galaxies, derived from ~ 1 Jy low frequency samples, such as the Parkes Selected Regions, or B3. While these objects have radio powers typically only a factor of $\sim 5 - 10$ less than their strange 3C cousins, they seem to be much better behaved: at a given redshift, they appear optically fainter, redder, with weaker emission lines, and much less prominent alignment effects (Peacock 1993, Thompson *et al.* 1994b). All this suggests that they may be much closer in their nature and properties to the genuine giant ellipticals, or at least progenitors thereof. Identifying and measuring sufficient statistical samples of sub-Jansky radio galaxies at high redshifts, and checking the dependence of their optical and IR properties on the radio power, may lead to a formulation of samples for which the K band Hubble diagram may again become a viable cosmological test.

An alternative approach may be to use the intercept of the fundamental plane correlations with the luminosity as one axis. This may be pursued in clusters out to $z \sim 1$, perhaps. While the differences in cosmological models are not as large at such redshifts, the *precision* of the standard luminosity defined by an ensemble of cluster ellipticals through the fundamental plane should be higher than for the "average" luminosities of BCMs or any radio galaxies.

5 Protogalaxy Searches

Discovery of primeval galaxies (PGs), usually defined operationally as the luminous ellipticals or bulges undergoing their first major bursts of star formation at large redshifts, remains one of the key challenges of modern observational cosmology. The subject was reviewed recently, e.g., by Djorgovski (1992a, 1994), Djorgovski & Thompson (1992), Pritchet (1994), and references therein. Most searches to date were based on the use of the Lyα line emission as a signature of star forming galaxies. Yet, despite considerable efforts, which have reached both sufficiently low limiting fluxes and covered a sufficient volume, no obvious population of Lyα luminous PGs has been found.

While there are now strong limits from the COBE FIRAS experiment (Mather *et al.* 1994) on the possible sub-mm background generated by completely dusty PGs, there is a real possibility that PGs may have been at least slightly dusty, which would considerably lower their Lyα line luminosities. If that was the case, then a good way to detect young, star forming galaxies at large redshifts would be through their Balmer or nebular oxygen emission lines. Useful estimates of the expected parameters can be found in Thompson *et al.* (1994a). In a nutshell, the expected Balmer and oxygen line fluxes from a PG powered by a star formation rates of $100 M_\odot$/yr are $\sim 10^{-16\pm1}$ erg cm^{-2} s^{-1}, for a range of redshifts of interest, and plausible cosmologies. The expected number densities of PGs are ~ 0.05 arcmin^{-2} for a redshift slice $\Delta z = 0.01$, which is a typical narrow band filter width.

A preliminary narrow band IR imaging search in 4 fields with the Keck (Pahre & Djorgovski 1995) improved the previous Palomar limits by an order of magnitude in both the area coverage and the limiting flux. While no obvious PGs were found, this was deepest IR PG search to date, reaching the limiting line fluxes of a few $\times 10^{-17}$ erg cm^{-2} s^{-1}. We are thus starting to reach the relevant portion of the parameter space where we may plausibly expect to see a population of IR PGs.

It may make sense to search around the known objects at high redshifts, e.g., quasars or radio galaxies, as they may well mark favorable sites of galaxy formation. Early dissipative mergers of gas-rich protogalactic fragments may be a common mechanism for both elliptical galaxy formation and the formation and feeding of the first AGN. A segment of one of our IR images is shown in Figure 4. There is an extremely red faint galaxy seen in projection next to the $z = 4.7$

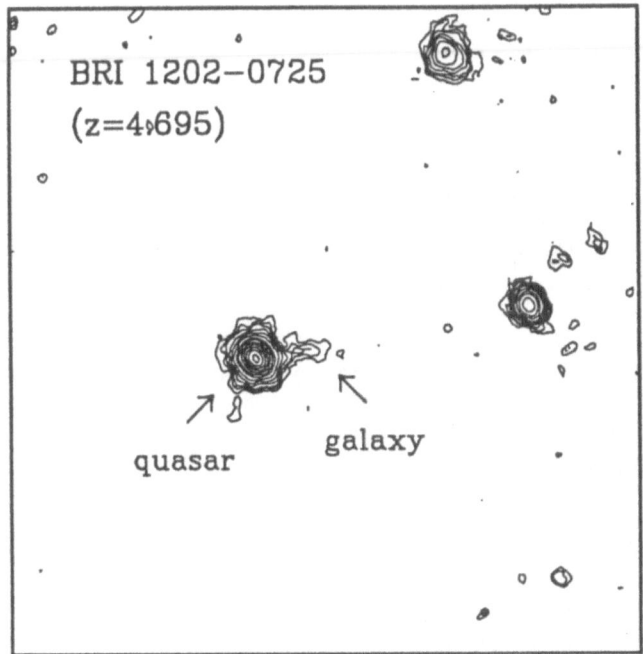

Fig. 4. A section of a K band image of the field of a $z = 4.695$ QSO BRI 1202–0725, obtained at the Keck. The field shown here is 15 arcsec square. An extremely red, faint galaxy is seen next to the quasar; its redshift is as yet unknown.

quasar BRI 1202–0725 (a similar case was seen in the case of $z = 4.9$ quasar PC 1247+3406 by Soifer *et al.* 1994). Sub-mm dust emission from this quasar has been detected by Isaak *et al.* (1994). It is tempting to speculate that we may be seeing an interacting or forming system at $z = 4.7$, perhaps similar to IRAS F10214+4724.

The author wishes to thank his collaborators, and in particular D. Thompson, J. Smith, M. Pahre, and R. de Carvalho, for many useful discussions. This work was supported in part by the NSF PYI award AST-9157412.

References

Bruzual, G. 1983, ApJ, 273, 105

Cowie, L., Gardner, J., Hu, E., Songaila, A., Hodapp, K.-W., & Wainscoat, R. 1994, ApJ, 434, 114

Djorgovski, S. 1992a, ASPCS, 24, 73

Djorgovski, S. 1992b, in G. Longo, M. Capaccioli, & G. Busarello (eds.), *Morphological and Physical Classification of Galaxies*, (Dordrecht: Kluwer), p. 337

Djorgovski, S. 1992c, ASPCS, 24, 19

Djorgovski, S. 1994, in *Mass-Transfer Induced Activity in Galaxies*, I. Shlosman (ed.), (Cambridge: Cambridge University Press), p. 452

Djorgovski, S., de Carvalho, R., & Han, M.-S. 1988, ASPCS, 4, 329

Djorgovski, S., & Thompson, D. 1992, in Proc. of the IAU Symp. #149, B. Barbuy and A. Renzini (eds.), (Dordrecht: Kluwer), p. 337

Djorgovski, S. *et al.* 1995, ApJ Letters, in press (January 1995)

Franceschini, A., Toffolatti, L., Mazzei, P., Danese, L., & De Zotti, G. 1991, A&ApSS, 89, 285

Franx, M 1993, ApJ, 407, L5

Hu, E., & Ridgway, S. 1994, AJ, 107, 1303

Hu, E., Huang, J.-S., Gilmore, G., & Cowie, L. 1994, Nat, 371, 493

Isaak, K., McMahon, R., Hills, R., & Withington, S. 1994, MNRAS, 269, L28

Koo, D., & Kron, R.G. 1992, ARAA, 30, 613

Kormendy, J., & Djorgovski, S. 1989, ARAA, 27, 235

Lilly, S.J. 1989, ApJ, 340, 77

Mather, J. *et al.* 1994, ApJ, 420, 439

McCarthy, P.J. 1993, ARAA, 31, 639

Metcalfe, N., Shanks, T., Roche, N., & Fong, R. 1993, Ann. N.Y. Acad. Sci., 688, 533

Pahre, M.A., & Djorgovski S. 1995, ApJ Letters, submitted

Peacock, J. 1993, in *First Light in the Universe: Stars or QSO's?*, Rocca-Volmerange, B. *et al.* (eds.), (Gif-sur-Yvette: Editions Frontieres), p. 115

Rigler, M., & Lilly, S.J. 1994, ApJ, 427, L79

Sandage, A. 1988, ARAA, 26, 561

Sandage, A., & Perelmuter, J.M. 1990, ApJ, 361, 1

Soifer, B.T. *et al.* 1994, ApJ, 420, L1

Songaila, A., Cowie, L., Hu, E., & Gardner, J. 1994, ApJS, 94, 461

Spinrad, H., & Djorgovski, S. 1987, in Proc. of the IAU Symp. #124, Hewitt, A. *et al.* (eds.), (Dordrecht: Reidel), p. 129

Steidel, C. 1993, ASPCS, 49, 227

Thompson, D., Djorgovski, S., & Beckwith, S. 1994a, AJ, 107, 1

Thompson, D., Djorgovski, S., Vigotti, M., & Grueff, G. 1994b, AJ, 108, 828

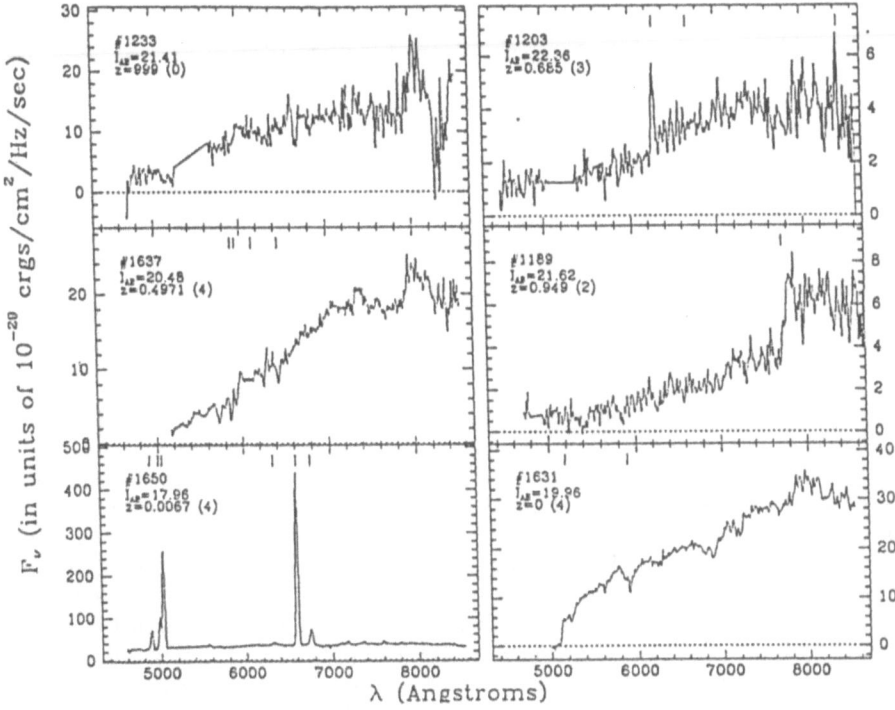

Fig. 1. Six spectra (8 hour integration) from the CFRS data base. Notes are indicated in parenthesis; note = 4,3 and 2 correspond to reliabilities higher th an 99.5%, 98%, and 82% respectively; note = 9 distinguishes the single emission line objects; note = 1 redshifts are only tentative, and note = 0 an absence of determination. These two last categories accounting for the incompletness (section 2).

2 Technical Details and Results

Sources have been selected from our deep I-photometry which was complete up to I=23, and no attempt has been made to separate star from galaxies. The multiplexing capability of the MOS at C.F.H.T. was optimized for our project, and up to 80 objects have been spectroscopically observed at the same time (Figure 1). For each individual I<22.1 object, 8 to 24 exposures of 1 hour have been taken with a low spectral resolution grism (45Å). Each spectrum has been extracted, combined and then analysed by three of us in an independent way. Comparison of the three analyses has led to a final redshift and the corresponding note of reliability (6 different notes, see Figure 1). Duplication of 8 X 1 hour exposure have been dedicated to about 200 objects, for which each 8 hour integration set has been indepe ndently analysed. This has allowed a (numerical) estimate of the reliability of our redshift assignment (Figure 1).

Deep Spectroscopy of 780 Galaxies: Results and Prospects for the VLT

F. Hammer[1], S. Lilly[2], O. Le Fèvre[1],
D. Crampton[3], L. Tresse[1]

[1] DAEC, Observatoire de Paris-Meudon, 92195 Meudon, France
[2] Astronomy Dpt., University of Toronto, Toronto, Canada
[3] Dominion Astrophysical Observatory, Victoria, Canada

Abstract. Two thirds of the available spectra of faint galaxies have been obtained at C.F.H.T. in the frame of a French-Canadian collaboration. It provides the first luminosity function of galaxies up to z=1. While the luminosity distribution of red galaxies is apparently similar to the local one, intrinsically bright L * galaxies show evidence of evolution of their luminosity since z=0.5. At lower redshift and luminosity range (z<0.3, L<0.5L*), blue galaxies are apparently more abundant than in the present day. We have pushed to its limit the capability of a spectrograph on a 4 meter telescope and are now awaiting the next step with the VLT. Indeed our knowledge of large scale structures is still too poor to accurately determine the luminosity function and to well understand the nature of the distant Universe. Only a large field spectrograph such as WFIS can provide the large data base required for deep surveys on large enough areas of the sky.

1 Goals

Deep counts of galaxies have revealed the presence of an overwhelming population of blue galaxies (Tyson, 1988), corresponding to a number excess of 3 to 5 time s more galaxies at B>23 than one can expect from the local luminosity function . These objects lie apparently at moderate redshift (z = 0.3 for B=23-24) as sho wn by spectroscopy of B-selected galaxies (Colless et al, 1990). However beyond z=0.2 the B filter samples the range below the 4000Å break at rest wavelengths, a spectral region sensitive to short-lived stars, implying that star forming galaxies are likely over represented in B-selected samples. More recently several tens of galaxies have been selected at redder wavelengths (Tresse et al, 1993; Lilly, 1993), in order to estimate the old stellar light from faint galaxies, which is somehow related to their stellar mass. Our two groups have chosen to merge in a wide French and Canadian collaboration (Canada France Redshift Survey, CFRS) which aims to gather a much larger s ample of 1000 faint objects. This sample is designed to study the luminosity function in detail (at various redshift and color ranges) and has been completed in 30 nights at C.F.H.T. within two years.

The resulting data base is made up of 996 spectra observed so far, plus the deep photometry done in B, V I and K bands. Among them 832 spectra were securely identi fied (85% completness) including 222 stars and 5 QSOs. Most of the 164 unidenti fied objects are galaxies having low surface brightnesses and/or belonging to the faintest luminosity range. The redshift distribution ranges from z=0 to 1.3 (Figure 2), but one can expect an increasing incompleteness beyond z=1 (see Le Fevre et al, this volume), while at low z the decreasing CCD efficiency below 4500Å hindered the identification of the 4000Å break of red absorbing-spectrum galaxies.

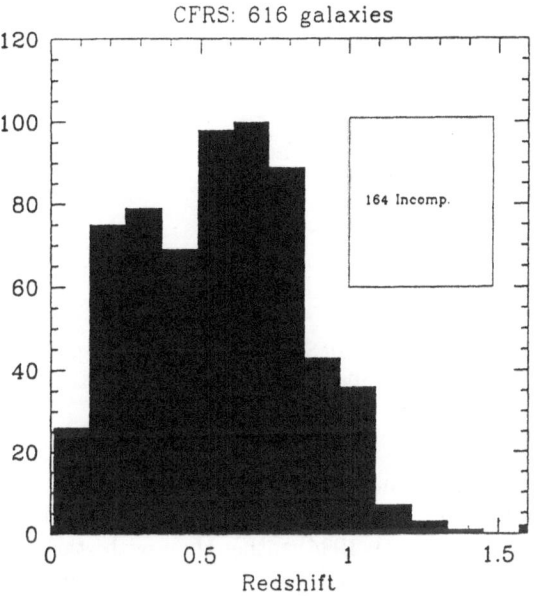

Fig. 2. Redshift distribution of the I<22.1 galaxies. Average z is 0.58.

The luminosity function has been derived after calculating the B absolute magnitude of each galaxy from the V and I photometry, through k-corrections derived from a grid of 14 galaxy SEDs (Bruzual and Charlot, 1990). Every point of the global luminosity function is above the local luminosity function determined by Loveday et al (1991). The former has however a poor meaning since it blends galaxies of different luminosities and different redshifts. Indeed, luminosity functions for galaxies redder than Sbc shows little or no discrepancies with the local one , while bluer and intrinsically bright galaxies are brighter by several tenths of magnitude beyond z=0.5 (Figure 3). This figure also shows that intrinsically faint blue galaxies were more numerous in the (recent) past than now, which confirms studies of faint B-selected samples (Cowie et al, 1991). Emission line spectra of z<0.3 galaxies have also been systematically analysed. Emission line ratios have been plotted in diagnostic diagrams from Veilleux and

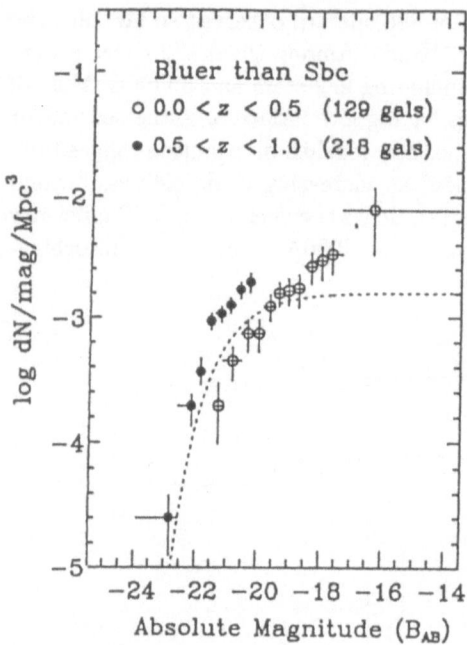

Fig. 3. Luminosity function for the galaxies bluer than Sbc in two redshift ranges. Each dot corresponds to 10 galaxies. Dotted line represents the local luminosity function from Loveday et al (1992).

Osterbrook (1987), in order to identify the nature of the ionization source. The results are that up to 50% of the faint blue galaxies (B=24) are associated with narrow emission line AGNs (Tresse et al, 1994) rather than with starburst galaxies as previously stated. This result can shed a new light on the nature of the overabundant population of blue galaxies.

3 Spatial Coverage and Structures

Galaxies have been selected in 5 10'X10' fields of high Galactic latitude (b>5 0 degrees). About 20% of all the I<22.1 galaxies in these fields have been spectroscopically observewd, while higher spatial coverage is reached within smaller areas (strips of 2'X10') due to our instrumental procedure. We have identified some clust ers of galaxies (about 1 per two fields) and several other structures revealed by a somewhat peaked redshift distribution in each individual field. One of such structure at z=0.985 has been already discussed by Le Fevre et al (1994) and corresponds, at this redshift, to an overdensity larger than 10. As shown in Figure 4, most of these structures have no central concentration and are apparently extended over scales larger than 10 arcmin (or 5 Mpc at z=0.5). In redshift space their extensions range from 1000 km/s to almost dz=0.1 which would be equivalent to 7 degrees at z=0.5 if it was projected on the sky plane. Our pencil beam survey reveals the presence of structures which might be similar in scales

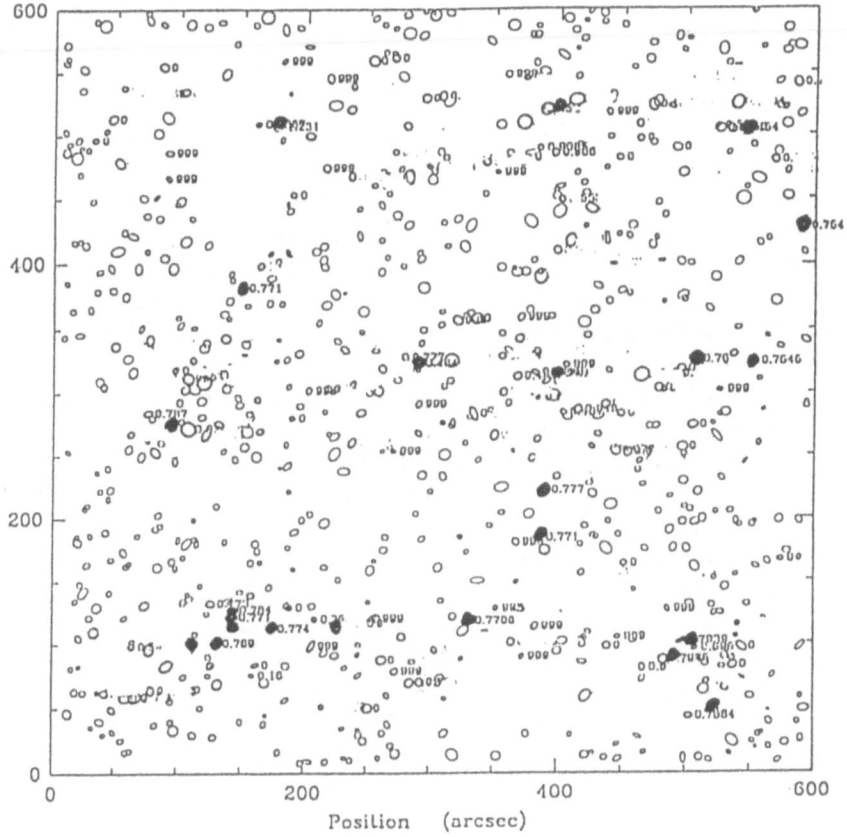

Fig. 4. I<22.1 galaxies in one 10'X10' field where we found a peak at z=0.7 7 in the redshift distribution; full dots represent galaxies belonging to the st ructure, the velocity dispersion over the field being 2000 kms^{-1}.

to the Great Wall, but lie at redshifts as high as 1 (40% of the present-day age of the Universe).

4 Conclusions: a Wide Field Spectrograph for the VLT ?

CFRS results have demonstrated the efficiency of a multi-object spectrograph to gather a large amount of data with cosmological interest. Our data base enables us to study in detail the evolution of galaxies from z=0.1 to z=1. Spectrophotometry combined with high-resolution imaging with HST will provide a new classification of galaxy type and its evolution from z=0.1 to z=1. It is the largest data base on faint galaxies that can be gathered from observations with an 4 meter telescope.

We are convinced that we need a much larger data base in order to sample a much larger volume of the Universe and better understand the role and the formation of structures at large scales. Structures are found to be preponderant in the nearby Universe and prevent us from an accurate determination of the local luminosity function (da Costa et al, 1994). On going projects are intended to tackle this issue by systematic spectroscopy of galaxies over very large areas (Sloan Digital Sky Survey: 1 million galaxies with B<19; 2df at the AAT: fiber spectroscopy of more than 100 000 galaxies with B=20). At the time of the VLT first light there will be little room for a fiber spectrograph to compete with these projects.

On the other hand, our pencil beam survey also demonstrates the influence of large scale structures, but much further away, i.e. up to z=1. A Wide Field Spectrograph (WFIS) for the VLT is presented by Vettolani et al (this volume) and would be ideally suited for deep studies of large areas of the sky. The CFRS project could be done in only one night and half with such an instrument on the VLT. This means that a deep survey over 10 sq degree gathering 20 000 galaxies at the same depth would become feasible in only 50 VLT nights. This would provide a unique data base for performing 3D correlations on scales as high as 100 Mpc, and to follow structure evolution from z=0.1 to z=1. A study of galaxy structures on a volume (4 million Mpc^3), similar to the Sloan project but much further away (z=0.2-0.3, 200 000 B=22 galaxies), could also be done in only 50 nights. A multi-slit wide field spectrograph will put European Astronomy in a good position with the forthcoming competition of the 21st century. The extremely large volume of the distant Universe. as well as the very large number of faint galaxies, should encourage us to pursue these goals, since one telescope cannot do the whole job. There is no instrument presently planned for the VLT which is able even to begin such projects.

References

Bruzual, G., Charlot, S.(1990): ApJ, 405, 538.

Colless, M., Ellis, R., Taylor, K., Hook, R. (1990): MNRAS, 235, 827.

Cowie, L., Songalai, A., Hu, E. (1991): Nature, 354, 460.

Le Fevre, O., Crampton, D., Hammer, F., Lilly, S., Tresse, L. (1994), Ap. J. , 423, L89.

da Costa, L., Geller, M. et al (1994): Ap. J. 424, L1.

Lilly, S. (1993): ApJ, 411, 501.

Loveday, J., Peterson, B., Efstathiou, G., Maddox, S. (1992): ApJ, 390,338.

Tresse, L., Hammer, F., Le Fevre, O., Proust, D. (1993): A&A, 277, 53.

Tresse, L., Rola, C., Hammer, F., Stasinska, G. (1994): in preparation.

Tyson, J. (1988): Astron. J., 96, 1.

Veilleux, S., Osterbrook, D., 1987: Ap. J. Supp., 63 295.

Deep Redshift Surveys on Faint Galaxies: Prospects for the VLT above z=1

O. Le Fèvre[1], D. Crampton[2], F. Hammer[1], S. Lilly[3], L. Tresse[1]

[1] DAEC, Observatoire de Paris-Meudon, 92195 Meudon Cedex, France
[2] Dominion Astrophysical Observatory, Victoria, Canada
[3] Astronomy Dept., University of Toronto, Toronto, Canada

Abstract. We present results from the Canada-France Deep Redshift Survey (CFRS). With more than 600 redshifts obtained for galaxies between $0 \leq z \leq 1.2$, and 210 galaxies at $z \geq 0.7$, this survey is an order of magnitude larger than any comparable survey. One of the main results is that the luminosity of the redder galaxies does not change much with time, but that blue galaxies have undergone a luminosity evolution of about 1 magnitude at $z \sim 1$. These results can be used to predict that about 30% of the field galaxies selected with $H \leq 21$, within reach of $\geq 8m$ telescopes, will be at redshifts in excess of 1. Only the $>8m$ telescopes like the VLT will be able to gather large samples of $z > 1$ galaxies and study the progenitors of present day normal galaxies, at look back times $<40\%$ the age of the universe. We emphasize the need to have efficient multi-slit spectrographs for the near-IR (up to 2 μm) for the VLT, in addition to multi-slit spectrographs in the visible.

1 The CFRS: the Largest Sample of Galaxies with $0 \leq z \leq 1.2$

For programs aiming at spectroscopic measurements, e.g. redshifts, of objects with a high projected density on the sky, the telescope collecting area is only one factor in the efficiency of observations, as multiplexing gains offered by multi-slit spectrographs can greatly enhance a telescope's performance.

The CFRS has been conducted to provide statistically meaningful information on the evolution of faint field galaxies with $17.5 \leq I_{AB} \leq 22.5$, with a projected density on the sky of 25000 gal/o^2, ideally suited for multi-slit spectroscopy. Observations were obtained at the CFHT with the MOS-SIS imaging spectrograph (Le Fèvre et al., 1994) and are described in a series of papers which will be shortly submitted for publication. More than a thousand redshifts have been measured with multi-slit masks with an average of 80 slits per mask. The unprecedented multiplexing gains at these faint magnitudes have been achieved thanks to a unique geometric arrangement of spectra on the detector, and are equivalent to using a 32m telescope with a single slit. The completeness of redshift measurement is $\sim 85\%$, and after identification of the galactic stars we have a sample of more than 600 galaxies spaning the redshift range $0 \leq z \leq 1.2$.

At redshifts larger than 0.7, more than 210 redshifts of galaxies have been measured. This is a sample an order of magnitude larger than any comparable deep survey and allows us for the first time to unambiguously establish the

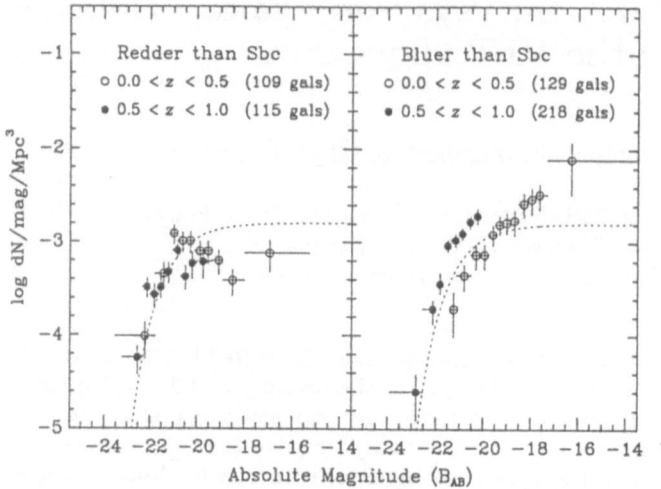

Fig. 1. Comparison of the luminosity functions computed from the CFRS for blue and red galaxy samples. The redder galaxies show little evolution with redshift (left), while the bluer population has brightened considerably by z~1, leading to an increased number at L* luminosities (right).

evolutionary properties of normal galaxies from z=0 to z~1.2. One of the main results of this survey is to demonstrate that, while redder galaxies do not seem to evolve much with epoch, the bluer galaxies have brightened considerably, by about 1 magnitude at z~1 (Fig. 1), implying an increase in the number of galaxies with luminosities ~L*. We were able to identify structures of galaxies at redshifts up to z=0.985 (Le Fèvre et al., 1994). Other results from this survey are given by Hammer et al. (this volume).

2 Predictions for the Observations at Fainter Limits

From the luminosity function of galaxies at z≥0.7 computed from the CFRS data, we can make predictions on the number of galaxies one would expect to detect in a sample of galaxies selected at a fainter flux limit. Fig. 2 shows the predictions for a sample of field galaxies selected with H≤21, chosen as a practical limit for the deepest surveys with one of the VLT 8m telescopes (see §4). Two cases are presented: (i) in the lower panel, the luminosity evolution is supposed to flatten to a fixed 1.3 magnitudes at and above z=1.3, which corresponds to the observed luminosity evolution in the CFRS; this is a "worst case" scenario as one would expect galaxies to continue to brighten at earlier epochs; (ii) in the upper panel, the evolution in luminosity of the galaxies has been extrapolated from what is observed in the CFRS. A fit of the increase in luminosity with redshift (reproducing a 1 mag. brightening at z~1), indicates that 2.5 mag. of evolution at z~2 is a possible extrapolation.

The predictions in these two cases show that 25% to 38% of the galaxies selected at H≤21 should be at redshifts ≥1. Obtaining a sample of galaxies at these faint fluxes would then provide invaluable information on the evolution of galaxies with $z \geq 1$, at look-back times <40% the present age of the universe.

3   Overcoming the z∼1 Barrier

As galaxies are redshifted, there are several problems that complicate their i-dentification: (i) the observed luminosity is decreasing as the distance modulus increases. As I_{AB}=22.5 is the practical limit for deep surveys on 4m telescopes, one needs the larger collecting area of the 8m telescopes to go to fainter magnitudes and larger redshifts; (ii) an increasingly large fraction of the flux is emitted in the near infrared, and most of the useful spectral lines used for redshift identification are disappearing from the visible window at z>1; (iii) the quantum efficiency of CCD detectors is low in the red. Table I gives the redshift range corresponding to the most-used spectral lines for redshift determination, observed in the visible and near-IR domain. One can note that above z=1, the number of lines in the visible domain is becoming small, and the paucity of strong lines in the spectrum of galaxies below [OII]3727Å makes the measurement of the redshifts of galaxies at z>1 extremely difficult. To overcome this difficulty one has to explore the near-IR, where most of the strong lines will be located.

Table 1. Redshift range of common spectral features

Feature	0.4-1 μm	1-1.8 μm
Lyα	2.3-7.2	7.2-13.8
[OII]3727Å	0.07-1.7	1.7-3.8
4000Å break	0-1.5	1.5-3.5
4304Å=G band	0-1.3	1.3-3.2
Hβ	0-1.1	1.1-2.7
[OIII]5007Å	0-1.0	1.0-2.6
Hα	0-0.5	0.5-1.7

4   Need for Efficient Multi-Slit Spectrographs in the Near-IR and Visible for the VLT

Deep redshift surveys to study the evolution of field galaxies and the distribution of galaxies in large scale structures require large numbers of galaxies to establish the basic evolutionary properties of galaxies at different epochs.

To go deeper than the current deepest surveys, and collect large samples of galaxies, one has therefore to (i) increase the collecting area, (ii) use the near-IR domain for faint-object spectroscopy, AND (iii) use the large multiplexing gains offered by multi-slit spectroscopy. The VLT 8m units address the first point, and new instruments need to be defined to overcome the limitation to long slit capabilities of the current planned near-IR instrumentation of the VLT.

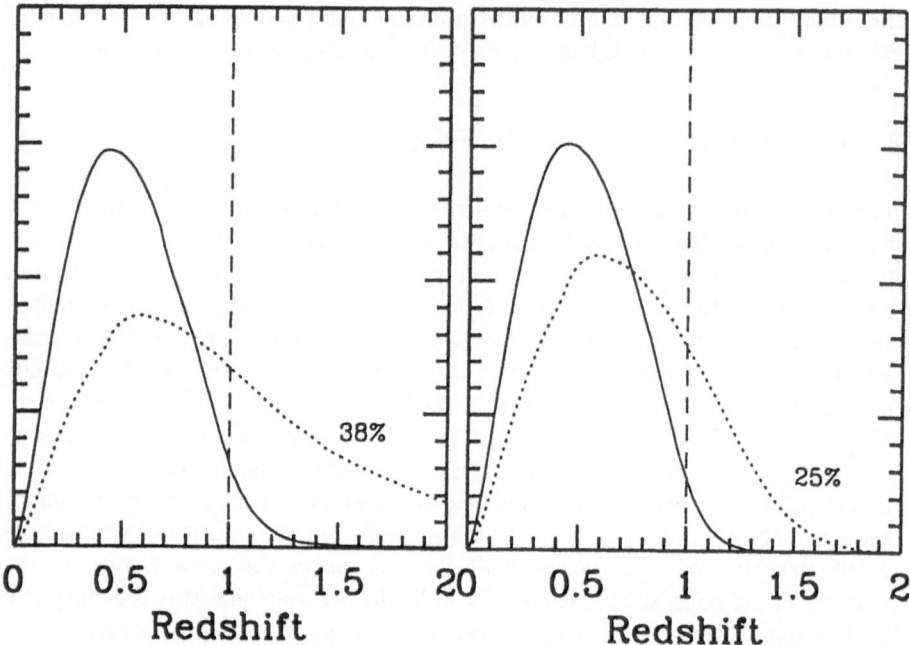

Fig. 2. Predictions for the redshift distribution of a sample of galaxies selected at H≤21. The full line shows the observed distribution for the CFRS survey (I_{AB} ≤22.5). The dashed line in the left panel shows the prediction for luminosity evolution of 1 mag. at z~1, as observed, and 2.5 mag at z~2 as predicted by a fit of the luminosity evolution observed in the CFRS, 38% of galaxies would be at z>1. The right panel shows the prediction if the luminosity evolution observed in the CFRS flattens to stay as observed at 1.3 mag above z=1.3. In this "worst case" situation, the fraction of galaxies at z>1 would be 25%.

A pilot project to study the evolution of field galaxies beyond z=1 with the VLT would require observation of several hundred field galaxies at H=21. With the current state of near-IR detectors, one would need 4hrs of integration on an 8m to obtain spectra with S/N~10. A single slit IR spectrograph, as currently planned for the VLT first generation of instruments, would therefore require several hundred nights to complete this project, clearly an unrealistic prospect.

To overcome this limitation, and widen the VLT field of investigation to the most distant normal galaxies beyond z=1, we therefore propose NIRMOS, a near-IR multi-object spectrograph, as part of a second generation of VLT instruments (see Le Fèvre et al., this volume). Its capabilities would allow the redshifts of about 400 galaxies with z>1 to be obtained in 5 nights in a sample of ~1200 galaxies with H<21. We also note that an efficient multi-object spectrograph operating in the visible such as the proposed WFIS (Vettolani et al., this volume), would also be an assest to quickly complete the redshift measurements.

The possibility to obtain such large and faint galaxy samples is a truly exciting prospect for our understanding of galaxy formation and evolution as well as the evolution of large scale structure, and would only become reality for European astronomy if near IR multi-slit spectrographs are available on the VLT.

References

Le Fèvre, O., Crampton, D., Felenbok, P., Monnet, G., 1994, A&A, 282, 325
Le Fèvre, O., Crampton, D., Hammer, F., Lilly, S. J., Tresse, L., 1994, ApJ, 423, L89

VLT
Observations with Large Natural Gravitational Telescopes

Yannick Mellier

LAT-URA285; Observatoire Midi-Pyrénées, 14, avenue Edouard Belin, 31400 Toulouse, France

Abstract. The coming of the 4-8 meter ESO VLT telescopes is an opportunity to start new challenging programmes not feasible with 4 meter telescopes. Many programmes related to gravitational lenses should make important breakthroughs in extragalactic astronomy and cosmology with the unique combination of a VLT and a gravitational telescope provided by the universe itself. In this paper I present some programmes related to arc(let)s and lensed quasars, which should benefit greatly from the VLT instrumentation.

1 Towards Very Large Gravitational Telescopes (VLGTs)

One decade of observations of gravitational lensing phenomena has confirmed early theoretical predictions that they are indeed formidable tools for astrophysicists. Gravitational lensing deflects, magnifies and multiplies *all* images of a distant source to some level, depending on their position behind the deflector. On account of their magnification capabilities, lenses can be used as *gravitational telescopes* (GT) for observing distant/faint sources which would be otherwise undetectable. Gravitational lens studies also provide the mass distribution instead of the light distribution; and since they work on all mass scales, they can weigh any gravitational systems from stars (10^{-3} $M_\odot$) to large scale structures (10^{16} $M_\odot$). Finally lenses can also provide information on cosmology.

Although the first observations where mainly dedicated to galaxy-lens effects on distant quasars, intensive studies are now performed on rich cluster-lenses on distant background sources. The large angular cross sections of rich clusters of galaxies give a new class of lenses with a useful *lens field* that can easily reach 10 arcmin or more. In constrast to quasars, all the sources in this field will present a detectable shear induced by the gravitational potential, providing a unique opportunity to map the distribution of dark matter or to discover new magnified distant objects. In this respect, cluster lenses act as *natural giant telescopes*. Fig. 1 (right) shows that there are basically three lensing regimes: strong; medium; and weak.

Strong and medium regimes induce visible distortions (shear) which may be spectacular. Giant arcs are rare events, but they probe accurately the inner mass of clusters of galaxies. On-going studies of arcs give robust mass-to-light ratios within cluster cores (≈ 100 kpc) and have clearly demonstrated that their core radii are smaller than what was believed from previous analyses.

Weak lensing perturbs all the galaxies behind the lens, but induces almost undetectable ellipticity in individual galaxies. The gravitationally-induced shear is computed statistically from the ellipticity and orientation averaged over a large number of galaxies (30 to 100). This demands that very faint limiting magnitude ($B \approx 28$) be reached, in order to have a high density of sources (10^5 gal./deg^2/mag.). The weak shear analysis is an efficient technique for measuring the mass distribution of clusters, even at a distance as large as $3h_{50}^{-1}$ Mpc. We now have compelling evidence that the total mass of rich clusters of galaxies is about three times their virial mass, and the assumption that light traces mass is incorrect on large scales. A formidable result is that the mass-to-light ratio of clusters reaches 1000 in the outermost regions, leading to a value of Ω close to 1 ! No similar work has yet been done on poor clusters and groups.

All lensing regimes can be also used to study evolution of galaxies. Optical and infrared photometry, and spectroscopy of highly magnified arcs gives the spectral energy distribution of distant galaxies and their redshift distribution. This can also be done statistically by using the radial distribution of sheared galaxies in the image plane as a low resolution spectrograph (see Sect. 3.2). Both techniques give consistent values of the averaged redshift of these sources, which seems to be below 1 (Kneib et al 1994, Bonnet et al 1994, Smail et al 1994). But results are preliminary and we do not have yet the dispersion, nor the shape, of the distribution.

Schneider et al. (1992), Blandford & Narayan (1992), Surdej & Refsdal (1994), and Fort & Mellier (1994) have reviewed the present status of gravitational lensing and discussed the outlook. They largely stressed the importance of observations of the background sources to address some key issues where we expect significant advances within the next decade.

- What is the *amount* of dark matter (DM) on scales ranging from 10^3 M$_\odot$ to 10^{16} M$_\odot$? In particular in poor clusters, groups and galaxies.

- What is the *distribution* of DM? How does this distribution vary with scale?

- What is the *relative distribution* of mass and light?

- What is the *mass spectrum* of gravitational structures?

- What is the *nature* of DM?

- What is the *shape of the redshift distribution* of faint sources?

- When did galaxies and clusters form? How do they *evolve*?

- What are the values of Ω_M, Ω_Λ and H$_0$?

Experienced observers in these fields are aware that the 4-m class telescopes are pushed to their limits to observe arcs and to obtain spectra of the brightest ones. It is clear that the VLT will improve the situation because it has three advantages: (1) on account of its collecting power. It can go as deep as as 4-meter telescopes in shorter exposure times. The gain is in fact larger than expected from

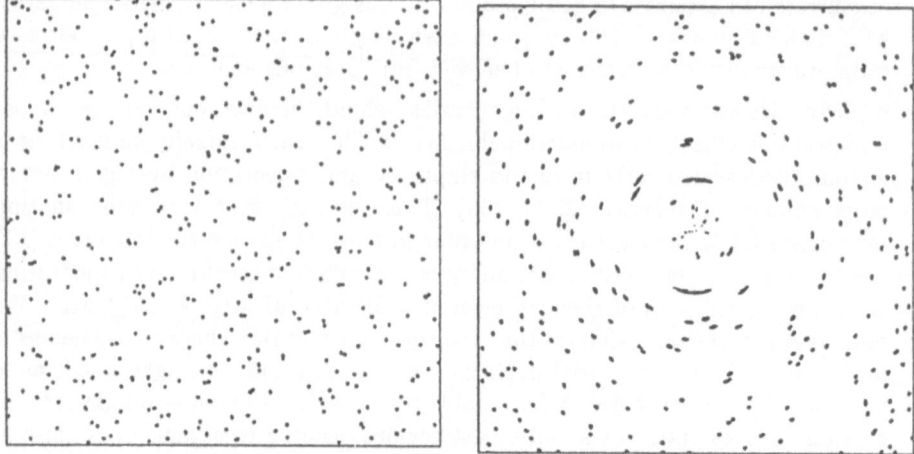

Fig. 1. A randomly distributed population of galaxies at z=1 (left panel) lensed by an isothermal sphere at z=0.5 (right panel). Arcs and small ellipses show the strong and medium lensing region. Outside, the individual shear is imperceptible and must be averaged (From Bonnet 1995, PhD Thesis)

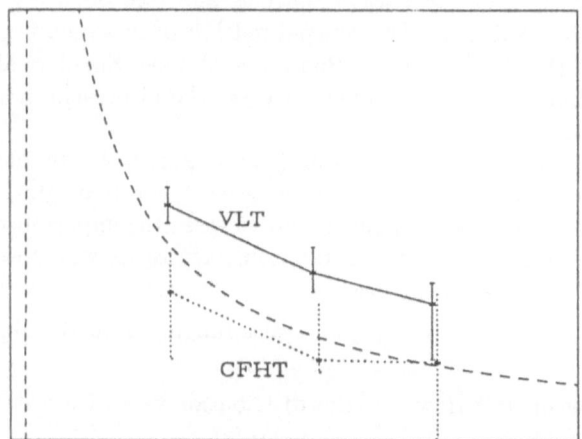

Fig. 2.. Shear profile induced by an isothermal sphere (dashed line), and comparison of what we would infer from observations on 4m (CFHT) and 8m (VLT) in similar conditions (seeing 0.7", sampling 0.2"/pix, exp. time: 4 hours). The curves have been offset with respect to the ideal isothermal model for clarity

just telescope size, because short exposures increase the probability of having very good image qualitity on deep images; (2) it can obviously go deeper than 4-m telescopes during long exposures; (3) the VLT has 4 different telescopes and a large diversity of instruments which allow an optimal multiplex instrumental configuration for a given scientific programme tobe choosen. In this paper we review a few astrophysical programmes that will benefit from the VLT and its instrumentation. Our main purpose it to stress that the use of VLT with the gravitational telescope (GT), will provide the universe's largest instruments, the Very Large Gravitational Telescopes (VLGTs)!

2 Dark Matter

2.1 The Mass-Profile of Rich Clusters of Galaxies

The discrepancy between the Ω values found from large-scale galaxy flows and from cluster analysis needs a careful estimate of the mass distribution of clusters of galaxies at their periphery ($>> 1h_{100}^{-1}$ Mpc). Mass profiles are also important to understand the dissipative processes which occur during the non-linear regime of cluster formation. The first tentative measurement of the weak shear in Cl0024+16 shows that the mass profile on scales larger than $2\ h_{100}^{-1}$ Mpc is compatible with an isothermal sphere (Bonnet et al. 1994). However, as shown by Bonnet et al. (1994) and in Fig. 2, the uncertainties are still too large to exclude a de Vaucouleurs profile or even another power law. Two approaches can improve the measurements. One is to go to a larger distance, where the two shear profiles separate easily (Bonnet et al. 1994). This demands very large C-CD arrays covering a typical angular scale of 20'×20'. The other possibility is to observe the same field at a fainter limiting magnitude in order to detect a larger number of galaxies. This will increase the signal within each bin and therefore decrease the error bars shown in Fig.2a. Both collecting power and image quality are crucial for this programme and need a VLT on a very good site, with image quality similar to the best obtained at the NTT. Deep galaxy counts in the I-band minimize the atmospheric refraction and reach a typical number density of 5×10^5 gal/deg^2/mag at I=26, instead of 2×10^5 gal/deg^2/mag at V=27.5 obtained by Bonnet et al. (1994). From the imaging capabilities described in the ESO document, FORS should reach this limit within 1 hour. However, a larger field of view would be advantageous since the typical angular scale to be mapped is 10'. An instrumental project like WFIS would fulfill this requirement.

2.2 Shear Map and Evolution of Distant Clusters of Galaxies

The discovery of high-z rich clusters is crucial for understanding the formation of structures in our Universe. It is currently believed that they are rather young gravitational systems which virialized at low redshift (≤ 2), and we expect to observe their evolution even at moderate redshift, by looking at the evolution of weak shear on a sample of clusters uniformly distributed with redshift. The shear efficiency should decrease with increasing redshifts of the clusters. Preliminary deep investigations of the field of the double quasar Q2345+007 show that the shear field found by Bonnet et al (1993) could be a lensing effect by a cluster at z about 1.5 (Fischer et al. 1994, Mellier et al. 1994):- first evidence for the presence of a massive, very distant cluster.

As the cluster redshift increases, the background sources become fainter and their number density may also change, which makes the measurement and the interpretation of the shear evolution non-trivial. Furthermore, since low redshift clusters show a large spread in characteristics, a statistically significant number of clusters is required. An ideal mapping requires a typical scale of 1 h_{50}^{-1} Mpc ($\approx$ 1' at z=1, and 5' at z=0.2). Efficient shear by a cluster at z =1 concerns galaxies with redshift of 1.5–2, which may have strong emission in the near-infrared, rather than in the optical. Therefore, this programme needs ultra-deep observations in the R, I and K bands (limiting mag. R=26, I=25 and K=23). The 7 hours/cluster exposure time on a 4-m telescope will be reduced to only 2 hours by using FORS and ISAAC. This opens the possibility of surveying a large sample of clusters. Note that a project like NIRMOS is best-suited for this programme since ISAAC has a small field of view.

2.3   Weak Shear around Galaxies

Weak shear measurement could be also extended to poor clusters of galaxies, groups, and even single massive galaxies (Fort, Mellier & Bonnet in preparation). This is certainly a major step toward understanding the mass spectrum of gravitational structure and we should not dismiss these systems in our future prospect. Nevertheless, we prefer to concentrate on galaxies only because the problem of weighing their massive haloes is still an open question.

Present-day observations of galaxies do not provide a complete description of their mass profile and extent. Galactic dark halos could have non-isothermal profile which could extend to very large distance. Because of the strong cutoff of their light component, it is not possible to probe the total mass by using current ground-based observations. An alternative is the statistical measurement of the weak shear amplitude at a typical distance of 100 kpc from the center.

At z=0.1, 150 h_{100}^{-1} kpc is equivalent to 1.7'. Because the shear is weaker than for clusters, a large sample of *sosie* galaxies is needed for each class of galaxies whose halo we want to characterize. The signal-to-noise ratio can be increased by selecting a sample of galaxies with similar morphology and co-adding their individual shear field. This is a typical programme for FORS, if we observe 10 galaxies in each category with 0.1"/pixel, and 6h shift-and-add exposures.

3 Sources

3.1 Spectral Energy Distribution and Redshift of Arcs

Thanks to magnification, redshifts of $\sim10\%$ of arcs have already been obtained, with the aim of producing an unbiased sample of distant field galaxies and to study their spectral history (Smail et al. 1993). Redshifts of arcs also provide the scaling of lenses and give their total mass. However, only the brightest arcs are observable with 4-m telescopes and almost all the redshifts successfully measured are based on emission-lines. This may bias the sample toward star-forming galaxies, and restricts these galaxies to a small redshift window at optical wavelengths, where [O II]3727Å is visible. A deeper spectroscopic survey of arcs is expected with the VLT, especially if we use the "Va-et-Vient" spectroscopic technique (Cuillandre et al. 1994), but, because many arcs could be at redshift larger than 1.5, spectroscopy must be done in both the optical and the infrared.

This is a typical program for the FORS and NIRMOS low resolution spectrographs. ISAAC could be used also, but multi-object spectrographs (such as MOS) have more versatile capabilities. Typical spectroscopic exposure times on giant arcs with EMMI are between 2 and 10 hours! By using FORS, the same redshifts would be obtained in 1 to 5 hours, or equivalently FORS should reach 24.2 mag/arcsec^2 within 1.5 hours (brightness of the arc in A370).

3.2 Redshift of Arclets from Lens Modelling

Kneib et al. (1994) attempted to give the most probable redshift of each individual arclet in A370. Almost similar attempts have been made by Bonnet el al. (1994) and Smail et al. (1994a). Their results show that the averaged redshift of galaxies with magnitude between 25 to 27 could be obtained this way while spectroscopy is impossible (see Fig.3). If the "lensing-redshifts" were spectroscopically confirmed for the brightest arclets, it will open a new technique to study the distance of faint galaxies and therefore to extend the analysis of faint blue galaxies beyond $B = 25$. The goals are to understand their formation and their evolution and to compare these galaxies with those at $B \approx 22 - 25$, whose nature is still elusive (Ellis 1993). It could be that a new population of more distant galaxies is present at fainter luminosity.

Before starting a massive analysis, we need first to confirm the lensing-distance by obtaining the redshift of some bright arclets. By looking at the magnitude of these objects, it is clear that we can hardly use 4 m telescope. We estimated that FORS will reach B=25, R=300 and S/N=5 in 6 hours! The MOS capabilities are also needed to measure the redshift of several arclets simultaneously. Furthermore, it is possible that many arclets are at redshift above 1 and a NIRMOS instrument will be also useful. We should reach a typical magnitude K=21 in 6 hours, with an expected $(B - K) = 4$ value for galaxies.

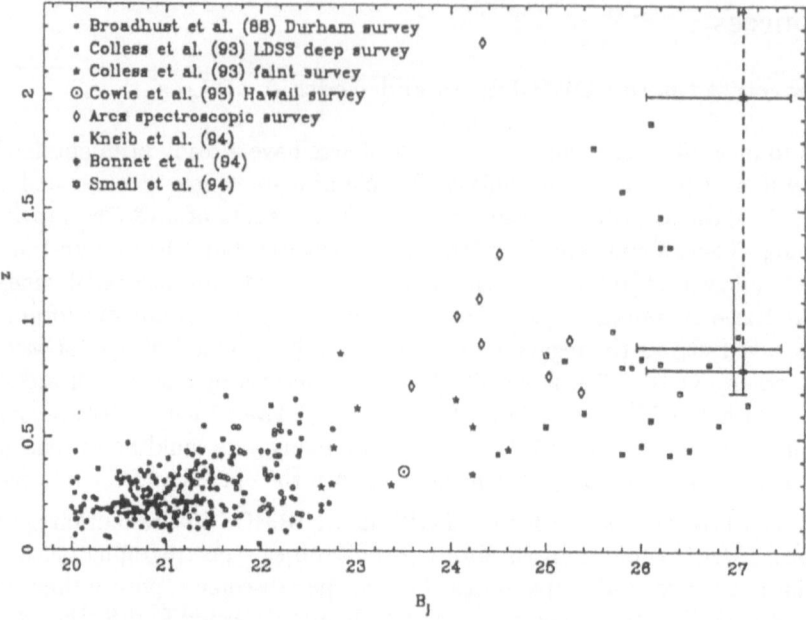

Fig. 3. Magnitude redshift diagram. The overlap between spectroscopic surveys and "lensing redshifts" is at $B = 25$, a magnitude range accessible to the VLT (from Fort & Mellier 1994)

3.3 Gravitational Pairs and the Discovery of a New Class of Objects?

When small lensed sources lie close to a caustic line, they form a gravitational pair of reversed images separated by a critical line. Pairs are as faint as arclets and were reported by Fort et al. (1988), Smail et al. (1994b) and Kneib et al. (1994), while Fort & Mellier (1994) reported a first systematic investigation in three clusters. But the scientific interest has been stressed by Miralda-Escudé & Fort (1993), who discussed the fact that most lensed pairs appear as point-like systems and suggested that perhaps they are a new class of extragalactic object. They argued that by going deeper, the number density of background sources becomes large enough to have a full mapping of the critical lines "drawn" by the lensed pairs.

It will be important to get a spectroscopic confirmation of the gravitational origin of lensed pairs. With FORS or NIRMOS, the major step consists in doing ultra-deep imaging up to B=29, and even higher. Miralda-Escudé & Fort (1993) extrapolated deep counts and estimated that about 1000 gal/arcmin^2 should be observed at B=29.5. FORS is then needed and will reach this magnitude in about 8 hours. This is clearly not feasible on 4-m telescopes.

3.4 Testing the Tully-Fisher Relation at Large Distance

Soucail & Fort (1992) and Narasimha & Chitre (1993) have pointed out that the elongated shape along a magnification axis also magnifies the rotation curves of galaxies in this direction. This can be used for testing the Tully-Fisher relation on distant galaxies, or even for testing lens models by observing the 2-D radial velocity pattern in giant arcs. Rotation patterns must be observed with an integral field spectroscopy and can be done provided the magnification is large enough in order to properly sample the galaxy.

This program could be performed with the ARGUS mode implemented on FUEGOS. The velocity accuracy imposes R~3000 in order to have enough details in the velocity map, and a signal-to-noise ratio of about 20. The presence of the [O II]3727Å emission line makes the observations easier, and in that case it requires about 4 hours on a VLT.

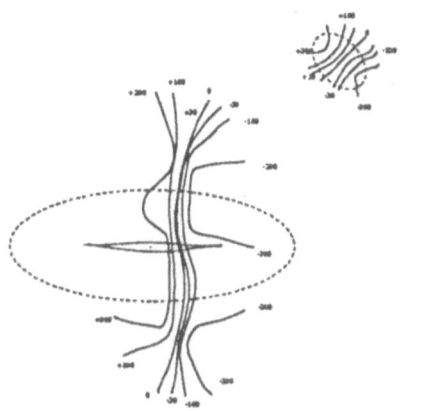 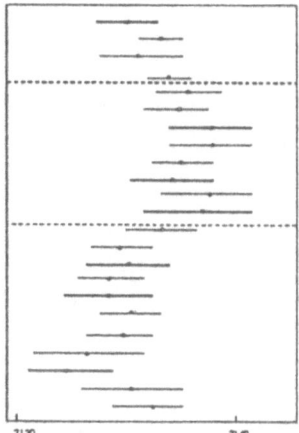

Fig. 4. Left panel: simulation of a 2-D magnified rotation map of a spriral galaxy as it would be observed with ARGUS on FUEGOS. The right panel shows the wavelength variation along the straight arc in A2390 observed by Pelló et al. (1991). It is interpreted as the rotation curve observed in three merging images.

Spatial variation along the magnification axis could be done with long slits for straight arcs (Pelló et al. 1991), and curved slits otherwise. FORS and NIRMOS should be capable of observing a large sample of arcs and testing the Tully-Fisher relation for galaxies at redshift larger than 1.

4 Cosmology

4.1 Testing the Cosmological Constant with Lensed Quasars

Gravitational lensing was always thought of as an ideal tool to test the cosmological models of the universe. Historically, Press and Gunn (1973) suggested using the angular distribution of doubly-imaged quasars for measuring the cosmological parameters, but the *number* of double image QSO's is a better statistical approach because it strongly depends on Λ (Turner, 1990). The first attempts only gave marginal results because complete catalogs have too few quasars, and selection biases are still misunderstood. In fact, the separation between each cosmological model increases with quasar redshift. Therefore, this programme demands complete and deep catalogs which include faint (distant) quasars.

Mathez, and Zamorani (these proceedings) proposed spectroscopic surveys on unbiased samples of Seyferts and quasars. These programmes could be extremely powerful to obtain good statistics on the number of lensed quasars and should put interesting constraints on Λ. The MEDUSA mode of FUEGOS is the only instrument on the VLT which has a large enough field of view for this program. It is capable to going as deep as R=23 in one hour and could survey about 1000 quasars, with more than 10 expected at z> 3.

4.2 An Ideal Case of Double Giant Arcs

Ideally, giant arcs produced by two galaxies at different redshifts would have radial positions with respect to the cluster center which depend on the cosmological parameters, the cluster redshift and the source redshifts (Fort & Mellier, 1994). Consequently, if, by chance, we could measure the redshifts of these two arcs, we should be able to compute Ω. The probability of observing two giant arcs bright enough to be able to get their redshift is almost zero on present day telescopes. However, for larger telescopes the probability increases because we can go deeper to detect fainter giant arcs, and we can also perform spectroscopy of much fainter arcs. If the arcs are bright enough, we should be able to measure the two redshifts with FORS and ISAAC, but if they are too faint, optical and infrared photometry with FORS and ISAAC would give constraints on their redshift. More sophisticated approaches can be developped by using arcs and arclets simultaneously. But in all cases it is necessary to find, from a very large survey, a simple cluster which can be easily modelled with the best suited multiple image configuration. As we expect about 1000 clusters in the whole sky, this programme requires a systematic survey of arcs. In any case, this programme, although fascinating, is marginal even for the VLT.

5   Summary and Conclusions

We have described a few scientific programmes which should lead to an important step in cosmology and extragalactic astrophysics by using an exceptional telescope association, namely the VLT and the Gravitational Telescope. It is clear that the imaging and spectroscopic capabilities of FORS and ISAAC will push observations for the faint arc(let)s much deeper. The ARGUS mode of FUEGOS should allow measurement of the radial velocity maps of distant, magnified spiral galaxies.

However, the first instrumentation set for the VLT does not cover all the scientific requirements given in this presentation. The first need is for a near-infrared multi-object spectrograph. A project like the proposed NIRMOS could fullfill our scientific goals because of its large field of view compared to ISAAC and its MOS capability. The second need is a visible wide field imaging capability to map the large scale structures by using the weak shear technique. The proposed WFIS with its MOS capabilities could improve the VLT efficiency because of its large field of view compared to FORS.

Nevertheless, it is a remarkable point that the VLT has already a large number of designed instruments covering the whole wavelength range from the UV to the infrared. In this respect it will be a unique and very competitive instrument for gravitational lensing studies of very faint objects.

Acknowledgements

It is a pleasure to acknowledge B. Fort for numerous stimulating discussions on gravitational lensing, cosmology and faint object observational techniques. They have led to most of the scientific programmes proposed in this presentation. I also thank J.-P. Picat, G. Mathez and H. Bonnet for helpfull discussions and advice, and J.-C. Cuillandre and R. Pelló for reading the manuscript.

References

Blandford, R. D., Narayan, R. (1992): ARA&A **30**, 311

Bonnet, H., Fort, B., Kneib, J.-P., Mellier, Y., Soucail, G. (1993): A&A, **280**, L5.

Bonnet, H., Mellier, Y., Fort, B. (1994): ApJ **427**, L83

Brainerd, T. G., Smail, I., Mould, J. R. (1994): preprint Fort, B., Prieur, J.-L., Mathez, G., Mellier, Y., Soucail, G. (1988): A&A **200**, L17

Cuillandre, J.-C., Fort, B., Picat, J.-P., Soucail, G., Altieri, B., Beigbeder, F., Dupin, J.-P. , Pourthié, T., Ratier, G. (1994): A&A, **281**, 603.

Ellis, R. S. (1993): In *Sky Surveys: Protostars to Protogalaxies*, ed. by B. T. Soifer (PASP Conf. Series 43) pp. 165.

Fischer, P., Tyson, A. J., Bernstein, G. M., Guhathakurta, P. (1994): ApJ, 431, L71

Fort, B., Mellier, Y. (1994): A&AR **5**, 239

Kneib, J.-P., Mathez, G., Fort, B., Mellier, Y., Soucail, G., Longaretti, P.-Y. (1994): A&A **286**, 701

Mellier, Y., Dantel-Fort, M., Fort, B., Bonnet, H. (1994): A&A, 289, L15

Miralda-Escudé, J., Fort, B. (1993): ApJ **417**, L5

Narasimha, D., Chitre, S. M. (1993): A&A **280**, 57

Pelló, R., Le Borgne, J.-F., Soucail, G., Mellier, Y., Sanahuja, B. (1991): ApJ **366**, 405

Press, W. H., Gunn, J. E. (1973): ApJ **185**, 397

Refsdal, S., Surdej, J. (1994): Rep. Prog. Phys. **56**, 117

Schneider, P., Ehlers, J., Falco, E. E. (1992): in *Gravitational Lenses* (Springer Verlag, Berlin)

Smail, I., Ellis, R. S., Aragón-Salamanca, Soucail, G., Mellier, Y., Giraud, E. (1993): MNRAS **263**, 628

Smail, I., Ellis, R. S., Fitchett, M. (1994): preprint (1994a)

Smail, I., Couch, W. J., Ellis, R. S., Sharples, R. M. (1994): preprint (1994b)

Soucail, G., Fort, B. (1991): A&A **243**, 23

The ESO Instrumentation Department (1994): *Instrumentation for the ESO VLT*, ed. by A.F.M. Moorwood (ESO, Garching)

Turner, E. L. (1990): ApJ **365**, L43

Integral Field Spectroscopy of Selected Extragalactic Objects with FUEGOS

Jean Surdej[1], Christian Vanderriest[2], Marie-Christine Angonin-Willaime[3], Jean-François Claeskens[4], Paul Felenbok[3]

[1] Space Telescope Science Institute[2], 3700 San Martin Drive, Baltimore, MD 21218, USA
[2] Canada-France-Hawaii Telescope Corporation[3], P.O. Box 1597, Kamuela, HI 96743, USA
[3] Observatoire de Meudon, Place Jules Janssen, F-92195 Meudon Cedex, France
[4] European Southern Observatory, La Silla, Casilla 19001, Santiago 19, Chile

Abstract. Combining the large collecting area of the VLT, optimal seeing conditions (typically < 0.7") and the integral field spectroscopic capabilities (ARGUS mode) of FUEGOS (also known as the Multi-Fibre Area Spectrograph), it will become possible to address several new and important scientific issues in the field of gravitational lensing, physical studies of quasar host galaxies, etc. After a brief introduction, we recall some of the well known effects caused by macro- and micro- lensing as well as a few related applications of great astrophysical and cosmological interest. We then review the expected instrumental performance of FUEGOS in its ARGUS mode and the type of observational data that can be retrieved. Finally, we describe typical challenging FUEGOS/ARGUS observations of gravitational lens systems that could be successfully carried out at the Nasmyth focus of VLT No 3.

1   Introduction

It is well known that superb direct imagery and spectroscopic observations can be achieved, under seeing conditions as good as 0.5", with telescopes like the CFH 3.6m in Hawaii, and also the ESO/MPI 2.2m and the NTT at La Silla. Seeing monitor campaigns carried out at ESO, and also elsewhere, have confirmed the occurence of such unusual conditions. The atmospheric seeing in Paranal is known to remain better than 0.5" during approximately 16 % of the clear nights (periods of at least one hour; see Sarazin 1990). Furthermore, the median seeing over the last years in Paranal has been measured to be 0.66"; with an exceptional average seeing of 0.32", recorded during one full, best night. Therefore, astronomers should be given the technical possibility to obtain with the VLT, under such very good seeing conditions (FWHM < 0.66"), first class scientific

---

[2] Member of the Astrophysics Division, Space Science Department of the European Space Agency; on leave from the Fonds National de la Recherche Scientifique, Institut d'Astrophysique de l'Université de Liège, Belgium
[3] on leave from the Observatoire de Meudon, France

observations of faint extragalactic objects (not only direct imagery but also spectroscopy), with optimal angular resolution sampling. In subsequent sections, we shall describe several extragalactic projects, including observational studies of selected gravitational lens systems for which major scientific achievements could be made thanks to the large collecting area of the VLT combined with a high angular resolution, bi-dimensional, spectroscopic mode. Let us first recall some of the basic features of gravitational lensing.

2   Macro– and Micro–Lensing Effects

Because of the deformation of the space time near a massive object (e.g. a galaxy lens), we know that a spherical wavefront originating from a distant source (cf. a quasar) will get distorted in the vicinity of the galaxy and that if an observer happens to lie very close to the line connecting the source and the lens, the former will see multiple images of the background quasar. For a typical galaxy deflector and cosmological redshifts for the source (QSO) and the lens, angular separations between the lensed images are typically of the order of 1 or several arcsec (see Refsdal and Surdej, 1994 for a recent review on *gravitational lensing*). We also know that because the light trajectories have different geometrical lengths, and also because time retardation varies along the different trajectories, a time delay Δt will become measurable between the lensed images by the observer, provided of course that the quasar source undergoes photometric light variations. We know from the work of Refsdal (1964) that Δt is inversely proportional to H_o, and it is therefore straightforward (at least in principle) to derive H_o from Δt if we dispose of a good model for the lens. Such a good lens model may be derived if we are able to measure the source redshift, the lens redshift, the positions of the lensed images with respect to that of the deflector, their flux ratio(s) (via the ratio(s) of emission-line fluxes, see below), the velocity dispersion of the deflector and, if possible, the mapping of that velocity dispersion across the deflector (cf. Falco et al. 1991 for the particular modelling of 0957+561 A and B).

There is however one more difficulty. When a foreground galaxy (the "macrolens") produces multiple images of a background quasar, these images are seen through rather dense parts of the galaxy and there is a good chance that one or several "macro-images" are affected by micro-lensing (Chang and Refsdal 1979). The "micro-lens" is a star (or several stars) of the galaxy, acting as a magnifying lens with a very small "field of view", which produces a more or less intricated network of micro-caustics (cf. Kayser 1992 and Wambsganss 1993 for recent reviews on *micro-lensing*). When the source crosses this network, its different components are differentially amplified, according to their sizes and locations. There will thus result a differential amplification of the different components of the quasar. For instance, in the spectrum of a micro-lensed quasar image, the optical continuum will be more amplified than the Broad Line Region (BLR) which has a larger extension. Due to relative proper motions, this phenomenon varies on a time scale of a few months or years and produces characteristic light curves (and very likely slightly variable spectroscopic line profiles for the broad emission-lines). Of course, the shape of these curves depends on the size of the source. A spectroscopic monitoring of the lensed QSO images will thus allow the structure and size of the continuum source to be probed, as well as the distribution in size (with an angular resolution of the order of 10^{-6}") and velocity of the BLR clouds.

High resolution spectroscopy of the multiple (2-4) lensed images of a background quasar should also allow one to set interesting constraints on the size and structure of the Ly-α and/or heavy absorbing element clouds located along their lines-of-sight (cf. Smette et al. 1992). It should even be possible from such observations to set constraints on the deceleration parameter q_o of the Universe.

So we see that gravitational lens systems, consisting of several variable macro-lensed images and of a deflector with angular separations of the order of one second of arc, constitute unique laboratories to probe very important cosmological and astrophysical parameters.

3 Description of the ARGUS Mode of FUEGOS

One of the astronomer's dreams would of course be to obtain individual spectra for all contiguous elements covering the (very narrow angular) field of compact gravitational lens systems, or fuzz around distant quasars, etc. It will in fact be possible to achieve such observations in the ARGUS mode of the spectrograph FUEGOS which should normally be installed at the Nasmyth focus of VLT No 3 (see Felenbok et al. 1994 and Felenbok 1994, in these proceedings, for a complete and detailed description of this instrument). We should like to stress here that somewhat similar instruments (cf. SILFID described in Vanderriest and Lemonnier 1988, TIGER in Courtès et al. 1988, ARGUS of MOS at CFH in Vanderriest, 1994) have already been used successfully at the focus of 4m telescopes.

In the FUEGOS spectrograph, 670 optical fibres forming the ARGUS bundle (this is the maximum number of fibres that can be accommodated across the 2048 × 2048 CCD format; pixel size of 24 microns) are closely packed together on the axis of the telescope focus to provide a good spatial sampling for small extended objects (cf. Fig. 3 in Felenbok 1994, these proceedings). The ARGUS fibre tips are provided with the same type of micro-lens array (array of hexagonal micro-lenses) as in the MEDUSA design. At the input extremity, they optimise the field aperture filling factor, while at the other end they transform the fast fibre output beam for the f/10 spectrograph collimator. To satisfy the scientific requirements, three different spatial samplings are provided by means of three optical relays, which can be selectively positioned in front of the ARGUS entrance (angular field of 5" × 5" with 0.2" sampling, or 12" × 12" with 0.45" sampling, or 18" × 18" with 0.7" sampling). As in the MEDUSA mode, spectral resolutions of R = 1500 (covering a range of 3500 Å), R = 5000, R = 10000 and R = 30000 will be accessible in the spectral range 3700 – 9000 Å. The ARGUS mode is provided with an atmospheric dispersion corrector (ADC) which can be inserted, when necessary, in the incoming telescope beam.

Of course, observations of gravitational lenses in this mode should only be carried out under excellent seeing conditions, i.e. in a service observing mode. Handling of the recorded spectra will make it possible to reconstruct images, by inverse geometrical transformation, in the QSO source continuum at any chosen wavelength – or range of wavelengths –, in specific emission- lines, to measure the velocity dispersion of the deflector, to produce maps of the radial velocity dispersion across the lens, maps of the radial velocity, of emission-line ratios, etc. (cf. Adam et al. 1989, Fitte and Adam 1993). Further analysis of these images should enable one to decompose the observed mirage into its individual lensed images plus deflecting galaxy by adequate fitting techniques.

4 FUEGOS/ARGUS Observations of Selected Extragalactic Objects

4.1 Gravitational Lens Systems

For most of the presently known gravitational lens systems, the distant ($z_L >$ 0.5) deflecting galaxies are generally much fainter ($R_L \simeq 19$–23) than the lensed QSO images. The information available on most of these lenses is thus dramatically poor. Two-dimensional spectroscopy will allow optimal separation of the faint light of the lens from the glare of the lensed quasar images. It will be possible to sum up the signal from all the fibres illuminated by the galaxy and subtract the quasar contamination. Let us note that thanks to the 'image slicer' effect, the spectral resolution will be kept constant for each individual image. We would recommend to derive (D) or to confirm (C) the lens redshift z_L and/or to measure the velocity dispersion (VD) for the following lens galaxies (see Angonin-Willaime et al. 1993 and Refsdal and Surdej 1994 for more information on the individual objects): UM673 ($R_L \simeq 19$ and $z_L \simeq 0.49$, C, VD), PG1115+080 ($R_L \simeq 19.8$ and $z_L \simeq 0.29$, C, VD), 2016+112 ($i_L \simeq 22$ and $z_L \simeq 1.01$ and 2?, D and C), MG0414+0534 ($R_L \simeq 22.4$ and $z_L \simeq 0.47$, C, VD), PKS1830-211 ($V_L \simeq 23$, D), B0218+357 ($r_L \simeq 20$ and $z_L \simeq 0.68$, C, VD), 2237+0305 ($R_L \simeq 14.5$ and $z_L \simeq 0.04$, mapping of the VD), 1654+1346 ($r_L \simeq$ 18.7 and $z_L \simeq 0.25$, C and VD). In order to have an approximate idea about the requested exposure times, we have estimated for the case of PG1115+080 that, using a spatial sampling of 0.4" per pixel and a spectral resolution R = 1500, it should be possible to obtain in 5 hours a good spectrum for the lensing galaxy characterized by a signal-to-noise ratio S/N $\simeq$ 20. From the same exposure, it will also be possible to extract individual spectra for the 3 lensed QSO images (A, B and C), characterized by a S/N $\simeq$ 100–500, depending on their apparent brightness. Using a spatial sampling of 0.2" per pixel, it should even be possible to record in just 2 hours individual spectra (S/N $\simeq$ 200) for the two merging macro-images A1 and A2, separated by only 0.5".

In addition, very high signal-to-noise and high resolution spectra (with a sampling of 0.2" per pixel) of the individual lensed QSO images for UM673, PG1115+080, 2237+0305, H1413+117 (see Angonin et al. 1990 for similar observations obtained with SILFID at CFH, sampling of 0.33" per pixel), B1422+231,

1009-025, HE1104-1805, etc. will allow the structure and size of the continuum source to be probed, as well as the distribution in size and velocity of the BLR clouds (see discussion in the previous section). These data will also be used to constrain the transversal size of (Ly-α and metallic) absorption line clouds (cf. Smette et al. 1992).

4.2 Other Extragalactic Objects

On account of space limitations, we shall only enumerate hereafter additional types of faint extragalactic objects for which 2-D spectroscopic observations with FUEGOS would also be very useful:

i. bright giant luminous arcs such as A370, Cl2244-02 (mapping the velocity dispersion across the lensed galaxies)

ii. the galaxies detected near (i.e. < 3") Highly Luminous Quasars (redshift and velocity dispersion measurements of those galaxies ($20 \lesssim R \lesssim 23$) to independently estimate the magnification bias and to test our understanding on the cosmic distribution, luminosity function and evolution of galaxies in the Universe)

iii. the faint (R > 21) environment of distant luminous quasars (extended haloes?, cooling flows in a surrounding cluster?)

iv. the host galaxies around low redshift quasars

v. kinematics (2-D velocity field) of the central regions of galaxies

vi. resolved structures in the inner regions of AGNs and merging galaxies (mapping velocities, excitation parameters, abundances...)

vii. etc.

References

Adam, G., Bacon, R., Courtès, G., Georgelin, Y., Monnet, G., Pécontal, E.: 1989, *Astron. Astrophys.* **208**, L15

Angonin, M.-C., Remy, M., Surdej, J., Vanderriest, C.: 1990, *Astron. Astrophys.* **233**, L5

Angonin-Willaime, M.-C., Hammer, F., Rigaut, F.: 1993, review paper in the proceedings of the 31st Liège International Astrophysical Colloquium 'Gravitational Lenses in the Universe', p. 85, Eds. Surdej, J., Fraipont-Caro, D., Gosset, E., Refsdal, S. and Remy, M. (University of Liège, Belgium)

Chang, K., Refsdal, S.: 1979, *Nature* **282**, 561

Courtès, G., Georgelin, Y., Bacon, R., Monnet, G., Boulesteix, J.: 1988, in the proceedings of 'Instrumentation for Ground-Based Optical Astronomy: Present and Future', p. 266, Ed. Robinson, L.B.

Falco, E.E., Gorenstein, M.V., Shapiro, I.I.: 1991, *Astrophys. J.* **372**, 364

Felenbok, P., Cuby, J.-G., Lemonnier, J.-P., Baudrand, J., Casse, M., Andre, M., Czarny, J., Daban, J.-B., Marteaud, M., Vola, P.: 1994, SPIE: Instrumentation in Astronomy V III, **2198**, 115

Felenbok, P.: 1994, these proceedings

Fitte, C., Adam, G.: 1993, in the proceedings of the 31st Liège International Astro-physical Colloquium 'Gravitational Lenses in the Universe', p. 445, Eds. Surdej, J., Fraipont-Caro, D., Gosset, E., Refsdal, S. and Remy, M. (University of Liège, Belgium)

Kayser, R.: 1992, in 'Gravitational Lenses', (Proceedings, Hamburg 1991), Lecture Notes in Physics **406**, p. 143

Refsdal, S.: 1964, *Monthly Notices Roy. Astron. Soc.* **128**, 307

Refsdal, S., Surdej, J.: 1994, *Reports on Progress in Phys.* **57**, 117

Sarazin, M.: 1990, ESO-VLT Report N0 62

Smette, A., Surdej, J., Shaver, P.A., Foltz, C.B., Chaffee, F.H. et al.: 1992, *Astrophys. J.* **389**, 39

Vanderriest, C., Lemonnier, J.-P.: 1988, in the proceedings of 'Instrumentation for Ground-Based Optical Astronomy: Present and Future', p. 304, Ed. Robinson, L.B.

Vanderriest, C.: 1994, CFH Information Bulletin, **31**, 22.

Wambsganss, J.: 1993, review paper in the proceedings of the 31st Liège Internation-al Astrophysical Colloquium 'Gravitational Lenses in the Universe', p. 369, Eds. Surdej, J., Fraipont-Caro, D., Gosset, E., Refsdal, S. and Remy, M. (University of Liège, Belgium)

Perspectives in the Study of Large Scale Structure and Galaxy Evolution with the VLT

G. Vettolani

Istituto di Radioastronomia CNR, Bologna Italy

Abstract. Inferring from present knowledge, the role of the large telescopes in the study of large scale structure and galaxy evolution is outlined. The instrumentation necessary to attack these problems, as well as examples of possible programs, are also discussed.

1 Introduction

Our present view of the galaxy distribution is mainly the product of the observational efforts over the last decade for systematic mapping of the local Universe (i.e. $z \leq 0.05$) over the whole sky. These surveys (see Giovanelli and Haynes 1991, for a review) have convincingly demonstrated some important properties of the distribution of galaxies, such as:

a. galaxies are clustered on scales less than $10h^{-1}$ Mpc in dynamical systems of different richness (from poor groups to rich clusters);[1]

b. large, underdense regions (voids) have been detected with maximum sizes of the order of $50h^{-1}Mpc$;

c. the "field" galaxies around the voids form structures which are connected and bidimensional, with sizes up to $100h^{-1}Mpc$.

At intermediate depth ($z \leq 0.2$), redshift surveys, which are still in progress, have been confined to small areas (few tens of square degrees), but they have already confirmed the validity of the picture of the large scale distribution of galaxies derived from shallower surveys. Furthermore they have shown the beginning of the interplay between large scale structure and galaxy evolution (see e.g. Vettolani et al 1994).

As the timescales ($\leq 10^9$ years) for evolution of the stars responsible for the optical light are short by cosmological standards, the samples of faint galaxies, besides probing the geometry and structures and their evolution, contain much information on the evolution of the stellar population in the galaxies.

However, due to the large amount of telescope time needed to observe faint galaxies, redshift surveys over large areas have never been accomplished and surveys have been limited to some hundreds of galaxies up to $b_j \leq 22 - 23.5$ in small pencil beams of a few square arcminutes, thus probing galaxy evolution to $z = 0.3 - 0.6$ (see Koo and Kron 1992 for a review).

---

[1] $h = H/100$, where H is the Hubble constant.

Despite the efforts of many groups and the large investment of telescope time, many possible issues about the properties of large scale structure and galaxy evolution remain, however, largely unsolved. Concerning the large scale galaxy distribution, we have little or no information on the distribution function of the dimensions of structures and voids, their relative density contrast, their topology and statistical properties and the evolution with cosmic time of these properties. Concerning the stellar population evolution with cosmic time, any quantitative measure is still lacking.

In order to determine these properties, one has to study statistically representative samples of large numbers of galaxies covering large areas and volumes.

For surveys of galaxies with $b_j \leq 20 - 21$, a fundamental role is played by the 4 meter class telescopes, if properly equipped with efficient multi-fiber spectrographs with a large field of view and hundreds of fibers. The best example of this instrumentation is the 2dF Spectrograph at the Anglo–Australian Telescope, with 400 fibers over a two degree field of view (Taylor 1994). Surveys of fainter galaxies are the domain of the very large telescopes as the gain they provide in the number of photons hitting the detectors is just enough to make surveys of faint galaxies feasible in a reasonable amount of time. It is, therefore, easily predictable that the large telescopes will have a strong impact on the study of the formation and growth of the diverse structures, as well as constraining galaxy evolution (see Section 2).

In Section 3 the instrumental constraints and the needed instrumentation are discussed and finally in Section 4 some examples of possible surveys are given.

2 Driving Science

In the following I briefly describe the possible role of large telescopes in two nowadays very challenging problems: the nature of faint blue galaxies responsible for the galaxy counts excess; the study of the evolution of galaxy clustering.

2.1 Counts of Faint Blue Galaxies

Counts of galaxies in the blue band as a function of magnitude show a large excess ($\simeq$ a factor 5 at $b_j = 24$) over what is predicted by extrapolating the number of galaxies per unit volume as measured locally (through the galaxy luminosity function) to distant volumes and large lookback times. These predictions are hampered by our poor knowledge of galaxy luminosity function for different morphological types, the appropriate cosmological K-corrections at large redshifts and the amount of the evolutionary corrections which take into account the evolution of the stellar population in galaxies. These simple passively evolving models totally fail to predict the counts at faint magnitudes. Conversely K band counts, i.e. counts of galaxies selected on the basis of the luminosity in the older stellar population, show almost no excess over prediction.

One could explain the blue counts if these blue faint galaxies were distant galaxies with young stellar populations (hence the blue color) or a nearby very low luminosity population. However, the fact that their redshift distribution is similar to the predicted distribution from the same model which fails to predict the counts and that does not show a a tail at high or very low redshifts, requires a more complex scenario.

Many solutions have been put forward as for example:

a. non conservation of the galaxy numbers due to mergers;

b. a population of dwarf galaxies disappearing at short lookback times;

c. non zero cosmological constant.

What we do know is that some physical process involving a substantial fraction of the galaxy population was acting at redshifts around 0.5 and beyond. Identifying this process (or processes) is really challenging. Observing deeper samples (hence larger redshifts and lookback times) would permit testing of some of the hypotheses which have been put forward, such as the different shapes of the redshift distributions in faint samples.

1.2 Correlation Function

The modelling of the formation and growth of the diverse structures observed in the Universe can be strongly constrained by the measure of galaxy clustering, as quantified by the two-point spatial correlation function. This function $\xi(r)$ measures the joint probability of finding galaxies in the volume elements δV_1 and δV_2 separated by the distance r.

At small scales $\xi(r)$ evolves in time under the action of gravity while the signal on large scales strongly reflects the form of the initial spectrum of the perturbations in the Universe.

As its measure requires the knowledge of the galaxy distribution in three dimensions, we have very little information on the evolution of its shape and amplitude with cosmic time, due to the essentially 1-dimensional nature of the deep surveys. In principle the spatial correlation function can be indirectly determined from the angular correlation function. The procedure is quite similar to the counts modelling, in the sense that it requires integration along the line of sight to all galaxies in the sample contained within the appropriate volume of space. Again this procedure requires the knowledge of the luminosity function, its dependence on the morphological types , the amount of luminosity evolution that galaxies have undergone with cosmic time etc. Given the difficulties in the modelling of galaxy counts, it is clear that any determination of $\xi(r)$ and of its evolution from the angular function suffers from large uncertainties.

The determination of the spatial clustering properties at large lookback times would also enormously help to solve the puzzle of the excess of faint blue galaxies. The blue galaxies are vigorously forming stars, therefore, once separated from the more normal population through for instance the width of the [O II] line, their correlation function should clearly show if they are an unclustered new population of which there is no trace in the near samples, or if they have similar clustering properties to the more normal galaxies, which would favour a merger hypothesis.

Table 1. Cumulative distribution of faint stars and galaxies

(1) mag (b_j)	(2) N Galaxies per square arcmin	(3) B-R	(4) N stars
19.5	0.014		
20.0	0.036	1.53	0.19
20.5	0.08		
21.0	0.16	1.40	0.32
21.5	0.34		
22.0	0.58	1.43	0.50
22.5	1.05		
23.0	1.89	1.23	0.78
23.5	3.48		
24.0	6.12	1.19	0.90
24.5	10.57		

3 Survey Instruments for the VLT

3.1 The Guidelines

We stress that, at variance with a large fraction of astronomical research, *the study of large scale structure and galaxy evolution can be attacked only by studying complete statistical samples of many hundreds of objects.*

This fact has an immediate implication in the required instrumentation for faint galaxies surveys: the basic need is to simultaneously observe as many objects as possible in the largest possible field of view. Therefore, the comparison and choice among different instruments and designs is driven by this extremely simple concept: at a given magnitude limit, the advantage is defined by the ratio between the number of observable objects per unit area per unit time and the number of existing objects in the cosmos in the same area.

Table 1 gives the relevant numbers of objects (galaxies and stars) in magnitude bins (cumulative distribution), adapted from Jones et al (1991). Basically, one should try to match the number of slits to the density of objects at some interesting magnitude. If we assume that $b_J \sim 23$ is such a magnitude (see e.g. the example on large scale structure in Section 4.), the required slit density is of the order of 1.8 per square arcmin. The high density of objects per square arcmin calls for a mask system, which can accommodate more slits than a system of movable slits as in FORS.

3.2 Constraints

The requirement of a large field of view, with as large as possible number of slits, calls for the Nasmyth Focus of one of the VLT telescopes, with its unvignetted field of view of ≥ 20 *arcmin* diameter.

The constraints on resolution are posed essentially by the required accuracy in velocity measurement, which is typically of the order of $50 - 100$ km/s. The accuracy of emission line measurements is constrained by the signal-to-noise on the continuum (which depends on resolution, R, as the square root of $\lambda \times R^{-1}$) and which, in any case, for faint objects, calls for low resolution.

Spectral resolutions between 250 and 2000 seem quite well adapted to most of the programs, with the lowest value well tailored for quasar surveys or very faint galaxies and the highest value for accurate redshift measurements in clusters of galaxies.

The number of slits is constrained by the detector dimensions given the slit length. It is absolutely necessary to sample as well as possible the sky outside the object in order to get the best possible sky subtraction. Sky subtraction errors are in fact suppressive, in the sense that, for sky limited observations, the maximum attainable signal to noise of the spectrum is essentially defined by the error in the sky subtraction and, beyond a certain limit, it does not increase any longer with the the exposure time. This forbids the use of shorter slits. An acceptable length for the slits is between 10 and 15 arcsec.

3.1 A Spectrograph in the Visible

A design of a spectrograph fulfilling most of these requirements is the Wide Field Imaging Spectrograph (WFIS) described by Delabre et al (1994) which consists of four identical channels with a field of view of $7 \times 7 arcmin$ each. Assuming slits of 10–15 arcsec length, a total number of 100-120 spectra, or more depending on field geometry, overlap etc, can be obtained in a single exposure in the four CCDs, corresponding to a slit density of 0.4-0.6 per square arcminute. Therefore, complete sampling of all galaxies brighter than $b_j = 23$ would be obtained with 3 to 4 exposures. With the parameters of the spectrograph described in Delabre et al (1994), exposure times of the order of 1 hour will provide $S/N = 10$ in the continuum. Such a signal to noise is more than adequate to obtain reliable redshift determinations.

3.2 A Spectrograph in the Infrared

When dealing with faint galaxy samples, a further point which should be taken into account is that a "visible" spectrograph has a limit for cosmological studies of galaxies at $z \simeq 1.2$, when [O II]3727Å and CA II H and K lines disappear from the bandpass. Furthermore, quantitative studies requiring detection of lines in the red (in the restframe), such as evaluation of star formation rates (given by the $H\alpha$ flux) are not possible with a spectrograph working in the visible even at quite modest redshifts. Therefore, a spectrograph working in the visible, as

WFIS, should be complemented with a near-infrared multi-object spectrograph, such as NIRMOS (Le Févre 1994), and vice versa. In the same scientific context, NIRMOS will survey the cosmos at larger redshifts, and will allow direct comparison of spectral properties in the same restframe range 3700 to 6000 $\mathring{A}$ among galaxy samples at small redshift (i.e. nearby galaxies where the bulk of our knowledge is), medium redshift samples available through spectroscopy in the visible and high redshift galaxy samples. This comparison is essential in order to understand properly the evolution with cosmic time of the stellar population in galaxies.

4   Examples of Possible Programs

Assuming the availability at the VLT of a wide field multislit spectrograph working in the visible domain, such as WFIS, we give in the following two examples possible programs: the first one dealing directly with large scale structure and evolution; and the second with galaxy clusters.

Examples for similar instrument working in the infrared are given by Le Févre et al. elsewhere in this volume.

This is certainly not an exhaustive list of objectives: a large number of studies other than cosmological ones can also be accomplished, including stellar cluster dynamics, extragalactic globular clusters, etc.

4.1   Large Scale Structure

We start by describing a program which, for its size, is daring to propose, but which we believe should be considered for two main reasons: first, because it directly compares to the most ambitious program ever conceived in cosmological research, namely the SLOAN survey (1991; second, because it gives directly the flavour of how easily we will be able to accomplish programs with scientific objectives which are not conceivable without such an instrument on the VLT.

At the beginning of next century, the SLOAN survey will be accomplished with a fully dedicated 2.5 meter telescope. After a planned five year photometric survey, all galaxies (plus stars) brighter than $V \simeq 18$ magnitude in the northern sky will have been observed spectroscopically. This survey will have a typical depth (at the maximum of the selection function) of $\sim 240h^{-1} Mpc$, corresponding to an explored volume of $\simeq 8.3 \times 10^6 Mpc^3$, over 1.8 steradians.

A redshift survey at $b_j = 23$ has a depth (at the maximum of the selection function) of $\simeq 900h^{-1} Mpc$ (depending on q_0). Therefore it is easy to show that ~ 1800 fields ($15 \times 15 arcmin$) would give the *same* explored volume. These can be arranged in a way as to maximize the information on large scale structure transverse to the line of sight. Observing these fields with a galaxy sampling of one third, this would call for some 200 observing nights: a large but possibly feasible program!

A program like this would allow the determination at $z \simeq 0.3$ of the properties of the large scale structure with approximately the same statistical power as the SLOAN survey at $z \simeq 0.1$. A comparison of the results of these two surveys would immediately give us interesting and possibly exciting results on the evolution with cosmic time of large scale structure. Furthermore, a survey like this would have the enormous advantage, over the SLOAN survey, of tackling directly the problem of galaxy evolution (on which SLOAN tells nothing) over a lookback time 30% of the age of the Universe.

4.2 Galaxy Clusters

Clusters of galaxies are the more massive bound structures which have ever formed in the Universe, possibly through the hierarchical growth of subunits. A comprehensive dynamical study of a galaxy cluster, sampling its luminosity function well below M_*, requires a few hundred radial velocities of galaxies with accuracies of the order of $50 km s^{-1}$. This has been accomplished, or is in progress, for several nearby clusters with fiber spectrographs at 4 meter telescopes. At intermediate redshifts (0.3 − 0.5), extensive studies of a few galaxy clusters have been mostly motivated by the study of the Butcher–Oemler effect, i.e. an increasing fraction of blue, starburst and/or poststarburst galaxies, with respect to nearby clusters. More recently, comprehensive dynamical studies of a few clusters have led to the discovery of arcs and arclets due to the presence of a rich cluster in the line of sight of a background faint galaxy.

These studies are at the borderline of 4 meter class telescopes, requiring very long integrations per object (or per field in the case of multislit spectrographs), and therefore the number of clusters with useful dynamical information is quite limited.

At $z \simeq 0.3 - 0.4$ a 15 arcmin field corresponds to $\simeq 4 - 5$ Mpc and is extremely well suited to study the whole velocity field of a rich cluster and its surroundings. Assuming that 80 galaxies could be observed in 2-3 hours, 1 night of observing time would be enough to get details of the velocity field of a galaxy cluster up to $z \simeq 0.4$.

The high end of the cluster luminosity function can be then investigated, and the new observations would provide accurate estimations of the dynamical mass and its distribution over each whole cluster. These masses can be compared to the mass derived from X-ray gas models, as well as to the mass estimation from arcs and arclets. Spectroscopy of the latter could be done during the same time and greatly helps to estimate the cluster dark matter and its distribution. This data would provide unprecedented constraints on the largest clumps of dark matter in the Universe and hence on mass related cosmological parameters. Furthermore these observations would allow a detailed study of the Butcher–Oemler effect in relation to the cluster dynamical properties, as this effect probably arises from infall of field galaxies into the cluster.

References

Delabre,B., D'Odorico,S., Vettolani,G. (1994): in SPIE Conference "Instrumentation for the 21st Century", Kona, Hawaii, in press

Giovanelli, R., Haynes,M.P.: 1991, Ann. Rew. A. A., 29, 499.

Jones,L.R., Fong,R., Shanks,T., Ellis,R.S., Peterson,B.A.: MNRAS 249, 481 (1991).

Koo, D. and Kron R. 1992 Annual Review A. A. 30, 613

Le Fevre, O., Delabre, B., felenbok, P., Hammer, F., Tresse, L., Vettolani, P., Mellier, Y., Picat, J. P., Lilly, S. J.: 1994, these proceedings

Taylor K. 1994: AAO Newsletter No. 69

Vettolani, G. et al 1994 in IAU Symposium 161: "Astronomy from Wide Field Imaging", H.T. Mac Gillivray Edt. page 687

The Nature of Compact Objects at R=23

G. Mathez[1], Y. Mellier[1] and J.P. Picat[2]

Observatoire Midi-Pyrénées,
[1]URA285, 14, Av. E. Belin, F-31400 Toulouse
[2]URA1281, 9, Rue du Pont de la Moulette, F-62500 Bagnères de Bigorre

Abstract. In order to check current samples for completeness and to minimize redshift biases, we suggest searching for quasars by taking spectra of *all* stellar objects.

1 Scientific Objectives

Large and complete samples of QSO's are needed to understand the intrinsic quasar phenomenon and to study the statistics of the whole quasar population. The latter aspect deals with the quasar cosmological evolution and the use of quasars in classical cosmological tests: distribution of redshifts, maximum redshift, quasar-quasar correlation and statistics of lensed quasars. It is also essential to achieve the construction of these samples with the fewest and the simplest selection criteria, since many unexpected biases, never easy to understand, can affect such a construction.

Since the very first QSO's were found by identification of optical counterparts of radiosources, it was predicted (Schmidt 1970) that most quasars do *not* emit in radio. Later on, this was directly established from radio studies of optically detected QSO's. Since that time, other spectral windows have been used to discover new quasars, namely in the X-ray and Gamma-ray ranges. However, even a non optically detected quasar requires optical or near IR spectroscopy to measure its redshift. The resulting catalogue suffers a double flux limitation: one in radio or X-ray for the detection, and one in the optical range for spectroscopy. It is to avoid this drawback that many purely optical searches for QSO's have been performed.

The programme described below is aimed at the construction of a deep sample of QSO's with the fewest possible selection criteria, thus the fewest possible biases in the redshift distribution. Incidentally, it could serve to improve our knowledge on the faint end of the stellar luminosity function in the Milky Way.

1.2 Optical Surveys of QSO's

The very first purely optical systematic quasar survey was performed through the selection of BSO's (Blue Stellar Objects) showing infrared excess with respect to main sequence stars (Braccesi et al. 1968). The most popular way of finding QSO's then turned out to be a photometric pre-selection of UVX stars (UVX for UV excess), followed by spectroscopic search for strong emission lines. But the UV excess comes from emission lines in the U band, and as the bluest strong emission line is Lyα (1216 Å), UVX QSO's have redshifts limited to $z < 2.3$. This method is likely to bias the samples towards those objects from which we receive a significant part of their energy in the blue. Webster (1994) concluded from the IR study of a complete radio-selected quasar sample, that current optical surveys (in UV or B) could miss a significant fraction of red QSO's due to dust-reddening.

A second optical way to detect quasars is by using an Objective Prism, Grens or Grism to search for emission lines in low dispersion spectra over a given field. This method is efficient, but it applies mainly at high redshift z>1.8, and biases are difficult to quantify, depending on the equivalent widths of the main emission lines.

Multiband optical photometry has also been used to distinguish QSO candidates from Main Sequence stars. The method consists in assuming a standard QSO spectrum, and, as the redshift varies, to follow the corresponding "evolutionary" track in the various color-color planes. The main drawbacks of this method are that it relies on the existence of a "universal" QSO spectrum, and on a modelling of its evolution (continuum+lines). It also requires very precise (δ mag $\simeq 0.01$) narrow band photometry in six or more bands. Finally, it is sensitive to time variations in spectra (continuum+lines) and it is time consuming, far more than UVX searches.

Other selection criteria (e.g. light variations, lack of proper motion) have been used to find quasars, but with marginal success. Each of the main optical methods induces specific redshift biases.

1.3 Pushing the Redshift Limits in Complete Samples

The net result is that most of the complete large samples so far available are UVX QSO's, thus limited to objects emitting preferentially in the blue and of redshift z<2.2. There are however good reasons for pushing the redshift beyond this limit.

At redshift less than z=2.2, there is compelling evidence that quasars show a strong evolution, either of their mean density (Schmidt 1968), or of their mean luminosity (Mathez, 1976,1978; Boyle et al. 1987; but see Kassiola & Mathez, 1990). However, the cosmological evolution of the quasar luminosity function, whatever its precise form, seems to reverse somewhere between z=2 and z=3 (Osmer, 1982), so the question is to what extent is this inversion of tendency explained by the transition between two selection criteria ?

Quasars are essential for cosmology because they lie in galaxy nuclei. Both the quasar-quasar correlation function and their maximum redshift are key constraints for models of galaxy formation (especially for its epoch and duration). Cosmological tests do require high-z objects. Accordingly, quasars are the most promising class of objects, through the distribution of their absorption redshifts (Rauch, 1994), the typical scale of their spatial distribution (Deng, Xia and Fang 1994), or the distribution of the 3D orientations of the quasar-quasar separations (Phillipps 1994). The QSO redshift distribution has also been used, provided that the evolution is modelled (Schade and Hartwick, 1994), or can be bypassed in some way (Mathez et al., 1994). Finally, there is presently a renaissance of the cosmological constant. Its influence increases drastically with increasing redshift, making QSO's a unique tool to investigate such a possibility.

2   Observational Programme

The programme we propose consists in the *systematic spectroscopy* of objects selected in a photometric catalogue, with *no other criteria than compactness and a limiting magnitude R=23*. It might be achieved within a reasonable timescale thanks to the collecting power of the 8-m telescope.

2.1   Similar Previous Surveys

Similar surveys were already done on 4-m telescopes: Koo et al. (1986) made a pre-selection of candidates brighter than B=22.5, based on several criteria namely *color, variability and astrometry*. Schade (1991) made a preselection based on the *compactness* of objects brighter than V=23.1. Since they were essentially interested in galaxies, Colless et al. made (1991) a *random* selection to B=23.5, and Le fèvre et al. (1994) made their search *without any preselection* other than I<22.1.

2.2   Observing Strategy

There are about 7500 compact objects up to R=23 per sq. deg. From the results of previous surveys we can reasonably extrapolate the following statistics in one average FUEGOS Field (1/7 sq. deg.): 1040±70 stars; 25±10 quasars (1 σ); and 40±15 AGN. To get 1000 QSO's, about 6 sq. deg., or 40^{+25}_{-10} (±1σ) FUEGOS Fields) must be surveyed, which requires a total of about 60 nights. In the process, 1800 AGN spectra and 40,000 star spectra will be obtained.

It should be feasible to implement FUEGOS with 200 fibres without increasing too much the cost. The optimal number of fibres devoted to the sky acquisition is the square root of the total number. We assume below that 185 fibres are useful, 15 being reserved for the sky spectrum, because sky subtraction will be in any case crucial. Under these hypotheses, an average of 6 exposures will be necessary in each FUEGOS field.

Preliminary UBVRIK photometry allows *all compact objects brighter than R=23* to be included in a catalogue. There are two advantages to operate this selection in the R frame. First, the galactic envelopes extend farther in red light, so there will be fewer doubtful compact objects and less spectra to be taken. This point is crucial since galaxies dominate over stars at these magnitudes. Second, selection in the red will induce less discrimination against high-z quasars.

We assume that the full spectral range 4700 – 9000 Å will be available with a resolution R≃600. In these conditions, the S/N ratio $\simeq$ 5 to 7 will be achieved in the emission lines after about 90 min. exposure time. In case of ambiguous spectra (no emission line, or emission line falling on a sky feature), these objects will be observed with FORS, (see Table 1 for a comparison of these two instruments estimated from the ESO booklet), allowing the sky correction to be more secure. Fig. 1 shows the two allowed redshift windows for the 6 strongest emission lines in QSO's corresponding to the respective spectral ranges of FUEGOS (5000 – 9000 Å) and NIRMOS (8000 – 18000 Å). Only one emission line is expected with FUEGOS for redshifts $z\simeq1$ (MgII) or $z>4.8$ (Lyα). Objects with a single emission line will benefit from spectroscopy with NIRMOS to determine their redshift. Redshifts up to 10 are thus measurable with Lyα 1216 Å *and* CIV 1549 Å .

Table 1. Comparison between FORS and FUEGOS, giving the total exposure time to survey all compact objects brighter than R=23 over 1 sq. deg.

	Field Area $('^{-2})$	Obj. /Pose	Obj. /Field	Poses deg.$^{-2}$	Total Time deg.$^{-2}$ (hours)
FORS					
19 slits	46	19	100	410	310
FUEGOS					
80 Fibres	530	70	1100	110	160
FUEGOS					
200 Fibres	530	185	1100	40	60

3 Conclusions

The star spectra, being obtained with too low a resolution, would have little or no interest for galactic dynamics. However they might provide a large database for spectral classification of dwarf stars.

The main interest of this programme would be to offer an unbiased estimate of the respective contributions of stars and QSO's, and of the fraction of dust-reddened QSO's, to the counts at R=23. It will provide an objective test of the efficiency of current color-based selection criteria of QSO's and an objective estimate of the completeness of current QSO samples. The result will lead to the first homogeneous, complete and large sample of QSO's without obvious bias in redshift, allowing progress in the understanding of QSO evolution and in the determination of the QSO redshift distribution, especially their maximum redshift.

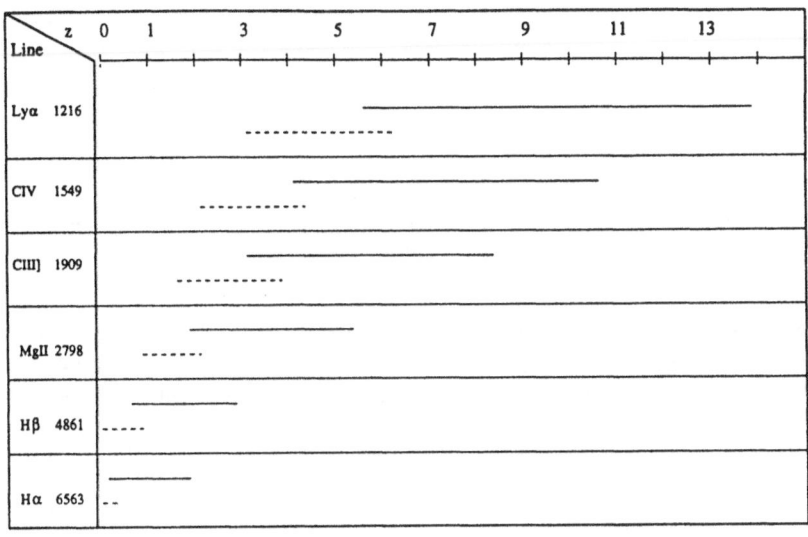

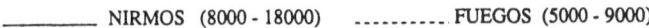

——————— NIRMOS (8000 - 18000)   ·········· FUEGOS (5000 - 9000)

Fig. 1. The two allowed redshift windows for the 6 strongest emission lines in QSO's corresponding to the respective spectral ranges of FUEGOS (5000 – 9000 Å) and NIRMOS (8000 – 18000 Å).

References

Boyle, B.J., Jones, L.R., Shanks, T. (1991) MNRAS 251, 482

Braccesi, A., Lynds, R., Sandage, A., (1968), ApJ 152, L105

Colless, M., Ellis, R., Taylor, K., Shaw, G. (1991) MNRAS 253, 686

Deng, Z., Xia, X., Fang, L.Z. (1994) ApJ 431, 506

Kassiola, A., Mathez, G. (1990) A & A 230, 255

Le fèvre, O., Hammer, F., (1994), this conference

Mathez, G., (1976), A & A 53, 15

Mathez, G., (1978), A & A 68, 17

Mathez, G., et al. (1994), *in preparation*

Osmer, P.S., (1982) ApJ 253, 28

Phillipps, S., (1994), MNRAS 269, 1077

Rauch, M., (1994) this conference

Schade, D., (1991), AJ 102, 869

Schade, D., Hartwick, F.D.A., (1994) ApJ 423, L85

Schmidt, M., (1968), ApJ 151, 393

Webster, R., (1994), XXIInd IAU General Assembly, The Hague.

Faint AGNs, Evolution and the X-Ray Background

Gianni Zamorani

Osservatorio Astronomico, Via Zamboni 33, I-40126 Bologna, Italy Istituto di Radioastronomia, Via Gobetti 101, I-40129 Bologna, Italy

Abstract. I briefly describe three possible VLT observational programs, each of which will clarify different aspects of the evolution of Active Galactic Nuclei. Programs like these will allow us to start giving some answers to simple, but important questions such as: Do the faint AGNs evolve in the same way as the more luminous ones? Is there a continuity in the properties of low and high luminosity AGNs? Are simple unified schemes consistent with the observations? Are AGNs short- or long-lived? What is the evolutionary history of single AGNs?

1 Introduction

In this paper I will briefly describe three possible observational programs, each of which will clarify different aspects of the evolution of Active Galactic Nuclei (AGNs). The first program is a spectroscopic survey of faint AGNs, aimed at studying the evolutionary properties of low luminosity AGNs; the second one, optical identification of a complete sample of faint infrared selected sources, is aimed at studying the ratio between type 2 and type 1 (i.e absorbed and unabsorbed) AGNs as a function of luminosity and/or redshift; the third one, high spatial resolution imaging and spectroscopy of a sample of nearby galaxies, is aimed at determining the frequency and mass distribution of black holes in normal nearby galaxies. The combined results of these programs will have an important impact on our phenomenological description of the evolutionary properties of AGNs as a whole, on the problem of the production of the X-ray background (XRB) and, finally, on our knowledge of the evolutionary history of the active nuclei.

2 Evolution of the Luminosity Function of AGNs

In the last decade the number of AGNs in complete samples has increased almost exponentially. Just to give an example, this number for $z > 0.3$ has increased from ~ 100 in 1985, to ~ 1000 in 1990, to probably more than 3000 today. In parallel with the increase in number, there has been also an improvement in the limiting magnitude of the faintest of such complete samples, from $m_B \simeq 20.0$ to $m_B \simeq 22.5$, and in the redshift coverage, with the inclusion of a number of samples selected at high redshift (i.e. $z \geq 2.2$), where the usual UV excess selection criterion is not efficient any more. This increased sampling of the luminosity -

redshift plane has allowed more detailed fits of the global evolutionary properties of AGNs; while at low redshift (z ≤ 2.0) most of the data are consistent with a pure luminosity evolution of the luminosity function (Boyle et al. 1991; but see Hewett et al. 1993 for recent evidence of systematic changes in the shape of the bright end of the luminosity function). At higher redshift the situation is still less clear, also because of the more complex selection and incompleteness functions of the existing samples (Warren et al. 1994).

Table 1. Existing Complete Samples of Faint AGNs

	MZ	KK
Selection:		
Multicolor	yes	yes
Variability	yes	yes
Proper motion	no	yes
X-ray	yes	no
Area (deg^2)	0.3-0.7	0.3
m_{lim} (spectra)	≤ 22.5	≤ 22.5
N_{AGNs}	64	35
N_{AGNs} with $m_B ≥ 21.5$	26	15
AGN surface density (obj./deg^2):		
$m_B ≤ 21.5$	73 ± 13	69 ± 15
$21.5 < m_B ≤ 22.0$	49 ± 12	38 ± 11
$22.0 < m_B ≤ 22.5$	40 ± 13	48 ± 13

Despite the large amount of effort and telescope time invested in this field, the behavior of the faint end of the luminosity function is still relatively poorly known. This is due to the fact that the number of AGNs in complete samples at faint magnitudes is still small today (i.e. probably less than 100 with $m_B > 21.5$). Most of these faint AGNs in complete samples are part of two different samples, studied by Marano, Zamorani and collaborators (MZ; see Zitelli et al. 1992) and Koo, Kron and collaborators (KK; see Trevese et al. 1994). Table 1 gives a few characteristics of the two samples, such as the techniques for the selection of the candidates, the areas surveyed, the limiting magnitude of the spectroscopic survey and the total number of spectroscopically confirmed AGNs. It is interesting to note that the assembling of each of these two samples, including both the selection and the spectroscopic confirmation, has required observing runs over about 10 years at 4m class telescopes.

As clearly seen in Table 1, both samples have been selected by using a variety of selection techniques. For this reason, the current estimates of incompleteness for the two samples are extremely small, of the order of $\simeq$ 5-10%. Not only the areas, the limiting magnitude and the number of objects are comparable in the two samples, but also the surface densities in different magnitude ranges are in excellent agreement with each other (see Table 1). In comparing the surface densities of the two samples, I should note that the surface density for the MZ sample in the faintest magnitude bin should be considered as a lower limit, because of some residual incompleteness in the spectroscopic observations in this magnitude range. The similarity between the two samples, the high level of completeness and the excellent agreement in the derived surface densities allow us to merge the two samples together (99 AGNs) in order to estimate what we should expect from deeper surveys with VLT.

Firstly, which class of AGNs do we expect to find at fainter magnitudes? No correlation between redshift and apparent magnitude is shown by these AGNs in the magnitude range $20 \leq m_B \leq 22.5$. As a consequence, there is instead a strong correlation between absolute and apparent magnitude, in the sense that the proportion of intrinsically fainter AGNs increases significantly at fainter magnitudes. This is shown in Fig. 1, where, somewhat arbitrarily, we have divided the sample of AGNs into Seyferts and quasars at $M_B = -23.0$ ($H_0 = 50$ km s^{-1}Mpc^{-1}). The ratio between quasars and Seyferts decreases from 7.4 for $m_B \leq 21.0$ to 1.0 for $m_B > 21.0$. We can therefore expect that, at magnitudes fainter than 22.5, most of the AGNs will be in the luminosity range typical of the local Seyfert galaxies.

Secondly, what surface density of AGNs do we expect at fainter magnitudes? Rather than integrating a model for the luminosity and evolution functions, I prefer to extrapolate to fainter magnitude the observed counts with a slope similar to what is observed in the magnitude range $20.0 < m_B < 22.5$. In this way, I obtain an estimate of $\sim$ 190 AGNs per deg^2 for $22.5 < m_B < 23.5$ and $\sim$ 165 AGNs per deg^2 for $23.5 < m_B < 24.0$. Figure 2 shows a color - color diagram for stellar objects brighter than $m_B = 23.5$. These data, derived from four CCD fields of 6 × 6 arcmin each, have been recently obtained at the ESO NTT telescope. A few objects are clearly visible outside the locus occupied by main sequence stars. The triangles are spectroscopically confirmed AGNs, 5 out of 6 brighter than $m_B = 22.5$, while the asterisks are stars. In the magnitude range $22.5 < m_B < 23.5$ there are 9 AGN candidates, for 8 of which no spectroscopy is available yet. If all of these turn out to be AGNs, the corresponding surface density would be 225 ± 75 AGNs per deg^2. Since at these magnitudes the contamination from galactic stars in the region of the color - color diagram occupied by AGN candidates should be negligible and, in addition, some AGN candidates may have been classified as extended and, therefore, would not appear in this diagram, we conclude that the estimates for the surface density of faint AGNs given above can be considered reliable and, possibly, even lower limits to the real value.

With this information I can now outline a possible VLT observational program with the already approved Multi-Fiber Area Spectrograph (MFAS). With MFAS one can obtain 80 or more spectra simultaneously over a field of view of

30 arcmin, corresponding to 0.2 deg². Table 2 gives a schematic summary of an AGN spectroscopic survey optimally tuned for this instrument on the VLT. As shown in the table, ~ 150 hours of VLT time would produce more than 1,000 AGNs in a magnitude range where very few data are available today. Obviously, deep photometric work for the selection of the candidates on 4m class telescopes is required. This work should start soon, in order to be ready with a reliable and complete list of candidates when the VLT becomes available.

Table 2. A Spectroscopic Survey for Faint AGNs with MFAS

m_{lim}	# of fields	Exposure time per field (hours)	# of AGNs per field	Total # of AGNs
$m_B \leq 22.5$	15	2	33	500
$22.5 < m_B \leq 23.5$	10	5	38	380
$23.5 < m_B \leq 24.0$	6	12	33	200

With this sample of $\sim 1,000$ AGNs we will be able to study the continuity of properties between Seyfert galaxies and quasars by obtaining a good and reliable determination at all redshifts of the faint part of the luminosity function. At the same time, we will quantitatively determine whether low luminosity AGNs do evolve similarly to the higher luminosity ones. Both these topics are important also for models which explain the production of the XRB as integrated emission from AGNs, since intrinsically faint AGNs are supposed to contribute a significant fraction of the XRB. In addition to these "statistical" studies, deeper, higher signal-to-noise observations of a subsample of these 1,000 AGNs will extend significantly, at each redshift, the range of luminosity for detailed analyses of physical relationships which may depend on both luminosity and redshift (e.g. Baldwin effect, metallicity and evolution in the Eddington ratio).

An essential requirement for all these projects is the completeness of the sample to be analyzed or, at least, a quantitative understanding of the possible incompleteness. At $22 < m_B < 24$ the relative surface density between galaxies, stars and AGNs in high galactic latitude fields is approximately 40:4:1. How many AGNs may be hiding among the stars with "normal" colors and/or among the objects classified as extended? The first part of the question can be answered with the same MFAS observations because, as seen in Table 2, the expected number of AGNs per single exposure (~ 35) is less than a half of the total number of available fibers. An answer to the second part of the question can be obtained much more efficiently with an instrument like the Wide Field Imaging Spectrometer (WFIS), not yet approved, but currently under study. With ~ 100 slits in ~ 200 arcmin² and its higher efficiency, such an instrument is much more suited for observations of objects of high surface density as are faint extended objects. It is likely that with WFIS there will be a number of spectroscopic surveys of all the extended objects in a limited area of the sky and in the magnitude range of interest here. We will therefore use the results from

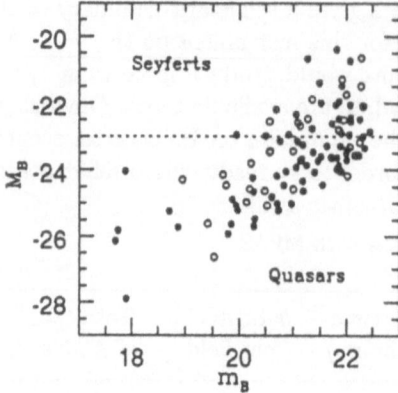

Fig. 1. Absolute versus apparent magnitude for the AGNs in the MZ (filled circles) and KK (open circles) samples. A dividing line between Seyferts and Quasars is indicated at $M_B = -23.0$.

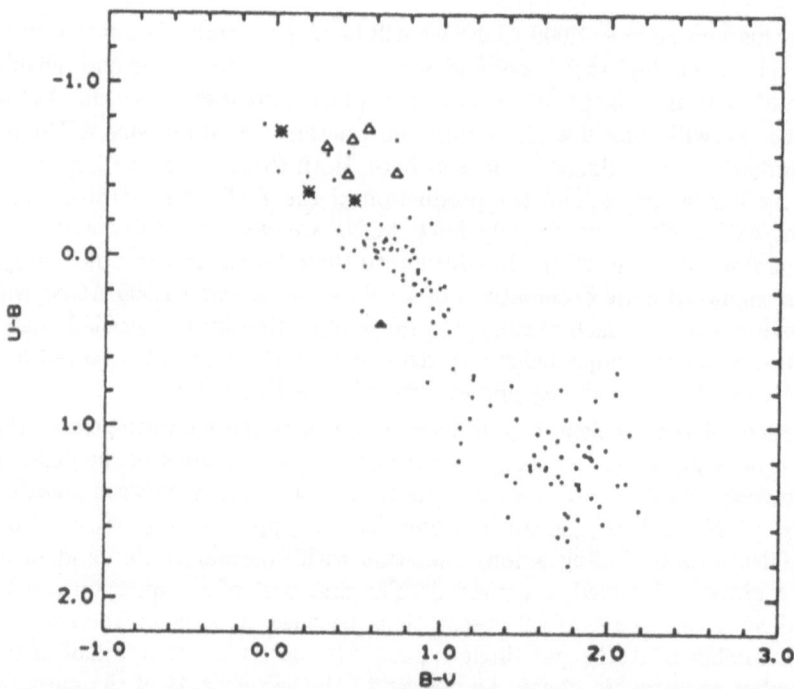

Fig. 2. U-B vs. B-V diagram for stellar objects brighter than $m_B = 23.5$ (see text for details). These data, derived from four CCD fields of 6×6 arcmin each, have been recently obtained at the ESO NTT telescope. The triangles are spectroscopically confirmed AGNs (open triangles for $z < 2.2$, filled triangle for $z > 2.2$), while the asterisks are stars.

these surveys for a statistical estimate of the possible losses in our selection of the AGN candidates.

3 Evolution of the Ratio Between Type 2 and Type 1 AGNs and the XRB

The problem of the origin of the extragalactic XRB has attracted much renewed attention in recent years following, in particular, the results obtained with the X-ray satellites GINGA and ROSAT (see Fabian and Barcons 1993 and Zamorani 1994 for recent reviews). It has been shown in a number of papers that, using the most recent spectral X-ray data for AGNs, it is possible to obtain good fits to the XRB spectrum. However, on this basis alone, it is difficult, if not impossible, to discriminate between the various proposed models. In a recent paper, Comastri et al. (1994) have discussed a self-consistent AGN model for the synthesis of the XRB in the framework of a simple AGN unified scheme, taking into account also all the additional observational constraints in the X-ray band, as, for example, the observed source number counts in the soft and hard X-ray bands, the redshift distributions and the spectral characteristics of different types of AGNs (i.e. type 2 (absorbed) and type 1 (unabsorbed)). Figure 3 shows the observed XRB spectrum compared with the Comastri et al. model. As clearly seen in the figure, a key parameter for this model, and for other similar models as well (see, for example, Madau et al. 1994), is the ratio between absorbed and unabsorbed AGNs. With some reasonable assumptions, this ratio, derived from the X-ray data, can be converted into the ratio between Seyfert 2 and Seyfert 1 galaxies, derived from optical spectroscopic data. In the Comastri et al. model the resulting ratio between Seyfert 2 and Seyfert 1 is in the range 2.4-3.7, in reasonable agreement with the value 2.3±0.7 found by Huchra and Burg (1992) for a complete sample of optically selected Seyfert galaxies.

Some important questions with respect to the class of models described above are: is the ratio between type 2 and type 1 AGNs a function of luminosity; is it constant with time, i.e. are the evolutionary properties of these two classes of AGNs the same? Unfortunately, the existing determinations of this ratio are all based on extremely local samples of relatively low luminosity AGNs, and therefore do not allow any conclusive answer to these questions. It may be surprising to note that no complete sample of Seyfert 2 galaxies exists for redshifts higher than 0.05, but this is due to the objective difficulty in finding these elusive objects. In the optical band, the absorption in the torus around the nucleus not only hides the broad emission lines, but also, reddening the emission from the nucleus, makes it difficult to use the standard UV-excess selection criterion. In the soft X-ray band the deep ROSAT surveys, which are finding a large density of sources (Hasinger et al. 1993), comparable to the expected number of optical AGNs at $m_B \sim$ 23.5-24.0, are biased against the detection of these objects because of the strong photoelectric absorption at low X-ray energy. As seen in Fig. 3, deep X-ray surveys with good spatial resolution at energies greater than 10-20 keV would in principle be able to provide the sample of distant, absorbed

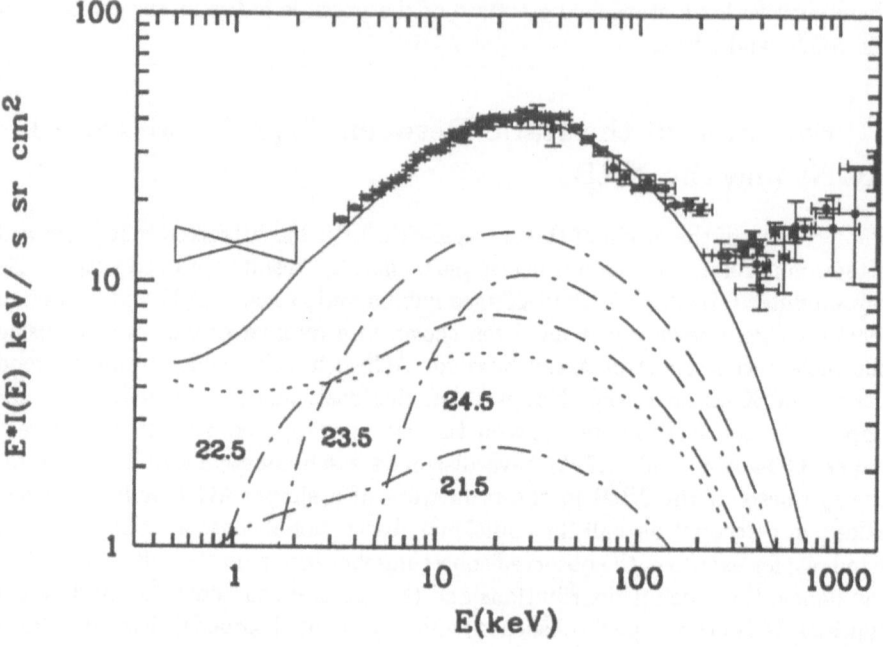

Fig. 3. The observed XRB spectrum compared with the Comastri et al. model. The solid line represents the best fit of the model, the dotted line represents the contribution of unabsorbed AGNs, while the absorbed AGNs are indicated by dot-dashed lines. The labels are the logarithms of the corresponding neutral hydrogen column densities.

AGNs we are looking for. Unfortunately, no such mission is foreseen for many years.

Probably the best wavelength window for a systematic search for these objects is the mid-infrared which is usually dominated by circum-nuclear dust or the non-thermal nuclear emission (Danese et al. 1992). While the IRAS sensitivity was rather poor at these wavelengths, the forthcoming ISO mission has an optimal sensitivity for such a program. Deep surveys of a couple of well studied fields are currently planned in the ISOCAM central program. In one of these fields (the "Marano" field, accessible from the south), this survey is expected to cover ∼ 540 arcmin² with a sensitivity limit of 25 and 60 μJy at 6.7 and 15 μm, respectively. Using models of AGN evolution derived from the optical band and the available information on the optical-IR nuclear colors of AGNs, the detection of ∼ 50 type 1 AGNs with a typical blue magnitude in the range 22.5-23.0 and a typical redshift in the range 1.0-3.0 is predicted (Franceschini et al. 1991 and Cesarsky et al. 1993). These objects are expected to be about 10% of the total number of infrared sources at these fluxes. On account of the absence of any information on the evolution of type 2 AGNs, no detailed prediction exists for these objects. In the framework of the AGN unified scheme, a similar or even higher number of type 2 AGNs would be expected; the redshift distribution should also be similar, while the typical optical magnitude should be somewhat fainter, by 1.0-1.5 magnitudes, because of possible effects due to obscuration of the nucleus. With an expected surface density of ~ 0.2 AGNs per arcmin^2 and a total surface density of ~ 1 infrared source per arcmin^2, this deep ISO selected sample would be a perfect sample for VLT follow-up spectroscopy. Five to ten deep exposures on these ISO sources with the multi-slit instrument WFIS should be enough to give a quantitative answer to the question on the evolution of the ratio between type 2 and type 1 AGNs and, at the same time, to test the validity of the above described models for the production of the XRB. Quite a small VLT time investment for the solution of an important and much debated scientific issue!

4 The Black Hole Mass Function in Nearby Galaxies

In the last ten years or so there have been many attempts to understand the physical reason for the strong evolution of the AGN population. The observed evolution of the luminosity function should be reproduced by physical models of evolution of single objects. Two main classes of models have been explored. In the first, a single generation of active objects is active continuously over a time-scale $\geq 10^9$ yr; in the second, the active lifetime of each AGN is $\ll 10^9$ yr, but many generations flare and fade, producing the observed luminosity functions. Detailed investigations by Cavaliere and Padovani (1988, 1989) have shown that the continuous models, implying very high masses (typically larger than 10^{10} $M_\odot$) for the central black holes in local AGNs and a much lower than observed ratio of the bolometric to the Eddington luminosity, appear to be at variance with the available observational data. The many generation models, on the other hand, predict remnant black holes, of significantly lower mass, in the center of a significant fraction of all galaxies. Recently, Haenelt and Rees (1993), assuming that AGNs are short-lived and constitute the first phase of the formation of the galaxies in the potential well of a dark matter halo, have been able to reproduce the observed dependence of the AGN luminosity function on redshift and to predict the mass function for the remnant black holes in normal galaxies (see their Fig. 8). Obviously, this mass function depends on the details of the model and should be tested with observational data.

As described in a number of papers in these proceedings, VLT with its excellent spatial resolution will be able to study in detail, both in imaging and in spectroscopy, the inner regions of local AGNs, improving on the already exciting results which are currently being obtained with the Hubble Space Telescope (see, for example, the mass determination of M = $(2.4 \pm 0.7) \times 10^9$ $M_\odot$ within 0.25 arcsec from the nucleus of M87 recently obtained by Harms et al. (1994) with HST FOS spectroscopy). For the scientific aim described in this Section we would be interested not in studying in detail the physical nuclear properties of a few particularly interesting local AGNs, but in determining the black hole mass function for a statistically significant sample of nearby galaxies. With this aim in mind, a possible VLT program could be organized as follows:

1. selection of a large (> 100) sample of galaxies of various luminosities and morphological types. A correct selection of the sample is crucial, because such a sample should obviously be free of any possible bias, in order to allow a correct interpretation of the final results;

2. VLT observations:

   a) infrared imaging with CONICA. With these observations, taking advantage of the excellent infrared spatial resolution ($\sim$ 0.05 arcsec), we will look for a signature of the presence of a central black hole, searching, for example, for indications of a stellar cusp or for the presence of a disk-like structure around the galaxy nucleus (see Ford et al. 1994 for HST high resolution images of the central regions of M87);

   b) bi-dimensional imaging and spectroscopy with MFAS in the ARGUS mode. With a spatial resolution of $\sim$ 0.2 arcsec and a spectral resolution of $\sim$ 5,000 we should be able to study in detail the inner velocity fields and the velocity dispersion as a function of position around the nucleus;

3. combined analysis of the imaging and spectroscopic observations will allow an estimate of the black hole mass, or setting upper limits for it in each galaxy and, finally, derivation of the mass function of black holes for the entire sample, to be compared with the predictions of various AGN evolution models. In this way, we should be able to set strong physical constraints on these models.

5 Conclusions

Evolution is probably one of the most striking features resulting from studies of AGNs. In this paper I have briefly described three observational programs with the VLT which should allow us to start giving some answers to simple questions such as: do the faint AGNs evolve in the same way as the more luminous ones; is there a continuity in the properties of low and high luminosity AGNs; are simple unified schemes consistent with the observations; are AGNs short- or long-lived; what is the evolutionary history of the single AGNs?

The questions are simple, but the answers we will obtain with these VLT programs will have a profound impact on our knowledge of the physics of these, still poorly understood, objects.

References

Boyle B.J., Jones L.R., Shanks T., Marano B., Zitelli V., Zamorani G. 1991, in Proceedings of "The Space Distribution of Quasars", Astronomical Society of the Pacific Conference Series, D. Crampton ed., Vol. 21, p. 191.

Cavaliere A., Padovani P. 1988, ApJL, 333, L33.

Cavaliere A., Padovani P. 1989, ApJL, 340, L5.

Cesarsky C., Chase S.T., Danese L., Desert X., Franceschini A., Harwit M., Hauser M., Koo D., Mandolesi N., Puget J.L. 1993, Proposal for ISO Guaranteed Time.

Comastri A., Setti G., Zamorani G., Hasinger G. 1994, A&A, in press.

Danese L., Zitelli V., Granato G.L., Wade R., De Zotti G., Mandolesi R. 1992, ApJ, 399, 38.

Fabian A.C., Barcons X. 1992, ARA&A, 30, 429.

Ford H.C., Harms R.J., Tsvetanov Z.I., Hartig G.F., Dressel L.L., Kriss G.A., Davidsen A.F., Bohlin R., Margon B., 1994, ApJL, in press.

Franceschini A., Toffolatti L., Mazzei P., Danese L., De Zotti G. 1991, A&A Suppl., 89, 285.

Haenelt M.G., Rees M.J. 1993, MNRAS, 263, 168.

Harms R.J., Ford H.C., Tsvetanov Z.I., Hartig G.F., Dressel L.L., Kriss G.A., Bohlin R., Davidsen A.F., Margon B., Kochar A.K. 1994, ApJL, in press.

Hasinger G., Burg R., Giacconi R., Hartner G., Schmidt M., Trumper J., Zamorani G. 1993, A&A, 275, 1.

Hewett P.C., Foltz, C.B., Chaffee F.H. 1993, ApJL, 406, L43.

Huchra J., Burg R. 1992, ApJ, 393, 90.

Madau P., Ghisellini G., Fabian A.C. 1994, MNRAS, in press.

Trevese D., Kron R.G., Majewski S.R., Bershady M.A., Koo D.C. 1994, AJ, in press.

Warren S.J., Hewett P.C., Osmer P.S. 1994, ApJ, 421, 412.

Zamorani G., in Proceedings of "Extragalactic Background Radiation: a Meeting in Honor of Riccardo Giacconi", Cambridge Univ. Press, in press.

Zitelli V., Mignoli M., Zamorani G., Marano M., Boyle B.J. 1992, MNRAS, 256, 349.

Part 10

**High Resolution Imaging
and Interferometry**

High-Resolution Imaging with the VLT at Optical Wavelengths

Gerd Weigelt

Max-Planck-Institut für Radioastronomie, Auf dem Hügel 69, D-53121 Bonn, Germany

Abstract. With the 8 m VLT telescopes and speckle methods it will be possible to achieve unprecedented angular resolution for many classes of astronomical objects and to obtain unprecedented astrophysical information. For example, at $\lambda \sim 4000$ Å, diffraction-limited images with a resolution of $\lambda/D \sim 0.010$" (10 mas) can be obtained. Optical long-baseline interferometry with the VLT interferometer will yield the fantastic resolution of 1 milli-arcsec. We discuss scientific objectives, examples of previous speckle observations, and computer simulations of optical long-baseline interferometry in the multi-speckle mode.

1 Scientific Objectives

Diffraction-limited images can be reconstructed by the Knox-Thompson method (Knox and Thompson 1973), speckle masking method (Weigelt 1977; Weigelt and Wirnitzer 1983; Lohmann et al. 1983), and other techniques. Diffraction-limited autocorrelations can be obtained by speckle interferometry (Labeyrie 1970). At $\lambda \sim 4000$ Å, the diffraction-limited resolution of a telescope with diameter $D = 8$ m is $\lambda/D \sim 0.010$" (10 mas). Since the diameter of the VLT mirrors is 3.3 times larger than the diameter of the Hubble Space Telescope mirror, speckle imaging can yield 3.3 times higher resolution than the HST at any given wavelength, if the object is bright enough. The limiting magnitude is about 18th magnitude for objects consisting of a small number of resolution elements (point sources) in nights of very good seeing (see Sect. 3). The signal-to-noise ratio in the reconstructed image is inversely proportional to the third power of seeing. Because of ! this strong seeing dependence, we plan to apply the VLT adaptive optics system, with or without Laser guide stars, in order to improve the SNR in the reconstructions.

In addition to high angular resolution, spectral information can be obtained by various speckle spectroscopy techniques, for example, O(x,λ)-projection speckle spectroscopy (Grieger and Weigelt 1990, 1992) and high-resolution objective prism (slitless) speckle spectroscopy (Weigelt et al. 1986; Afanasyev et al. 1992). A spectral resolution of 1 Å to 0.1 Å can be obtained.

The following objects are examples of important candidates for speckle imaging and speckle spectroscopy. Of course, the list of projects is not complete. Most of the objects discussed in the following sections have already been resolved by speckle observations with 3 m class telescopes, demonstrating the feasibility of such projects. More detailed discussions of most of the projects can be found in review papers, for example, by Appenzeller (1979, 1988), Davis (1979), McAlister (1979, 1988), Ulrich (1979, 1981, 1988), and Refsdal and Surdej (1992, 1994).

(1) Mira stars, red giants and supergiants: wavelength dependence of the diameter, shape, extended atmospheres, limb darkening, bright surface features, dust envelopes, companions. Mira stars show spectacular variations of angular size with wavelength which are related to TiO opacity. This wavelength dependence of the diameter was observed for the first time using speckle interferometry (Labeyrie et al. 1977). For Mira and χ Cygni very interesting wavelength dependence of the structure and size of the atmosphere was found (Bonneau et al. 1982; Foy 1988). Studies of the exploratory models of Scholz and Bessell (Bessell et al. 1989; Bessell and Scholz 1989; Scholz 1993) show that monochromatic radii observed in suitably chosen filters, and at suitably chosen phases, are very sensitive diagnostic tools for the investigation of the structure of a Mira photosphere. Accurate monochromatic radius measurements, combined with conventional colour and line profile observations, may pin down the photospheric parameters and discriminate between different current pulsation models. Speckle observations with single VLT telescopes can provide the basic observ! ations for a quantitative model analysis of the photospheric structure of a large number of Mira variables. No quantitative analysis of any Mira photosphere exists to date.

Observations of α Ori show time-varying bright features on the surface (Buscher et al. 1990), possibly due to convective hot-spots. Surface structure observations are important for our understanding of large-scale stellar convective processes and for the interpretation of stellar spectra. The goal of speckle observations with the VLT is to search for similar features on many objects, and to measure the wavelength dependence of the diameter due to variations in TiO opacity which, similar to Miras, can be used for investigating the photospheres of non-variable M type giants (see Scholz 1985). The important applications of angular diameter measurements (e.g., determination of the absolute emergent flux distribution at the stellar surface and effective temperatures) have been discussed by Davis (1979). Finally, several observers have shown that interferometric observations at visible wavelengths can contribute to our understanding of these objects by providing measureme! nts of circumstellar dust disks (Ricort et al. 1981; Roddier and Roddier 1983) and companions.

(2) Spectroscopic binaries: luminosities, masses, distances, MLR. Binary star studies play a fundamental role in observational astrophysics as they provide the only direct means for measuring stellar masses. McAlister (1988) discussed the importance of very accurate speckle observations of close binaries and of the determination of magnitudes and colors of the individual components. For double-lined spectroscopic binaries, the luminosities as well as the masses of the individual components can be determined in a fundamental manner and in this way new accurate points can be added to the empirical mass-luminosity relation (McAlister 1988).

(3) Pre-main sequence (PMS) stars: mass determinations with close binaries, circumstellar envelopes and disks. During the past few years many IR speckle observations of circumstellar envelopes or disks around various PMS stars have been reported (e.g., Beckwith et al. 1984; Zinnecker et al. 1987; DeWarf and Dyck 1993; Leinert et al. 1991, 1994). At optical wavelengths only a few preliminary observations exist. Images with the highest possible resolution of about 10 mas are important to test present models of the PMS star envelopes and highly collimated mass outflows. For mass determinations it is necessary to determine the orbit of close binaries with separations of 10 mas to 100 mas in nearby star forming regions. The distance of the nearest star forming regions is about 150 pc. At this distance semi-major axes of, for example, 3 AU correspond to 20 mas separation and short enough periods.

(4) LBVs and Be stars. Eta Car is an extremely luminous, eruptively unstable Luminous Blue Variable (LBV) with spectacular shell ejections. As the most extreme known LBV, Eta Car is uniquely critical for studies of the LBV phenomenon (Davidson 1989). Eta Car has a surprising sub-arcsec structure near its central star. Three objects at separations 0.11", 0.18", and 0.21" have been detected by speckle observations (Weigelt and Ebersberger 1986; Hofmann and Weigelt 1988). Speckle observations at different wavelengths, speckle spectroscopy, and speckle polarimetry with 10 mas resolution are required to study the nature of Eta Car B to D, and of fainter circumstellar structures. By comparing the images obtained at different epochs, it will be possible to derive the proper motion and the date of the ejection of the objects.

Visible speckle imaging in the continuum and in emission lines, speckle spectroscopy, and speckle polarimetry can also contribute to our understanding of Be stars by providing measurements of the circumstellar envelopes. Quirrenbach et al. (1994) resolved for the first time the envelope of the Be star ζ Tauri. The morphology of the 10 mas $\times$ 3 mas Hα image is most easily interpreted as a disk seen almost edge-on.

(5) Very massive LMC stars and clusters. Speckle observations allow us to reveal the multiplicity of the most massive LMC stars (e.g., Weigelt and Baier 1985; Pehlemann et al. 1992). The results will set new upper limits for stellar masses and better define the shape of the upper IMF. Observations in strong emission lines will examine the number ratio of Wolf-Rayet to normal O-type stars which is related to the evolutionary state of the stellar groups.

(6) Seyfert galaxies: structure, ionisation, and kinematics of the NLR, starburst regions, intermediate region between NLR and BLR. Previous speckle observations have resolved the NLR and circumnuclear starburst regions in several AGN down to scales of $\sim$ 10 pc (e.g. Ebstein et al. 1989; Hofmann et al. 1989; Afanasyev et al. 1992; Mauder et al. 1992, 1994). Most of the NLRs turned out to be clumpy aggregates, often confined in linear or cone-like structures. This confirms earlier results derived from spectroscopic investigations (e.g. Wagner and Appenzeller 1988) that the NLRs consist of individual clouds whose physical and dynamical properties differ from each other. It is necessary to obtain high resolution images in the light of different diagnostic lines in order to investigate the structure and ionization conditions of the clouds. For example, observations of objects which have Extended NLRs are required in order to: study the distribution of the clouds on small scales; determine the size spectrum of the individual clouds; search for possible correlations with the dynamical state of the ENLR and the luminosity and spectrum of the central source. The velocity of the individual clouds can be measured by speckle spectroscopy with about 1 Å spectral resolution (Hα and [O III]) to study the kinematics of the NLR, and to test present AGN models. Ulrich (1979, 1981, 1988) discussed the intermediate region between the broad line region and the narrow line region and the importance of high-resolution (10 to 100 mas) observations for our understanding of the mass loss from the broad line region and the origin of the gas in the narrow line region.

(7) Gravitational lenses: mass of the lensing galaxy, Hubble constant, micro-lensing and the physical structure of quasars, statistical gravitational lensing and the cosmological density of compact objects. Several speckle groups have resolved gravitational lenses of magnitude 16 to 17 with telescopes of 1.5 m to 3.6 m diameter (e.g., Triple QSO PG1115+08; Hege et al. 1981; Foy et al. 1985; Weigelt et al. 1986). An extremely important application of gravitational lenses is the determination of the mass of the lensing galaxy. In this way very distant galaxies otherwise not accessible can be investigated. Measurements of the time delay of the brightness fluctuations of the individual lensed QSO images will allow the Hubble Constant to be determined (Refsdal 1964) if the QSO is intrinsically variable and if a reasonable mass model of the lensing galaxy can be derived. Furthermore, individual stars in the lensing galaxy may induce recognizable micro-lensing brightness variations for one (or more) of the macro-lensed QSO image(s). Speckle imaging, polarimetry and spectroscopy of micro-lensed images offer a unique possibility to retrieve physical information concerning the structure of the QSO central source and emission-line region (see Refsdal and Surdej 1994 for a review on this subject). Finally, a survey for multiple lensed QSO images with angular separations in the range 0.01"-1" will directly enable the setting of values (or limits) on the cosmological density of compact objects in the mass range $10^7 - 10^{11} M_\odot$ (Surdej et al. 1993).

2 Examples of Speckle Masking and Speckle Spectroscopy: η Car, R136, NGC 3603, NGC 1386, and NGC 7469

Figure 1 shows speckle masking observations of the core R136 of the central starburst cluster in the giant H II region 30 Doradus (Pehlemann et al. 1992), the Seyfert galaxy NGC 1386 (Mauder et al. 1992), the starburst ring of the Seyfert galaxy NGC 7469 (Mauder et al. 1994), and speckle spectroscopy of the close double star GC 6771 (separation 0.10"; Grieger and Weigelt 1990).

Figures 2 (top) shows a comparison of speckle masking observations ($\lambda \sim$ 8000 Å; Hofmann and Weigelt 1988) and HST observations ($\lambda \sim 5500$ Å; Weigelt

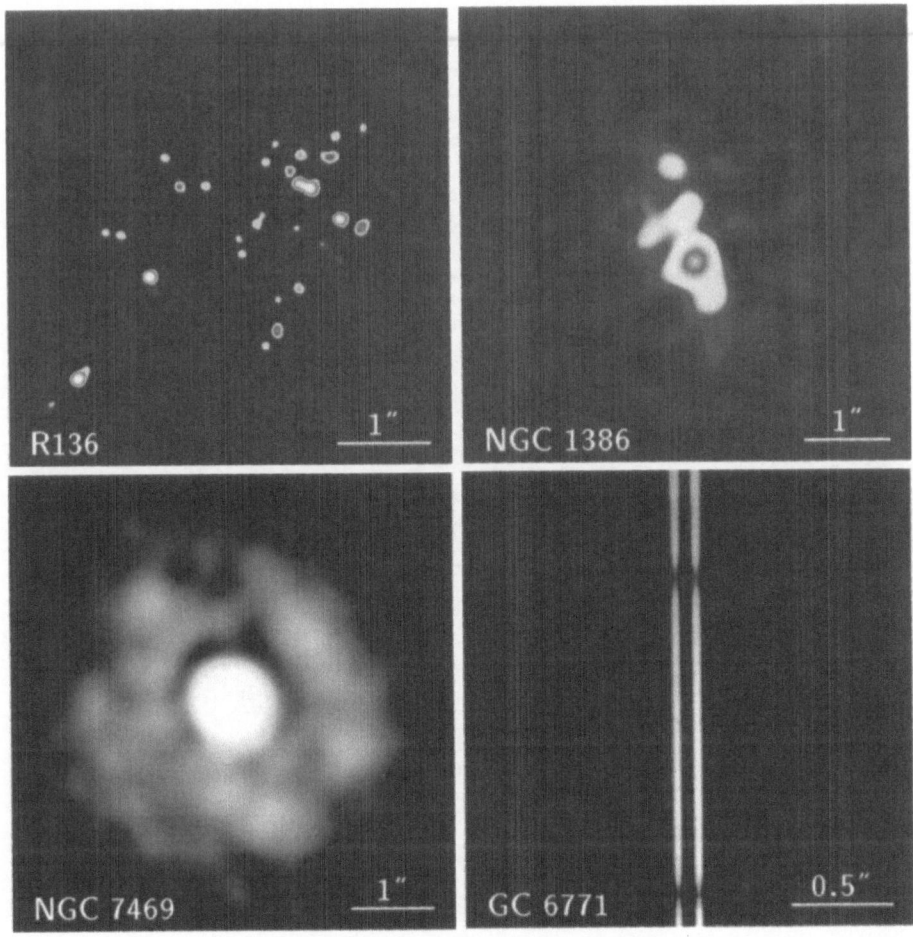

Fig. 1. Speckle masking observations of R136, NGC 1386, NGC 7469 and speckle spectroscopy of the close double star GC 6771 (separation 0.10").

et al. 1995) of the circumstellar material around Eta Carinae. The 550 nm HST image looks more diffuse than the 800 nm speckle image since scattering by dust is stronger in the visible.

Figures 2 (bottom) shows a similar comparison for the central starburst cluster in the giant H II region NGC 3603 (Baier et al. 1988). Both speckle reconstructions were obtained and published before the launch of the HST.

420

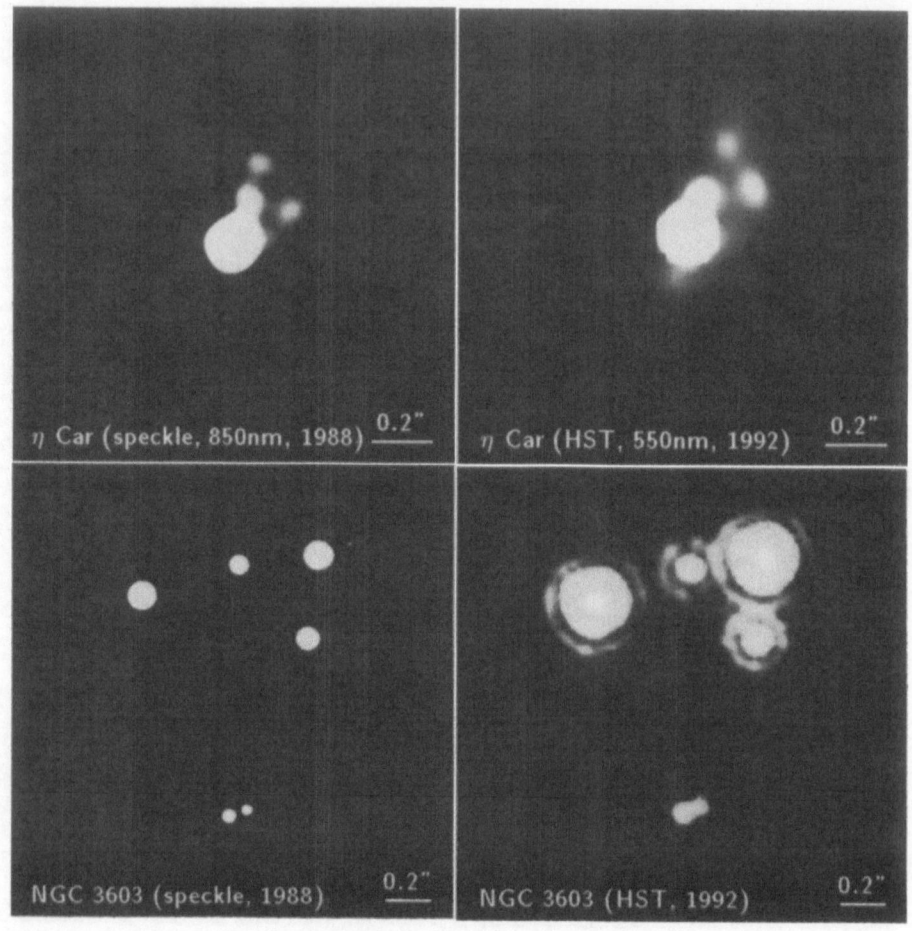

Fig. 2. Comparison of speckle masking observations and HST observations. (Top, left) Eta Car, speckle masking, filter RG830, 1988; (top, right) Eta Car, HST FOC, filter F550M, 1992; (bottom, left) NGC 3603, speckle masking, filter RG610, 1988 (the separation of the double star at the bottom is 74 mas); (bottom, right) NGC 3603, HST FOC, filter F550M, 1992.

3 Limiting Magnitude of Speckle Imaging

The computer experiments reported by Hofmann and Weigelt (1993) show, for example, that about 10 000 speckle interferograms are required for $D/r_0 = 10$ (D = telescope diameter, r_0 = Fried parameter) and with 25 to 100 photoevents per speckle interferogram. In this computer experiment the object was a cluster of 5 stars. For 25 photoevents per interferogram, the mean photometric error of the stars in the speckle masking reconstruction was about 15%, for 100 photoevents per interferogram the photometric error was about 5% (see also Hofmann et al. in these proceedings). Smaller errors are obtained if more than 10 000 speckle interferograms are reduced. 10 000 speckle interferograms correspond to only about 500 s observing time for a frame rate of 20 frames per s. A count number of 25 photoevents/frame corresponds to about magnitude 16 to 18 for an 8 m telescope and typical values for quantum efficiency, exposure time, and filter bandwidth.

4   Optical Long-baseline Interferometry with the VLT

At optical wavelengths all interferometers with large telescopes produce interferograms consisting of several speckles with fine interference fringes in each speckle (so-called "fringed speckles"– see Fig. 3). The number of speckles is approximately equal to the number of turbulence cells (diameter $\approx$ Fried parameter r_0) in front of the telescopes. For the VLTI and $\lambda \approx 5000$ Å the average speckle diameter is about 10 milli-arcsec and the width of the finest interference fringes is about 1 milli-arcsec. Imaging with this type of interferogram is called optical long-baseline interferometry in the multi-speckle mode.

Each r_0-subpupil of an array of large telescopes can be regarded as an individual r_0-telescope with a different phase error. For example, for optical VLTI observations with $r_0 = 40$ cm the number of r_0-subpupils or r_0-telescopes with different phase errors is about $4 \times 20 \times 20 \approx 1600$. Because of the high density of r_0-telescopes for large single-dish telescopes, or optical arrays of several large telescopes, each baseline with the same length and orientation between two r_0-subpupils exists many times. In the above VLTI example each baseline exists about 400 times. The advantage of speckle masking is the fact that it can directly be applied to the multi-speckle interferograms described above, in spite of the very high redundancy of the baselines of r_0-telescopes (see discussion in Reinheimer et al. 1993). Conventional phase closure methods require a non-redundant exit pupil. A non-redundant exit pupil can be obtained if a non-redundant pupil mask with many r_0-holes is inserted into the pupil. A disadvantage of this technique is the light loss. In speckle masking observations with coherent arrays, non-redundant pupil masks are not required. Therefore, speckle masking is an extension of the classical phase closure methods to highly redundant arrays (such as optical large single-dish telescopes and interferometers of large telescopes) and to faint objects (Roddier 1986; Cornwell 1987, 1989). Figure 3 shows a computer simulation of optical long-baseline interferom! etry with the VLT in the multi-speckle mode (see also Reinheimer et al. in this proceedings volume).

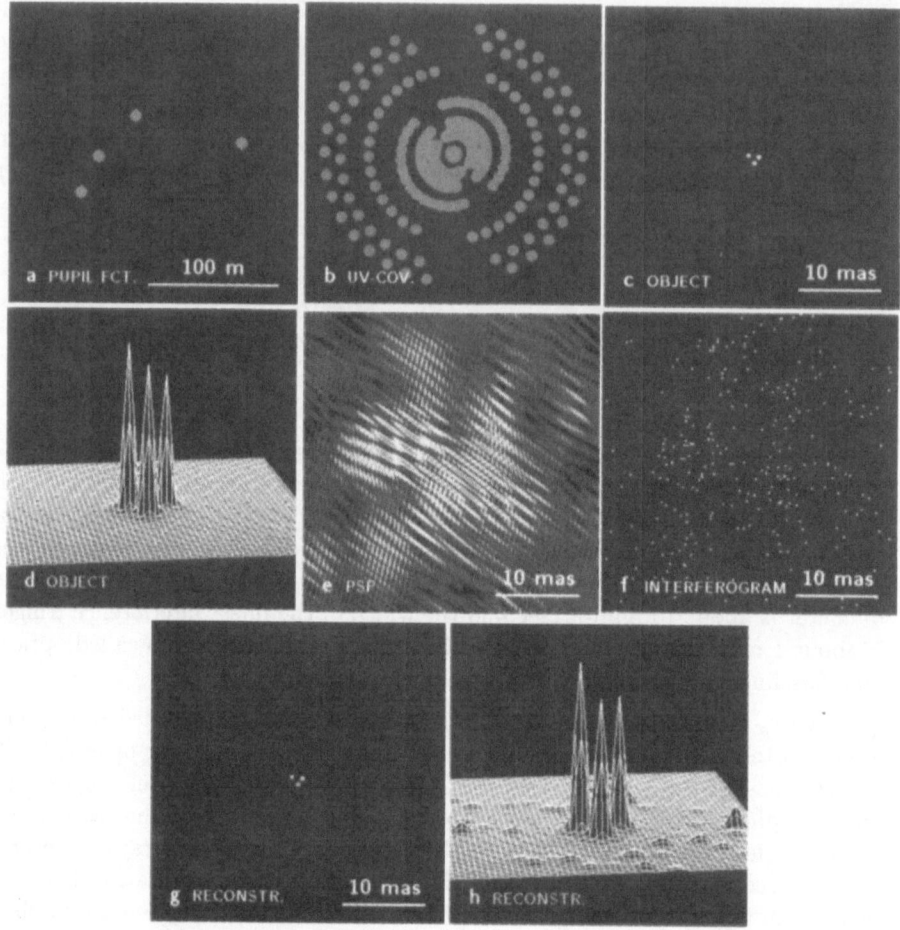

Fig. 3. Computer simulation of optical long-baseline interferometry in the multi-speckle mode. (a) VLT pupil function, (b) uv-coverage (9.3 h aperture synthesis, geographic latitude –24°, declination –70°), (c,d) object, (e) one of the 48 000 computer-generated multi-speckle interferograms, (f) one of the 48 000 interferograms with photon noise (300 photoevents/interferogram), (g,h) diffraction-limited image reconstructed from the 48 000 noise-degraded interferograms by speckle masking and the building block method (from Reinheimer et al. 1993).

Acknowledgements

I thank I. Appenzeller, M. Scholz, J. Surdej, S. Wagner, and H. Zinnecker for many interesting and helpful discussions.

References

Afanasyev, V.L., Balega, Y.Y., Orlov, V.G., Vasyuk, V.A., 1992, A&A, 266, 15

Appenzeller, I., 1979, in: ESA/ESO workshop on Astronomical Uses of the Space Telescope, eds F. Macchetto, F. Pacini, M. Tarenghi, p. 47

Appenzeller, I., 1988, in: NOAO-ESO Conf. on High-Resolution Imaging by Interferometry, ed. F. Merkle, p. 19

Baier, G., Eckert, J., Hofmann, K.-H., Mauder, W., Schertl, D., Wehorn, H., Weigelt, G., 1988, The Messenger (ESO) 52, 11

Beckwith, S., Zuckermann, B., Skrutski, M.F., Dyck, H.M., 1984, ApJ 287, 793

Bessell, M.S., Brett, J.M., Scholz, M., Wood, P.R., 1989, A&A 213, 209

Bessell, M.S. and Scholz, M., 1989, IAU Coll. 106, 67

Bonneau, D., Foy, R., Blazit, A., Labeyrie, A., 1982, A&A 106, 235

Buscher, D.F., Haniff, C.A., Baldwin, J.E., Warner, P.J., 1990, MNRAS 245, 7p

Cornwell, T.J., 1987, A&A 180, 269

Cornwell, T.J., 1989, Science 245, 263

Davidson, K., Humphreys, R.M., 1986, A&A 164, L7

Davidson, K., 1989, IAU Colloq. 113, p.101

Davis, J., 1979, IAU Coll. No. 50, High Angular Resolution Stellar Interferometry, eds. J. Davis and W.J. Tango, p. 1-1

DeWarf, L.E., Dyck, H.M., 1993, Astron. J. 105, 2211

Ebstein, S.M., Carleton, N.P., Papaliolios, C., 1989, ApJ 336, 103

Foy, R., 1988, in Instrumentation for Ground-Based Optical Astronomy.Present and Future. ed. L.B. Robinson (Springer Verlag), p. 345

Foy, R., Bonneau, D., Blazit, A., 1985, A&A 149, L13

Grieger, F., Weigelt, G., 1990, SPIE 1319, 440

Grieger, F., Weigelt, G., 1992, ESO Proc. High Resolution Imaging by Interferometry II, p. 481

Hege, E.K., Hubbard, E.N., Strittmatter, P.A., Worden, S.P., 1981, ApJ Letters 248, L1

Hofmann, K.-H., Mauder, W., Weigelt, G., 1989, in: ESO Proc. Extranuclear Activity in Galaxies, p. 35

Hofmann, K.-H., Weigelt, G., 1988, A&A 203, L21

Hofmann, K.-H., Weigelt, G., 1993, A&A 278, 328

Knox, K.T., Thompson, B.J., 1974, ApJ Lett. 193, L45

Labeyrie, A., 1970, A&A 6, 85

Labeyrie, A., Koechlin, L., Bonneau, D., Blazit, A., Foy, R., 1977, ApJ 218, L75

Leinert, Ch., Haas, M., Lenzen, R., 1991, A&A 246, 180

Leinert, Ch., Richichi, A., Weitzel, N., Haas, M., 1994, Near-Infrared speckle observations of Herbig Ae/Be stars, in: Proc. Nature and Evolutionary Status of Herbig Ae/Be, ASP Conf. Ser.

Lohmann, A.W., Weigelt, G., Wirnitzer, B., 1983, Appl. Opt. 22, 4028

Mauder, W., Appenzeller, I., Hofmann, K., Wagner, S., Weigelt, G., Zeidler, P., 1992, A&A 264, L9

Mauder, W., Appenzeller, I., Wagner, S., Weigelt, G., 1994, A&A 285, 44

McAlister, 1979, IAU Coll. No. 50, High Angular Resolution Stellar Interferometry, eds. J. Davis and W.J. Tango, p. 3-1

McAlister, H., 1988, in: NOAO-ESO Conf. on High-Resolution Imaging by Interferometry, ed. F. Merkle, p. 3

Pehlemann, E., Hofmann, K.-H., Weigelt, G., 1992, A&A 256, 701

Quirrenbach, A., Buscher, D.F., Mozurkewich, D., Hummel, C.A., Armstrong, J.T., 1994, A&A 283, L13

Refsdal, S., 1964, MNRAS 128, 307

Refsdal, S., Surdej, J.: 1992, "Gravitational Lensing", Invited discourse during the XXIst General Assembly of the International Astronomical Union (Buenos Aires, July 1991), IAU 'Highlights of Astronomy' Vol. 9, 3-32, J. Bergeron (ed.)

Refsdal, S., Surdej, J.: 1994, Reports on Progress in Physics 57, 117

Reinheimer, T., Hofmann, K.-H., Weigelt, G., 1993, A&A 279, 322

Ricort, G., Aime, C., Vernin, J., Kadiri, S., 1981, A&A 99, 232

Roddier, F., 1986, Optics Commun. 60, 145

Roddier, C., Roddier, F., 1983, ApJ 270, L23

Scholz, M., 1985, A&A 145, 251

Scholz, M., 1993, IAU Coll. 139, New Perspectives on Stellar Pulsation and Pulsating Variable Stars, eds. J.M. Nemec and J.M. Matthews, p. 201

Surdej, J., Claeskens, J.F., Crampton, D., Filippenko, A.V., Hutsemèkers, D., Magain, P., Pirenne, B., Vanderriest, C., Yee, H.K.C.: 1993, Astron. J. 105, 2064

Ulrich, M.H., 1979, in: ESA/ESO workshop on Astronomical Uses of the Space Telescope, eds F. Macchetto, F. Pacini, M. Tarenghi, p.261

Ulrich, M.H., 1981, in: ESO Conf. on Scientific Importance of High Angular Resolution at Infrared and Optical Wavelengths, eds. M.H. Ulrich and K. Kjär, p. 411

Ulrich, M.H., 1988, in: NOAO-ESO Conf. on High-Resolution Imaging by Interferometry, ed. F. Merkle, p. 33

Wagner, S., Appenzeller, I., 1988, A&A 197, 75

Weigelt, G., 1977, Optics Commun. 21, 55

Weigelt, G., Baier, G., 1985, A&A 150, L18

Weigelt, G., Ebersberger, J., 1986, A&A 163, L5

Weigelt, G., Baier, G., Ebersberger, J., Fleischmann, F., Hofmann, K.-H., Ladebeck, R., 1986, Opt. Eng. 25, 706

Weigelt, G., Wirnitzer, B., 1983, Optics Lett. 8, 389

Weigelt, G., Albrecht, R., Barbieri, C., Blades, J.C., Boksenberg, A., Crane, P., Davidson, K., Deharveng, J.M., Disney, M.J., Jakobsen, P., Kamperman, T.M., King, I.R., Macchetto, F., Mackay, C.D., Paresce, F., Baxter, D., Greenfield, P., Jedrzejewski, R., Nota, A., Sparks, W.B., 1995, "HST Observations of Eta Carinae", Revista Mexicana de Astronomia Y Astrofisica, in press

Zinnecker, H., Perrier, Ch., Chelli, A., 1987, IAU Symp. 115, p. 71

From Planets to Galaxies: Adaptive Optics Revolution and VLT Interferometry

P. Léna

Université Paris VII & Observatoire de Paris, 92195 Meudon, France

Abstract. Three years of scientific operation of adaptive optics on the 3.6 meter ESO telescope at La Silla have produced a great wealth of astronomical results on a variety of objects, and demonstrated the power of this new observing method to reach with high sensitivity the diffraction-limited resolution of ground-based large telescopes. Extrapolating the gained experience to the VLT 8-m telescopes is now clearly understood. Mastering adaptive optics techniques also leads to confidence of their use in the VLT Interferometric mode (VLTI). The results obtained on the 3.6-m telescope indicate the need to obtain the full VLTI resolution (1 to 10 mas) in order to understand a great diversity of phenomena only observable at such resolution.

1 Introduction

Adaptive optics (AO) was considered as a nice dream without practical applications as close as ten years ago. But through the foresight of the VLT proposers, its prototype development at La Silla has led, since 1990, to a full demonstration of its applicability and to a wealth of new results. Thus putting European astronomers in the forefront in this field, hidden until then in classified military work.

The primary goal of AO is to correct the detrimental effects of atmospheric turbulence on an image, in order to improve the angular resolution, the dynamic range and the contrast over the sky background hence enhance the signal-to-noise ratio on unresolved sources and reduce confusion in crowded fields. In addition, a number of potential impacts on classical observing techniques such as spectroscopy, coronography or polarimetry may be expected. In Sec. 2 we outline the demonstrated capabilities of AO on 4-m class telescopes. From such a demonstration new problems are discovered and solutions proposed to optimize the design of AO systems on future telescopes, especially the VLT.

It is interesting to notice that the ESO AO developements were strongly pushed as early as 1983-1984 by the discussions arising from the potential use of the VLT 8-meter telescopes as an interferometric array. Roddier & Léna (1984) showed that large telescopes do not bring any sensitivity gain over smaller ones, unless the wavefronts are corrected for turbulence effects on each telescope. On the contrary, if such a correction, even partial, is obtained, huge sensitivity improvements over smaller telescopes may be obtained by interferometric combination. Dyck & Kibblewhite (1984) had rightly argued that it was pointless to combine large telescopes coherently, as no sensitivity gain resulted, but these authors failed to see the potential of AO.

Although no interferometric combination of AO-corrected images coming from different telescopes has yet been achieved in practice, it is entirely accepted that interferometric combination of telescopes with diameter D significantly larger than the coherence size of the turbulence, i.e. the Fried parameter $r_0(\lambda)$, must include AO correction for maximum efficiency. The larger the ratio $D/r_0(\lambda)$, the more important it is to implement the correction, as discussed in Sec. 3 in the specific case of the VLTI (VLT Interferometric mode).

2 Astrophysical Results with Adaptive Optics

The VLT Prototype adaptive optics system is an evolving instrument, with its successive versions - ComeOn (1989-1991), ComeOnPlus (1992-1994) and Adonis (1995) - equipping the 3.6-m telescope at La Silla. Detailed descriptions of the system and performance may be found in Rigaut et al. (1991), Beuzit et al. (1995), while a number of specialized papers have adressed the new problems arising from AO use for astronomical observations. They are reviewed by Beckers (1993) and further discussed in a dedicated SPIE Conference (Ealey & Merkle, 1994) where the reader will see the explosive growth of the field.

An overview of the to-date scientific results was given by Beuzit et al. (1994) and Léna (1994). These results cover imaging in the near-infrared, namely 1.25 to 5 μm, where the current ESO system is the most efficient and where diffraction-limited performance are achieved. Although there is no point in repeating here the complete list of results, it is of interest to extract a few typical cases, especially with the goal of demonstrating how interferometric observations represent a natural and necessary continuation of AO imaging with single telescopes.

Solar System objects. The ease of observing such objects is due to their intrinsically bright magnitude, most of the time brighter than $m_V \approx 9$. The planets or their satellites provide easy referencing for the correction, as discussed in Saint-Pé et al. (1993). Images of Ceres, Pallas and Titan have been obtained with typically 0.15" resolution, i.e. some tens of pixels in the image. The planet Jupiter has been observed (Drossart et al. 1994), using Galilean satellites as convenient reference sources, as they provide a scanning reference for the extended planet surface. Currently these images are unique as obtained with AO on widely extended astronomical objects. (Saturn images may have also been obtained in July 1994 by the Starfire Range group, Albuquerque).

Stellar Astronomy. One obvious case for AO imaging is multiple stars. Yet the systematic analysis of assumed "supermassive" stars in the LMC (Heydari-Malayeri & Beuzit 1994) shows, not only the possibility of accurate mass determination through precise, unconfused photometry (spectral type determination will become possible when AO is coupled to spectrographs), but also a fundamental impact on the understanding of the upper mass limit of stars.

At the other end of the mass spectrum, a systematic study of close binaries has been undertaken. With higher sensitivity and better use of telescope time than with speckle techniques, orbital data for binaries with 0.1-0.2" separation can be obtained (Duquennoy et al. 1994), the accuracy of positioning reaching 50 milli-arcsec (mas) or better, further improved by blind deconvolution (Thiébaut & Conan 1994). A long term program is underway, but already from a list of 30 sources, more than half have been resolved.

The analysis of the Initial Mass Function for a complete sample (to $m_K \approx 16$-19) in an active star forming region is facilitated by the AO sensitivity gain and confusion removal. It has been demonstrated on the R136 region in the LMC (Brandl et al. 1994) and will be applied to closer Galactic clusters of newborn stars.

Circumstellar Environments. This includes star formation, young objects, accretion and proto-planetary discs, reflection nebulae, bipolar flows, stellar winds, novae, sporadic ejections of matter, etc. Observing these galactic objects appears especially suitable for AO, as the presence of the star provides a natural reference, usually bright enough ($m_V \leq 13$) to study the environment within 1-2" (ca. 10-100 a.u. at 10 to 100 pc distances), in a perfect isoplanatic field. In addition, infrared observations in J,H,K,L,M bands are sensitive to cooler dust and the decreasing optical depth allows deep investigation. The study of light polarization on a small scale gives hints on scattering close to the illuminating object.

A broad selection of young, low mass stellar objects (T Tauri and Herbig Ae/Be stars) is under systematic study, with the general aim to search for circumstellar disks and remnants of the parent molecular cloud. Such disks and the suspected occurence of associated bipolar jets become observable on nearby (≤ 100 pc) stars, all of them being bright enough for proper referencing.

Observations of η Carinae (Rigaut et al. 1994) illustrate the performance of adaptive optics applied to a complex object as compared to previous speckle techniques. The variability of this star has been interpreted by massive ejection of dust and there is probably a thick circumstellar disk. AO images reveal symmetrical structure close to the central object (equatorial plane at position angle 35° and jet like structures at 140°). The core is well resolved. Detailed analysis shows hot (1000 K), resolved dust close to the central source and puts constraints on the mass loss and radiation field.

Images of the FU Orionis star Z CMa (Malbet et al. 1993) illustrate the discovery of a circumstellar disk around a close binary, previously separated by speckle work and confirmed by AO observations. Magnitudes of T Tau and FU Ori stars are in the range $7 \leq m_V \leq 13$, and therefore fully adequate for referencing.

Herbig Ae-Be stars are pre-main sequence stars of intermediate mass presenting a strong near-infrared excess whose origin is still controversial: accretion disk or circumstellar envelope(s)? This is further complicated by the fact that some of the IR excess may also originate from the presence of a low mass companion. To clarify this point, a long-term survey of a large sample is being undertaken with AO multicolor infrared imaging. From a first run carried out in December 1993, 15 stars were observed in J,H,K bands and 7 clearly resolved into binaries (Bouvier et al. 1994). Companions do exist, but their faintness can not explain the total infrared excess.

At the other end of the stellar evolution sequence, ejection of matter through stellar winds deserves close investigation, as it is the main source of replenishment for the interstellar medium. A systematic study is needed to trace the complete evolution sequence from the stage of late Asymptotic Giant Branch stars until the young planetary nebulae. As most of these objects are bright and present a large infrared excess, images close to the diffraction limit allow derivation of the departure from spherical symmetry, the dust grain size and chemical composition, and hence the mass loss history. An example is given by the post AGB star the Frosty Leo, where evidence for a disk seen edge-on was obtained (Dougados et al. 1992). Adaptive optics imaging shows the central star, but also a lack of symmetry which may be interpreted as the presence of a companion, the mass of which can even be surmized (Beuzit, Rouan et al. 1994). A similar AO approach by Roddier et al. (1994) confirms the double structure of the star.

The protoplanetary matter around stars is attracting great interest. The combination of a coronographic approach and AO is able to investigate distances as close as 0.1" from the star at 1μm. The main problem here is dynamic range, and a number of tests are in progress to maximize it. By imaging the star HR4796, we have detected, 4.9" away, a nearby star 10.5 magnitudes less bright (a factor of 10^4) at 4σ, with the use of a coronograph (Léna 1994). An obvious application is the study of the brightness distribution near β Pictoris and similar stars. Observations of β Pic with coronographic masks of 2" and even 0.8" diameter allow the disk to be followed very close to the star (Beuzit & Lagrange-Henri 1994). For distances of ca. 10pc, radii as close as 1 a.u. may be investigated. The star 68 Oph where the presence of a disk was announced, and later possibly denied, was observed with AO in April 1993. No sign of circumstellar material was found down to a limiting magnitude of $m_K \geq 17$ arcsec^{-2}, an indication of the current (and still improvable) sensitivity of AO.

Galaxies. Extragalactic AO imaging is limited by the possibility of adequate referencing. No systematic study of chance coincidences of adequate stellar references close to galaxies within an isoplanatic field has yet been undertaken to our knowledge, but the task is not difficult. Several tens of infrared luminous galaxies, most of them identified by IRAS, present sufficiently bright nuclei ($m_V \leq 14$) to be observable with AO. In addition, the star formation is close to the nucleus and its study requires high resolution. Three programs have currently started.

Star Formation in Active Galaxies. The SAb galaxy NGC 7469, known as a Seyfert-I galaxy, presents an active nucleus surrounded by a region of star formation. Its triggering process is not known and several hypotheses have been formulated such as tidal effects from a companion, presence of a bar, etc. AO observations demonstrate the existence of a structured ring (Lai et al. 1994)) which was mapped in J,H and K bands. Preliminary analysis indicates that the main emission comes from red supergiants (M0-M5) and that fairly high starburst activity ($L_* = 10^{11}$ $L_\odot$) is consistent with the luminosity measured by other methods. Here is a good example of the AO capabilities on relatively faint objects, as the magnitude of the nucleus is $m_V = 12$.

Seyfert Nuclei. The brightest nucleus is the center of NGC 1068. With AO the stellar content of the circumnuclear environment is determined by imaging at 0.45" resolution at $\lambda=2.2\mu m$ (Marco et al. 1994). After applying a CLEAN deconvolution, three bright spots appear within the nucleus. Their spatial and angular distribution are fully compatible with recent images made in [O III] with the HST and a map at $\lambda =10.3\mu m$ (Cameron et al. 1993).

Remote and Possibly Primeval Galaxies. As they are not sufficiently bright to provide adequate referencing, a different strategy is used: stars are selected in fields which are expected, from a variety of criteria, to be rich in remote galaxies. They are systematicaly mapped as far from the star as allowed by the isoplanatic field (usually 20-30", depending on seeing, amount of expected correction and wavelength). Two galaxies were observed in K band in the cluster J1836.3CR at a redshift z =0.42. The resolution is 0.4" and the galaxies are clearly resolved (Sams et al. 1994). Their integrated magnitudes are K=15 and K=18. The (V-K) colors indicate the brighter source to be an elliptical and the fainter a spiral galaxy bluer by 0.4±0.3 magnitude. This program is pursued systematically to determine the colors and morphological types. This new possibility to classify remote galaxies, combined with HST observations, is important for the understanding of evolution.

In summary, AO during three years of preliminary work has demonstrated its productivity. With minor improvements, mainly on the wavefront sensor sensitivity, reference objects as faint as $m_V\approx16\text{-}17$ will be within reach. The use of asteroids as moving references (Ribak & Gendron 1994) greatly increases the number of observable objects, at least in a band ±30° from the ecliptic. The concept of "modal control optimization" (Gendron & Léna 1994a, 1994b) has been implemented to adapt the system behaviour to a particular set of observing conditions (magnitude of the reference, angular distance to the imaged object, height and time dependence of the turbulence and seeing) and to maximize the final Strehl ratio **S** (viz. the sharpness of the corrected image).

The current system may reach $S\geq0.3$ (i.e. 30% of the energy in the diffraction-limited core of the image) for $\lambda\geq1\mu m$ with references brigher than $m_V\approx13$, in a 15" isoplanatic field; these values being degraded in case of fast seeing. Careful deconvolution, based on a stable AO PSF, provides cleaned images with a dynamic range of 10 (close to the first dark ring of the Airy patttern) to 10^4 at typically 3" from the center. It is worth mentioning the natural complementarity between AO performances above $\lambda 1\mu m$ and the Hubble Space Telescope imaging capabilities at shorter wavelengths, since the achieved resolutions are almost identical (ca.100 mas or slightly better).

We refer to Léna (1994) for a more extensive discussions of various techniques coupled to AO: coronography, spectrography, polarimetry and light extraction with single mode optical fibers. As there is no current or foreseen effort around the VLT to implement, even on an experimental basis, laser stars for artificial reference sources, contrary to the large efforts currently undertaken in other large telescope projects, we shall not discuss here the potential gains such method would provide to AO observing.

3   From Adaptive Optics to the VLT Interferometric Mode

We recall that the VLTI concept contains two separate configurations which can work either independently or combined: VISA (VLT Interferometric Sub Array), a coherent array of three (or more) movable 1.8-m telescopes; and VIMA (VLT Interferometric Main Array), a coherent array formed by the four 8-m fixed telescopes positioned on a trapezoidal optimized configuration (see The VLT Interferometer Implementation Plan, 1989). The Hybrid mode combines coherently VISA and VIMA to realize the full VLTI potential. The scientific capabilities of either have been carefully assessed (for a complete VLTI summary and references, see von der Lühe et al. 1994).

If one considers any of the scientific problems described above, the resolution gain of 2.2 provided by an AO equiped 8-m is significant, but far from sufficient. Current diffraction-limited resolution above $1\mu m$ provides typically 10 x 10 pixel images of objects whose size is in the $1''$ range, such as satellites and asteroids, cometary nuclei, circumstellar disks and bipolar flows, galactic nuclei. The ca. 100 meters typical VLTI baseline (maximum 190 m) will lead to 1-10 mas resolution, sufficient to resolve volcanic activity on Io, details on Titan's surface, accretion disks on young stellar objects, bipolar flows and clumpy ejection of matter, trace of protoplanet formation through the gaps left in circumstellar disks (such as β Pic, HL Tau, etc.) and Seyfert nuclei structure. These few examples are entirely outside the reach of the diffraction-limited resolution achievable by a 10-m class single telescope. Once the resolution is achieved, two essential questions must be adressed: (i) are these astronomical objects bright enough to provide cophasing of separate telescopes, ensure proper fringe detection, accurate phase and visibility measurements; (ii) will the u-v spatial frequency plane be properly filled to provide images with a sufficient number of pixels?

Limiting Magnitudes and Cophasing. The question of limiting magnitude has been well studied (Beckers et al 1992). It is independent of telescope size and therefore common to VISA and VIMA. The cophasing limiting magnitudes are identical to the ones encountered in AO, the so-called "bright object" case. In other words, any object sufficiently bright to be either partially (Strehl ratio $S \ll 1$ or significantly ($S \approx 0.3$-1) AO corrected on a single telescope, can also be observed with long integration times (in order to get a high SNR on the visibility) in a coherent mode combining several telescopes. As already demonstrated by the AO scientific results, this first simple case opens a great wealth of objects to investigation. One should note that only the simplest zero-field configuration for the interferometer is at this stage required.

Beyond this, several schemes have been proposed to use an off-axis reference, by giving field-of-view to the combined interferometric focus, in order to maintain the cophasing on a sufficiently bright off-axis object, while performing long integration on the faint source of interest. The original VLTI design aims at a large ($8''$ for VIMA, 3.5 to $8''$ for VISA) FOV for any combination (VISA or VIMA), while the reduced (so called "Better Science"or more recently "The

Alternative") VISA configuration will temporarily operate with zero field. It has also been proposed to consider the injection of the reference off-axis source into the zero-field interferometer with single-mode optical fibers. Such a scheme could be implemented in the "Better Science" current VISA configuration and greatly expands the number of observable sources.

In summary, limiting magnitudes of VISA or VIMA with zero-field will be identical to the ones achieved with ComeOnPlus, but remain improvable with a better wavefront sensitivity as explained above. Much fainter objects will become observable with VISA or VIMA when using off-axis cophasing, in very similar manner as AO does when using a "bright" off-axis reference to image faint galaxies in the example given in Sec.2. Indeed, the sensitivity gain of VIMA (8-m telescopes) over VISA (1.8-m) is then in direct proportion to the areas, a key consideration when it was decided to make the 8-m coherent combination an integral part of the program.

A side issue, but fundamental for VIMA, is the legacy to extrapolate the current AO mastering on a 3.6-m telescope to the 8-meters. This point has been deeply studied (Hubin et al. 1994) and leaves no doubt. In addition, the operational experience gained with ComeOnPlus data acquisition, real-time use and a posteriori data treatment will help to quickly exploit the currently decided AO system at one Nasmyth focus of UT2. When AO is implemented at Nasmyth, it was suggested (Mariotti et al. 1994) that two equipped foci on separate telescopes could be coherently coupled with single-mode fibers, on the basis of a fully demonstrated technology of guided optics for interferometric combination of telescopes (Coudé du Foresto & Ridgway 1991, Coudé du Foresto 1994).

Another open question is the possible need to equip the VISA telescopes with AO. Their smaller size would only require a limited number of corrected modes (less than 10 to operate beyond 1μm, a few tens for a useful correction below 1μm) and greatly improve over the multi-speckle interferometric combination.

Imaging Capabilities. In an image obtained with a particular u-v sampling, the final number of pixels is directly equal to the number of baselines and associated measured visibilities. This is why the ultimate VLTI configuration should have the capability to combine the four 8-m telescopes, to include in this combination a number, possibly larger than three, of auxiliary 1.8-m telescopes, and to exploit Earth rotation in supersynthesis. This point was stressed as early as 1987 (see "The Interferometric Mode of the VLT", 1987) and the VLTI imaging potential is described by von der Lühe et al. (1994).

Let us emphasize here the imaging capability of the "Better Science" version of VISA (called "The Alternative" in the most recent documents), considered as the first step of the program. With only four operative stations for the movable auxiliary telescopes, 3x3 pixel images will be available at the 1-10 mas resolution in the so called "snapshot" mode (short exposure), and 10 x 10 pixel images with supersynthesis. With these limited capabilities, three remarks may be made: (i) images of barely resolved objects at this resolution will be of high value, as they do not require a large number of pixels; (ii) images of complex objects will be meaningful only if they are made as a follow-up to AO observations carried at

≈50 mas resolution; (iii) images of very complex and extended objects will be out of reach in this first VLTI phase. Points (i) and (ii) are, as demonstrated at radio frequencies, already sufficient on almost every class of the above described objects, to leave ample space for imaging with VLTI.

We simply conclude here that adaptive optics, rightly considered ten years ago as a mandatory condition for a proper interferometric use of large telescopes, has now demonstrated its feasibiity. Performances announced for the VISA, VIMA and Hybrid modes of the VLTI are sustained by actual results. In addition and even more important, the study at ca.100 mas resolution of a great diversity of objects with a diffraction-limited 3.6-m telescope, complemented by the HST images at the same resolution, demonstrates the scientific need to increase the resolution to the 1-10 mas range.

Acknowledgments

The AO results presented here have been communicated, often prior to publication, by numerous graduate students and colleagues, all involved at some degree in the development of AO with the ComeOn system. It is a pleasure to thank them all. The author is also indebted to Prof. Genzel and his group, for a thorough cooperation on the imaging part of the program, as well as for scientific collaboration. Finally, the ESO involvement through N.Hubin and M.Faucherre has been invaluable.

References

Beckers, J., 1993, Ann.Rev.Astr.Astrophys., 31.

Beckers, J.M., Braun, R., di Benedetto, G.GP., Eckart, A., Faucherre, M., Foy, R., Genzel, R., Léna, P., von der Lühe, O., Koechlin, L., Merkle, F., Weigelt, G., 1992, ESO VLT Report 65, Garching.

Beuzit, J.L., Brandl, B., Combes, M., Eckart, A., Faucherre, M., Heydari-Malayeri, M., Hubin, N., Lai, O., Léna, P., Perrier, C., Perrin, G., Quirrenbach, A., Rouan, D., Sams, B., Thébault, P., 1994, The Messenger, 75, 33.

Beuzit, J.L., Thébault, P., Perrin, G., Rouan, D., 1994, A.&A., in press.

Beuzit, J.L., Rousset, G. et al, 1995, in prep.

Beuzit, J.L., Lagrange-Henri, A.M., 1994, in prep.

Bouvier, J. et al., 1995, in prep.

Brandl, B., Sams, B., Eckart, A., et al, 1994, in prep.

Cameron, M., Storey, J.W., Rotaciuc, V., Genzel, R., Verstraete, L., Drapatz, S., Siebenmorgen, R., Lee, T.J., 1993, Ap. J., 419, 136.

Coudé du Foresto, V., 1994, IAU Symposium on Very High Angular Resolution Imaging, eds. Robertson, J. .G. & Tango, W.J., 158, 261.

Coudé du Foresto, V., Ridgway, S., 1994, in High Resolution Imaging by Interferometry II, eds. Merkle, F. & Beckers, J., ESO, Garching.

Dougados, C., Rouan, D., Léna, P., 1992, A.& A., 253, 464.

Drossart, P., Forni, O., Beuzit, J.L., Baines, K., Orton, G., 1994, Bull.AAS, in press.

Duquennoy, A, Mayor, M., Perrier, C., Mariotti, J.M., 1995, these proceedings.

Dyck, H.M., Kibblewhite, E.J., 1986, PASP, 98, 260.

Ealey, M.A., Merkle, F., 1994, in Adaptive Optics in Astronomy, Proc. Hawaii Conf., Vol. 2201, S.P.I.E.

European Southern Observatory, 1989, The VLT Interferometer Implementation Plan, VLT Report 59b.

Gendron, E., Léna, P., 1994a, A.&A., in press.

Gendron, E., Léna, P., 1994b, A.&A., submitted.

Gendron, E., Ribak, E., 1994, A.&A., in press.

Heydari-Malayeri, M., Beuzit, J.L., 1994, A.& A., 287, L17.

Hubin, N., Théodore, B., Petitjean, P., Delabre, B., 1994, in Adaptive Optics in Astronomy, S.P.I.E., 2201, 34.

Lai, O., Eckart, A., Rouan, D, et al, 1994, in prep.

Léna, P., 1994, in Adaptive Optics in Astronomy, S.P.I.E., 2201, 1099.

Malbet, F., Léna, P., Bertout, C., 1993, A.&A., 271, L9.

Marco, O., Blietz, M., Alloin, D., 1994, in prep.

Mariotti, J.M., Coudé du Foresto, V., Perrin, G., Zhao, Peiqian, Léna P., 1994, submitted to ESO.

Ribak, E.N., Gendron, E., 1994, A.&A., 289, L47.

Rigaut, F., Gehring, G., et al, 1994, in The Carinae Nebula: a laboratory for stellar evolution, Revista mexicana A.&A., in press.

Rigaut, F., Rousset, G., Kern, P., Fontanella, J.C., Gaffard, J.P., Merkle, F., Léna, P., 1991, A.&A., 250, 280.

Roddier, F. et al, 1994, Private communication.

Roddier, F., Léna, P., 1984, Journ. Optics (Paris), 15, 363.

Saint-Pé, O., Combes, M., Rigaut, F., Tomasko, M., Fulchignoni, M., 1993, Icarus, 105, 263.

Sams, B., Brandl, B., Beckers, J., Genzel, R., Léna, P., 1994, in prep.

Thiébaut, E., Conan, J.M., 1994, J.O.S.A., submitted.

Von der Lühe, O., Ferrand, D., Koehler, B., Neng-hong, Z., Reinheimer, T., 1994, in Amplitude and Intensity spatial Interferometry, S.P.I.E., 2200.

A Strategy for High Angular Resolution Astrophysical Programmes at the VLT in the Visible

Renaud Foy

Observatoire de Lyon, CNRS URA 300, 69561 Saint-Genis-Laval, France

Abstract. For certain astrophysical programmes we consider the trade-off between requirements for the spatial resolution, the spectral resolution and the signal-to-noise ratio in the restored images. This discussion takes into account the capabilities of the monolithic aperture VLT and the diluted aperture VLTI, as well as those of the HST.

1 Introduction

Owing to the rapid progress in the field of imaging techniques - speckle interferometry, Michelson interferometry, adaptive optics, image restoration-, the VLT should become one of the very best tools in the world for high angular resolution imaging. Scientific programmes relevant to these capabilities will span a very wide range of astrophysics if the high angular resolution instrumentation is adequately defined. Both the astrophysical programmes, and the specifications of the instruments, have to take into account the resolution of the Very High Angular Resolution Camera (VHARC) with a single 8 meter VLT unit telescope, of the VLTI and also of the HST as well as of the image restoration algorithms.

It seems that, for a decade or so, adaptive optics at the focus of large telescopes will be restricted to partial compensation of the deformed incoming wavefront in the visible. Typically the spatial resolution will range around 0.1", at a limiting magnitude $m_v \lesssim 13$ which will be extended towards faintest objects with the help of an artificial laser guide star. Spatial resolutions down to the diffraction limit, i.e. $\approx 0.01"$ at $4000\AA$ require the use of interferometric methods. From present experience, the limiting magnitude of speckle interferometry reaches 16 to 18 when the seeing is better than $\approx 0.5"$. Fainter limiting magnitudes are to be expected from coupling adaptive optics and laser guide stars with speckle interferometry, but this mostly remains to be investigated.

The VLTI will offer spatial resolutions down to 0.001" or less at $H\alpha$.

To set up an efficient strategy for astrophysical programmes based on these capabilities, these numbers should be compared with the performance of the HST. A diffraction limited 2.40m telescope has a resolution of .047" in the V band. The spatial resolution of the current wide field imaging camera of the HST is limited by the sampling to $\approx 0.1"$. Direct ground-based imaging, using adaptive optics in partial compensation mode, also reaches this spatial resolution (Roggemann et al. 1994; Connan 1994). Resolution enhancement, beyond the diffraction limit, performed with deconvolution algorithms can be applied to any image, whatever the telescope. Nevertheless, the signal-to-noise ratio close to the cut-off frequency is much higher in the case of a diffraction limited telescope, making it more efficient to run such algorithms.

Figure 1 shows where a few astrophysical targets are located in the spectral resolution versus spatial resolution plane, together with the limiting spatial resolutions of the above mentioned imaging telescopes. In the following, we will consider two typical programmes, showing that a continuous range of spatial resolution, with high spectral resolution, is highly desirable.

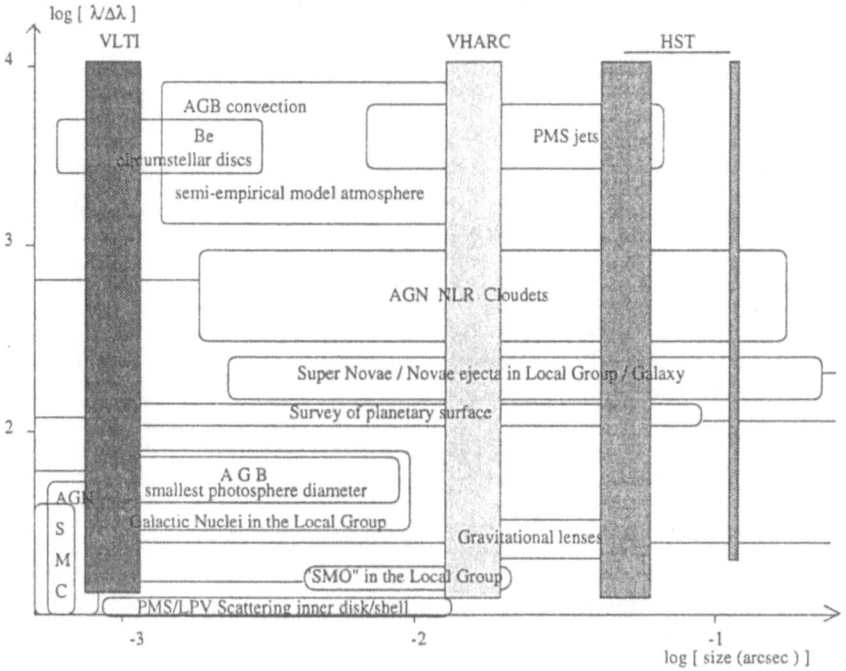

Fig. 1. Domains of astrophysical programmes in the spectral resolution versus spatial resolution plane, compared with the spatial resolution of the VLTI (dark strip), the VHARC (light strip) and the HST (gray strip). The strip widths span the visible spectral range. The thin strip on right shows the resolution of the HST currently limited by the sampling of the WFC.

2 The Limiting Magnitude

The signal to noise ratio Λ depends on the noise propagation in image restoration algorithms. Therefore for the sake of simplicity, let us consider the formula for the case of the power spectrum at spatial frequency ν derived from Roddier (1981):

$$\log \Lambda (\nu) = 7.53 + \log \Delta\lambda + \frac{1}{2} \log T + \log \rho - \frac{1}{2} \log \Delta v + \frac{5}{2} \log r_0 - \frac{m}{2.5} \quad (1)$$

where $\Delta\lambda$ stands for the spectral bandwidth in nm, T for the observing time in seconds, r_0 for the Fried parameter at the wavelength of the observation (in metres), Δv for the standard deviation of wind velocities, in metres per second, weighted by the structure constant of the refractive index, and ρ for the total efficiency of the atmosphere-telescope-instrument-detector. With $\Delta\lambda = 0.1nm$, $r_0 = 0.2m$, $\Delta v = 10$ m/s, $\rho = 0.03$, and $T = 3600$ sec, which are typical parameters for the VHARC, one gets $\Lambda(\nu) \approx 10$ at $m = 8.9$.

Limiting magnitudes for the 1.8 metre telescopes of VISA will range from 8 to 14, depending on the observing mode (ESO Interferometry Panel 1992).

3 Late Stages of Stellar Evolution

Long Period Variables (LPV's) have a very complex atmospheric structure, owing to several factors:

- *the extension* of the atmosphere is huge, up to $\approx 10R_*$ (Bonneau et al. 1982; Wilson et al. 1992), which makes the plane-parallel approximation no longer valid.
- *several molecule species* contribute heavily to the absorption, e.g.: TiO is the dominant spectral feature in the visible for M stars, or C_2 for carbon stars. Therefore the optical depth in the atmosphere is rather sensitive to the dissociation equilibrium, i.e. to the temperature and the pressure.
- *dust grain formation* has to occur somewhere in between the photosphere and the circumstellar shell. The formation process is probably related to the vertical profile of molecular density and to the shockwave propagation. The atmosphere structure is strongly affected in the grain formation region.
- *shockwaves* propagate from the inner photosphere to the outermost atmospheric layers during the light cycle. It raises a lot of questions, about the pre-shock and the post-shock, its possible effect on the atmosphere extension, and its possible relation with dust grain formation.
- *mass loss* puts these stars among the most important contributors to the ISM recycling. It is related to at least the two previous items.
- *convection* is also a major factor in the atmosphere structure, e.g.: through the large scale inhomogeneities it could produce due to the amplitude of the temperature distribution at the base of the photosphere (Scwharzschild 1975).

- *an asymmetry* of the outermost atmosphere has been reported (Karovska et al. 1991); this is also shown in Fig. 2 (Thiébaut 1994). Hence 1-dimensional atmosphere models are not appropriate.
- *pulsation* is probably related to the generation of shockwaves; the atmospheric structure undergoes large temperature fluctuations with time.
- other mechanisms, such as *tidal* interaction or *mass exchange* from companions, generally negligible in the case of other types of stars, may play a significant role in the atmospheric structure of LPV's.

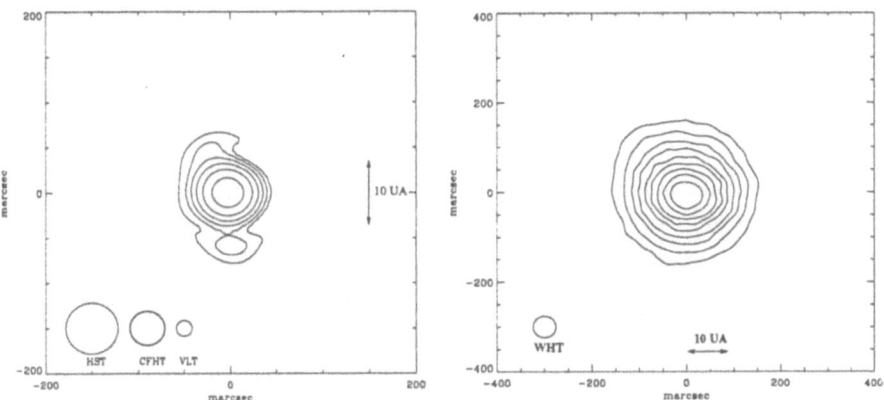

Fig. 2. Image of T Tau in Hα at the resolution of the CFHT (Thiébaut 1994).

Fig. 3. Image of χ Cyg at 496 nm (TiO) at the resolution of the WHT (Thiébaut 1994).

3.1 Probing the Geometrical Depth

Due to the strong absorption coefficient of TiO, its spectrum dominates the visible spectrum within the temperature range 1000-3500K (for an illustration, see the spectrum of TU Oph in Warner et al. (1988). The atmospheric layers $\tau_{TiO} = 1$ for different bands span a large range of altitudes (Bonneau et al. 1982; Labeyrie et al. 1977). The measurement of the angular extension of the atmosphere within bands centered on TiO features will allow one to build semi-empirical atmosphere models.

i/ Spatial resolution requirements

The very closest LPV's have diameters of the outermost layers within these TiO bands of ≤ 0.1", but the *innermost layers* remain unresolved, requiring the resolution power of an *8 metre telescope*. In the case of the base of the *photosphere*, which could be observed at $1.04\mu m$ or in the blue-UV spectral domain, the resolution of the *VLTI* is required: the diameter of Mira in the 400-420nm range is $\approx 0.025"$ and that of χ Cygni is $\approx 0.015"$; both these stars are the closest LPV's, and both have peculiarities (Mira is a binary undergoing mass exchange, and χ Cygni is a M-S type star). Other LPV's are more distant, and they require baselines $\gtrsim 15$ m which will be provided by the VLTI. Let us recall that the radius of the photosphere is a fundamental parameter for modelling both the atmosphere and the circumstellar shell.

ii/ *Polarimetric imaging requirements*

Whether the absorption is pure absorption, scattering, or both, is an open question, which is crucial to the correct modelling of the atmosphere of LPV's. The polarisation is known to vary abruptly with the spectral features (Boyle et al. 1986). Observing the geometrical structure within different polarisations should bring new constraints in this field. If the extension of the atmosphere turns out to be insensitive to polarisation, then one should expect that scattering does not contribute significantly to the TiO absorption.

iii/ *Spectral resolution requirements*

TiO band heads have a typical width of $\Delta\lambda \lesssim 2\mathring{A}$: this sets the lower limit of the resolution when imaging LPV's within TiO features: $\lambda/\Delta\lambda = R \geq 2500$.

Thus the strategy of such observations is that they should be made:

- with a single VLT unit or with the VLTI, depending on the extension of the region to study;
- for different bands characterized by different intensities and/or band systems, in order to address different atmospheric layers,
- for different phases of the light cycle, in order to model the behaviour of the atmospheric components with time, and
- for different objects in order to address different kinds of LPV's (e.g.: Population I versus Population II stars, stars pulsating in different modes, ...)
- in parallel with spectrometry and spectropolarimetry for flux calibration.

Other molecular bands or atomic lines are of major interest from the point of view of probing the geometrical depth:

- in χ Cygni, light in the CaI 4227$\mathring{A}$ line, as well as in the Hδ and Hγ lines, is polarized as compared with the polarization in the local continuum (Boyle et al. 1986). Why? What is the possible geometrical cause(s) for this polarization? A spectral resolution of $R \approx 4000$ is best for such observations.
- unlike the TiO bands for the closest LPVs, the C_2 bands in carbon LPVs are formed in atmospheric regions which are not resolved with 4 meter class telescopes. Is this due to a distance effect, or due to a different line formation process? Investigating these features will require a single unit 8 meter telescope, or the VLTI.
- similarly, ZrO bands are a dominant feature of S-type LPV's, and their polarisation has a markedly different behaviour from that of TiO bands.

3.2   What is the Size of Convective Eddies?

The role of convection in the atmosphere structure of LPV's is not understood. It could produce polarisation. It could be related to the generation of shockwaves. A major parameter characterizing the convection is the mean size of eddies, which is completely unknown today. It may range in between a half and a few hundredths of the stellar radius (Scwharzschild 1975).

i/ *Spectral resolution and polarimetry requirements*

The TiO spectrum is strongly temperature sensitive. Therefore the mean s-cale of convective eddies should be measured within bandwidths centered on TiO bands to enhance the contrast, with the assumption that the surface distribution of the intensity of these lines is related to the underlying surface distribution of temperature. The spectral resolution required is the same as above : $R \approx 2500$, again with polarimetric capabilities.

ii/ *Spatial resolution requirements*

Very few objects will be accessible with an 8m telescope, since the minimum diameter of Mira is ≈ 0.025" and that of Betelgeuse is ≈ 0.05" (Wilson et al. 1992). Nevertheless they should first be observed with this spatial resolution, because it will consume much telecope time to image such "large" surfaces with an interferometer with as few apertures as the VLTI at the required dynamic range. Almost all the other objects will require the VLTI.

3.3   What Do Shockwaves Look Like?

We know that shockwaves are produced in the outer layers of LPV's. Indeed, emission lines are observed: Hydrogen lines and atomic lines, e.g.: TiI, FeII. These lines may have P Cyg profiles (Ferlet & Gillet 1984). Hα and Hβ profiles show two emission bumps ≈ 1Å apart, the blue one being brighter (Gillet et al. 1983). Hγ and Hδ are single asymmetric lines. Also, the radial velocities of low and high excitation lines of CO at $2.2\mu m$ are interpreted in terms of a velocity field which traces the propagation of shockwaves (Mailllard 1974; Hinkle et al. 1982). The physical process which gives rise to these shocks is not known: are they produced by pulsations or by the convection, or are they Alfvèn waves or acoustic waves, or something else entirely ?

i/ *Spatial resolution requirements*

The shock at Hγ and Hδ has been resolved with the CFHT in the case of χ Cygni (Foy 1988): the major contributor to the emission flux is a region of size 0.040". Since χ Cygni is the second closest VLP to the Sun, a 8m telescope, or the VLTI, is required to image the shocks for most of the other LPV's in the solar neighborhood, and to follow the shock propagation. The shape of the shock could be important for understanding the pulsation mode of the star (Tuchman 1991). The extension of the postshock region in not known. It is not resolved in the Hδ line in χ Cygni with the CFHT; we will probably need to work the VLTI with long baselines to be able to measure it. Another open question is whether the shock wave is homogeneous or not, and if yes, is there a relation with the stellar surface inhomogeneities possibly caused by convection? This requires a spatial resolution similar to that required to study the convection.

ii/ *Spectral resolution requirements*

Imaging the objects with narrow bandwidths across the Hydrogen emission line profiles will provide information about the velocity field, which is crucial for understanding the physical mechanism of the line formation: is the double peak profile due to a geometrical effect (an expanding shell) or to absorption by far away material absorbing in the central region of the line profile (Gillet 1988a; Gillet 1988b). The spectral resolution required to isolate a bump is $R \lesssim 1\text{Å}$; it is $R \lesssim 0.5\text{Å}$ to isolate the central depression.

3.4 Is There Enough Flux?

The amplitude of the light curve of Mira: $3 \lesssim m\,(o\text{Ceti}) \lesssim 8$ or of χ Cygni : $5 \lesssim m\,(\chi\text{Cygni}) \lesssim 13$ has to be compared with the limiting magnitude derived in Sect. 2. The answer is *"Yes! There are enough photons!"*. Even taking into account a loss of 80% in strong TiO absorption bands, such bright stars are observable over the whole light cycle. From the absolute magnitude versus period relationship (Foy et al. 1975), every VLP closer than ≈ 1 kpc to the Sun will be observable at maximum light with this signal to noise ratio, which will allow stars to be observed with very different physical parameters such as e.g.: the pulsation mode or the chemical composition.

4 Pre-Main Sequence Stars

Pre-main sequence stars, like T Tau stars, or FU Ori stars have a complex environment; they often show bipolar jets which are probably fed by circumstellar discs. Their spectral energy distribution is characterized by both an IR and U-V excess (Bertout 1989). Large scale jets are known. HST observations of DG Tau have revealed a jet as close as 0.25" to DG Tau (Kepner et al. 1993). A jet extends from 0.3" from T Tau to HH1555 30" away westward, in [O I] and [S II] lines (Böhm & Solf 1994). The diffraction limited image of T Tau in Hα (Thiébaut 1994) shows an extension at a much smaller scale, which could be the cross section of the westward jet (Devaney et al. 1994). The study of the fine structure of the jets, and possibly the study of the scattered light of the inner disk would help to understand the production mechanism of the jet. In particular, it is important to know whether the jet is continuous, or whether there are knots or other structures along it, and to have a measure of the cross section of the jet to probe the efficiency of the collimation process.

4.1   Spectral Resolution Requirements

The typical full width of the Hα line is $\approx$ 400km/s, i.e.: $\approx$ 10$\mathring{A}$. A reciprocal spectral resolution of $\Delta\lambda \approx 3\mathring{A}$ seems to be the minimum possible for one to be able to derive a 3-dimensional Hα map of the jet. Three-dimensional mapping is crucial to determine the velocity field. The physical process of the Hα emission is not known. Observing the base of the jet, where the line seems to be partly formed, will allow one to check whether its formation involves a shock mechanism or not. If it does, then the map obtained within the forbidden lines [O I], [S II] and [N II] should ressemble that in Hα. These lines have an equivalent width of a few $\mathring{A}$. This requires that the spectral resolution is again $\approx$ 3$\mathring{A}$ (in this case, one does not attempt to measure the velocity field).

4.2   Spatial Resolution Requirements

From the very recent results cited above, observations in Hα, [O I] or [S II] will require the resolution of a single 8m to better resolve the extension close to T Tau, and later that of the VLTI, particularly for measuring the width of the collimated jet close to the star. In addition a resolution of $\approx$ 0.1" is required for intermediate scale imaging. The latter may be provided either by the HST or by adaptive optics working in partial compensation mode.

It is worth noting that mapping the inner disc to measure the amount of scattered light and derive grain properties will be more efficient as one moves toward shorter wavelengths. This supports the *use of the VLTI in the blue* spectral range. Of course, due largely to telescope vibrations, the contrast of the fringe pattern will decrease at the same time. But there is no discontinuity which would forbid observing at shorter wavelength than a given threshold: it is better to have observations with a lower contrast than no observations at all!

4.3   Is There Enough Flux?

The brightest pre-main-sequence stars in Taurus have magnitudes ranging $\approx$ 9-11 at maximum light. From Sect. 2, the answer is again "Yes there is!", they can be observed with an 8m telescope with a spectral reciprocal resolution of 1$\mathring{A}$. Applying (1) with $\Delta\lambda = 3\mathring{A}$, leads to the limiting magnitude $\approx$ 16: thus these stars at minimum light, or fainter objects, will be also observable and with the VLTI as well.

5   Conclusions

I have restricted the discussion in this communication to only two topics. We have addressed in the same way several other ones, leading to similar conclusions which are summarized below.

Within the coming few years, additional observations should be carried out to prepare for high angular resolution imaging with the VLT/VLTI. Spectroscopy, polarimetry and speckle interferometry with 4 meter class telescopes, to optimize the wavelengths and the bandwidths of the observations, are all required.

Also, one has to further analyse the comparison of the performance with respect to the HST for the high angular resolution, in particular addressing the following items:

i/ **Spectral resolution** The spectral resolution which is today available at the HST is not flexible. Neither the central wavelength nor the bandwidth of the filters can be tuned continuously. It should be possible to have such a capability in the future "Advanced Camera" (Brown 1993); I suggest equipping this camera with a Courtès monochromator (Tallon et al. 1995). If this flexibility is not available with the Avanced Camera, then it will strengthen the need for the VHARC.

i/ **Polarimetry** There is today no facility for high angular resolution imaging in polarized light at the HST.

ii/ **Field of view** The advantage of the HST is huge, since it is not affected by the anisoplanatism due to the atmosphere.

iii/ **Dynamic range** The advantage of the HST is again huge, since the transfer function is not degraded by the atmosphere. With respect to the examples given in this communication, this is particularly worth considering for imaging intermediate scale jets, or the outermost atmospheric layers.

iv/ **Spatial resolution** Diffraction limited images at given wavelength will be sharper with the VHARC by a factor 8m/2.4m = 3.3, which is a critical factor in many cases, in particular for the above mentioned programmes, and as it appears in Fig. 1. It should also be noted that the minimum base of the VLTI is 8 metres; so one will need observations at the diffraction limit of a single 8 m telescope to avoid gaps in the low frequency domain of the so-called (u,v) plane, for which the weight is high in image restoration algorithms.

Whatever the imaging technique, direct imaging with the HST or with adaptive optics, speckle imaging with the VHARC or Michelson imaging with the VLTI, the spatial resolution can be enhanced by running deconvolution algorithms. Direct images should be more enhanced because the transfer function is not degraded at high spatial frequencies by the atmosphere. A priori it does not seem possible to retrieve the spatial resolution of the VHARC with deconvolved HST images. But theoretical computations and numerical simulations, confirmed later on by observations, have to be run to understand better how these imaging techniques compare.

To check new concepts and astrophysical programmes for the VHARC, for the VLTI, and for the adaptive optics, our programme of development includes

- the construction of a speckle imaging camera (the SPID), providing a spectral resolution of 1Å, for astrophysics, and including a wavefront sensor for self-referenced speckle holography (Tallon et al. 1995);

- the development of new image restoration methods, in particular merging bispectrum analysis and wavefront sensing.
- the extension of these methods to the case of a diluted aperture, of which the interest is considerable, because it breaks the limitation of the signal to noise ratio $\propto D^{\frac{1}{6}}$ due to the spatial multiple mode of the VLTI (Tallon & Tallon-Bosc 1992).
- the participation in the study of the new combiner REGAIN for the astrophysical exploitation of the GI2T (Mourard et al. 1994).
- the R&D for the polychromatic artificial star of which the purpose is to provide full compensation for the tilt with full sky coverage with a laser guide star (Foy et al. 1994).
- the coupling of speckle interferometry with adaptive optics operated in partial compensation mode.

There are a lot of items to work on in order to be able to take full scientific advantage of the high angular resolution at the VLT. It requires a comprehensive approach for instruments, data processing and scientific programmes. From the point of view of astrophysics there is a strong link from the VLTI to adaptive optics and the HST, through the VHARC. The spatial resolution of the VLTI is at least one magnitude higher than that of the HST, so that there is no competition. **With respect to the HST**, a visible adaptive optics camera has comparable resolution, and **the VHARC has 3.3 times higher spatial resolution**. They will compete very favourably with the HST if they provide a **spectral resolution** which is **both high and flexible** and **polarimetry capabilities**. This is a crucial specification to achieve a high angular resolution instrumentation which maximises the benefits of the complementary performances of the HST and of the VLT.

References

Bertout, C. (1989): Ann. Rev. Astron. Astrophys. **27** 351

Böhm, K.-H., Solf, J. (1994): ApJ **430** 277

Bonneau, D., Foy, R., Blazit, A., Labeyrie, A. (1982): A&A **106** 235

Boyle, R. P., Aspin, C., Mclean, I. S., Coyne, G.V. (1986): ApJ **164** 310

Brown, R. A. (1993): "The future of Space Imaging" STScI Baltimore

Connan, J.-M. (1994): thesis Université Paris XI

Devaney, N., Thiébaut, E., Foy, R., Blazit, A., Bonneau, D., Bouvier, J., de Batz, B., Thom, Ch. (1994): A&A, submitted

ESO/VLT Interferometry Panel (1992): "Coherent combined instrumentation for the VLT Interferometer", ESP VLT Report No. 65

Ferlet, R., Gillet, D. (1984): A&A **133** L1

Foy, R., Heck, A., Mennessier, M. O. (1975): A&A **43** 175

Foy, R. (1988): in "Instrumentation for Ground-Based Optical Astronomy, Present and Future", ed. L. Robinson (Springer-Verlag : New-York), p. 345

Foy, R., Migus, A., Biraben, F., Grynberg, G., McCullough, P. R., Tallon, M. (1994): A&A in press

Gillet, D. (1988a): A&A **190** 200

Gillet, D. (1988b): A&A **192** 206

Gillet, D., Maurice, E., Baade, D. (1983): A&A **128** 384

Hinkle, K. H., Hall, D. N. B., Ridgway, S. T. (1982): ApJ **252** 697

Karovska, M., Nisenson, P., Papaliolios, C., Boyle, R. P. (1991): ApJ Letters **374** L51

Kepner, J., Hartigan, P., Yang, C., Strom, S. (1993): ApJLetters **415** L119

Labeyrie, A., Koechlin, L., Bonneau, D., Blazit, A., Foy, R. (1977): ApJ Letters **218** L75

Maillard, J.-P. (1974): Highlights of Astronomy **3** 269

Mourard, D., Blazit, A., Bonneau, D., Merlin, G., Tallon-Bosc, I., Vakili, F., Ménardi, S., Rebattu, S., Hill, L., Rousselet, K., Boit, J.L., Le Merrer, J., Lasselin-Waultier, G., Saisse, M., Pouiliquen, D., Joubert, M. (1994): SPIE 2000 58 in press

Roddier, F. (1981): Proc. ESO Conf. "Scientific Importance of High Angular Resolution at Infrared and Optical Wavelengths", ed. M.-H. Ulrich et K. Kjär, (Garching : ESO), p. 423

Roggemann, M. C., Caudill, E. J., Tyler, D. W., Fox, M. J., von Bokern, M. A., Matson, C. L. (1994): Appl. Opt. **33** 3099

Schwarzschild, M. (1975): ApJ **195** 137

Tallon, M., Tallon-Bosc, I. (1992): A&A **256** 715

Tallon, M., Baranne, A., Belkine, I., Foy, R., Chatagnat, M., Dubet, D., Kohler, B., Lacroix, D., Robert, D. (1995): these proceedings

Thiébaut, E. (1994): Thesis, Université Paris VII

Tuchman, Y. (1991): ApJ **383** 779

Warner, B., Fairall, A. P., Overbeek, M. D. (1988): Ap&SS **143** 211

Wilson, R. W., Baldwin, J. E., Buscher, D. F., Warner, P. J. (1992): MNRAS **257** 369

Narrow-Angle Astrometry with the VLT Interferometer

O. von der Lühe[1], A. Quirrenbach[2], and B. Koehler[1]

[1] European Southern Observatory, Garching, Germany
[2] Max-Planck-Institut für Extraterrestrische Physik, Garching, Germany

Abstract. We discuss the merits of a narrow-angle (up to 40 arcsec) astrometric mode for the VLT Interferometer Sub-Array (VISA), which could perform with a precision of the order of 10 micro-arcsec. We present a concept for an astrometric instrument at the combined focus of VLTI.

1 Introduction

The interferometric mode of the ESO Very Large Telescope, the VLT Interferometer (VLTI; Beckers 1991; von der Lühe et al. 1994), is conceived as a pure imaging array, astrometry has never been considered as part of its concept. The precision, to which vector baselines (i. e., the separation between the points of intersection of azimuth and altitude axes) must be known, increases with the separation between the sources whose relative position is to be measured. The large 8m Unit telescopes cannot be optimised for astrometric requirements, so astrometry over wide angles will forever remain impossible with VLTI. The precision requirements are substantially relaxed when astrometry is performed over narrow angles only. Shao and Colavita (1992) pointed out the scientific value of narrow angle astrometry (NAA), where fields of a few dozen arcsec are used, and explained the advantage of large baseline interferometers over filled pupil telescopes in this regime. Their considerations led us into analysing whether (and how) NAA capability can be implemented with VLTI. We found that the already existing extended field-of-view capabilities of VLTI offer many advantages.

Why should one want to do narrow-angle astrometry with an interferometer? The most severe contributor to ground-based astrometric error other than instrumental effects is atmospheric turbulence. Due to the lateral correlation of atmospheric optical delay (isoplanatism), the atmosphere behaves differently over small angles than over large angles. The transition between these regimes depends on the baseline. The dependency of the astrometric error σ on the baseline B (either the diameter of a single-dish telescope or the separation of elements in an interferometer) and on field angle θ is given in Table 1, for a given height h of the turbulent atmospheric layer above the observatory.

The astrometric error due to turbulence as a function of reference star separation angle is shown in figure 1 for several baselines. These curves assume an integration time of 30 minutes and were obtained using measured $C_n{}^2$ (structure constant for the refractive index of air) and wind data from Paranal.

Table 1. Dependency of narrow angle astrometric error (σ) on baseline (B) and field angle(θ).

Observing Instrument	Single dish telescope with $B \ll \theta h$	Two-element interferometer with $B \gg \theta h$
Dependency of σ on B Dependency of σ on θ	independent $\sigma \propto \theta^{1/3}$	$\sigma \propto B^{-2/3}$ $\sigma \propto \theta$

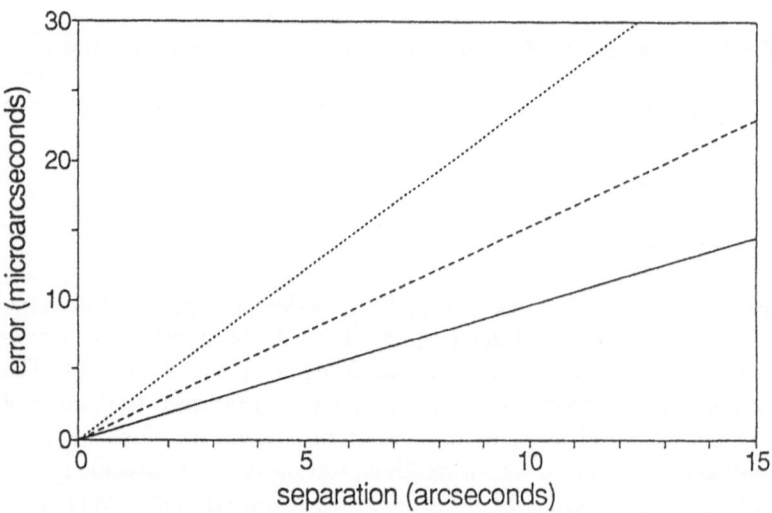

Fig. 1. Astrometric error due to turbulence at Paranal, as a function of angular separation. Baselines are 50 m (dotted), 100 m (dashed) and 200 m (solid).

Potential applications of narrow-angle astrometry are discussed by Quirrenbach (1995). Many target objects are intrinsically bright and can serve for adaptive optics wavefront control and for fringe tracking. A suitable position reference can therefore be much fainter. Shao and Colavita estimated a 90% probability to find an $m_K = 17$ star within a 17 arcsec field radius. Many program sources will therefore have a suitable position reference star sufficiently close by.

2 How Can Narrow-Angle Astrometry Be Done with VLTI?

The goal is to exploit narrow-angle astrometric capabilities of VLTI with a precision comparable to the limit set by turbulence. A useful target is 10 micro-arcsec, which results in the following technical requirements. The accepted field should be ideally 30 arcsec, and at least 15 arcsec. Vector baselines must be known to overall precision of $\frac{\partial B}{B} \approx \frac{\partial \theta}{\theta}$, where ∂B denotes the precision in baseline (all vector components), θ the angular separation of the sources of interest, and $\partial \theta$

denotes the desired astrometric precision. For baselines of 100 m we arrive at $\partial B \approx 50 \mu m$. Differences in optical path between the two sources must be known to the 5 nm level. The sensitivity should be sufficient to exploit reference sources of $m_K \approx 15 \cdots 18$ within a $15 \cdots 30$ arcsec field.

2.1 Unvignetted Field

The current unvignetted field of VLTI is 8 arcsec with Unit telescopes and 3.5...8 arcsec with Auxiliary Telescopes, depending from which station an Auxiliary telescope operates and on the position of the delay line carriages. This field is quite difficult to modify for the Unit telescopes. A simple addition to the coudé relay optics of the Auxiliary Telescopes (AT) substantially increases their unvignetted field. A two mirror beam expander reduces the pupil demagnification from a factor 100 to a factor 20 and produces 90mm parallel beams. A suitable axial position of the expander results in a transferred pupil much closer to the delay line. The unvignetted field for the AT varies between 15 ... 40 arcsec with this modification. Figure 2 shows the size of the unvignetted field as a function of the optical path from the AT station to the center of the delay line tunnel, for both the current concept and the beam expander concept, as well as for the near and far delay line positions. The field is particularly large when stations are used which are close to the beamcombiner lab. In this case, the delay line carriages will never approach the far position. The field will remain in the 15 arcsec range if stations far from the tunnel center are used.

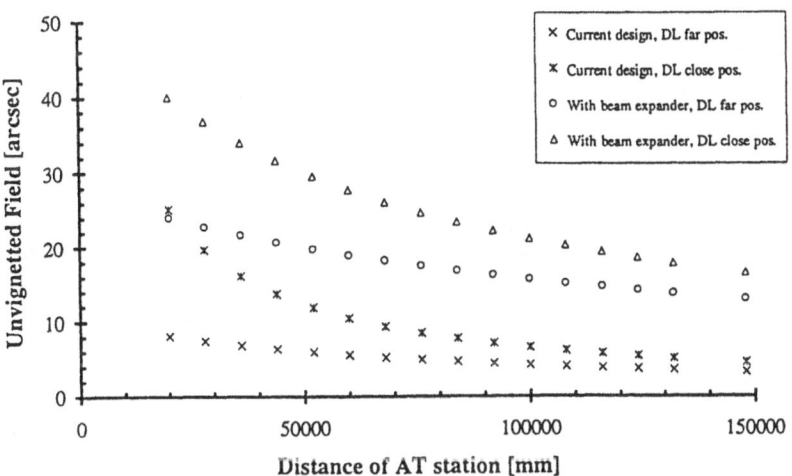

Fig. 2. Unvignetted field without and with beam expander.

2.2 Precision of Vector Baselines

The main requirement is to achieve a precision of $50\mu m$ on all three coordinates of a baseline over periods of time corresponding to an astrometric observation - typically two hours. This can be achieved by a metrology system (complex and expensive) or by repeated astrometric solutions (time consuming). The observation of six position references for the astrometric solution would take some 30 minutes, which appears acceptable. Maintaining the required mean precision does not appear a great challenge on a 2 hr timescale given that Auxiliary Telescopes are already built for stability.

2.3 Differential Optical Path

Instrumental optical path difference (OPD) between the two field positions (program source and position reference) must be *known* to a precision of 5 nm. Both light beams share all optics in the train up to the point where beam mixing occurs - this is a feature of the already existing VLTI design and a great advantage. One of the main contributors to differential OPD is the quality of all optical surfaces which are not at a transferred pupil - these need to be monitored with high precision. Repeated characterisation of the optical surfaces as well as calibration with narrow-angle distance references, and the corresponding data calibration procedures, will be required. The variable curvature mirror at the delay line cat's eyes may pose a particular problem. A detailed error budget needs to be established.

2.4 Sensitivity

Beckers (1993) estimated a limiting magnitude of $m_K = 20$ for observing conditions which apply to NAA - i.e., presence of a near-by bright reference which is used for image tracking and fringe co-phasing. Even if one allows for reduced sensitivity of an astrometric instrument, there is sufficient margin to conclude that sensitivity is not a severe limitation. The sensitivity limit for the program source is $m_K = 13$ if visible light is used for fine guiding, and $m_K = 9$ if fine guiding is done at the observed wavelength.

3 Instrumentation for Astrometry

Narrow-Angle Astrometry with the VLTI will require a dedicated astrometric instrument. We have established some major requirements which include simultaneous measurement of the white light (central) fringe on both program and reference sources, sharing of optics for both sources up to the point of beam mixing to avoid differential delays, capability to accept beams from up to three telescopes, and the operation in the near IR.

A possible concept for such an instrument is based on a design by Colavita (1994) and is presented in Figure 3. It consists of three major units.

1. A source separator whose purpose is to take the larger fraction of light of the program source (Fig. 3a). The separation is achieved in the image plane with a field mirror whose central ~2 arcsec is not coated. Partially reflected light of the program source which is centered on the mirror, and all of the reference source light is directed towards the beam combiner. The transmitted light of the program source is directed towards image tracking and co-phasing sensors. Properly designed reflectivity of the central part of the mirror helps in equalising the intensities of both program and reference sources which improves fringe detectibility. Another advantage of this design is that no field de-rotation is required.

2. A "beam switcher" which is used only if three telescopes are combined (Fig. 3b).

3. A two-element beamcombiner which includes a high precision differential delay line in one of the two arms and an OPD modulator for synchronous detection in the other one (Fig. 3c). The differential delay line splits the pupil and introduces a delay into one half of the beams. The delay is adjusted such that zero OPD is measured for the program source in one half pupil, and for the position reference in the other half pupil. The required differential stroke for a source separation of 20 arcsec and for a 100m baseline is 5mm and needs to be measured to a precision of 5nm. Separate pairs of detectors receive the light of each half pupil after beam combination by the beamsplitter. The wide detected bandwidth requires the use of a compensator plate. The extra reflection in one of the beams is needed to maintain an even difference of beam turnings.

4 Conclusions

We have investigated the possibility to include narrow-angle astrometry into the existing concept of VLTI. Doing so would substantially increase the scientific scope of VLTI. Preliminary analyses indicate that NAA may be possible with the VLT Interferometer Sub-Array (1.8m Auxiliary Telescopes) with only minor extra effort. A dedicated astrometric beam-combining instrument which is currently not in the program would be required.

References

Beckers, J. M. (1991): J. Optics (Paris) 22, 73
Beckers, J. M. (1993): *VLT Interferometer: Science Objectives and Requirements*, VLT Programme Review 1993
Colavita, M. (1994), Astron. Astrophys. 283, 1027
Quirrenbach, A. (1994): these proceedings
Shao, M., Colavita, M. (1992): Astron. Astrophys. 262, 353
von der Lühe, O., Ferrand, D., Koehler, B., Zhu, N., Reinheimer, Th. (1994): in *Amplitude and Intensity Spatial Interferometry II*, J. B. Breckinridge (Ed.), SPIE Proc. Vol. 2200, p. 168

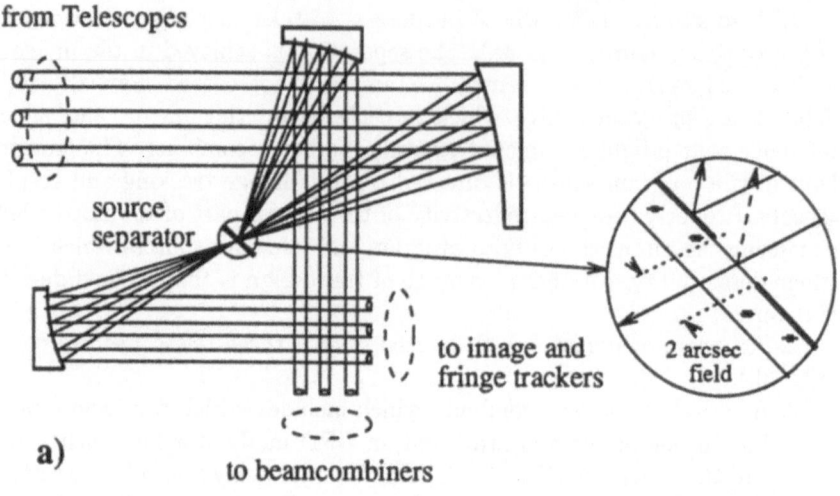

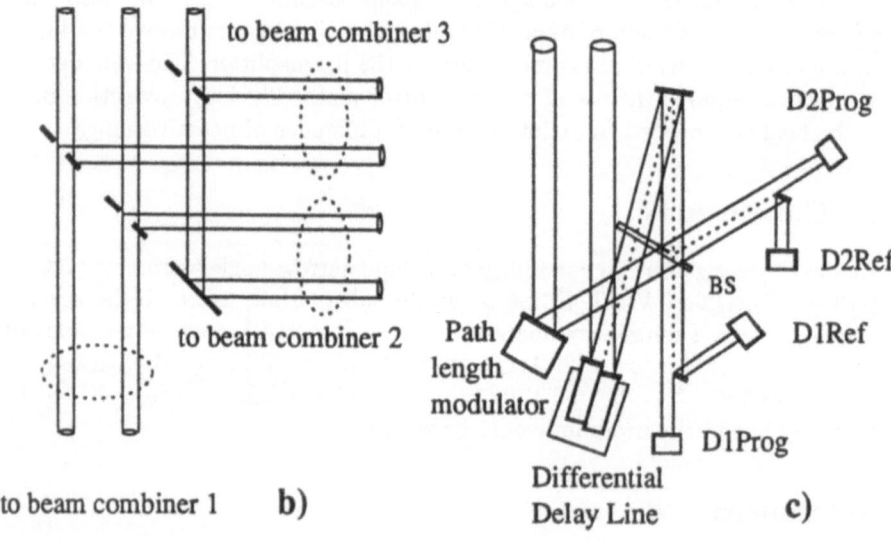

Fig. 3. Astrometric instrument concept; a: source separator, b: beam switcher, c: beam combiner.

Part 11

**Strategies for Use
of the VLT**

Flexible Professional Observing at the VLT

L. Woltjer

Observatoire de Haute Provence, F-04780 Saint-Michel l'Observatoire, France

Abstract. Some typical programmes for the VLT are outlined. All require a very flexible scheduling. This fact as well as considerations of observing efficiency and quality lead to the conclusion that the VLT should be largely operated by professional observers rather than by visiting astronomers.

1 Typical Scientific Programmes

In this paper I shall discuss a few typical scientific programmes for the VLT and the consequences that such programmes would have for the mode of operation of the VLT.

1.1 Gravitationally Lensed Quasars

When the light from a quasar passes by concentrations of matter on the way to the observer, the gravitational lensing effect may lead to the appearance of multiple images. In principle, such images should have strictly identical spectral distributions. In practice, differential absorption along the slightly different light paths, differential magnification of different components of the quasar, differential effects associated with the time variation of the quasar, and the small differences in light travel time to the various images may make it more difficult to differentiate between gravitationally lensed images from a single quasar and images from quasars close in space - for example in a small group of galaxies. Differential absorption could be largely eliminated by observing in the IR, while an appropriately spaced sequence of observations could establish the effects of the variability. The latter is particularly important as it gives information about the gravitational effects of the concentrations of matter along the way and, if these have a simple structure, possibly about the geometry of the Universe (H_O, q_O). Lensing by mass concentrations like (clusters of) galaxies gives time delays between images of the order of years, but microlensing by stars in such galaxies may lead to more rapid variations associated with the motion of the star. Moreover, microlensing may "resolve" some of the quasar components.

The separation of the images is of the order of an arcsec in typical cases. Good spatial resolution is therefore required for accurate data. Spectral resolutions of 10^3 are desirable to study differential amplification of continuum and emission line regions. Observations need to be spaced at weekly to monthly intervals which creates scheduling problems for most telescopes. The gravitational lens community therefore has proposed to build a special telescope devoted to lensed objects. A change in scheduling procedures might be a better solution, since major gains would result if the spatial resolution (in the near IR) and spectral resolution of the 8-m telescopes could be combined.

1.2   Mapping of the Broad Line Region in Quasars and Seyferts

The BLR generally appears to be ionised by a variable compact source of ionising radiation. Because of light travel time effects in the BLR (a month - years) the emission lines fluctuate with time delays. Measurements of these delays give information about the distribution of the gas. From accurate liné profiles and their variations information on the kinematics may also be obtained. High signal-to-noise profiles at resolutions better than 10important in many AGN, observations in the IR seem particularly important, the more so since in the IR numerous coronal lines occur (high ionisation stages of Si, etc.) which give information about the far uv radiation field.

Daily - monthly observations are needed in typical cases, leading again to the need for a rather fluid scheduling.

1.3   X- and Gamma-Ray Sources

The first decade of the VLT will coincide with the operation of two major European space missions: XMM (1999 launch), a large X-ray collector for (modest resolution) imaging and deep spectroscopy, accompanied by an optical monitor (V ¡ 24.5); and INTEGRAL, a gamma-ray mission with good sensitivity, which among other things will make repeated scans of the Galactic plane. Both missions should detect various targets of opportunity, with a need for rapid follow-up by the VLT, as well as many variable AGN and X-ray binaries (with sometimes unpredictable high and low states), which should be observed simultaneously in the optical / IR. Again, a very fluid schedule is needed to execute such observations.

1.4   Supernovae, Novae, etc.

While novae are the classical targets of opportunity, other categories of variable object, such as supernovae and cataclysmic binaries, may well be much more numerous. Some observations may be preprogrammed at particular moments, others may have to be inserted in the observation programme at the last moment. In any case it is clear that none fit into the traditional way of running a telescope, with observers obtaining a few nights according to a schedule made a year before. Equally if a long time series of observations of an object is to be obtained, continuity is essential; if the night planned for the observation is cloudy, it should be observed the next night, displacing part of the programme of that night.

2 Efficiency of Telescope Use

The VLT is an expensive telescope ($\sim$ 1 MDM / week). Consequently it is necessary to use it with maximum productivity. The efficiency of the VLT is determined by:

a The site (atmosphere and other factors)
b Telescope optics
c Efficiency of instrumentation
d Overheads.

Once established (a) is fixed, while concerning (b) one could still wonder whether one or two telescopes should be optimised for IR use. Almost everywhere (see also the discussion on supernovae and cepheids as standard candles), interstellar absorption is a major hindrance which prevents observation in some objects and makes the results more uncertain in others. In the IR the problem largely disappears. Now that IR detectors with 10^6 pixels are around the corner, it is clear that the domain of 1 - 2.5 Cμm is particularly promising, the more so since this is also most suitable for the study of much of the Universe at a redshift of a few.

Concerning (c), there seems to be a general belief that multiobject capabilities are the essential factor. However, it should be noted that with a field of 50 arcmin^2 this is only an advantage for objects with an area density in excess of 300 degree^{-2}. While useful in clusters and in cosmology, there are many other areas of astronomy which do not benefit from multiobject capability. Maximum throughput remains an essential factor in instrument building.

The overhead issue is of major importance. Factors to be considered are:

1 Time used for technical work
2 Faults
3 Calibration
4 Experience of observer
5 Sequencing of observations.

Concerning (1), proper organisation of the work to be done at the telescope is probably the main factor. With the VLT one cannot bring only half-finished instruments to the telescope, as was not infrequently done at La Silla.

Faults frequently occur when instruments are exchanged. With three foci per telescope it should be possible to limit the instruments available per telescope to three. If these were kept in a constant configuration for periods of one or two years at a time, much of the problem should disappear. It also reduces the time needed for calibration. Especially now that the detector situation is beginning to stabilise - at least in the optical - it would be best to exclude "improvements" during one or two year periods; this also much helps the construction of a usable archive. To further reduce loss of time due to faults, a proper analysis of the needed component reliability and related factors would also be of interest. Of course, this does not mean that one would wish to go to space-qualified instruments, since then the financial and construction time advantages of ground-based facilities would be lost.

Concerning (4), it hardly seems possible to continue the present practice of having ever more complex instruments used by inexperienced observers. But any visiting astronomer who obtains 3 or 4 nights per year will necessarily be inexperienced. The efficiency of the use of telescope time and the confidence in the data obtained would probably be much increased if professional observers would actually make the observations. Such professional observers would not necessarily have to be astronomers, but they should have some understanding of the nature of the results to be obtained and enough understanding of the insttuments to deal quickly with minor problems.

Finally the sequencing of observations could be rationalised by having professional observers. Not only could they organise the observations according to atmospheric conditions, they also would make it possible to obtain the types of observation outlined in the earlier part of this paper.

Albrecht and Raimond (1992) have studied the archives of the 2.5-m INT at La Palma. From their data it may be inferred that of the time with good weather conditions, 8% is "Technical and Reserved", 7% is lost due to Faults, 25% is Acquisition or Idle, and 60 calibration. While such figures might be somewhat different for the VLT, depending upon the average length of an observation, they clearly indicate that major gains in observing efficiency should still be obtainable, in particular by reducing the acquisition time.

Our conclusion is that the great majority of the programmes should be executed by professional observers. This does not mean that under no circumstance should Visiting Astronomers have access to the telescopes to bring innovative, more special purpose instrumentation. Also there are programmes where on-line access to the data would be important in order to avoid making exposures longer than needed. In such a case, a data link with European institutes might be all that is needed. But in most cases, the actual observing and calibrating should be left to a professional observer.

An important question is how to find professional observers. Obviously a certain understanding of physics is a prerequisite. It may well be that an astronomical background would be an advantage. Probably an essential aspect is that the professional observer knows that he will be judged by his skills in observation, rather than by the papers he may write.

Can ESO afford professional observers? Undoubtedly the answer is positive. With few Visiting Astronomers the infrastructure is simplified. If the frequency of faults is significantly reduced by unchanging instruments (for periods of 1 - 2 years) and by professional handling, the reduction in the technical support needed should make the cost of professional observing acceptable. The consequent reduction in the total number of persons on site and of the related infrastructure would be an important benefit, also from the point of view of maintaining optimal observing conditions.

References

Albrecht, M.A., Raimond, E., 1992, in "Astronomy from Large Databases II", eds. A. Heck & F. Murtagh, ESO, Garching, p. 173.

The VLT: max (4 × 8; 1 × 16; 1/X × 64)?

Klaas S. de Boer

Sternwarte, Universität Bonn, Auf dem Hügel 71, D-53121 Bonn, Germany

Abstract. The VLT is planned as 4 identical telescopes with each 4 focal stations, a Cassegrain, a Coude, and two Nasmyth foci. The 4 telescopes can also be used together in an interferometric mode. Several questions are posed about what will be the most important science the VLT can help us achieve. This centers around questions of what can and what cannot be done with the VLT, as well as what can or should be adjusted in the VLT programme.

Introduction

When preparing for this Workshop I realised the programme had little about what the VLT is and what it can or cannot do for astronomy. This contribution is not one presenting a science programme. It is not a contribution in which I say "Dear Santaclaus, give me the instrument for the science I want to do". Even though the DG Ricardo Giacconi often said to the STC: "Tell us what you want and we'll do it", I fear (and know) it is not that simple. I will present some collected facts (and thus repeating existing data) and come with questions which I will not answer, but for which I hope discussion will bring insight. With that, I hope to make clear what is realistic and what is not realistic to expect from the VLT.

There are 3 questions I want to ask about the VLT:
is it 4 telescopes for coherent interferometry, VLT-I?
is it 4 telescopes each with the same instruments to work fast?
is it 4 telescopes with 4 foci each allowing for 16 different fixed instruments?
Let us address some of the technical issues.

Gains Due to Mirror Size

First, where are the largest gains expected due to the size of the VLT mirrors? The basic facts have been summarised elegantly by Beckers (1992), which I repeat in simplified form (telescope diameter D):
"bright" objects (photon noise dominates detector noise): D^2
"faint" objects (detector noise dominates photon noise): D^4
sky limited observations: D^2
high time resolution - short exposure ("bright" or "faint"): D^2 to D^4
So the real gain due to mirror size is for "faint" objects.

Quantum Efficiency, QE

The most likely detectors for the VLT will be CCDs. CCDs are, due to their material, either fit for the visual or for the infrared.

In the visual good CCDs have QE of better than 90% in the red ($\lambda \approx 700$ nm) and normally below 40% for $\lambda < 400$ nm. With coating the blue part may be made more sensitive, but never more than in the red. Thinning will possibly give a QE of $> 80\%$ all through the visual.

In the infrared (2 to 5 or 10 μm) the QEs are now near 50% but that may be improved to larger values.

Mirror Coatings

The mirror coatings to be used are either Aluminium, Silver, or Gold. Reflectivities for each are summarised in Fig. 1. They all work quite well in the infrared ($>95\%$) but not all are equally good in the visual. Gold is not so good at $\lambda < 600$ nm while Al has a dip in the reflectivity near $\lambda = 800$ nm. Silver is very poor in the UV but it suffers from very strong ageing, unless it is protected.

After just 2 reflections (M1 and M2, Cass focus) or 3 reflections (M1 to M3, Nasmith focus) the efficiency is still quite good. However, on the way to the combined focus almost 20 reflections are involved, so that in the combined focus less than 30% remains at visual wavelengths (or less, depending on the chosen coating combinations) while in the IR about 55 to 60% of the light remains.

Given these reflectivities, one wonders what the use is of bringing the light to the combined focus in an incoherent manner; working at the Cass focus with 4 exposures is more efficient.

For VLT-I (coherent beam combining) these losses must be accepted (although they should be minimised) in order to obtain the interferometric spatial resolution. Yet, the number of photons available will be much less than the number collected by the area of the 4 individual telescopes.

Imaging at Different Wavelengths

In the visual there should be baffles to shield the sky from diffusely illuminating the detectors. On the other hand, in the infrared the detector should NOT "see" any part of the telescope. These requirements are in conflict with each other. Will there be removable baffles? Or is it better to create wavelength-range specialised telescopes?

Since the VLT units will have approximately f/15 (a value commonly used for telescopes, and thus for instrumentation), the large mirror size (8 m) means that the imaging scale is reduced. Therefore the VLT must have bigger detectors than common telescopes, preferrably larger than 4000^2 pixels of some 20 μ size.

The effects of the intrinsic angular (spatial) resolution have been reviewed by Wampler (1986). I repeat his figure as Fig. 2 but add the expectation of resolution due to Adaptive Optics. Also I have made the lines giving the diffraction limits fat for those space instruments which at certain wavelengths will do better than the VLT unit telescopes.

One conclusion may be that for pure imaging the HST will, in the visual, be better than the VLT. The VLT, in its turn, will be better than the HST ever can be in the IR, and also be better than ISO (ecxept, of course outside the atmospheric windows). This leads to the suggestion to make the VLT telescopes infra-red specialised telescopes.

Limits to VLT-I?

With coherent beam combination the VLT will become effectively a telescope with a spatial resolution of that of a 100-m telescope. However, the u-v-plane coverage can only be modest (maximum about 10%) and the capacity to do imaging will depend on the number of photons available for each point in the u-v-plane. This means that there will be clear limits at the faint end for sources to be mappable.

16 Foci or 4?

With 4 unit telescopes each having 4 foci one may consider having 12 different instruments. This possibility deserves some comments. Should there be:
- many different instruments

pro: many different possibilities for measurements, simultaneous multiwavelength observations possible;
contra: slow progress in large programmes since (like on La Silla) just one instrument is available for a given task;
- few multiple instruments

pro: cheap instruments, fast progress for larger programmes, easy scheduling;
contra: limited possibilities, no exceptional science possible.

The VLT will have, in fact, something in between these two extremes. On UT1 3 instruments will be available (FORS1, ISAAC, CONICA) of which FORS1 and ISAAC will be available soon after first light. The instruments for UT2 are still not fully decided (except UVES), while UT3 will have FORS2 and MFES.

Conclusions?

Overlooking all information it seems one could conclude the following.

- Optimise telescopes for the UV+vis *or* for the IR.
Consequences for the design of M2 etc? Consequences for coatings, coating plant?

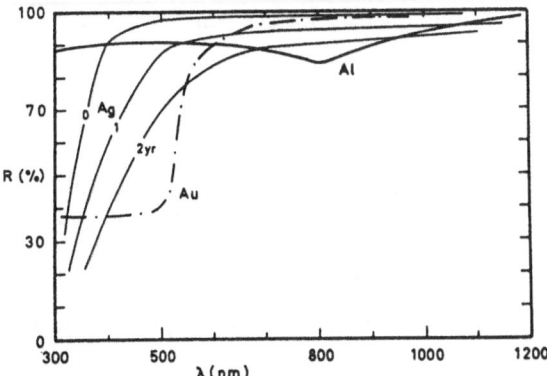

Fig. 1. The reflectivities of 3 materials commonly used in astronomy are shown as a function of wavelength and in relation with ageing. Note the dip near 800 nm for Al, and the severe ageing of Ag (fresh, 1, or 2 yr old) in the UV and visual. Polarisation effects have been ignored. The diagram is adapted from Pulker (1984) and Giordano (1990). **Note:** On the way to the combined focus there will be about 20 reflections!

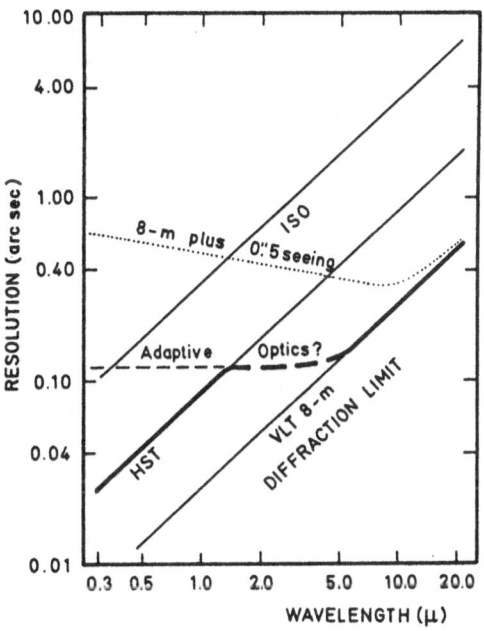

Fig. 2. The diffraction limits of the VLT, HST, and of ISO are shown as a function of wavelength (figure adapted from Wampler, 1986). The expected spatial resolution of the VLT unit telescopes due to Adaptive Optics (AO) is also indicated. The heavy line indicates which telescope is best at what wavelength (for ISO this is not relevant on account of its short mission duration 1995-1997)

- Optimise at most 1 telescope for the UV+vis.

The power of the VLT is in the IR. Consequences for the combined focus?

- No incoherent beam combining!

In particular the losses after the many reflections show this option brings no gain.

- No VLT-I in the visual, only in the IR! In particular the losses after the many reflections show that VLT-I in the visual will be of very limited use.

- All science on "bright" objects is to be barred from the VLT.

The largest gain of the VLT lies with "faint" objects. There is an important role for La Silla here.

- Imaging by unit telescopes of the VLT is in the visual always seeing limited. Consequence: for imaging always use HST! And use the VLT in the visual only for spectroscopy.

The title of this contribution was a question too. Are there answers?

Will the VLT be 4×8? Yes, with 12 available foci.

Will the VLT be 1×16? No, never. It will never have the diffraction limit of a 16 m telescope; moreover, the light losses after beam combining are such that the effective collecting area is at most equivalent to that of a 10 m telescope.

Will the VLT be $1/X \times 64$? The expression was not quite correct. It will be a 128-m telescope in VLT-I with just a 1/64 of the equivalent collecting area, only for a certain fraction of the time, and not before 10 years from now.

In order to make good choices, the scientific priorities have to be found. Personally, I believe they lie in IR imaging.

Two more suggestions are based on the exciting results Garth Illingworth showed from the Keck telescope, as well as the results expected from ISO. Many known sources will have been studied in great detail by others by the time the VLT will be ready (even by the time UT1 is operational). We therefore need to start NOW the work of finding southern sources fit for the unique capabilities of the VLT. We should take care to finish the DENIS 1 m La Silla IR survey and keep the new sources for ESO use. And ESO should start to search for faint southern Quasars (e.g. with a deep Schmidt survey) in order to have objects available for the ESO community to do unique work as of the availability of VLT UT1.

Acknowledgements

I should like to thank many colleagues who shared their ideas for this contribution, in particular Joe Wampler, who also helped with the source material.

References

Beckers, J.M. (1992): in "High resolution spectroscopy with the VLT", M.-H.Ulrich (ed), ESO Conf. No. 40, p.1-9

Giordono, P. (1990) ESO WP N° T 119-30

Pulker, H.K. (1984): "Coatings on glass", Elsevier, Amsterdam, p. 362

Wampler, E.J. (1986): in "The VLT", S. D'Odorico & J.P. Swings (eds), ESO, p 3-13

References

Synopses of Panel Discussions

Part 12

Synopsis of Panel Discussions

Synopsis of Panel Discussion I: Interferometry

Panel members:
- O. von der Lühe (Chair)
- I. Appenzeller
- R. Genzel
- P. Lena
- J.-M. Mariotti
- S. Ridgway
- J. P. Swings
- G. Weigelt

von der Lühe opened the discussion by briefly describing the essentials of VLTI. The three chief advantages over other planned interferometry systems were seen as:

1. High sensitivity (4 × 8m telescopes)
2. Unvignetted field upto 8" (larger for the small telescopes array - VISA)
3. Use in the southern hemisphere

The VLTI main array mode (VIMA) with coherent beam combination of four unit telescopes would not have good u-v plane coverage leading to missing data. The u-v plane coverage is better for VISA. A hybrid mode with 4 unit telescopes plus 4 small telescopes would offer high sensitivity plus good u-v coverage. The limiting magnitude depends on fringe tracking and whether the reference source is the program source or another in the field, so cannot necessarily be given in advance.

In the wake of the delays to interferometry resulting from budgetary pressures, a revised implementation plan has been developed. The plan includes a design and development phase which covers the next four years. This phase would be followed by the construction of VLTI as originally foreseen when full funding would become available. VLTI could be commissioned during the years 2003 to 2005 at the earliest. In order to prepare for the case of early additional funding of VLTI, be it extra contributions from the member states or the contributions of new ESO members, an alternative plan for the next 5-6 years has been developed consisting of the following:

- 3 auxilliary telescopes on 4 stations giving baslines 40-50m and fair u-v coverage
- 2 delay lines with 30m stroke
- 1 fixed cat's eye
- 1 interferometric instrument

and with a projected date of ~2000. In the second phase, the remaining delay line, and the remaining complement of auxiliary telescope stations, as well as a second instrument, would be constructed and integrated. At the same time, two unit telescopes would be equipped with coudé optical trains, enabling their use with the interferometric focus. The targeted end date for this phase would be ~2005. The remaining capability of the original VLTI could be added later.

Three instruments were under consideration by the former VLT Interferometry Panel (VLT Report No. 65):

an image plane multi-axial instrument for 0.45-2.5 μm with imaging and spectroscopic modes (spectral resolution $\sim 10^4$ in the optical and 100 in the IR);

a pupil-plane Fourier transform spectrometer for the 1.25-2.5 μm region with resolutions $\sim 10^4$;

a thermal IR imager for 4.6-12 μm with resolution ~ 100.

A. Leger stressed the implementation of a differential astrometric capability, targetted towards planet detection, which was so far not included in the plans.

von der Lühe replied that this had not yet been considered in the design concept and it was still not clear how feasible it would be. It could probably have a field up to 40" and would appear to require a relatively small effort to implement (see also the contribution by von der Lühe, Quirrenbach, and Koehler, these proceedings).

R. Giacconi mentioned that, in the reduced plan for interferometry, only the first two unit telescopes would have an operational coudé mode. He suggested that a a working group be set-up to specify a narrow angle astrometric mode and a proposal be made to the ESO Scientific and Technical Committee for its support.

Appenzeller enquired about commissioning of VLTI.

von der Lühe replied that the technical commissioning problems would be worked out step-by-step on the small telescopes over a period of 2 years using the auxilliary telescopes, which would be diffraction limited in the 1-2.5 μm region with fast guiding. Technically, this spectral region would be the best compromise for early successful operation.

A. Leger stated that in the development plan, science priorites should be put first.

Genzel said that the plans had already been cut down and that if a new mode (viz. narrow angle astrometric) was planned it would lead to more delays.

R. Giacconi said that it was important to design-in potential for the astrometric mode.

K. S. de Boer asked about interferometry in the visual.

von der Lühe replied that it was foreseen to operate VLTI in the visible, but not in the UV. However, technical considerations would make it attractive to start operations in the near-infrared. One concern is the polarization characteristics of coatings for the reflecting components which are known to be a cause of fringe contrast loss. Simple coatings like protected silver are excellent in the red and the IR but not for the UV.

G. Setti asked how long it would take to tune up the interferometer.

Lena answered. It was only planned to interferometrically combine 2 unit telescopes in 2005. Upto this time experience would be gained on VISA such that with a flip of a mirror on the 8m, the system could be working. However full adaptive optics (AO) is absolutely required on the 8m telescopes first. The first AO system is planned for Nasmyth in $\sim$1999, then it is needed at coudé. The system will be such that it will not distinguish the beam from an 8m telescope or another telescope.

von der Lühe added that no 8m time would be needed at first but that later it would be needed for testing the system with 2 unit telescopes.

M. Rich asked about daytime testing but von der Lühe said that this may be ruled out.

P. Lagage would have preferred mid-IR interferometry and asked about the scientific justification for near-IR.

Mariotti replied that things are cleaner in the near-IR, but that of course any group would be welcome to try a 10 μm camera at the combined focus.

Appenzeller mentioned the importance of the 10 μm region for study of warm dust in many environments. It should not be excluded from VLTI even if it required baffling and fast chopping.

von der Lühe asked how one can do chopping, while maintaining an optical path difference between interferometer arms of a fraction of a wavelength.

U. Kaufl stated that it should be bourne in mind that the sky is very bright at 10 μm and that the limiting magnitude for interferometry would be about 12 mag. (still 10 mags. fainter than the sky). There are other methods of doing interferometry in the IR, such as heterodyne techniques but the sensitivity may be less.

O. Lefevre was worried that the budget for VLTI might freeze out other instruments.

R. Giacconi stated that the baseline instrument budget allows for one new instrument per year at present, with an expected duration of construction of 4 years per instrument. Additional money is clearly needed for VLTI but the development is in parallel with VLT instrumentation.

K. S. de Boer asked about the consequences for VLTI of optimized telescopes for the IR or visual.

Genzel said that this would not be a problem as only single 8m telescopes would be optimized for a particular waveband.

Lena stated that of couse it was not just the primary but many other mirrors in the chain which needed to be optimized for interferometry.

A. Moorwood asked about polarization of mirror M3 and would silver be a problem.

G. Illingworth reported on experience in using silvered mirrors at UKIRT and IRTF. Bare silver coatings are good for about a year and there had been no polarization problems. Only the UV throughput is sacrificed.

A. Leger mentioned exposure times for the two modes of interferometric use: imaging - resolve x pixels across an object - may perhaps require several telescope configurations and at least a few nights per target; astrometric - fewer parameters to determine but of course dependant on the accuracy required - typically one measurement per night would be possible.

R. Cayrel stated that as scientists we were all interested in the shape of objects but he doubted that many wanted very detailed information.

Genzel replied that radio VLBI had started in the 1960's as an experimental curiosity, but was now routine.

R. Giacconi said that Come-on+ was still in the domain of the experts but he hoped that it would later be a general user instrument.

Lena stated that a ~5 year training period has been required to implement an AO system. But now 1 technician can get it working for the observer. Perhaps in five years it will be a push-button system but this may not be the best method of operation and service observing might be preferable.

P. Quinn asked about incoherent beam combinination.

von der Lühe said that there was no proposed instrument but the losses would be ~2 for VLTI in the near-IR and ~3 times in the visual compared with a coherent combined focus.

G. Djorgowski thought that long baseline interferometry in the visual was still a dream as 1000's of actuators would be required for AO as compared to the 100's in the IR. Come-on+ was a prototype for full correction in the IR with a feasible number of ~250 actuators. AO could be seen as pre-cleaning of the PSF. Keck would be going for the simple option of 1 baseline in the near-IR.

Lena stated that if Come-on+ on the ESO 3.6m was scaled up to 8m it was currently quite feasible and would have the same density of actuators.

R. Giacconi considered that there was little to be gained by only a small improvement in technology applied to AO. The groundwork for interferometry had been done but we must plan now, go ahead and develop the technology. It will take a long time to get science, so we need to start now.

G. Djorgowski believed that there were a whole new set of problems in trying to implement interferometry on a large telescope as compared to a small telescope.

T. Bedding said that AO in the visual on a 1.8m telescope presents the same challenges as in the IR on an 8m.

Lena affirmed that no-one was doing AO 6 years ago so progress was great.

von der Lühe stated that in the visual there was a heavy sensitivity problem with adaptive optics. High bandwidth and small subapertures are required, which imply large photon fluxes of the reference sources. For coverage of a significant fraction of the sky, the use of laser beacons is required.

L. Vigroux said that 10 μm diffraction limited imaging was very simple in comparison with the visual, and there were many scientifically interesting programmes to be done.

Weigelt stated that in the visual, interferometry had already been achieved with the Mount Wilson Mk III interferometer. In the optical where we can count individual photons, AO is not required as speckle can be used as has been amply demonstated on AGN, gravitational lenses and star clusters. With long baseline interferometry a multi-speckle mode can be used. AO in the IR could be used for partial correction in the visual together with speckle.

Appenzeller stressed that the reality was still very far from all the simulations of interferometry and he was worried that there would be many sources of vibration leading to loss of fringes.

von der Lühe mentioned that environmental sources of error for VLTI were being intensely studied at ESO.

Mariotti emphasized the 10 μm window as an important domain. Initially 2×10^6 DM was put-up for interferometry instruments, so a (technical) choice had to be made among the 3 proposed interferometric instruments. With more money becoming available, three instruments of somewhat lower complexity could be built. He stressed that the biggest gains at 10 μm were to be made from the 8m telescopes (S/N proportional to Dia^4). He proposed that the 10μm instrument should be for the 8m telescopes. The 1.8m telescopes are suited to the near-IR.

A. Leger hoped for science in 10 years. The discovery of extra-solar planets will be a key field.

Appenzeller, beginning the closing remarks from the panel, emphasized the chances of interferometry to produce the most exciting science for the VLT. However there were associated the greatest uncertainties and a great amount of development work was required.

Ridgway considered comparisons with other programmes. There are some proto-types with limited success; some facilities working measuring angular diameters of stars and binary star separations; and some concepts for doing astrometry in the search for planets. The advantage of large aperture was great in allowing faint sources to be studied, going to the thermal IR and being able to use high spectral resolution. High spectral resolution may be very effective. The three instrumentation concepts each seem remarkably powerful and unique in the world at present. He saw a great community awareness in Europe for the potential of high angular resolution imaging, which does not exist in other regions, and a very active high resolution community with advanced thinking. He hoped however that Ridgways 2nd rule of 'lost funding after fringes found' would not prevail. Already there had been 'lost funding before fringes' (mirth).

Mariotti stressed the favourible response to interferometry with a number of new programmes arising from the meeting. This was the first meeting where science and interferometry techniques had been discussed, and many new ideas had arisen. He wanted to keep the multi-wavelength approach and to put 10-20 μm interferometry under consideration. He supported an astrometric instrument which should not be expensive and have a small impact on the overall programmmme.

Genzel believed that interferometry in the near-IR will be a revolution similar to that which had taken place in the radio. Interferometry was a natural progression in resolution from AO. He said that we need to get young poeple excited but also to give them a surer perspective. Interferometry needed to come up with real programmes.

Weigelt believed that visitor instruments in the visible had a role to play as a learning tool. Delays would set us at a disadvantage in relation to other projects such as the Mk III and the Big Optical Array (BOA - array of six 60cm telescopes for optical interferometry), which would have many baselines and be operational by the end of the year. We must be careful not to be too late.

Lena recapped that 8 years ago the revolutionary decision had been taken to go for AO on the 3.6m. Science was now being produced from this initiative. This success gives us strength to plan AO for the VLT. We are now at a similar juncture for interferometry with an 8-10 year timeframe. The revised plan for VLT interferometry makes sense, is technically feasible and can achieve science. There must continue to be a constant productive involvment of ESO and the community. The interferometry panel should be revived in order to put meat on the current plans.

von der Lühe concluded by saying that he found the discussion very enlightening and there were clearly many VLTI customers. New areas for consideration were astrometry and thermal IR. Many interferometry projects were on-going around the world such as ASEPS-0 and, in particular, the Keck interferometric mode. Some are operational and some will come on-line in the next few years. The VLT interferometry initiative must not be lost.

Synopsis of Panel Discussion II: Future Instrumentation

Panel members:
- L. Vigroux (Chair)
- P. O. Lagage
- O. Lefevre
- P. Magain
- A. Renzini
- G. Vettolani
- J. Wampler
- G. Wiedemann

Vigroux opened the discussion by calling attention to the booklet 'Instruments for the VLT', edited by A. Moorwood, where the accepted and proposed instruments were briefly described. These were not the only possibilities and there was probably room for 5 instruments in addition to those planned (excluding Coudé foci). He thought that consideration had to be given to operational constraints, such as instruments being run in 100% remote (service) observing mode or with visiting observers, in designing new instruments. There should be flexibility for new instruments.

Lagage described briefly the projects for which the proposed Mid-Infrared Imaging Spectrometer (formerly MIIS, now VISIR) was well suited, such as dust discs around YSO's, environment of planet formation, AGB mass loss, dust in external galaxies and AGN. In addition there were many spectral lines of interest in this band. VISIR (MIIS) had been mentioned in 14 talks at the Workshop. This would be an progression from TIMMI, would follow up on ISOCAM and be a unique instrument. It would be capable of 10-20 μm imaging, low resolution spectroscopy (R~500) for dust and a maximum spectral resolution in the range $15\text{-}20 \times 10^3$ for spectral line work.

Lefevre described the scientific goals for NIRMOS (Near-InraRed Multi-Object Spectrograph). At high redshift most of the observed flux from galaxies is in the near-IR. The instrument would use multi-apertures for good sky subtraction and be aimed at deep IR galaxy counts, surveys of galaxy evolution and large scale structure, nature of distant galaxies and QSO's and surveys for brown dwarfs. With a field of 4 × 5x5 arcmin, 30 slits per field, 0.8-1.8 μm capability (possibly extendible to 2.3 μm) at resolutions of ~200-3000, it is expected to be useful for galaxies to z~3. The field is currently open and the only other project competing was OSIS on the CFHT with a 4x4arcmin field and only 20 slits.

Magain outlined the Very High Resolution Spectrograph (VHRS) proposal. The aimed-for resolution was $3\text{-}6 \times 10^5$, coverage from UV to near-IR, with a very clean instrumental profile, a narrow spectral range per exposure and high stability. The scientific aims were many, including stellar oscillations, Beta-Pic type discs, thorium cosmochronology, measurement of isotope ratios from hyperfine splitting, cosmic microwave background temperature measurement as a function of redshift, and IGM abundances from absorption lines towards SN in Virgo.

Vettolani briefly described WFIS the wide field optical imaging spectrograph. This was primarily aimed at the collection of large complete samples in order to attack the problems of galaxy and cluster evolution. At B~25 there are ~16 objects per square arcmin and the four quadrant fields allow spectra of 120 objects to be recorded simultaneously at resolutions of 500 - 2000. This instrument would have about four times the spatial density sampling of FORS in the same exposure time.

Wampler, in describing a Wide Field Direct Imager, said that in very deep images (e.g. in Tyson fields) there are many faint galaxies and objects ~0.5 arcsec in size whose nature is not known. Deconvolution is required to exploit good seeing in order to obtain morphological information on such small images. For a Wide Field imager an Atmospheric Dispersion Corrector is needed and the instrument would need to be at the Cassegrain focus and require baffling of the telescope. Active optics adjustment could be implemented during readouts.

Wiedemann outlined the CRIRES (CRyogenic InfraRed Echelle Spectrograph). The resolution was to be aimed at 10^5 for the 1-5 μm region using a cross-dispersed echelle and a 1024x1024 array. Many programs were of interest from the Solar System, stellar atmospheres, star formation to galaxies, opening up new scientific areas. There might be possibilities of extension to 8-14 μm. The only competition in this field was a similar instrument being built for Kitt Peak 4.2m.

Vigroux mentioned two other future instruments which were not presented: a high resolution camera (VHARC) and a fast recording optical imager and spectrophotometric instrument (FRISPI).

Renzini started the discussion with some provocative questions. He questioned what was the driving science behind these instrumentation proposals. Was it deep observational cosmology as at Keck and Subaru? He questioned whether the instrumentation was adequate to compete in the field of observational cosmology with Keck and Subaru. One deficiency was that Subaru has a wide angle Prime Focus camera. He thought that WFIS could make a big contribution. He was however worried that with WFIS many targets must first be located and with the possible loss of the ESO Schmidt this requires smaller survey telescope(s), which has not been considered.

K. S. de Boer stated that the VLT with 4 telescopes and many instruments could service many interests at the same time in contrast with the more limited facilities of Keck and Subaru. However a priority list for science needed to be drawn up since everything can't be done at once. He believed that cosmology had the highest priority and that planet searches were going to be very important.

R. Genzel said that cosmology was only one field and that we should be careful about comparisons with Keck which served a small community. From studying stars in our own and nearby galaxies we can transfer that knowledge to distant galaxies, so the two aspects should not be seperated.

R. Giacconi doubted the use of an 8m telescope and a site such as Paranal simply for doing fast photometry. On the need for a wide field he considered that we should to do more than just examine the sky with FORS. There seemed to be many wish lists, but which was better or more practical had to be specified. In observational cosmology he asked what field is required to get the statistics needed.

Vigroux stated that for pulsar photometry imaging photometry was often required, and that a simple one or two-channel photometer would not be adequate.

Lefevre reiterated that to study the evolution of galaxies needs many thousands of spectra carefully statistically selected.

F. Hammer pointed out that in the local universe structure dominates and even out to z=1 structure influences the luminosity function. Large scale surveys such as Sloan (for 10^6 galaxies) and 2dF to z=1 would be best suited to determining the local structure-luminosity function.

S. Cristiani said that it would be difficult to do survey work with existing ESO facilities and thence to use the VLT effectively. It was pointless to do wide field work with the VLT. He stressed that MFIS and VISIR were required for galaxy surveys in assocation with wide field capabilities on smaller telescopes (eg. the LIGHT project). He suggested developing wide field 4m class telescope projects at ESO and sharing wide field capabilites with other facilities. He wanted FUEGOS (the Multi-object fibre spectrograph) to have ≥ 100 fibres for observational cosmology.

P. Felenbok in reply said that if you wanted to go to hundreds of fibres you would need two spectrographs in FUEGOS, which would be very expensive and above the ceiling for single instruments. FUEGOS was not aimed at large surveys although could be changed in scope to peform them.

M. Rich stressed the astrophysical importance of galaxy formation at $z \geq 1$. At such red-shifts sky suppression in the 7000 - 10000Å region becomes important for removal of night sky emission and telluric absorption. Coatings of optics in instruments were also of great importance for maximizing throughput.

I. Appenzeller stated that a wide field survey imager would be an advantage for the VLT but was not the best and most cost-efficient way of using an 8m telescope. It would be better suited to smaller telescopes. Single object work will always be of considerable importance.

Vettolani said that the Sloan survey, with a 2m telescope, will go to 19th mag., and others such as 2dF to 22 mag. (23 in 8 hours). These surveys give many objects with well defined sampling. However for galaxy formation at $z \geq 0.7$ an 8m is required for proper sampling to determine how structure evolves. Otherwise a 4m could be kept occupied for ~ 20 years.

R. Kudritzki suggested that for cosmology surveys we should put deep wide field survey instruments on all unit telescopes, but he questioned whether we wanted to do this. He thought the field was too small for this sort of survey work.

Lefevre countered that a 10 arcmin field is big enough. In CFHT deep surveys, in 8 hour exposures, spectra of I=22 mag. galaxies have been obtained. A bigger telescope and multi-object instrument would open up such limited science and go deeper.

L. Chincarini said that a dedicated 4m telescope could do the imaging required for deep galaxy surveys but an 8m was required for the spectroscopy.

Vigroux reiterated that an imaging depth of 25-26 mag. required to supply the spectroscopic surveys could easily be achieved with small telescopes. To go deeper and obtain morphological information a high resolution camera and or deconvolution are required with adequate sampling of the PSF.

P. Felenbok reminded everyone that the Nasmyth field of the VLT is 30arcmin, and that this is a substantial field to cover. Most instruments do not exploit all this field.

R. Kudritzki stressed that it was not only deep surveys that lead to cosmological discoveries but also high resolution spectroscopy for determining abundances in stars and isotopic abundances, e.g. Deuterium.

Renzini stated that at least one unit telescope should be dedicated competitively to cosmology. He agreed with Genzel that we need to study near-by stars in order to understand distant galaxies. He again stressed the need for wide fields on smaller telescopes.

A. Leger drew attention to the need for high resolution spectroscopy with the highest stability for planetary searches. A resolution of $\leq 10 \text{ms}^{-1}$ was at the photon noise limit but stability was more important than resolution.

R. Cayrel made a plea that global efficiency of instruments and telescopes should be given high priority. Great care should be taken not to loose photons.

U. Kaufl said that the Keck was showing that stellar spectroscopy was important on large telescopes. For the VLT there was the LMC, SMC, Galactic Centre and important star-forming regions observable, which would mean that stellar astronomy would be a strong contributor, more so than at northern hemisphere observatories.

J. Andersen posed the question on how to decide between what to do and what not to do among all the options. In particular consideration must be given to what should still be done on La Silla in the VLT era (e.g. in the form of preparation and follow up).

J. Lafon stressed that high angular resolution was essential for advance in modelling circumstellar envelopes.

Vigroux suggested that a simple mid-IR imager and an OH line suppressor should be considered among the suite of new instruments. There should also be room for small innovative instruments built by teams to use at the VLT.

R. Genzel pointed to the example of the British in accepting innovative flexible new instruments on their telescopes. This was a cost effective way to implement new technology. The VLT should allow space for novel instrumentation which could then be developed and, if succesful, offered to a larger community.

G. Weigelt said that the VLT should not just consist of monster instruments. The NTT for example could not easily accommodate other instruments. It should be part of the plan that small groups be allowed to develop cheaper specialised instruments which could be offered to the community. He quoted the SHARP camera as an example.

R. Giacconi closed by stressing that we had to be serious about running facilities properly with discipline and high efficiency. It was no good keeping them going with token manpower. It seemed often that we wanted all, but did not want to pay the price. There seemed to be a lack of priority among these suggestions for future instruments. ESO cannot do everything.